Applied Homogeneous Catalysis with Organometallic Compounds

Volume 3: Developments

Edited by
B. Cornils and W. A. Herrmann

Further Titles of Interest

B. Cornils, W. A. Herrmann, R. Schlögl, C.-H. Wong (Eds.)
Catalysis from A to Z
A Concise Encyclopedia
2000, ISBN 3-527-29855-X

A. Liese, K. Seelbach, C. Wandrey
Industrial Biotransformations
2000, ISBN 3-527-30094-5

R. A. Sheldon, H. van Bekkum (Eds.)
Fine Chemicals Through Heterogeneous Catalysis
2001, ISBN 3-527-29951-3

D. E. De Vos, I. F. J. Vankelecom, P. A. Jacobs (Eds.)
Chiral Catalysts Immobilization and Recycling
2001, ISBN 3-527-29952-1

K. Drauz, H. Waldmann (Eds.)
Enzyme Catalysis in Organic Synthesis
A Comprehensive Handbook in Three Volumes
Second, Completely Revised and Enlarged Edition
2002, ISBN 3-527-29949-1

Applied Homogeneous Catalysis with Organometallic Compounds

A Comprehensive Handbook in Three Volumes

Volume 3: Developments

Edited by
Boy Cornils and Wolfgang A. Herrmann

Second, Completely Revised
and Enlarged Edition

 WILEY-VCH

Prof. Dr. Boy Cornils
Kirschgartenstraße 6
D-65719 Hofheim
Germany

Prof. Dr. Dr. h.c. mult. Wolfgang A. Herrmann
Anorganisch-chemisches Institut
der Technischen Universität München
Lichtenbergstraße 4
D-85747 Garching
Germany

This book was carefully produced. Nevertheless, editors, authors and publisher do not warrant the information contained therein to be free of errors. Readers are advised to keep in mind that statements, data, illustrations, procedural details or other items may inadvertently be inaccurate.

Cover picture: Homogeneous catalysis in aqueous phase: the yellow catalyst solution separates spontaneously from the colorless phase consisting of butyraldehydes. The underlying molecular model symbolizes the water-soluble ligand of the organometallic complex. The picture was taken at the plant site of Celanese (formerly Ruhrchemie), Oberhausen/Germany (see Chapter 1 and Section 2.1.1).

Library of Congress Card No.: applied for

A catalogue record for this book is available from the British Library

Die Deutsche Bibliothek – CIP-Cataloguing-in-Publication-Data
A catalogue record for this book is available from
Die Deutsche Bibliothek

ISBN 3-527-30434-7

© Wiley-VCH Verlag GmbH, D-69469 Weinheim, 2002

Printed on acid-free paper

Composition: Hagedorn Kommunikation, Viernheim
Printing: betz-druck GmbH, Darmstadt
Binding: Buchbinderei Schaumann GmbH, Darmstadt

Printed in the Federal Republic of Germany

Contents

Volume 1: Applications

Volume 2: Developments

Volume 3: Developments (continued)

3

Recent Developments
in Homogeneous Catalysis
(continued)

3.2.7   Homologation

Helmut Bahrmann

3.2.7.1   Historical Background

The energy and oil supply crisis of 1973 focused the interest of the West and Japan primarily on the production of large-volume commodity chemicals on the basis of domestic raw materials such as coal, with syngas as the chemical building block. It was thought that syngas facilitates the transition from oil- to coal-based products. Regardless of what it is made from, it is an identical chemical species and it fits in with existing equipment and technology. It was reckoned that in the syngas building block the carbon itself would cost only half as much as in the ethylene building block [1]. The highest efficiency in the use of syngas is reached if the oxygen content of the carbon monoxide remains in the end-product. This is the case in the classical syntheses of methanol and in the Monsanto process for the formation of acetic acid [2] (cf. Section 2.1.2.1). In this connection, homologation of methanol to ethanol and subsequent dehydration could provide a new route to ethylene. Later on, when the use of tetraethyllead as an octane booster was scheduled to be discontinued for health reasons, a mixture of ethanol and higher alcohols became of interest as an octane booster and cosolvent for methanol in wet hydrocarbons, because methanol alone could separate from the hydrocarbons at lower temperatures and attack various parts of the fuel distribution system [3].

Originally, the homologation reaction, i.e., the enlargement (extension) of the carbon chain of oxygen-containing molecules by a $-CH_2-$ group according to eq. (1), e. g., was discovered by Wietzel, Eder, Vorbach, and Scheuermann during the period 1941–1943 [4]. They converted aliphatic primary alcohols into the next higher alcohols on reacting with syngas as a source for $-CH_2-$ at raised temperatures and pressures (eq. (2)).

$$R'OH \ + \ \big(CH_2\big) \ \xrightarrow{\ cat.\ } \ R'\text{-}CH_2\text{-}OH \tag{1}$$

$$ROH \ + \ CO \ + \ 2\,H_2 \ \xrightarrow{\ metal\ carbonyls\ } \ RCH_2OH \ + \ H_2O \tag{2}$$

The by-products consisted of higher acids, their esters, and hydroxy ethers. The results of this experiments remained unnoticed so that work in 1949 by Wender et al. [5] – after whom this reaction was named – was a kind of rediscovery. In 1952, the essential conclusions of the work of Wietzel et al. were confirmed in a fundamental paper by Ziesecke [6].

In 1956 Berty [7] introduced a iodine activator to the basic catalyst cobalt carbonyl; this made possible a noticeable increase in the reaction rate and represented the transition from the high-pressure syntheses (40–100 MPa) to the medium-pressure syntheses (18–40 MPa), a state-of-the-art which lasted until 1988.

A further improvement to the selectivity of the reaction, especially in the period 1975–1985, could be reached in the field of catalyst development. This development can be characterized by the transition of the original one-component catalyst cobalt to complex multicomponent catalyst systems [8]. The added catalyst compounds are activators, such as halogens or halogenides, and promoters, such as donor ligands of the Group 5 elements, as well as additional co-catalysts, such as ruthenium or nickel. In the homologation of methanol the transition metal ruthenium is especially useful for an in-situ hydrogenation to ethanol of the primary product acetaldehyde, whereas nickel facilitates the formation of acetaldehyde dimethyl acetal [9]. Some attention was also dedicated to the development of a process for the production of acetaldehyde [10, 11]. The state-of-the-art up to 1982 has been reviewed in [12]. Later, most of the activity has still been concentrated on the homologation of methanol. An ultimate breakthrough may have been achieved in 1988 by Moloy and Wegman when they changed the basic catalyst metal from cobalt to rhodium. They succeeded in the development of a novel rhodium–ruthenium–diphosphine–methyl iodide catalyst which enabled a low-pressure, (relatively) low-temperature homologation of methanol for the first time [47].

Table 1 includes new literature on catalyst systems and reaction conditions up to 1994. As can be seen from this table, a large range of different catalyst combinations, promoters, and solvents has been investigated. However, the overall conversion and selectivity remain unsatisfactory. It seems to be impossible to achieve both a high conversion of methanol and a high selectivity to ethanol at the same time. Thus, the results of all the efforts have not justified a commercial realization of the homologation process up to now.

3.2.7.2 Chemical Basics and Applications

Although the homologation reaction was originally restricted to aliphatic alcohols, its scope has been extended to a broad range of basic organic chemicals (eq. (3)).

$$R \left\{ \begin{array}{l} CH_2OH \\ CHO \\ COOH \\ COOR \end{array} \right. + \ 2 \ CO/H_2 \ \longrightarrow \ R \left\{ \begin{array}{l} CH_2\text{-}CH_2OH \\ CH_2\text{-}CHO \\ CH_2\text{-}COOH \\ CH_2\text{-}COOR \end{array} \right. + \ H_2O \qquad (3)$$

Within this scope, homologation reactions are all variants of enlarging the carbon chain in a oxygen-containing molecule by one C atom with the use of syngas.

Thus the homologation reaction can be used, for example, for the synthesis of acetaldehyde from methanol [48], propionic acid from acetic acid [47], or ethyl acetate from methyl acetate [50]. Styrene may be produced from toluene by oxidation to benzyl alcohol [51] and homologation to 2-phenylethanol, which in turn can be dehydrated to styrene. From the chemical point of view, the applications of homologation reactions are broad and useful. But, as mentioned before, low selec-

Table 1. Homologation of methanol – current state in the development of catalysts.

Press. [bar]	Temp. [°C]	Catalyst	Promoters	Remarks	Conversion [%]	Product/ selectivity [%]	By-product selectivity [%]	Ref.
	80–170	Co	I_2/Br_2, phosphine chelate	Addition of inert solvent		Acetaldehyde		[13]
	170–250	Co	Halogen containing complex phosphine chelate			Ethanol		[14]
	170–250	Co	Halogen containing complex phosphine chelate	Addition of water		Ethanol		[15]
350	260	Co	$(Co_2B)_5H_3$		9.5	Ethanol/74	Methyl acetate/6.8	[16]
320	260	$Co_2(CO)_8$	NaI, Na_2CO_3	CO and water conversion	42	Ethanol/29	Methyl acetate/14 1,1-dimethoxy-ethane/7	[17]
400	185	$Co_2(CO)_8$		Pretreatment with CO	23	Ethanol/56	Methyl acetate/5.5 Ethyl acetate/5.6	[18]
200	190	$[CpCo(CO)_2]$	I_2, PPh_3		67	Ethanol/36	Acetaldehyde/9.6 Acetic acid/13.7	[19]
200	185	$Co(OAc)_2$	I_2, Br_2, and AsR_3, SbR_3 or BiR_3	Addition of inert solvent	55	Acetaldehyde/33	Ethanol/3.3 Acetic acid/8.1	[20]
200	205	$Co(OAc)_2$	I_2, PBu_3	Addition of inert solvent	47	Ethanol/70	Methane/25.5 Acetic acid/6.8	[21]
200	190	$Co(OAc)_2$	I_2, Br_2, phosphine chelate		50.6	Ethanol/65	Acetaldehyde/1 1,1-Dimethoxy-ethane/10 Methane/18.4	[22]

Table 1. (Continued)

Press. [bar]	Temp. [°C]	Catalyst	Promoters	Remarks	Conversion [%]	Product/ selectivity [%]	By-product selectivity [%]	Ref.
200	185	Co(OAc)$_2$	I$_2$, PPh$_3$, RuCl$_3$		43	Ethanol/65		[23]
400	200	Co(OAc)$_2$	I$_2$, RuI$_3$		50	Ethanol/33.3	n-Propanol/3 Ethyl acetate/3 Acetaldehyde/2	[24]
275	180	Fe/Co	[Bu$_4$N]$^+$ [FeCo$_3$(CO)$_{12}$]$^-$ + MeI		42	Acetaldehyde/84	Methyl acetate/14	[25]
270	220	Ru/Co	CpRu(PPh$_3$)$_2$Co(CO)$_4$		54	Ethanol/80	Methane/1 Dimethyl ether/2 Methylethyl ether/3 Methyl acetate/2 Diethyl ether/3 n-propanol/5 Ethyl acetate/3	[26]
276	175	Co(acac)$_3$	PR$_3$, I$_2$, Ru		59	Ethanol/60	Dimethyl ether/5 Methyl acetate/15 Diethyl ether/14	[27]
276	200	Co^{II} *meso*-tetraaromatic phosphine	I$_2$		68	Acetaldehyde/62	Dimethyl ether/9.3 Ethanol/11.4 Methyl acetate/12.9	[28]
280	200	[Co(CO)$_3$L]$_2$ where L = AsR$_3$	Iodine, AsR$_3$, SbR$_3$		71	Acetaldehyde/53	Dimethyl ether/6.7 Ethanol/18.1 Methyl acetate/13	[29]
280	200	Co(CO)$_3$L]$_2$ where L = PBu$_3$	I$_2$, Ru(acac)$_3$		44	Ethanol/72	Dimethyl ether/3.5 Diethyl ether/3.0 Methyl acetate/9.6 Others/11.8	[30]

Table 1. (Continued)

Press. [bar]	Temp. [°C]	Catalyst	Promoters	Remarks	Conversion [%]	Product/ selectivity [%]	By-product selectivity [%]	Ref.
400	185	Co(OAc)$_2$	I$_2$, PPh$_3$, 1,2-bis(diphenylphosphino)-ethane			Acetaldehyde/53.3		[31]
200	250	CoS, Co$_2$S$_3$	PR$_3$	Addition of nitrogen-containing solvent	36	Ethanol/86.3	Acetaldehyde/2.1 Methyl acetate/3.2 Ethyl acetate/1	[32]
300	200	Co$_2$(CO)$_8$	CH$_3$I		80	Ethanol/42	n-Propanol/4 n-Butanol/2 Acetaldehyde/2	[33]
260	215	Co$_2$(CO)$_8$	NaI + CH$_3$I, Ru$_3$(CO)$_{12}$			Acetaldehyde		[10]
210	181	Co(OAc)$_2$	I$_2$, P(C$_6$H$_{11}$)$_3$ Ru$_3$(CO)$_{12}$		43	Ethanol/80		[34]
245	190	Co(OAc)$_2$	I$_2$		100	Ethanol/65.5	Acetaldehyde/1.7 Methyl acetate/23.2 Ethyl acetate/9.6	[35]
290	200	Co(OAc)$_2$	HI, PPh$_3$, and Ni(OAc)$_2$		58	1,1-Dimethoxy-ethane/80	Acetaldehyde/8.6 Methyl acetate/5–8	[9]
300	200	Co(OAc)$_2$	HI, PPh$_3$, W(CO)$_6$			1,1-Dimethoxy-ethane		[36]
		RuCl$_3$/ Co$_2$(CO)$_8$	I$_2$	Addition of ethers		Ethanol/80		[37]

Table 1. (Continued)

Press. [bar]	Temp. [°C]	Catalyst	Promoters	Remarks	Conversion [%]	Product/ selectivity [%]	By-product selectivity [%]	Ref.
350	185	Co/Pt	Halogen or halide add. of 1,2-bis(di-phenylphosphino-ethane)		72	Ethanol/69	Acetaldehyde and acetals/2 n-Propanol/4	[38]
550	185	Co/RuCl$_3$	NaI, O-containing phosphine chelate		42	Ethanol/76	Acetaldehyde and acetals/Esters/5	[39]
235	200	Co/RuO$_2$	Halide, Bu$_4$P$^+$ Br$^-$			Ethanol		[40]
15	250	Rh/Fe	Heterocyclic amines			Ethanol		[41]
270	220	Co/Ru	MeI, PPh$_3$	Variation of source of Ru	58	Ethanol/86		[42]
550	220	Co/Ru	NaI, sulfonated and carboxylated phosphines		56	Ethanol/69	n-Propanol/2 Hydrocarbons/9 Ethers/17 Esters/3	[43]
550	200	Co/Rh	Iodide, amine or amide, heterocyclic		32	Ethanol/62 Amide		[44]
70	180	Co$_2$(CO)$_8$	N-contg. compds., 1,1'-bis(diphenyl-phosphino)ferrocene	1,4-Dioxane	82	Ethanol/58	Acetaldehyde/7 n-Propanol/14 Methyl acetate/5	[45]
200	230	Co$_2$(CO)$_8$	Bu$_3$P	C$_6$H$_6$/OH-contg. compds.	28	Ethanol/84		[46]
70	140	Rh/Ru	Diphosphine/MeI			Ethanol/70–80		[47]

tivities and/or activities of the existing catalyst systems have still prevented broad research and development in this field. Most of the research is still restricted to the laboratory scale and to reactions with methanol.

3.2.7.3   Mechanism of Reaction

Despite the quantity of informations available, the material does not allow clear deductions to be made about the mechanism. However, it is obvious that the mechanism of the homologation reaction depends basically on the main catalyst metal which is used.

Under the drastic reaction conditions and the acidic influence of the classical cobalt catalyst [$HCo(CO)_4$], acetaldehyde and subsequent acetals are formed, which in turn may be hydrolyzed back to acetaldehyde or directly hydrogenated to ethanol (cf. eq. (4)).

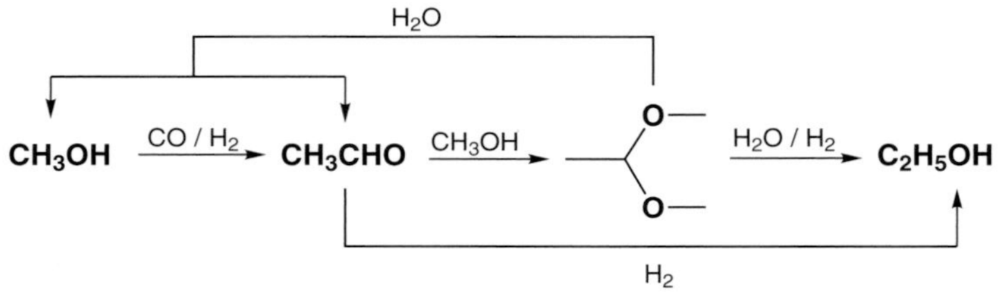

$$
\begin{array}{c}
H_2O \\
CH_3OH \xrightarrow{CO/H_2} CH_3CHO \xrightarrow{CH_3OH} \underset{H_2}{\overset{O-}{\diagup}\diagdown_{O-}} \xrightarrow{H_2O/H_2} C_2H_5OH
\end{array}
\qquad (4)
$$

In contrast, under the milder reaction conditions with rhodium, no acetals are observed (cf. eq. (5)).

$$
CH_3OH \xrightarrow{HI} CH_3I \xrightarrow{CO/H_2} CH_3CHO \xrightarrow{H_2} C_2H_5OH
\qquad (5)
$$

Similarly to the reaction with cobalt, the acetaldehyde intermediate formed will be further hydrogenated to ethanol. Overall, the Rh-catalized homologation mechanism resembles the Monsanto process with the exception that, as a result of the presence of hydrogen, acetaldehyde is now the main product and acetic acid definitely the only by-product. Some key catalyst components present at the end of the homologation reaction, such as Rh(diphosphine)COMe)I_2 and [$Ru(CO)I_3$]$^{4-}$ have been isolated and identified by Moloy et al. [49]. It may be assumed that the Ru complex is responsible for the intermediate in-situ hydrogenation to the high ethanol selectivity obtained.

More data are available from the cobalt catalyst system. Under reaction conditions the cobalt compounds will form the following equilibrium with syngas (eq. (6)).

$$
Co_2(CO)_8 + H_2 \rightleftharpoons 2\,HCo(CO)_4
\qquad (6)
$$

These cobalt carbonyl compounds may be involved in the primary step of the homologation reaction, the formation of a metal–alkyl complex. For this, nine different routes according to Scheme 1 are discussed in the literature [12c].

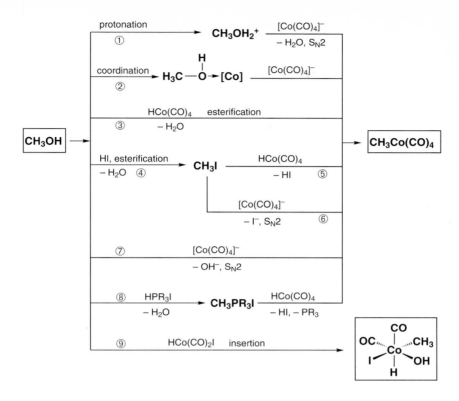

Scheme 1. Primary step of the homologation reaction: the formation of the metal–alkyl bond.

Additionally, the situation will be further complicated by the fact that, under the reaction conditions of the homologation reaction, $Co_2(CO)_8$ with methanol, halogen, halide, or phosphines may undergo various different disproportionation reactions, from which some compounds were identified by IR spectroscopy, e. g., $[Co(CH_3OH)_6]^{2+}$ $[Co(CO)_4]_2^-$ [52], $[Co(CO)_4]^-$ [53, 54], $CoX_2 + M^+[Co(CO)_4]^-$, $[Co(CH_3OH)_x(CO)_yI_z]^{n+}$, $[Co(CO)_4]_n^-$ [55], $[Co(CO)_3(L\frown L)]^+$, and $[Co(CO)_4]^-$ [56].

Nevertheless, from the nine different routes to form the key intermediate of the homologation reaction set out in Scheme 1, three remain the most convincing: the insertion mechanism; the S_N2 mechanism; and the phosphonium ion mechanism (cf. Schemes 2 and 3).

Scheme 2 outlines the insertion and the S_N2 mechanisms. In both cycles cobalt complexes are involved in the splitting-off of water. The key intermediate in the insertion mechanism is $HCo(CO)_2I$ and in the S_N2 mechanism the anion $[Co(CO)_4]^-$, resulting from one of the previously mentioned disproportionation reactions.

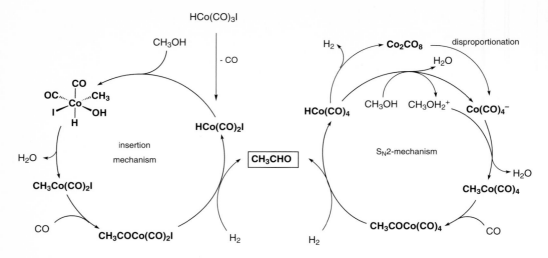

Scheme 2. Insertion and S$_N$2 mechanism of the homologation reaction with methanol.

As proposed by Keister [57] and supported by own investigations [58], phosphines are partly quarternized, so in the reaction mixture from the homologation of methanol in the presence of triphenylphosphine and hydroiodic acid methyltriphenylphosphonium iodide could be isolated and identified by IR spectroscopy, which shows a quantitative methylation of the intermediate formed [HPPh$_3$]$^+$I$^-$ with methanol. This fact suggests a phosphonium ion mechanism, which is proposed by the author and outlined in Scheme 3.

Within this mechanisms the phosphonium ions function as a methyl-group transfer agent and the critical step for the conversion of methanol – the splitting-off of water – is facilitated by taking place outside the direct catalyst cycle.

3.2.7.4   Technical Applications

So far, the homologation reaction has reached only the pilot-plant scale [58, 61]. Little information is available about the reaction in continuous operation. The only cobalt-catalyzed continuously conducted reaction led to a mixture of 20 different products. The yield of ethanol is low (16 mol %) [59]. By activation with iodine and variation of the space-velocity, the overall yield has been improved and the ratio of acetaldehyde/ethanol could be varied between 13:18 and 2:17 [60]. BP has described continuous homologation with the Co/I/PPh$_3$ catalyst system. The yield of ethanol reached only 25 mol % [11]. Semicontinuous work on the homologation reaction has been reported by the former Ruhrchemie AG [61].

Seven of the most convincing discontinuously developed catalyst systems were recycled nine times and conversion and selectivity were noted (cf. Table 2). The catalyst compounds were separated by a special distillation unit under CO/H$_2$

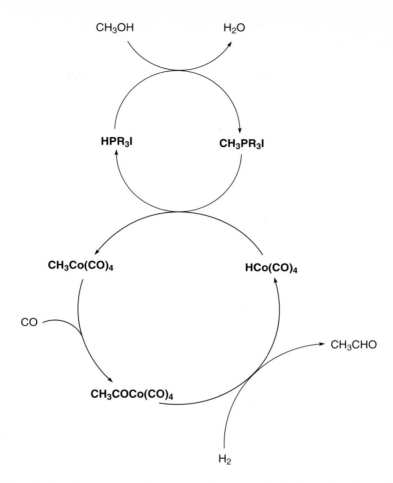

Scheme 3. Phosphonium ions as methyl-group transfer agents in the homologation of methanol.

pressure. The best results were reached with the system no. 1. During recycling, conversion of methanol decreased by 7 points and the selectivity to ethanol by 18 points. In further recycling experiments, fresh catalyst was added in such an amount that conversion and selectivity remained constant. It was found that 20–30 % fresh catalyst must be added in order to reach the steady state. The stability of the catalyst during the reaction and the recycling was too low for a technical application. Furthermore, owing to the insufficient methanol conversion, high energy and investment costs for separation and recycling of unconverted methanol would be required. Finally, the same is true for the separation of the different reaction products.

Table 2. Recycling behavior of selected Ruhrchemie AG catalysts.

No.	Catalyst system	Conversion [mol %] of methanol		Selectivity [mol %] to ethanol	
		Start	End	Start	End
1	Co, Ru, NaJ, 1,3-bis(diphenylphosphino)propane	59	52	80	62
2	Co, Ru, NaI, $C_6H_{11}P(CH_2CH_2COOH)_2$	48	21	80	10
3	Co, Ru, NaI, $PPh_3/HPPh_3{}^+I^-$	59	35	78	49
4	Co, Pt, NaI, 1,3-bis(diphenylphosphino)propane	59	32	76	62
5	Co, Ru, NaI, methyl-2-pyrrolidone[a]	49	<5	68	
6	Co, Ru, NaI, high-boiling solvent (polyglycol 1000)	45	40	51	32
7	Co, Ru, HI, high-boiling solvent (polyglycol 1000)	78	52	65	55

Reaction conditions: 550 bar, 3 h, $CO/H_2 = 1:2$, $CH_3OH/Co = 500$, NaI/M = 1:1, promotor/M = 2:1, Ru/Co = 0.1.
[a] Only one recycle possible.

3.2.7.5   Future Prospects

Despite considerable improvements in the 1980s with respect to the conversion of methanol and the selectivity to ethanol or acetaldehyde, both are still insufficient and prevent a technical application. This is also true for the other homologation reactions. It seems that the relatively severe reaction conditions of the classical cobalt catalyst system are inherently connected with this fact. One could speculate about whether the same effort and expenditure of research which was directed to the cobalt system were to be concentrated into the new rhodium system, with generally much milder reaction conditions and the potential for essentially higher selectivity, the state-of-the-art of the homologation reaction could be much brighter than it is. In this light, the homologation reaction is still waiting to be reinvestigated.

References

[1] J. I. Ehrler, B. Juran, *Hydrocarbon Proc.* **1982**, *2*, 109.
[2] J. F. Roth, J. H. Craddock, A. Hershman, E. Paulik, *CHEMTECH* **1971**, *10*, 600; D. Forster, *Adv. Organomet. Chem.* **1979**, *17*, 255.
[3] J. F. Knifton, J. J. Lin, D. A. Storm and S. F. Wong, *Catal. Today* **1993**, *18*, 355.
[4] BASF AG (G. Wietzel, O. Vorbach, A. Scheuermann), DE 843.876, 867.849, 875.346 (1941–1943).
[5] I. Wender, M. Orchin et al., *J. Am. Chem. Soc.* **1949**, *71*, 4160; I. Wender et al. *ibid.* **1951**, *73*, 2656; I. Wender et al., *Science* **1951**, *113*, 206.
[6] K. H. Ziesecke, *Brennstoff-Chem.* **1952**, *33*, 385.

[7] J. Berty, L. Marko, D. Kallo, *Chem. Tech. (Leipzig)* **1956**, *8*, 260.

[8] Commercial Solvents Corporation (A. D. Miley, W. O. Bell), US 3.248.432 (1961).

[9] Union Rheinische Braunkohlen Kraftstoff AG (J. Korff, M. Fremery, J. Zimmermann), DE OS 2.913.677 (1979).

[10] Rhône–Poulenc Industries (J. Gauthier-Lafaye), EP Appl. 22.735 and 22038 (1979).

[11] BP (W. J. Ball, D. G. Stewart) GB Appl. 2.053.915 (1980).

[12] (a) H. Bahrmann, B. Cornils, *Chem.-Ztg.* **1980**, *104*, 39; (b) H. Bahrmann, B. Cornils in *New Synthesis with Carbon Monoxide* (Ed.: J. Falbe), Springer-Verlag, Berlin, 1980; (c) H. Bahrmann, W. Lipps, B. Cornils, *Chem.-Ztg.* **1982**, *106*, 249.

[13] Agency of Ind. Sci. Tech., JA 56.020.536 (1979).

[14] Agency of Ind. Sci. Tech., JA 56.020.536 (1979).

[15] Agency of Ind. Sci. Tech., JA 56.025.122 (1979).

[16] Air Products and Chemicals Inc. (C. M. Bartish), US 4.171.461 (1978).

[17] Allied Chemical Corp. (M. Novotny, L. R. Anderson), US 4.126.752 (1978).

[18] Allied Chemical Corp. (M. Novotny), US 4.283.582 (1978).

[19] BP (G. M. Thomas), EP Appl. 29.723 (1979).

[20] BP (B. R. Gane), EP 1.936 (1977).

[21] BP (B. R. Gane), EP 1.937 (1977), EP Appl. 3.876 (1978).

[22] BP (B. R. Gane, D. G. Stewart), EP Appl. 10.373 (1978).

[23] B. R. Gane and D. G. Stewart, GB 2.036.730 (1977).

[24] Commercial Solvents Corp. (G. N. Butter), US 3.285.948 (1965).

[25] Exxon (G. Doyle), EP Appl. 27.000 (1980).

[26] Exxon (G. Doyle), EP Appl. 30.434 (1980).

[27] Gulf (W. R. Pretzer, T. P. Kobylinski, J. E. Bozik), US 4.133.966 (1977).

[28] Gulf (W. R. Pretzer, T. P. Kobylinski, J. E. Bozik), US 4.151.208 (1977).

[29] Gulf (J. E. Bozik, T. P. Kobylinski, T. W. R. Pretzer), US 4.239.704 (1978), US 4.239.705.

[30] BP (B. R. Gane), EP Appl. 1.937 (1977), EP Appl. 3.876 (1978).

[31] Mitsubishi Gas Chemicals, JP 52.136.111 (1976).

[32] Mitsubishi Gas Chemicals, DE-OS 3.016.715 (1979).

[33] Montedison (D. Paolo), IT Appl. 1.034.761 (1975).

[34] UCC (R. A. Fiato), US 4.233.466 (1979), US 4.253.987 (1980).

[35] UCC (W. E. Walker), US 4.277.634 (1980).

[36] Union Rheinische Braunkohlen Kraftstoff AG, BE 890.964 (1982).

[37] K. Kudo, *Nippon Kagaku Kaishi* **1982**, *3*, 462.

[38] Ruhrchemie AG (B. Cornils, C. D. Frohning, G. Diekhaus, E. Wiebus, H. Bahrmann), EP Appl. 53.792 (1980).

[39] Ruhrchemie AG (B. Cornils, C. D. Frohning, H. Bahrmann, W. Lipps), DE-OS 3.042.434 (1982).

[40] Texaco Development Corp. (J. F. Knifton, J. J. Lin), EP Appl. 56.679 (1982).

[41] Ethyl Corp. (D. C. Hargis, M. Dubeck), US 4.361.499 (1982).

[42] G. Doyle, *J. Mol. Catal.* **1983**, *18*(2), 251.

[43] Ruhrchemie AG (H. Bahrmann, B. Cornils, W. Lipps), EP 84.833 (1983).

[44] Ruhrchemie AG (H. Bahrmann, B. Cornils, W. Lipps), DE OS 3.330.507 (1983).

[45] Texaco Inc. (J. F. Knifton, J. J. Lin), US 4.476.326 (1984).

[46] Agency of Industrial Sciences and Technology (Y. Isogai), JP 62.242.636 (1987).

[47] UCC (W. R. Wegman, K. G. Moloy), US 4.727.200 (1988); K. G. Moloy and R. W. Wegman, *J. Chem. Soc., Chem. Commun.* **1988**, 820.

[48] Rhône–Poulenc Industries (J. Gauthier-Lafaye), EP Appl. 11.042 (1978).

[49] J. F. Knifton, *CHEMTECH* **1981**, *10*, 609.

[50] G. Braca, G. Branca, G. Valentini, *Fundam. Res. Homog. Cat.* **1979**, *3*, 221; Imhausen-Chemie-GmbH, DE-OS 2.731.962 (1979).

[51] Anon., *Chem. Eng. News* July 4, **1977**, 13.
[52] I. Wender, H. Sternberg, M. Orchin, *J. Am. Chem. Soc.* **1952**, *74*, 1216.
[53] W. R. Pretzer, T. P. Kobylinski, *Ann. N. Y. Acad. Sci.* **1980**, *333*, 58.
[54] Y. Iwashita, F. Tamura, H. Wakamatsu, *Bull. Chem. Soc. Jpn.* **1970**, *43*, 1520; F. Ungvary, A. Sisak, L. Marko, *J. Organomet. Chem.* **1980**, *188*, 373.
[55] P. S. Braterman, A. E. Leslie, *J. Organomet. Chem.* **1981**, *214*, C45.
[56] J. Ellermann, *J. Organomet. Chem* **1975**, *94*, 201; Andreetta et al., *Nouv. J. Chim. Ital.* **1978**, *2*, 436; A. Sacco, *Gazz. Chim. Ital.* **1963**, *93*, 698; D. J. Thornhill et al., *J. Chem. Soc. Dalton Trans.* **1974**, 6; R. L. Peterson et al., *Inorg. Chem.* **1973**, *12*, 3009.
[57] J. B. Keister, R. Gentile, *J. Organomet. Chem.* **1981**, *222*, 143.
[58] W. Lipps, H. Bahrmann, B. Cornils, Ruhrchemie AG, unpublished work.
[59] G. S. Koermer, W. S. Slinkard, *Ind. Eng. Chem., Prod. Res. Dev.* **1978**, *17*, 231.
[60] Mitsubishi Gas-Chem. Inc. (T. Asano, K. Ishida, T. Imai), JP 48.002.525 (1964).
[61] H. Bahrmann, W. Lipps, Ruhrchemie AG, Forschungsbericht T 85-054, *Technologische Forschung und Entwicklung,* Bundesministerium für Forschung und Entwicklung (BMFT), **1985**.

3.2.8   Homogeneous Electrocatalysis

Didier Astruc

3.2.8.1   Introduction

Redox processes have introduced powerful ways to activate substrates in inorganic [1] and organic chemistry [2]. Pairwise organometallic processes including homogeneous catalysis have also become an extremely rich field [3]. The combination of these two worlds should have a considerable impact on the advancement of catalysis [4]. In this chapter a summary of the catalytic aspects of redox processes is given, restricting the scope to homogeneous reactions although the heterogeneous aspects are also of great importance and can be well understood by using a similar and global approach [5]. Pairwise redox changes (oxidative addition and reductive elimination) are not discussed since the focus is on single-electron transfer processes. There are two major areas of homogeneous catalysis involving single-electron transfer:

(1) Electrocatalysis, also named *Electron-Transfer-Chain* (ETC) catalysis, whereby a reaction (mostly of non-redox type) is catalyzed by an electron (reduction) or by an electron hole (oxidation). Organotransition-metal complexes can carry an electron or an electron hole and, if they achieve this function without decomposition, they are electron-reservoir complexes [6].
(2) Redox mediation or catalysis whereby a redox reaction is mediated or catalyzed by a redox reagent. This type includes and is inspired by biological catalysis with metalloenzymes. Here again both states need to be stable in order to insure this function. Thus the redox mediator used must also be an electron-reservoir complex [7].

Each of these two areas can be further divided into two parts, depending on the nature of the redox change during the catalytic cycle: electron transfer versus atom transfer. The latter mode of redox change is necessarily an inner-sphere version of the electron-transfer reaction, whereas the first mode is frequently of the outer-sphere type [8].

Chain inorganic reactions have been reviewed several times [9–14] whereas redox catalysis is an extremely large area dealing with metal-catalyzed oxidations [15, 16] (cf. Section 2.4) and bioorganic catalysis [17, 18] (cf. Section 3.2.1). Many important references concern electrochemistry (heterogeneous electron transfer) and therefore are not cited here but interested readers can find them in [4].

3.2.8.2 Electron-Transfer-Chain (ETC) Catalyzed Reactions

3.2.8.2.1 Principles

Thermodynamically favorable reactions such as $A \longrightarrow B$ which suffer from kinetic limitations can be catalyzed by an electron or by an electron hole using, for instance, a redox carrier as an initiator. The reaction can be split into an initiation step and a chain propagation cycle. The latter can be distinguished as a "chemical" step and a cross redox step (Scheme 1).

overall: **A + C** → **B + D**

mechanism initiation: **A** → **A$^\pm$**

propagation:

    A$^\pm$ + C → **B$^\pm$ + D**   (chemical step)

    B$^\pm$ + A → **B + A$^\pm$**   (cross ET step)

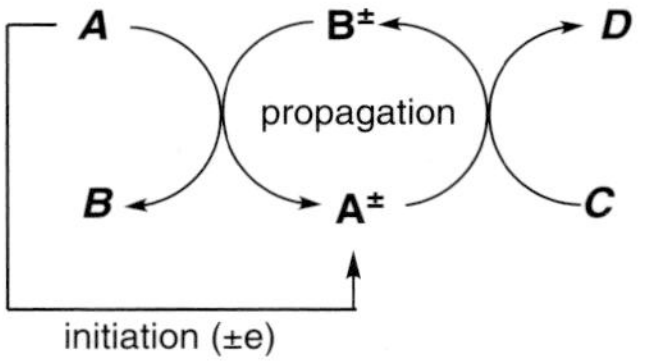

Scheme 1

The reaction works best if (1) the electron-transfer initiation step is exergonic i.e., the redox reagent is a good oxidant or a good reductant, depending on which is required (*vide infra*); (2) both propagation steps are exergonic (or if not, at least the overall propagation cycle must be exergonic, and side reactions of the endergonic step must be avoided).

Catalysis is obtained because the rate of the chemical step is often considerably more favorable at the odd-electron level than at the even-electron (closed-shell) level. Indeed, most of the time inorganic radicals react about 10^9 times faster than their isostructural diamagnetic analogs. For organic radicals, the rate enhancement is even larger [9].

3.2.8.2.2 Initiation

The initiation can be oxidation or reduction, but usually only one of these two possibilities is working. The reason is that the cross redox step should be exergonic to provide an efficient chain reaction. It is also easy to know if the redox step is exergonic, whereas this is not the case for the "chemical" propagation step. If the reaction product is more electron-rich than the starting material, the initiator should be a reductant in order to have an exergonic cross redox step; whereas it should be an oxidant if the reaction product is less electron-rich than the starting material. More precisely, the ergonicity of an electron-transfer reaction is given by the Rehm–Weller equation [19]:

$$\Delta G^0 \text{ (kcal mol}^{-1}) = 23.06 \ (E^0_{red} - E^0_{ox}) + 331.2 \ [(Z_{ox} - Z_{red} - 1)(f/\varepsilon d)] \tag{1}$$

where:
E^0 = thermodynamic redox potential,
Z = charge,
f = ionic-strength factor,
ε = dielectric constant of the solvent,
d = $r_{ox} + r_{red}$ (Å) = sum of the radii r_{ox} and r_{red} of the
        oxidant and reductant.

If the oxidant is a monocation ($Z_{ox} = 1$) and the reductant a neutral species ($Z_{red} = 0$), the electrostatic term is nil and the ergonicity is simply deduced from the respective values of the thermodynamic redox potentials. The electrostatic factor can be large, however, if the difference of charge is high, if the radii are small, and if the solvent has a low dielectric constant, e. g., THF ($\varepsilon = 7.5$), Et_2O ($\varepsilon = 4.2$) or hydrocarbons ($\varepsilon < 3$) [20].

These thermodynamic considerations can also be used for the cross redox step of the propagation cycle. Thermodynamic redox potential values E^0 for initiators can be found in Table 1.

3.2.8.2.3   Examples

Ligand Substitution Reactions

The first example of recognized ETC catalysis was published by Taube in 1954 [25]. Chlorine exchange in $[Au^{III}Cl_4]^-$ was initiated by the reductant $[Fe^{II}(CN)_6]^{4-}$ (eq. (2) and Scheme 2).

Footnotes to Table 1
[a] The redox potential of sandwich complexes can be finely tuned by modification of the number of ring substituents. (Only approximate values are given for the redox potentials). For a review of redox potential values, see [21]; for E^0 values of organometallic complexes, see [22]; for NAr₃, E^0 values range from 0.52 V to 1.72 V vs. SCE depending on the nature of Ar, see [23]; for transition-metal sandwich compounds, see [24].
[b] Abbreviations: bipy, bipyridine; Tol, *o*-tolyl; Cp, η-C_5H_5; TTF, tetrathiafulvalene; TMPD, tetramethylphenylenediamine; TCNQ, tetracyanoquinotdimethane; Cp, cyclopentadienyl; Cp*, pentamethylcyclopentadienyl; TBA, tetrabutylammonium.

Table 1. Redox reagents suitable for initiation of electrocatalysis.[a]

Reagent[b]	Redox potential E^0 vs. SCE at 25 °C (V)	Solvent
One-electron oxidants		
$[Ru(bipy)_3]^{3+}$	+ 1.32	MeCN
$(p\text{-Br-}C_6H_4)_3N^{\bullet+}$	+ 1.06	DMF
Ag^+	+ 1.11	DMF
NO^+	+ 1.27	DMF
$N(Tol)_3^{\bullet+}$	+ 0.76	THF
Fe_{3+}	+ 0.53	0.1 M HCl
$I_2(2e)$	+ 0.3	H_2O
$[FeCp_2]^{\bullet+}$	+ 0.40	MeCN
$[FeCp_2]^{\bullet+}$	+ 0.45	DMF
$TTF^{\bullet+}$	+ 0.30	MeCN
TMPD	+ 0.21	DMF
TCNQ	+ 0.127	MeCN
$[Fe(Cp^*)_2]^{\bullet+}$	− 0.12	MeCN
$TCNQ^{\bullet-}$	− 0.29	MeCN
O_2	− 0.82	MeCN
O_2	− 0.87	DMF
$NAr_3^{\bullet+}$		
One-electron reductants		
$Na^+(C_{10}H_8)^{\bullet-}$	− 2.54	THF
$Na\,[MnCp^*_2]$	− 2.17	DMF
$TBA^+Ph_2CO^{\bullet-}$	− 1.88	MeCN
$[Fe^I(Cp^*)(C_6Me_6)]^\bullet$	− 1.85	DMF
$[Fe^ICp(C_6Me_6)]^\bullet$	− 1.55	DMF
$[Fe^0(C_6Me_6)_2]$	− 1.50	DMF
$[Co^{II}(Cp^*)_2]^\bullet$	− 1.48	DMF
$[Fe^ICp(C_6H_6)]^\bullet$	− 1.25	DMF
$[Ru^0(C_6Me_6)_2](2e)$	− 1.0	MeCN
$[CoCp_2]^\bullet$	− 0.89	DMF

Footnotes see p. 1048.

$$[Au^{III}Cl_4]^- + 4\ ^{36}Cl^- \longrightarrow [Au^{III}\ ^{36}Cl_4]^- + 4Cl^- \qquad (2)$$

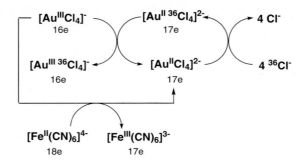

Scheme 2

Taube showed that several other reductants can work as well (Sn^{II}, Sn^{IV}, Au^{I}, V^{IV}), that the turnover number is 10^4 per equiv. initiator, and that the termination reaction is the dismutation of Au^{II} to Au^{I} and Au^{III}. Ligand exchange using this ETC catalysis technique works well in organotransition metal chemistry [6]. For instance, CO ligands are easily exchanged by P donors or isonitriles in neutral poly-nuclear metal–carbonyl complexes using reductive initiation [11, 26, 27] (Schemes 3 and 4) [28] whereas hydrocarbon ligands are replaced by P donors or solvent ligands using reductive initiation in mononuclear cationic complexes (Schemes 5 and 6). Likewise, the acetonitrile ligand of neutral mononuclear complexes can be replaced by a P donor using the oxidant ferricinium as the initiator [11, 29].

Oxidative versus Reductive Initiation:
Chelation of Monodentate Dithiocarbamate (dtc) Complexes

The comparison between oxidative and reductive initiation sheds light on the con-sequence of the choice of initiation mode and on the side reactions. The chelation of the monodentate dtc ligand corresponds to a carbonyl exchange (Scheme 7). Thus, in principle, an exergonic cross redox step would require a reductive initiation. However, the complex is not so easy to reduce, but easier to oxidize. Thus, oxidation (Scheme 7b) was first achieved using a catalytic amount of ferricinium salt (10 %). The yield of chelate complex obtained was not very high and was shown to depend on the counter-anion of ferricinium (38 % with $SbCl_6^-$). The reason for the poor ef-ficiency is the side reaction of the oxidized chelate complex (precipitation) which cannot react in an exergonic step with the monodentate complex. The relative suc-cess of the catalytic reaction is due to the fast chemical step at the favorable Fe^{III} oxidation level [36]. Reductive initiation requires a reductant as strong as naphthyl-sodium and works better (67 % yield, neither starting complex nor intermediate left) [37]. The chemical step, however, is not very selective due to the unfavorable attack by a sulfur ligand at the low Fe^{I} oxidation state level. Thus dtc loss also occurs, leading to the $[FeCp^*(CO)_2]^\bullet$ radical, which dimerizes [38] (Scheme 7a).

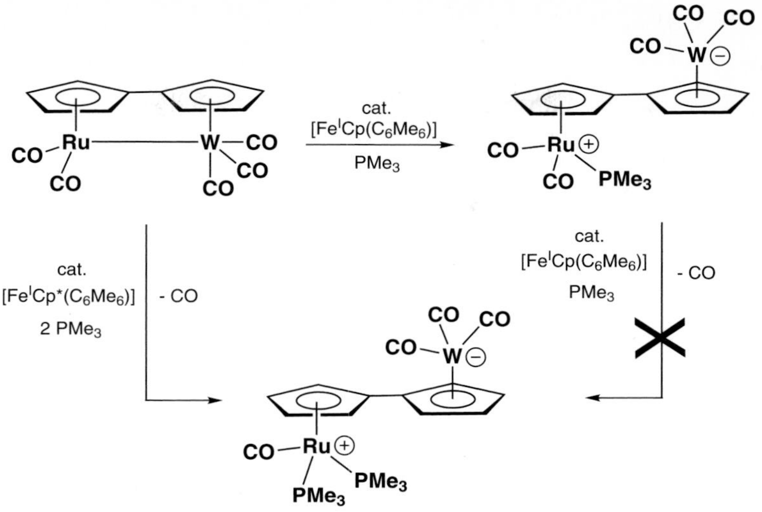

Scheme 3. Note how the strength of the reductant is important to initiate the reaction. Also the two metals are differentiated ("redox recognition"), which leads to regiospecific reactions [28].

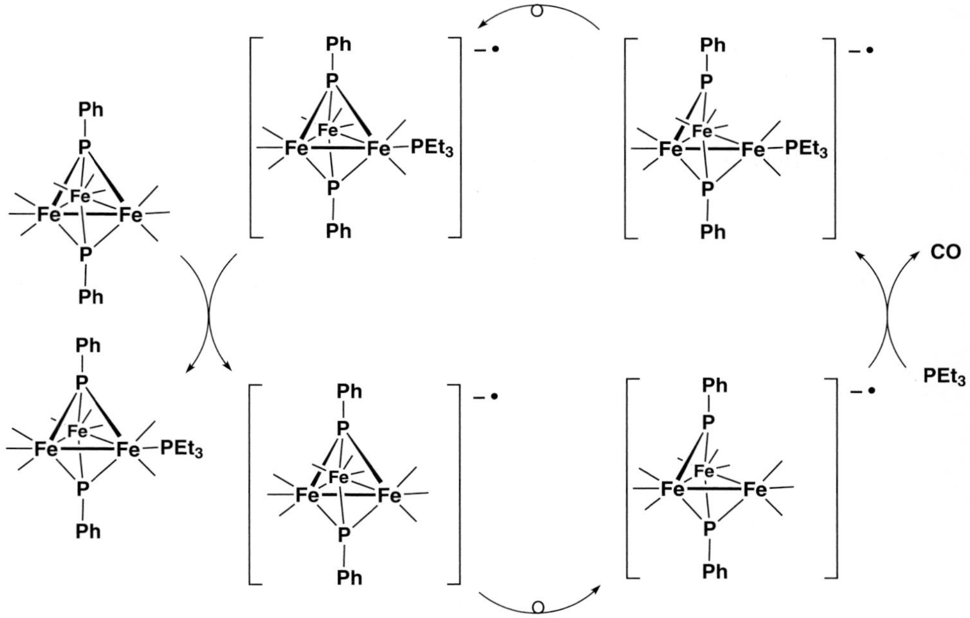

Scheme 4. Electrocatalytic exchange of CO by a P-donor in the bicapped cluster: the four radical anions participating in the propagation cycle have been characterized by EPR, by coupling with P ligand(s) in a stepwise sequence at low temperature [30].

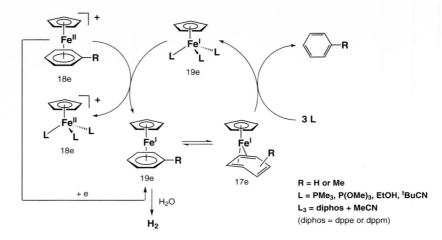

Scheme 5. The exchange of arene was first disclosed with solvent ligands [31]. Later, it was found that 19-electron complex [Fe^{I}Cp(arene)], generated from [Fe^{II}Cp(arene)]$^+$ and Na/Hg in THF, can catalyze the ligand exchange with P donors with turnover numbers of about 100 [32, 33].

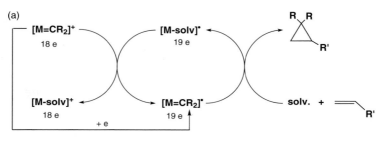

M = FeCp*(CO)$_2$; R = H, R' = Ph

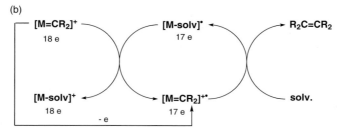

M = Fe(CO)$_2$PR'$_3$ (R' = Me, Ph); R$_2$C=CR$_2$ = tetrathiafulvalene

Scheme 6. Electrocatalytic carbene transfer to organic substrates: cyclopropanation of olefins (a) initiated by mild chemical reductants such as Zn powder [34], and carbene dimerization (b) initiated by mild oxidants such as O$_2$ from air [35].

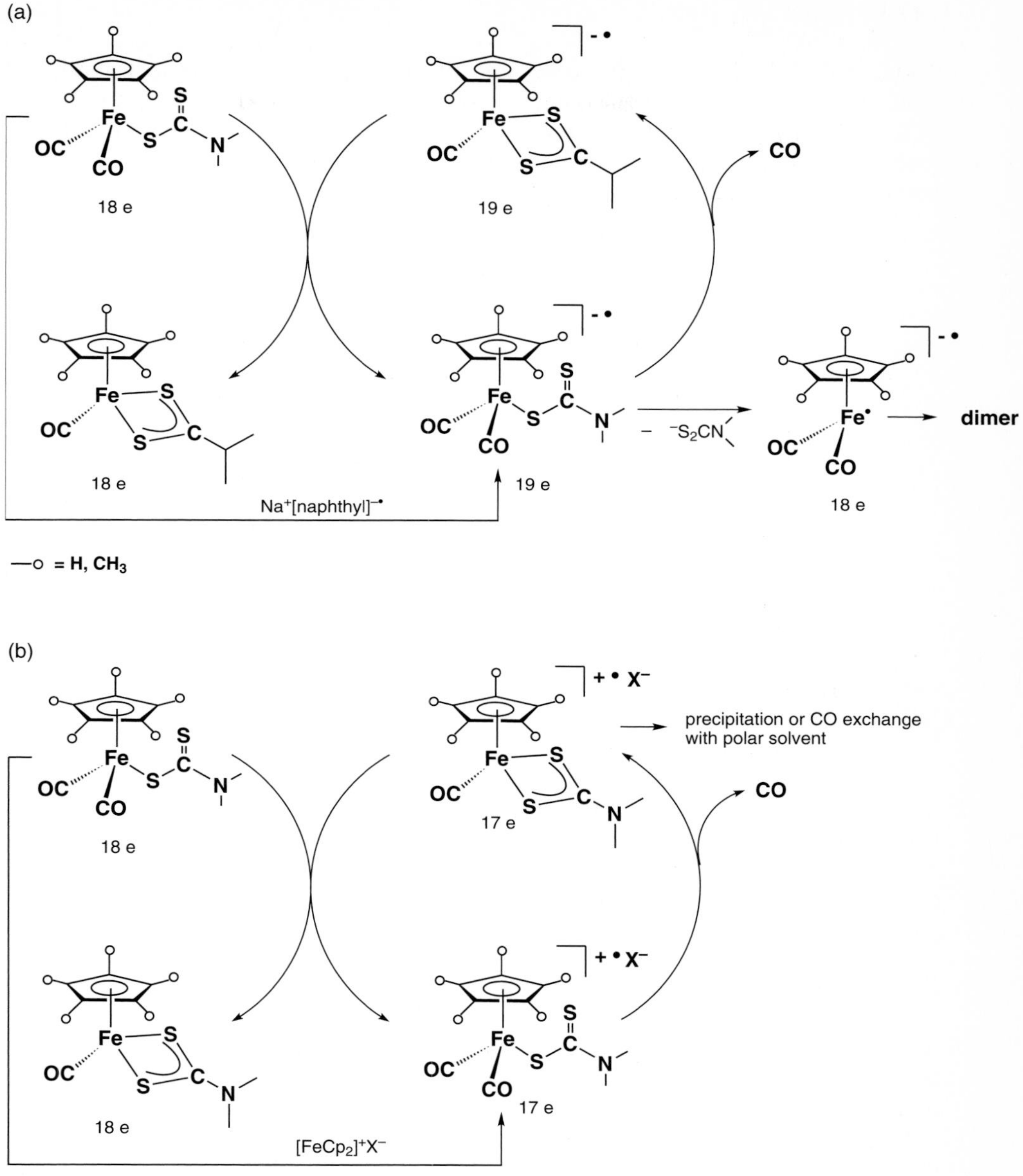
(a)
18 e
19 e
CO
18 e
19 e
Na+[naphthyl]−•
− −S2CN
dimer
18 e
—o = H, CH3
(b)
18 e
17 e
precipitation or CO exchange with polar solvent
CO
18 e
17 e
[FeCp2]+X−

Scheme 7

CO Insertion

The classic CO migratory insertion into the Fe–CH_3 bond can also be initiated either by an oxidant or a reductant, although the external ligand is not the same in the two cases. This difference of external ligand (CO versus PPh_3) makes the insertion reaction exergonic in both cases (eqs. (3) and (4); E_p^a = anodic peak potential; E_p^c = cathodic peak potential; $E_{1/2} = (E_p^a + E_p^c)/2$).

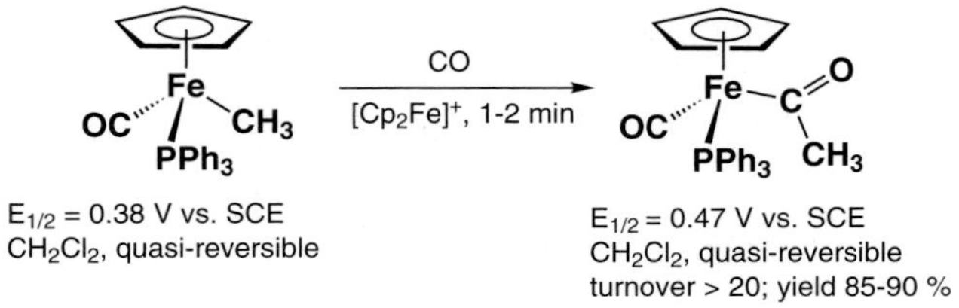

Fe
OC
CH3
PPh3
CO
[Cp2Fe]+, 1-2 min
Fe
C
O
OC
PPh3
CH3
E1/2 = 0.38 V vs. SCE
CH2Cl2, quasi-reversible
E1/2 = 0.47 V vs. SCE
CH2Cl2, quasi-reversible
turnover > 20; yield 85-90 %

(3)

Fe
OC
CH3
CO
PPh3
[cathode], -1.80 V vs. SCE
1 h, 20 °C
Fe
C
O
OC
PPh3
CH3
E1/2 = -1.90 V vs. SCE
irreversible
E1/2 = -1.95 V vs. SCE
irreversible
turnover 5-7; yield 85 %

(4)

Other Reactions

Most other inorganic reactions have been carried out using ETC catalysis: isomerization of octahedral complexes [39–41], disproportionation [42], metal–metal bond cleavage and formation [43, 44], CO extrusion in formyl complexes [11]. Although many studies involve electrochemical initiation, the use of a chemical oxidant is also often shown to work. It is possible to use a photoexcited state as the initiator given its enhanced redox power [45].

Coupling ETC Catalysis with Organometallic Catalysis

A major drawback in many catalytic reactions is the kinetic limitation to replace a ligand in the available precatalyst by the substrate in the coordination sphere of the transition metal. Heating or a long reaction time is required to overcome the induction period. If a suitable ETC-catalyzed ligand substitution reaction between the ligand and the substrate is coupled with the organometallic catalysis, it should be possible to enhance reaction rates and selectivities greatly and to reduce reaction temperatures and times. Such a system was designed for this purpose in the case of the W^0-catalyzed terminal alkyne polymerization [46] whose mechanism has been shown by Katz to proceed via metallacyclobutadiene intermediates [47]. Using $[W(CO)_3(NCMe)_3]$ the polymerization requires either heating to 100 °C, or one week of reaction at 20 °C. The replacement of an acetonitrile ligand by a less

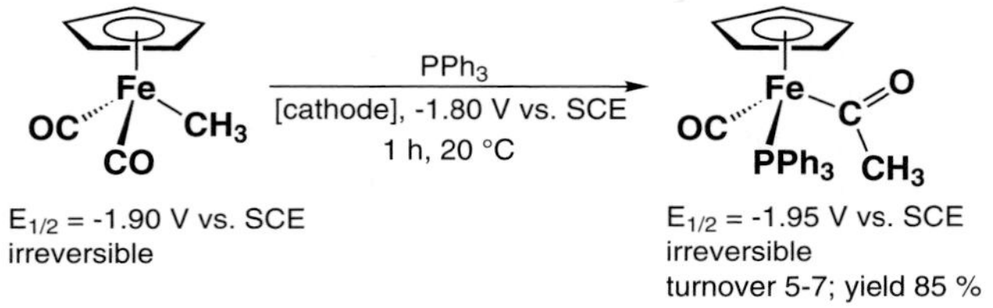

electron-donating alkyne ligand requires an oxidative initiation in order to provide an exergonic cross redox step. Thus ferricinium salt was used in milligram amounts in conjunction with W^0 catalyst since it must be present only in a catalytic amount with respect to the W^0 catalyst. Under these conditions, the reaction is fast at 20 °C, as expected (eq. (5) and Scheme 8).

$$\equiv\!\!-R \xrightarrow{\text{cat., init.}} \textbf{polyacetylenes} \quad \overline{M_w} \sim 25000 \tag{5}$$

cat. = [W(CO)$_3$(NCMe)$_3$];
init. = [FeCp$_2$]$^+$[PF$_6$]$^-$ (initiator);
R = alkyl, phenyl

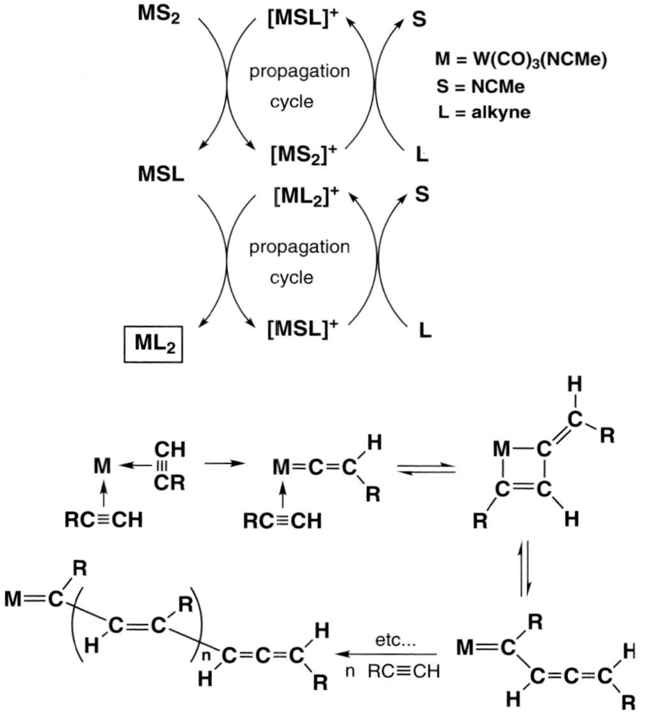

Scheme 8

3.2.8.3   Atom-Transfer-Chain (ATC) Catalysis

3.2.8.3.1   Halogen-ATC Catalysis

The particle transferred in the chain reaction is now a hydrogen or a halogen atom rather than simply one electron. The cross atom-transfer propagation step is exergonic if the bond formed is stronger than the bond broken. It was Taube again who reported the first recognized example in transition–metal chemistry. Radioactive ^{36}Cl was incorporated into $[Pt^{IV}Cl_6]^{2-}$ in water at 25 °C for a few minutes upon irradiation with diffuse visible light [25]. Taube's recognition of the chain mechanism for ligand substitution explained such reactions, which had in fact been found 120 years earlier [48] (Scheme 9). The termination step is the dismutation of the Pt^{III} intermediate to Pt^{II} and Pt^{IV}.

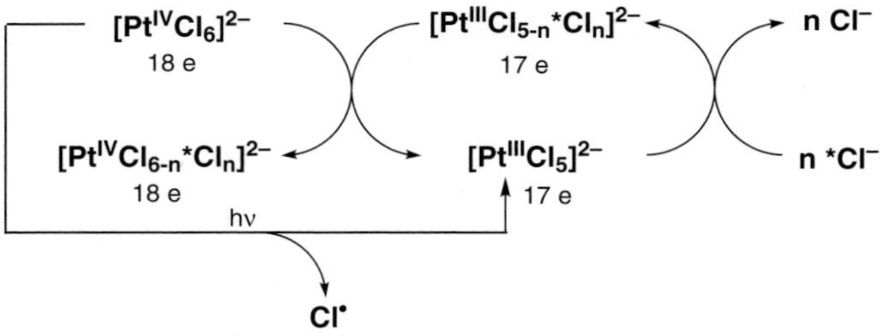

Scheme 9

The important oxidative addition reaction was found in 1972 by Osborn and co-workers to proceed according to an ATC mechanism in the case of the addition of certain halides to Ir^I. For instance, irradiation at 436 nm was proposed to induce reduction of EtI from the photoexcited state of Ir^I, generating $Et^\bullet$ and Ir^{II}, a process accelerated by addition of an electron-rich phosphine which produces a better reductant (eq. (6) and Scheme 10) [49].

$$EtI \; + \; [Ir^I(Cl)(CO)(L_2)] \; \longrightarrow \; [Ir^{III}(I)(Cl)(CO)(L_2)] \qquad (6)$$

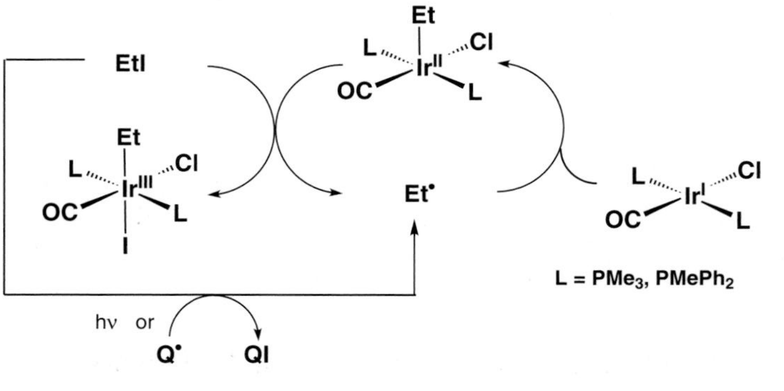

Scheme 10

Many classical reactions have been recognized to proceed according to this halogen-ATC catalysis mechanism; examples are the dehydroiodination of secondary iodides by Pt^0 [10], the bromination of $[Mn_2(CO)_{10}]$, and the exchange of carbonyls by phosphines or isonitriles [50].

3.2.8.3.2   Hydrogen-ATC Catalysis

In 1975, Brown [51] found that the substitution of CO by a phosphine in [Re(H)(CO)$_5$] proceeded according to a hydrogen-ATC mechanism. The reaction can be initiated, for instance, by irradiation at 311 nm of the dimer [Re$_2$(CO)$_{10}$] which generates the [Re(CO)$_5$]$^\bullet$ radical and can be inhibited by O$_2$ (eq. (7) and Scheme 11).

$$\text{H-M-CO} \ + \ \text{PR} \ \longrightarrow \ \text{H-M-PR}_3 \ + \ \text{CO} \tag{7}$$

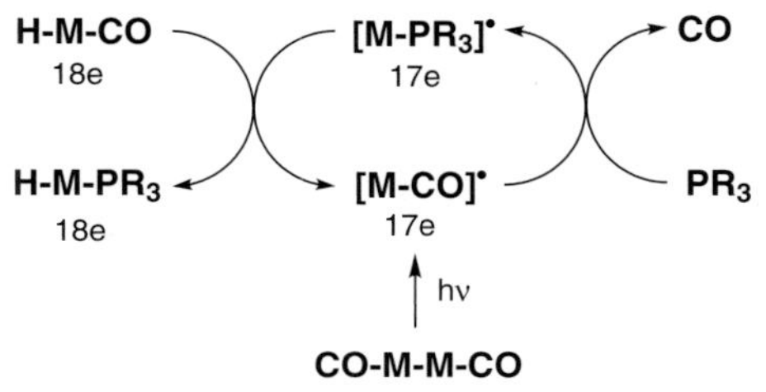

Scheme 11

This mechanism is quite general for this substitution reaction in transition metal hydride–carbonyl complexes [52]. It is also known for intramolecular oxidative addition of a C–H bond [53], heterobimetallic elimination of methane [54], insertion of olefins [55], silylenes [56], and CO [57] into M–H bonds, extrusion of CO from metal–formyl complexes [11] and coenzyme B$_{12}$- dependent rearrangements [58]. Likewise, the reduction of alkyl halides by metal hydrides often proceeds according to the ATC mechanism with both H-atom and halogen-atom transfer in the propagation steps [4, 53].

Finally, group-transfer chain catalysis involves the transfer of a group in the cross-propagation step; a few examples are known with transfers of groups such as allyl [59], cyclopentadienyl [60], and rhodium octaethylporphyrin [57].

3.2.8.4   Conclusions

Electrocatalysis and redox catalysis both involve catalysis using redox processes. Molecular engineering is a key feature in these areas of catalysis. It requires precise thermodynamic data for the investigated systems as well as imaginative designs, including the coupling of several types of catalysis in multicatalytic systems. Homogeneous systems are all the more useful as devices are complex because they are amenable to kinetic investigations, thus allowing a mechanistic approach. The latter, in turn, is useful for improving multicomponent systems which involve not only homogeneous processes but also heterogeneous ones including semiconductors and colloids. At this stage, molecular electronics and catalysis are clearly connected sciences. Such interdisciplinarity will spread in the future for the improvement of "catalytic engineering" to mimic the increasingly well-understood metalloenzymatic catalysis.

References

[1] H. Taube, *Electron-Transfer Reactions of Complex Ions in Solution*, Academic Press, New York, **1970**.

[2] D. R. H. Barton, S. I. Parekh, *Half a Century of Free-Radical Chemistry*, Cambridge University Press, Cambridge, **1993**.

[3] F. A. Cotton, G. Wilkinson, *Advanced Inorganic Chemistry*, 5th ed., Wiley, New York, **1988**.

[4] D. Astruc, *Electron Transfer and Radical Processes in Transition-Metal Chemistry*, VCH, New York, **1995**.

[5] D. Astruc in [4], Chapters 2 and 6.

[6] D. Astruc, *Angew. Chem.* **1988**, *100*, 662; *Angew. Chem., Int. Ed. Engl.* **1988**, *27*, 643.

[7] D. Astruc, *Acc. Chem. Res.* **1991**, *24*, 36.

[8] The distinction between inner-sphere and outer-sphere electron transfer was established by Taube; see Ref 1 (and for instance, Ref. 4, Chapters 1 and 7).

[9] M. Chanon, *Acc. Chem. Res.* **1987**, *20*, 214; M. Chanon, *Bull. Soc. Chim. Fr.* **1982**, II-197 and **1985**, 209; M. Chanon, *Chem. Rev.* **1983**, *87*, 425.

[10] J. A. Osborn in *Organotransition-Metal Chemistry* (Eds.: Y. Ishii, M. Tsutsui), Plenum, New York, **1978**, p. 69.

[11] J. K. Kochi, *J. Organomet. Chem.* **1986**, *302*, 389.

[12] M. I. Bruce, *Coord. Chem. Rev.* **1987**, *76*, 1.

[13] N. J. Coville in *Radical Processes in Organometallic Chemistry*, J. Organomet. Chem. Library, Vol. 22 (Ed.: W. C. Trogler), Elsevier, New York, **1990**, p. 108.

[14] M. Chanon, M. Julliard, J.-C. Poite (Eds.), *Paramagnetic Species in Activation, Selectivity, Catalysis*, Kluwer, Dordrecht, **1988**.

[15] R. A. Sheldon, J. K. Kochi, *Metal-Catalyzed Oxidation of Organic Compounds*, Academic Press, **1981**, New York.

[16] J. P. Collman, P. S. Wagenknecht, J. E. Hutchinson, *Angew. Chem.* **1994**, *106*, 1620; *Angew. Chem., Int. Ed. Engl.* **1994**, *33*, 1537.

[17] J. Reedjik (Ed.), *Bioinorganic Catalysis*, Dekker, New York, **1993**.

[18] L. Stryer, *Biochemistry*, 2nd ed., New York, Freeman, **1981**.

[19] D. Rehm, A. Weller, *Isr. J. Chem.* **1970**, *8*, 259.

[20] L. Eberson, *Electron Transfer in Organic Chemistry*, Springer, Berlin, **1987**, Chapter X.

[21] A. J. Bard (Ed.), *Encyclopedia of Electrochemistry of the Elements*, Vols. 1–14, Dekker, New York, **1980**.

[22] N. G. Connelly, W. E. Geiger, *Adv. Organomet. Chem.* **1989**, *23*, 1.

[23] E. Steckhan, *Top. Curr. Chem.* **1987**, *142*, 1.

[24] D. Astruc, *Chem. Rev.* **1988**, *88*, 1189.

[25] R. L. Rich, H. Taube, *J. Am. Chem. Soc.* **1954**, *76*, 2608.

[26] See [4], Chapter 7. For a recent example see: Y. Koite, C. K. Schauer, *Organometallics* **1993**, *12*, 4854; M. Hidai, Y. Misobe, *Chem. Rev.* **1995**, *95*, 1115; D. Sellmann, *Angew. Chem.* **1993**, *105*, 67; *Angew. Chem. Int. Ed. Engl.* **1993**, *32*, 64.

[27] For an in-depth study of ETC catalyzed substitution chemistry in $[Co_3(CO)_9(\mu_3\text{-}CR)]$ clusters, see K. Hinckelmann, J. Heinze, H. Schacht, J. S. Field, H. J. Vahrenkamp, *J. Am. Chem. Soc.* **1989**, *111*, 5078. Carbonyl substitution can also be autocatalytic (self-initiated); see W. Kaim in *Organometallic Radical Processes*, J. Organomet. Library, Vol. 22 (Ed.: W. C. Trogler), Elsevier, New York, **1990**, p. 173.

[28] D. S. Brown, M.-H. Delville-Desbois, R. Boese, K. P. C. Vollhardt, D. Astruc, *Angew. Chem.* **1994**, *106*, 715; *Angew. Chem., Int. Engl. Ed.* **1994**, *33*, 661.

[29] M. Tilset in *Energetics of Organometallic Species* (Ed.: J. A. M. Simoes), Kluwer, Dordrecht, **1992**, pp. 109–129.

[30] H. H. Ohst, J. K. Kochi, *J. Chem. Soc., Chem. Commun.* **1986**, 121; H. H. Ost, J. K. Kochi, *Inorg. Chem.* **1986**, *25*, 2066.

[31] C. Moinet, E. Román, D. Astruc, *J. Electroanal. Chem. Interfacial Electrochem.* **1981**, *241*, 121.

[32] J. Ruiz, M. Lacoste, D. Astruc, *J. Am. Chem. Soc.* **1990**, *112*, 5471.

[33] P. Boudeville, J.-L. Burgot, A. Darchen, *New. J. Chem.* **1995**, *19*, 179

[34] C. Roger, C. Lapinte, *J. Chem. Soc., Chem. Commun.* **1989**, 1598.

[35] D. Touchard, J.-L. Fillaut, H. Le Bozec, C. Moinet, P. H. Dixneuf in [14], p. 311.

[36] J.-N. Verpeaux, M.-H. Desbois, A. Madonik, C. Amatore, D. Astruc, *Organometallics* **1990**, *9*, 630.

[37] M.-H. Desbois, D. Astruc, *J. Chem. Soc., Chem. Commun.* **1990**, 943.

[38] D. Astruc, M.-H. Delville, J. Ruiz in *Molecular Electrochemistry of Inorganic, Bioinorganic and Organometallic Compounds*, NATO ASI Series, Vol. 385 (Eds.: A. J. L. Pombeiro, J. A. McCleverty), Kluwer, Dordrecht, **1993**, p. 277.

[39] R. D. Rieke, H. Kojima, K. Öfele, *J. Am. Chem. Soc.* **1976**, *98*, 6735; *Angew. Chem.* **1980**, *92*, 550; *Angew. Chem., Int. Ed. Engl.* **1980**, *19*, 538.

[40] N. G. Connelly, S. J. Raven, G. A. Carriedo, V. Riera, *J. Chem. Soc., Chem. Commun.* **1986**, 992.

[41] C. M. Arewgoda, B. H. Robinson, J. Simpson, *J. Chem. Soc., Chem. Commun.* **1982**, 284.

[42] D. R. Tyler in *Prog. Inorg. Chem.*, Vol. 36 (Ed.: S. J. Lippard), Wiley, New York **1988**, 125.

[43] S. L. Yang, C. S. Li, C. H. Cheng, *J. Chem. Soc., Chem. Commun.* **1987**, 1872.

[44] S. D. Jensen, B. H. Robinson, J. Simpson, *Organometallics* **1986**, *5*, 1690.

[45] D. P. Summers, J. C. Luong, M. S. Wrighton, *J. Am. Chem. Soc.* **1981**, *103*, 5238.

[46] M.-H. Desbois, D. Astruc, *J. Chem. Soc., Chem. Commun.* **1988**, 472; M.-H. Desbois, D. Astruc, *New J. Chem.* **1989**, *13*, 595.

[47] T. C. Klarcke, C. S. Yannoni, T. J. Katz, *J. Am. Chem. Soc.* **1983**, *105*, 1787.

[48] J. Herschel, *Philos. Mag.* **1832**, *1*, 58; M. Boll, *Ann. Phys. (Paris) 2*, **1914**, *5*, 226.

[49] J. S. Bradley, D. E. Connor, D. Dolphin, J. A. Labinger, J. A. Osborn, *J. Am. Chem. Soc.* **1972**, *94*, 4043; J. A. Labinger, A. V. Kramer, J. A. Osborn, *ibid.* **1973**, *95*, 7908.

[50] N. M. J. Brodie, A. J. Poe in [14], p. 345.

[51] B. H. Byers, T. L. Brown, *J. Am. Chem. Soc.* **1975**, *97*, 947; B. H. Byers, T. L. Brown, *ibid.* **1977**, *99*, 2527.

[52] T. H. Whitesides, J. Shelly, *J. Organomet. Chem.* **1975**, *92*, 215.

[53] T. L. Brown in *Organometallic Radical Processes*, J. Organomet. Chem. Library, Vol 22 (Ed.: W. C. Trogler), Elsevier, New York, **1990**, p. 67; N. J. Coville, *ibid.* p. 108.

[54] R. T. Edidin, J. R. Norton, *J. Am. Chem. Soc.* **1986**, *108*, 948.

[55] J. Halpern, *J. Am. Chem. Soc.* **1984**, *106*, 8319; J. Halpern in [14], p. 423.

[56] D. H. Berry, J. H. Mitstifer, *J. Am. Chem. Soc.* **1987**, *109*, 3777.

[57] B. B. Wayland, B. A. Woods, *J. Chem. Soc., Chem. Commun.* **1981**, 700.

[58] J. Halpern, *Science* **1985**, *227*, 869.

[59] M. Rosenblum, P. S. Waterman, *J. Organomet. Chem.* **1980**, *187*, 267.

[60] B. D. Fabian, J. A. Labinger, *J. Am. Chem. Soc.* **1979**, *101*, 2239; B. D. Fabian, J. A. Labinger, *Organometallics* **1983**, *2*, 659.

3.2.9   Homogeneous Photocatalysis

Andreas Heumann, Michel Chanon

3.2.9.1   Definitions

The actual meaning of the word *"photocatalysis"* is somewhat controversial [1, 2]. The aim of the contribution is to show how organometallic complexes have made it possible to invent novel chemical transformations without necessarily disentangling all the kinetic subtleties associated with this field. The definitions adopted will be the rather pragmatic ones proposed by Kirsch to represent the set of transformations involving photochemistry and more or less directly connected with catalysis [3]. However, other definitions are used elsewhere, and therefore references are cited in which other definitions have been adopted. Although stoichiometric transformations (e. g., [4]) are obviously interesting, nothing in their molecular description connect them with catalytic phenomena. If one thinks that the compulsory bottleneck of passing through very small quantities of electronically excited states is reminiscent of the very small quantities of catalysts used in an efficient catalytic transformation, one goes back to some old approaches of photochemistry where light was considered as a catalyst just because it was inducing a transformation.

Photocatalysis can be defined as "acceleration of a photoreaction by the presence of a catalyst" [3, 5]. The catalytic dimension may originate either because of the quantity of consumed photons or because of the quantity of one added substance [6]. The most restrictive situation corresponds to a transformation, photoinduced with a catalytic quantity of photons, provided that a catalytic quantity of an exogenous substance is added to the reaction mixture. The two other possibilities, where the catalytic dimension comes either from the number of consumed photons or from the presence of an added substance in catalytic quantities, are also well identified.

Several processes that are "catalytic" (a photon is not a substance) in photons and involve a catalytic quantity of one compound have been reported. Different labels were associated with such an overall situation: *electron transfer induced chain reactions* [7], *photoinduced catalytic reactions* [8], *or photogenerated catalysis* [9]. The main experimental observations which characterize such processes are:

(1) the presence of quantum yields greater than 1 (exceptions to this statement are known [10, 11]);
(2) the possible presence of an induction period;
(3) the possible continuation of catalysis after termination of the irradiation.

Salomon identified several types of generic situations among the transformations responding to these experimental criteria [9]. They are gathered in Scheme 1.

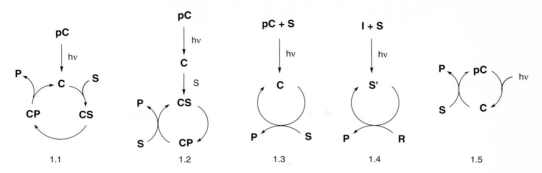

Scheme 1. Types of photocatalytic process (see also Scheme 4).

In this Scheme, pC stands for pro-catalyst, C for catalyst, CS for a complex between catalyst and substrate, CP for a complex between catalyst and product, I for an initiator, S′ for a structural variation of the substrate, R for an added reagent. In cases 1.1 and 1.2 the catalysis is based on a coordinative interaction between catalyst and substrate; in case 1.1 the product is released to regenerate C (for example by reductive elimination) whereas in case 1.2 the regeneration of CS results from a substitution of the complexed product by S. It should be clear that cases 1.1 and 1.2 do not exhaust the formal possibilities offered to photogenerated catalysis. One may actually imagine a photogeneration of catalyst from a selected pro-catalyst for any of the multiple catalytic cycles identified in homogeneous catalysis centered on transition metal complexes [12].

In case 1.3 the pro-catalyst interacts with the substrate (for example by photo-substitution) to generate the actual catalyst. Such a case is germane to *photo-induced electron transfer catalyzed reactions* [7] shown in case 1.4. Here, an excited state of a transition metal complex (designated as I) could interact with the substrate by a positive or negative electron transfer. Because of the activation induced by electron transfer [13] S would be transformed into S′, whose fate is to be regenerated one or several steps after yielding P.

In cases 1.1 to 1.4 the role of the photon is played outside the catalytic cycle, which explains why the quantum yield of P is usually greater than 1. In *stoichio-metric photogenerated catalysis* [9] (Scheme 1.5), also labeled *photoassisted* [8, 14], *photoenhanced* [15], or *photoactivated* [16], the role of the photon is inside the catalytic cycle; therefore every S → P transformation consumes a photon and the overall quantum yield for the production of P is smaller than 1. Scheme 1, case 1.5 should not be confused with sensitization because in the step S + C → P + pC both the substrate and the catalyst are in their ground state, which is not the case for photosensitization.

An example of case 1.1 is provided by the photosubstitution of CO ligands by L (L = PBu$_3$, PPh$_3$) in HRe(CO)$_5$ in the presence of catalytic quantities of Re$_2$(CO)$_{10}$ or Mn$_2$(CO)$_2$, for which a possible mechanism is shown in Scheme 2. This me-chanism is not the one proposed in the original report [17]: it illustrates the inter-vention of uneven catalytic cycles (contrasted with the ones involving species

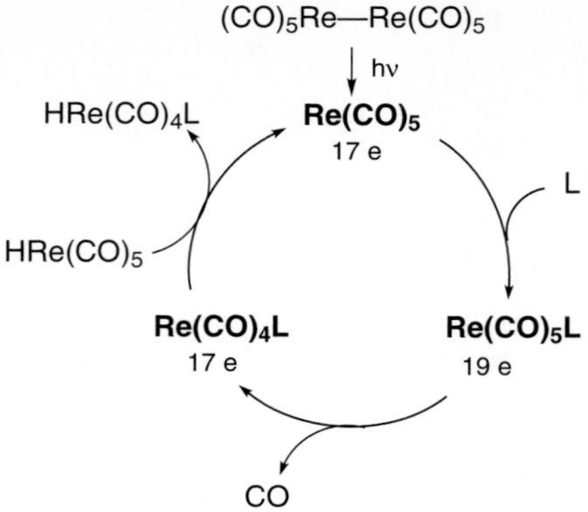

Scheme 2. Photosubstitution in rhenium carbonyl complexes.

with an even number of valence electrons) and the importance of 19 e species [18, 19].

The double bond migration or *cis–trans* isomerization of linear pentenes catalyzed by a variety of transition metal complexes ($Fe(CO)_5$, $Fe_3(CO)_{12}$, $Ru_3(CO)_{12}$) in the presence of irradiation illustrates the operation of case 1.3 [20, 21] (Scheme 3). Case 1.4, which covers photoinduced electron transfer

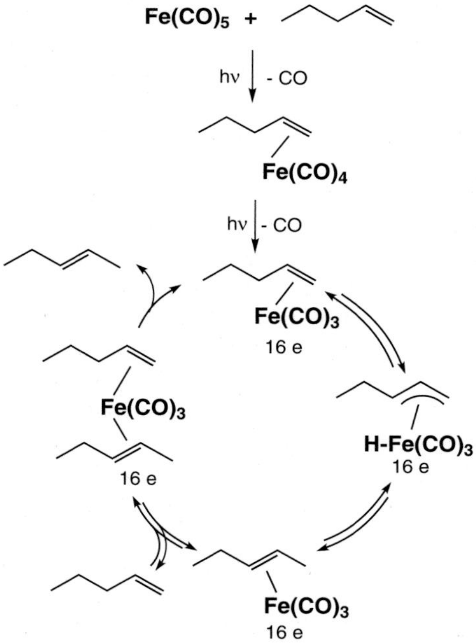

Scheme 3. Photoisomerization of linear pentene.

catalysis, has been dealt with thoroughly in [7]. Electrochemical studies [22–26] suggest that the photochemical scope of this type of process involving organometallics has been underexploited.

Case 1.5 would be illustrated by photoextrusion of a ligand from a 18e transition metal complex [27] creating an unsaturation, making possible the complexation of the substrate S to be activated.

The terms used to describe the class of processes that are noncatalytic in photons, depicted in Scheme 1, case 1.5, are rather confusing. Kutal proposed to cover all situations by the term *catalyzed photochemistry* [28]. This includes three cases:

(1) An excited state of the substrate S reacts with the transition metal complex, whereas S in the ground state would not react. Such a situation was labeled *catalyzed photoreaction* by Wubbels [1], Salomon [9], Mirbach [29], and Hennig [8].
(2) Case 1.5 in Scheme 1.
(3) Photosensitized reactions.

This combination may be justified because it is experimentally not trivial to distinguish the situation where photocatalysis results from an interaction between the excited state of the substrate and the catalyst, and the situation where photocatalysis results from a ground-state interaction between the catalyst and the substrate. It is apparently easier to distinguish experimentally a process that is catalytic in photons from a noncatalytic one, than to establish that the overall catalytic activity originates from an interaction between the excited state of the substrate (S) and a transition metal complex (C) and not from a ground-state interaction between S and C [30].

The other great class of processes connected to photocatalysis is photosensitization. Here the processes are no longer "*catalytic*" in photons but may involve catalytic quantities of sensitizer. Therefore the quantum yield of product formation is generally smaller than 1. If one relies on the most general definition of a photosensitized reactions as "a reaction in which a chemical species having absorbed light undergoes no practical change but some other species undergo a certain chemical reaction without absorption of photons" [31], then the sensitizer may sensitize the formation of a catalyst or an initiator. In these specific cases the quantum yield of product formation may obviously become greater than 1. The *formal* criterion to distinguish between $\phi <$ and $\phi > 1$ is to write the catalytic cycle and to check if $h\nu$ is operating inside or outside the catalytic cycle (see Scheme 1, cases 1.1–1.4 vs. 1.5). The difficulties in establishing experimentally the mechanism best fitting the formal scheme prevent the attribution of an unambiguous label to many of the identified transformations involving both light and transition metal complexes.

Several illustrations are necessary to clarify sensitized photoreactions. If we limit ourselves to the cases where the organometallic compound is the sensitizer, most of the situations are gathered in Scheme 4. The extension to photosensitized transformations of organometallics is straightforward [32].

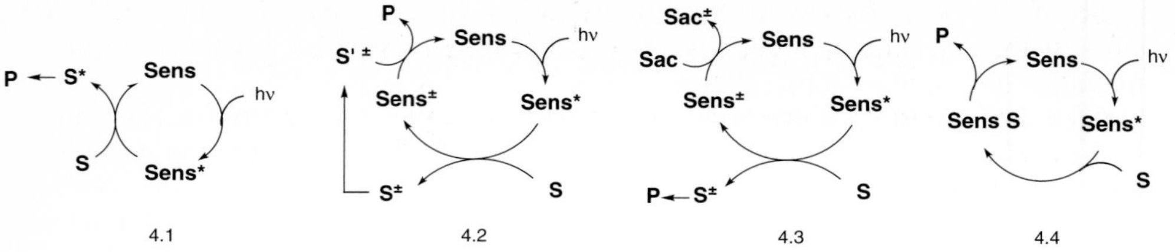

Scheme 4. Photocatalytic systems centered on sensitizers.

In Scheme 4, Sens stands for sensitizer, S for substrate, S′ for transformed substrate, Sac for sacrificial reagent (because 1 mol of Sac is consumed for 1 mol of S → P transformation). Case 4.1 corresponds to an energy transfer induced transformation of the substrate into product(s). Cases 4.2 and 4.3 correspond to a transformation resulting from an electron transfer between the sensitizer in its excited state and the substrate.

The difference between cases 4.2 and 4.3 is that in the first case the transformation of S into P is coupled with the regeneration of Sens (case 4.2), whereas in the second case (case 4.3) this regeneration has to be performed by an oxidant or a reducing agent purposely added to the reaction mixture. Finally, the excitation of the sensitizer may lead to an associative activation of the substrate toward the sensitizer and it is this substrate–sensitizer complex which evolves toward product(s) with regeneration of the sensitizer.

These schemes have been thoroughly explored with inorganic transition metal complexes, particularly with respect to the use of solar energy [33–35]. Far less has been explored in this direction using organometallic compounds as sensitizers or substrates.

Some references of reviews besides the ones already cited are given [1, 3, 5–9, 19, 23–25, 28, 31, 33]. Organometallic photochemistry [36] was excellently treated in [37] and may be compared with inorganic photochemistry to gain further inspiration [38–40]. A recent multiauthored book strongly overlaps with the subject matter of the present section, and should certainly be consulted [41]. Electron transfer reactions play a determinant role in many photocatalytic processes; several recent reviews and books may be cited on this topic [42–44]. The photochemistry of the M–CO bond [45] and the theme of photocatalysis by transition metal complexes [46] have recently been reviewed. Covalently linked donor–acceptor systems for mimicry of photosynthetic energy transfer have been discussed in [47]. Several special issues of *Coordination Chemistry Reviews* have been devoted to the photochemistry and photophysics of coordination compounds [48–50], and a special issue to photochemistry [51]. Further developments in photochemistry were the subject of a special issue of *Chemical Reviews* [52]. Practical considerations useful for designing photochemical experiments may be found in [53].

3.2.9.2 Synthesis and Activation – What *hv* Metal Catalysis Can Do Better?

In a review devoted to transition metal complexes in photocatalysis numerous examples of transformations [54] and for the selective activation of small molecules [55] have already been gathered.

3.2.9.2.1 C–C Bond Formation

Alkylation and Carbonylation

Allyl–allyl cross-coupling of allyl bromides (Structure **1**) and allyl sulfides (**2**) carrying homoallylic alcohol or ester functions takes place under irradiation with hexamethylditin (see eq. (1)). The reaction cleanly leads to 2,6-dienes (**3**) and no isomerization of allylic bromides is oberserved [56].

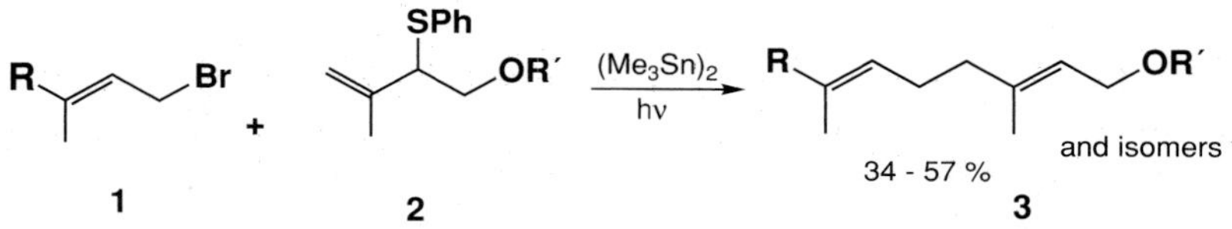

$$(1)$$

An interesting result of control of acyclic stereochemistry is reported by Nagano et al. [57], who showed that efficient 1,2-asymmetric induction can be achieved in radical-mediated allylation of diethyl (2S,3S)-3-bromo-2-oxo-succinates stereoselectively. In the Eu(fod)$_3$ (1.1 equivalent) photocatalyzed reaction of bromohydroxy compound (**4**) diastereoselectivity is reversed with respect to the simple photoreaction. On the other hand, substitution with silyl groups tends to enhance diastereoselectivity up to 8.6:1. The effect is still operative to a lesser extent with catalytic amounts of the lanthanide reagent (0.1 equivalents, *threo/erythro* [**5/6**] = 3:1) (eq. (2)) [57].

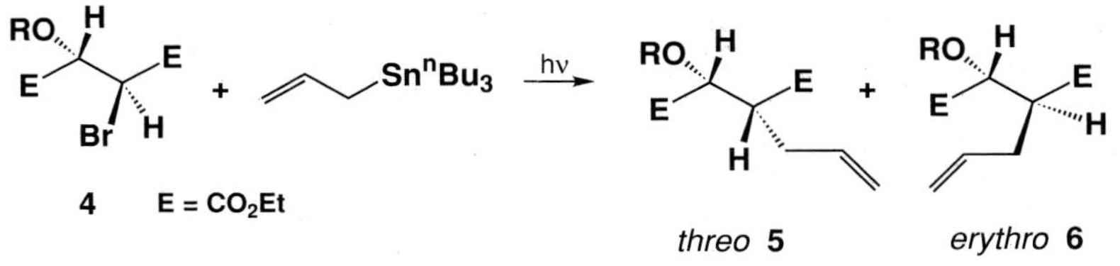

$$(2)$$

		threo 5		erythro 6
R = H		1	:	1.9
R = H	Eu(fod)$_3$	1.7	:	1
R = SiMe$_3$	Eu(fod)$_3$	8.6	:	1

Dimeric iron or manganese carbonyl complexes such as [CpFe(CO$_2$)]$_2$ and Mn$_2$(CO)$_{10}$, respectively, upon irradiation photocatalytically cleave carbon–halogen bonds. This leads to carbon-centered radicals which can be reduced to hydrocarbons or add to alkenes yielding saturated (**8**) and/or unsaturated (**9**) products (eq. (3)). This transformation is also suitable for cyclization reactions (intramolecular radical-trapping) [58].

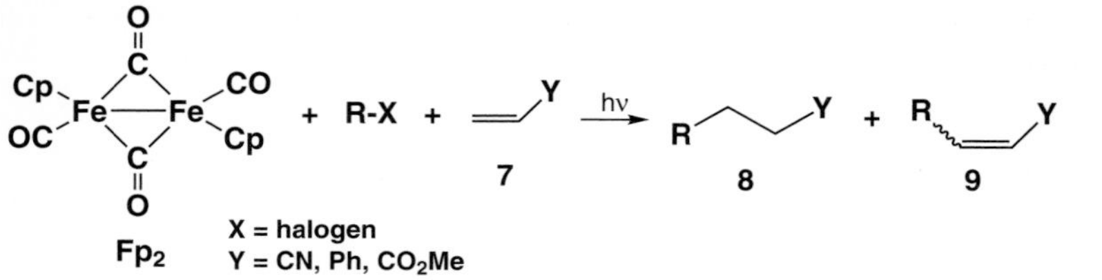

In the cobalt-catalyzed photochemical carbonylation of olefins, hydroformylation can be performed easily at ambient temperature (and high pressure) with high primary aldehyde selectivities (cf. Section 2.1.1) [59]. Under comparable conditions allylic amines are carbonylated to 2-pyrrolidinone, *N,N'*-diallylurea, and *N*-allyl-3-butenamide [60]. Photochemical methoxycarbonylation of olefins is possible at ambient conditions, i.e., at room temperature and atmospheric pressure [61].

In the photochemical activation of CO_2 the successful achievement of Ni catalysis is considered a milestone [62]. For the first time, a nickel–phosphine cluster can efficiently reduce CO_2 to radical anion $CO_2^{\bullet -}$ capable of carbon–carbon coupling reactions (cf. Section 3.3.4).

Photocyclization

In 1989 Curran and co-workers reported on a photocatalytically induced free-radical cyclization leading to various cyclic, bi-, or polycyclic carbocycles (fused and spiro) via isomerization of unsaturated iodides (alkenes, alkynes) [63]. This corresponds to the nonreductive variant of the tin hydride method. Under sunlight irradiation and in the presence of 10 mol% hexabutylditin, α-iodo esters, ketones, and malonates are efficiently transformed via an iodide atom transfer chain mechanism (eq. (4)).

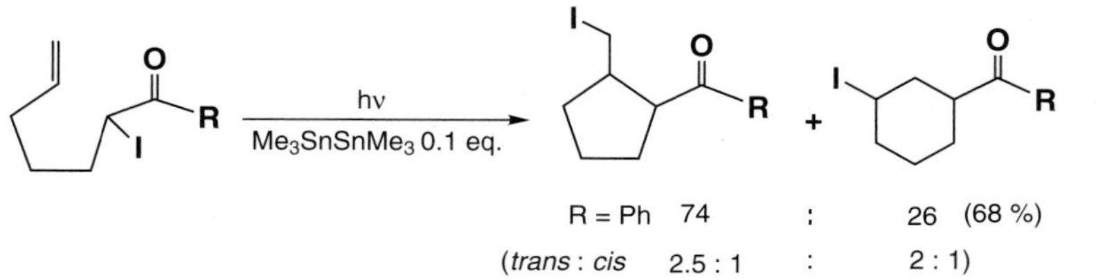

The γ-iodo carbonyl compounds can either be isolated or transformed *in situ* to deiodinated products or to lactones. Synthetic and mechanistic studies with hex-5-ynyl iodides showed the generality of the method [64]. When tested and compared under three sets of conditions, the photolysis with catalytic hexabutyltin showed the highest reactivity and selectivities. The tin additive plays the role of an iodine trap in a two-step radical chain reaction. The synthesis ofa capnellene (**10**) outlined in eq. (5) also shows the synthetic potential of this highly controlled radical reaction (a cascade or domino reaction [65, 66]).

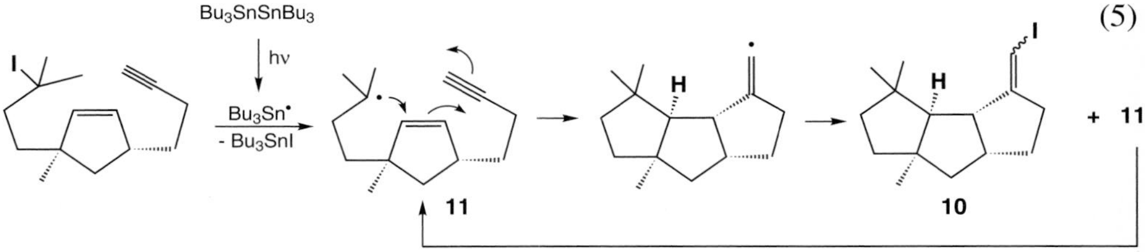

(5)

Curran's procedure has been used for ring closure reactions of various methylenecyclopropyl-substituted malonate iodides via 5-*exo*, 7-*endo*, or 8-*endo* cyclization [67].

The radical photoisomerization of iodoacetylenic esters (alkynes) represents a route to iodoalkylidene lactones [68]. Zinc has been added to reduce side reactions and to increase yields of the photolysis reaction. Bromoalkynyloxiranes are photocatalytically (tri-*n*-butyltin) cyclized to allenylidene tetrahydrofurans [69].

One may note that these radical reactions involve catalytic amounts of organometallic and catalytic amounts of light ($\phi > 1$) but, as is the case for some photoinduced electron transfer catalyzed reactions (cf. Section 3.2.9.1), they could be classified as chain processes rather as catalytic processes [70].

Photocyclization of methoxynaphthyl analogs of chalcone is reported to proceed via (unusual) electron transfer from excited vinyl arenes. Copper(II) gives an organocopper intermediate which evolves via a radical cation to a cyclized radical and the final naphthofuran **12** (eq. (6)) [71].

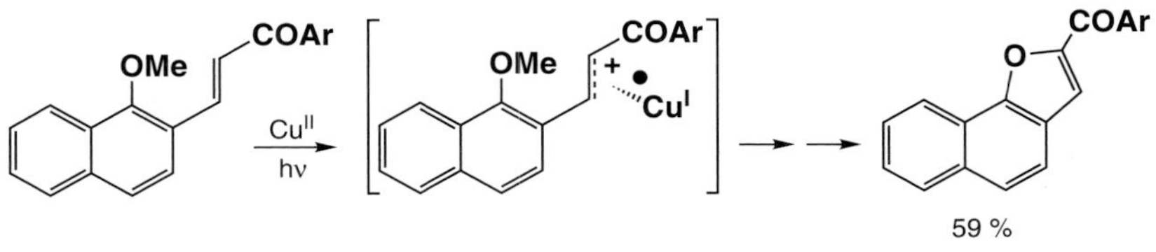

(6)

59 %

12

Palladium-catalyzed cross-coupling of alkyl halides is a challenging problem due to slow oxidative addition rates and rapid β-elimination of palladium hydride (Scheme 5).

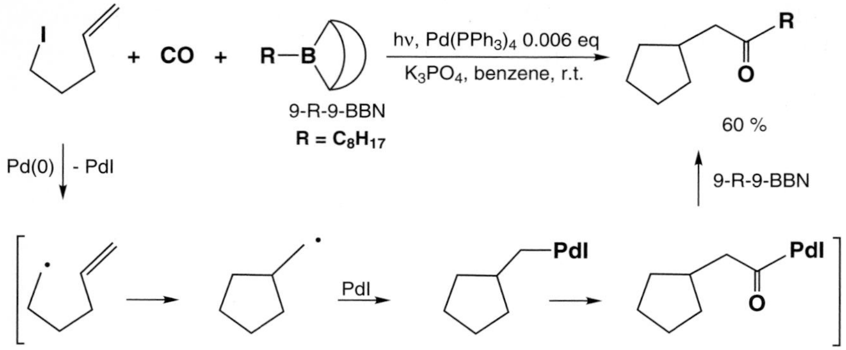

Scheme 5. Palladium-catalyzed photocatalytic carbonylative coupling with boranes [73].

Photocatalytic carbonylative coupling with 9-alkyl-9-borabicyclo[3.3.1]-nonanes (9-R-9-BBN), however, made it possible to transform alkyl halides to ketones [72]. Iodoalkenes or iodoalkynes are thus cyclized to five-membered rings [73]. The oxidative addition of iodoalkyl to palladium(0) proceeds via radicals allowing the ring closure to take place prior to the dual coupling with CO and the alkylboranes.

Photocycloaddition

Photodimerization and cross-cycloaddition of coumarins are improved by Lewis acids [74]. Similarly, photochemical [2+2] cycloadditions [75] of 1- and 2-naphthols [76] with ethylene are promoted by aluminum halides yielding the [2+2] adduct from the (complexed) enone form. According to the structure, substitution (e. g., methyl) vicinal to the OH group in 1-naphthol gives rise to ring-contracted indanone products. The formation of (ring-contracted) benzobicyclo[3.1.0]bicyclohexenone was already observed by irradiation (AlCl$_3$) of 1-naphthol without ethylene (Scheme 6) [77].

The [3+2] methylenecyclopentane annulation of [(trimethylsilyl)methylene]-cyclopropane dicarboxylates with unactivated and electron-rich alkenes (vinyl ether, vinyl thioether, or vinyl silyl ether) are efficiently photocatalyzed by butyl disulfide or bis(tributyltin) [78].

With the sequential [2+2] cycloaddition, *exo*-allylation, hydrohalogenation, and ring expansion, cycloalkenes and dichloroketene are transformed to *cis*-fused cycloheptanones. The photocatalytic step consists of radical alkylation (cycli-

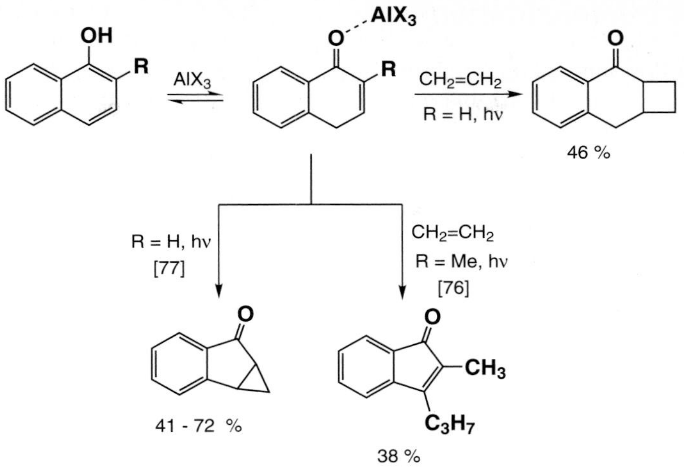

Scheme 6. [2+2] Photocycloaddition and ring contraction of 1- and 2-naphthols [76, 77].

zation) of the cyclobutanone and subsequent radical ring enlargement (Bu₃SnH) (eq. (7)) [79].

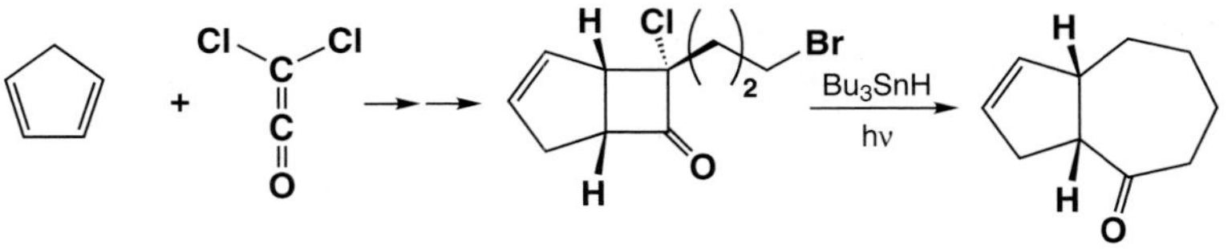

(7)

Curran's photocatalytically induced radical [4+1] annulation of phenyl isocyanide (**13**) and bromopyridone (**14**) represents the key step of the campto-thecin synthesis [80, 81]. The remarkable one-step synthesis of the tetracyclic heterocyclic system starts with photolytic cleavage of hexamethylditin to form the Me₃Sn radical, which then cleaves the C–Br bond in **14**. This new radical reacts with the isonitrile carbon to form **15** which yields the final **16** via two subsequent radical intermediates (eq. (8)) (cascade or domino reaction [65, 66]).

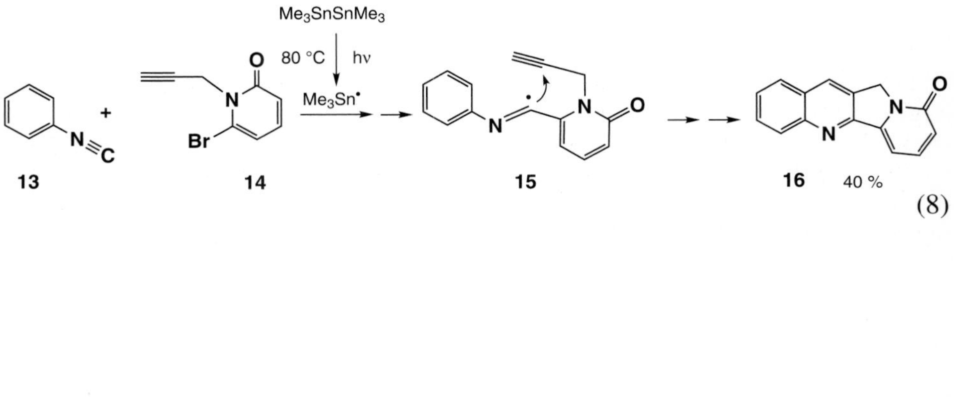

(8)

3.2.9.2.2 Photooxidation

The most important photocatalytical reactions are oxidation reactions which include the oxygenation of unsaturated systems, but also oxidations of saturated carbons with or without incorporation of oxygen (C–H activation). The photo-oxygenation of olefins in the presence of Ti^{IV}, V^{IV}, or Mo^{VI} catalysts leads *one-pot* to epoxy alcohols (**17**) via singlet oxygen (eq. (9)) [82].

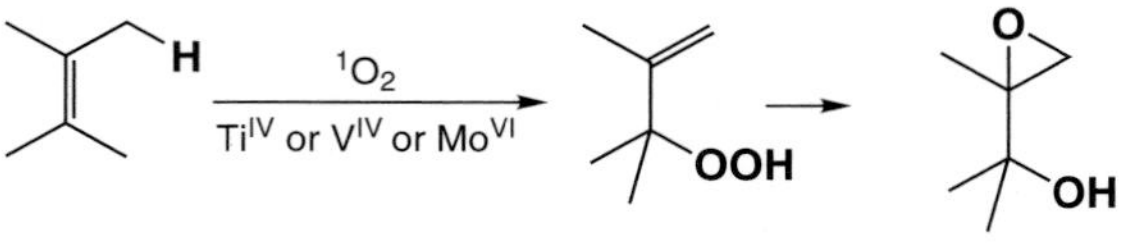

$$(9)$$

17

The advantage of this method lies in the fact that the peroxidic oxygen is generated *in situ*, does not accumulate, and transfers an oxygen atom to the allylic alcohol. Chemical yields and diastereoselectivities are good; the important pattern is the ene reactivity of the alkene with singlet oxygen. Chiral epoxides could be obtained with diethyl tartrate in a good enantiomeric excess of 72 % *ee*. The reaction has been successfully extended to vinylsilanes (oxyfunctionalization at the allylic site) [83], halogen substituted alkenes [84], and hydroxyvinylstannones which after TPP reduction predominantly yield *erythro* diols [85]. The same diastereoselectivity is observed in the singlet-oxygen ene reaction with chiral allylic acetates [86]. However, an allylic hydroxy group directs, via coordination of the incoming electrophilic oxygen, to *threo* 1,2-dioxygen products. Application of photocatalytic conditions to these hydroperoxy homoallylic alcohols leads to epoxy alcohols with unusually high diastereoselectivities [87]. Besides its synthetic interest, this transformation illustrates the difficulty of easily defining a borderline between organometallic and inorganic photocatalysis.

Bergman [88] reported on the Mo^{VI} oxo complex-catalyzed epoxidation of olefins by alkyl hydroperoxides (e. g., *t*-butyl hydroperoxide, TBHP) (eq. (10)). The active Cp^*MoO_2Cl catalyst is generated by irradiation of $Cp^*Mo(CO)_3Cl$ in the presence of dioxygen.

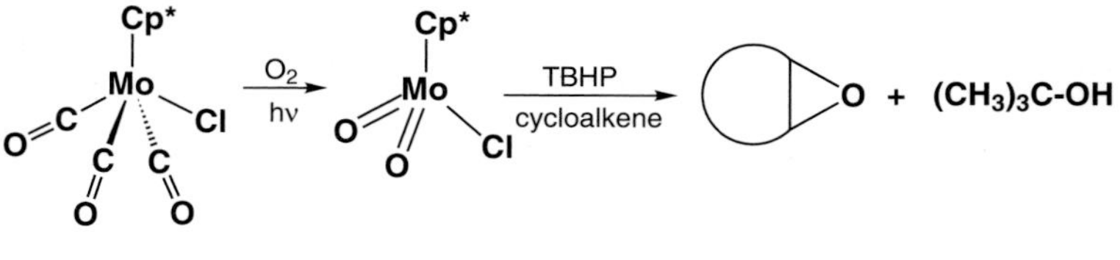

$$(10)$$

The combination of a (tetraarylporphyrinato)Fe^{III} photocatalyst and molecular oxygen transforms strained alkenes to (preferentially) epoxides, whereas unstrained olefins lead to allylic oxygenation products [89]. The use of water-soluble metal porphyrin complexes (Mn^{III}, Fe^{III}) facilitates the separation of substrates and products in aqueous solvent systems [90]. Copper(II) chloride induces chemo- and regioselectivity in the photooxychlorination of olefins (eq. (11)) [91].

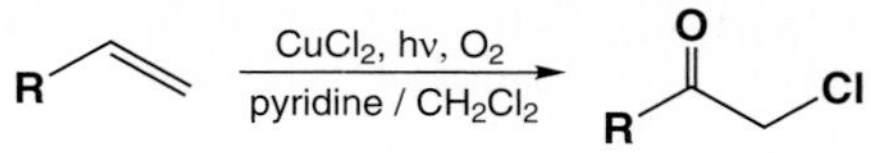

$$R\diagup\diagdown \xrightarrow[\text{pyridine / CH}_2\text{Cl}_2]{\text{CuCl}_2,\ hv,\ O_2} R\diagdown\overset{\displaystyle O}{\overset{\|}{C}}\diagup\diagdown Cl \qquad (11)$$

3.2.9.2.3 C–H Activation [92]

Organic substrates (alkanes alkenes, alcohols) are also photooxidized by *trans*-dioxo Ru^{VI} and Os^{VI} complexes [93]. The interest in these catalysts may lie in the transformation of cyclohexane to cyclohexanone and cyclohexanol in reasonable yields. The presence of alcohol, ester, and ketone functional groups is tolerated in the catalytic functionalization [94] with polyoxometallates and Pt^0 as co-catalyst [95].

Rh^I catalyzes the photochemical dehydrogenation of alkanes with high efficiency [96]. Cyclooctane was transformed with quantum yields up to 0.10 and turnover numbers as high as 5000 [97] (Scheme 7). *trans*-Rh(PMe$_3$)$_2$(CO)Cl was shown to be the only significant photoactive species in solution. The active catalyst Rh(PMe$_3$)$_2$Cl is formed by photoextrusion of carbon monoxide from the rhodium carbonyl complex, a process that delivers the energy needed for the thermodynamically unfavored dehydrogenations.

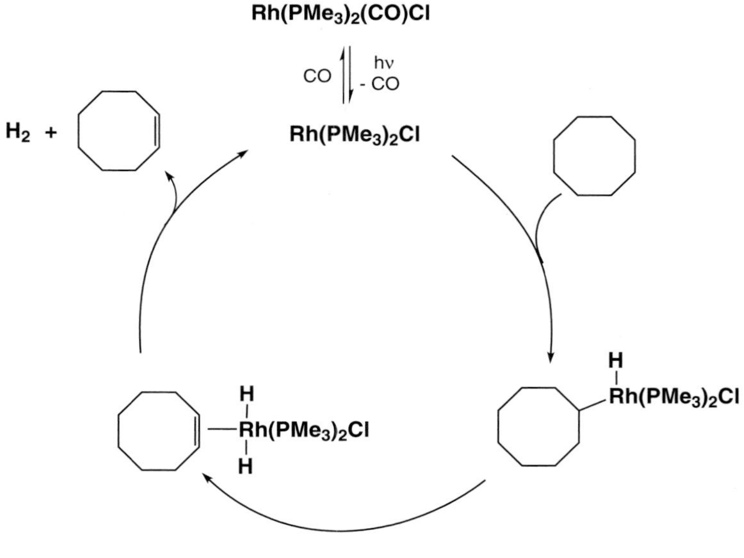

Scheme 7. Photochemical dehydrogenation of cyclooctane [97].

The same photocatalyst system permits the observation of the insertion of alkynes into C–H bonds [98]. Isonitrile insertion into aromatic C–H bonds with (C$_5$Me$_5$)Rh(CNCH$_2$CMe$_3$)$_2$ gives aldimines in low yield [99]. Photoinduced electron transfer between pyridine derivatives and alkyltin reagents leads to alkylation in α and γ positions of the pyridine [100].

The asymmetric coupling of 2-naphthol to optically active 1,1′-bis(2-naphthol)-derivatives (**18**; eq. (12) [101]) has now been realized photocatalytically with C_3-symmetric Δ-[Ru(menbpy)$_3$]$^{2+}$ (menbpy = 4,4′-di-(1R,2S,5R)-(-)-menthoxy-carbonyl-2,2′-bipyridine) as a photosenzitizer and [Co(acac)$_3$] as an oxidant (16.2 % *ee*) [102].

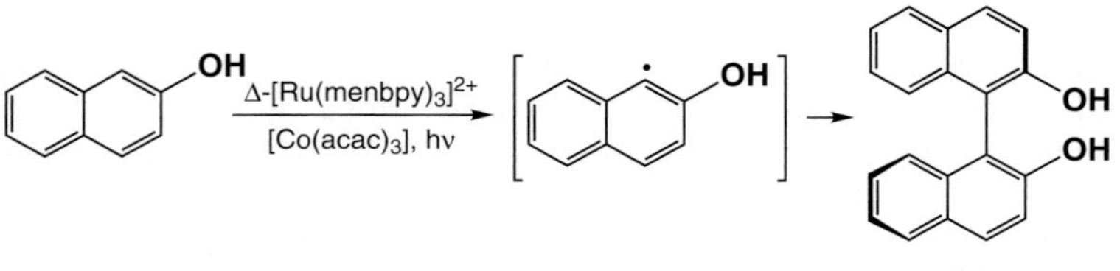

(12)

(R)-**18**

Mercury photosensitized ($^3P_1$-excited state) dehydrodimerization of hydrocarbons [103] has been developed into a useful organic synthetic method by using a simple reflux apparatus in which the radical reaction products are protected from further transformation simply by condensation (vapor-pressure selectivity) [104]. The selectivity of C–H cleavage increases from primary to tertiary carbons (350:1) and the method permits the formation of highly substituted C–C bonds (eq. (13)). One limitation for product formation is the appearance of four sets of obligatory 1,3-*syn* methyl–methyl steric repulsions (e. g., 2,3,4,4,5,5,6,7-octamethyloctane).

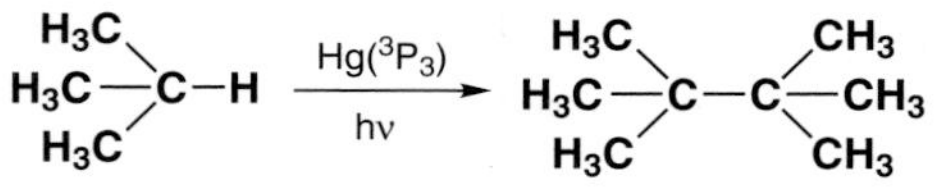

(13)

The reaction may proceed as homo- or cross-dehydrodimerization [105] and takes place with a wide range of substituted substrates such as higher alcohols, ethers, silanes, and partially fluorinated alcohols and ethers, but also with ketones, carboxylic acids, esters, amides, and amines [106]. Besides the formation of 1,2-diols from saturated alcohols, unsaturated substrates are also dimerized under hydrogen to form 1,n-diols other than the 1,2-isomers [107]. The regioselectivity of the diols is controlled by the formation of the most stable radical, which then dimerizes.

3.2.9.2.4   Photoreduction and Photocleavage

Photocatalytic reductions may concern organic unsaturations (C=O, C=C, etc.), or inorganic CO_2, or bicarbonate. Even carbon tetrachloride is very efficiently photoreduced to chloroform by alcohols with *meso*-tetra(2,6-dichlorophenyl)porphyrin [108]. Intermolecular hydrogen transfer is catalyzed by cobalt–phosphine complexes [109]. In this reaction photoirradiation generates the active hydride species "CoH[PPh(OEt)$_2$]$_3$" for the reduction of ketones with secondary alcohols. The efficient (TON up to 59) and chemoselective reduction of aldehyde carbonyls has been described with $Rh(PMe_3)_2(CO)Cl$ using cyclooctane as a hydrogen source (eq. (14)) [110].

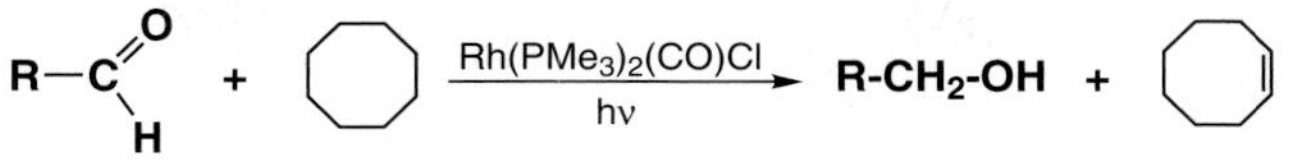

$$R-C\underset{H}{\overset{O}{<}} \; + \; \text{(cyclooctene)} \quad \xrightarrow[h\nu]{Rh(PMe_3)_2(CO)Cl} \quad R\text{-}CH_2\text{-}OH \; + \; \text{(cyclooctadiene)} \tag{14}$$

The challenging photochemical reduction of carbon dioxide to formate is catalyzed by Ru^{II} [111] (cf. Section 3.3.4). For example, with the 2,2′-bipyridine–ruthenium(II) complex the active species is formed by photolabilization. Water renders the system more efficient with quantum yields up to 15 %. Methanol is the photoproduct when CO_2 is reduced with TiO_2 in propene carbonate/2-propanol [112]. In a more sophisticated system, containing deazariboflavin (dRFl, **19**) as photosensitizer, N,N'-dimethyl-4,4′-bipyridinium (MV^{2+}) as primary electron acceptor, and sodium oxalate as sacrificial electron donor, in the presence of a Pd colloid stabilized by β-cyclodextrin (Pd-β-CD), bicarbonate is reduced to formate [113] (Scheme 8).

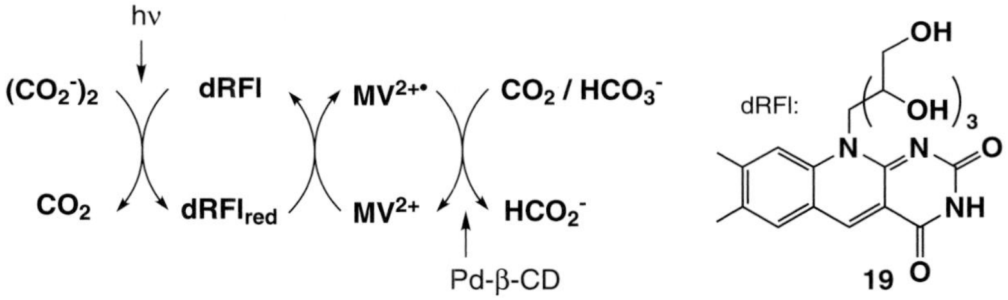

Scheme 8. Reduction of bicarbonate in the presence of a Pd colloid stabilized by β-cyclodextrin [113]

The mechanism of photocatalytic hydrogenation has been studied (by IR) with norbornadiene (nbd) and Group 6 metal carbonyls with respect of the role of H_2 [114] and the role of the diene [115]. In a subsequent study [116], *mer*-$[Cr(CO)_3(\eta^4\text{-norbornadiene})(\eta^2\text{-ethylene})]$ was found to be a key compound in the understanding of the photocatalytic diene hydrogenation (eq. (15)).

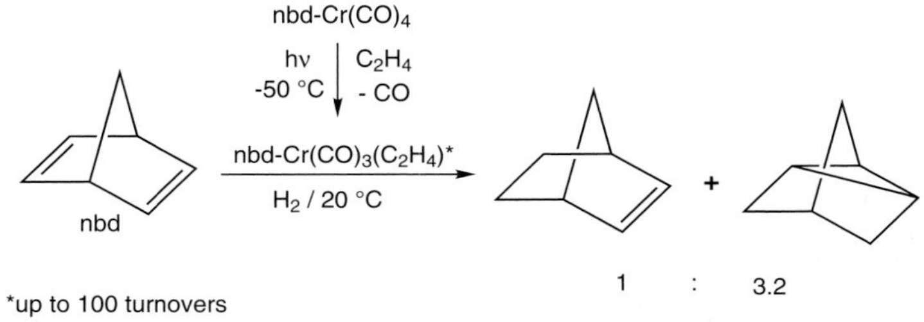

$$\tag{15}$$

*up to 100 turnovers

Photocatalytic cleavage of 1,2-diols [117] or 1,2-diphenylethane-1,2-diols [118] with Fe^{III} porphyrin (Fe^{III} (tmpyp)) leads to aldehydes and small quantities of the corresponding acids (eq. (16)).

$$\text{PhHC}\!-\!\text{CHPh} \; + \; 2\,Fe^{III}(tmpyp) \xrightarrow{\;h\nu,\,Ar\;} 2\,PhCHO \;+\; 2\,Fe^{III}(tmpyp) \;+\; 2\,H^+ \quad (16)$$
$$\underset{\text{OH \quad OH}}{\big|\quad\big|}$$

3.2.9.2.5 Isomerization

Iron carbonyl complexes are efficient in photoisomerizations of 2-alkenylphosphoramides to 1-alkenylphosphoramides [119] and of unsaturated alcohols to ketones (eq. (17)) or aldehydes, respectively [120].

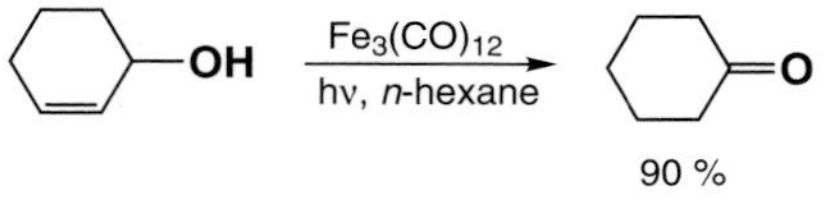

$$\text{(cyclohexenyl)}\!-\!\text{OH} \xrightarrow[\text{h}\nu,\,n\text{-hexane}]{Fe_3(CO)_{12}} \text{(cyclohexanone)}\!=\!\text{O} \qquad (17)$$
$$90\,\%$$

3.2.9.2.6 Polymerization

Photopolymerization with transition metals [121] has been used for the formation of homopolymers and block copolymers from norbornene (nbn) and phenylacetylene with $W(CO)_6$ (eq. (18); cf. Sections 2.3.3 and 3.3.10.1) [122].

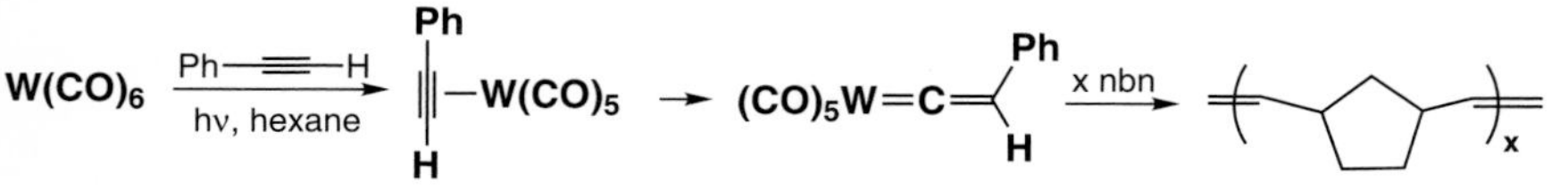

$$W(CO)_6 \xrightarrow[\text{h}\nu,\,\text{hexane}]{Ph\equiv\!\!\!\equiv H} \overset{Ph}{\underset{H}{|\!|}}\!-\!W(CO)_5 \;\longrightarrow\; (CO)_5W\!=\!C\!\!\overset{Ph}{\underset{H}{\diagdown}} \xrightarrow{x\,nbn} \text{(polymer)}_x \qquad (18)$$

3.2.9.3 Conclusion: What Photochemical Techniques Can Provide in Mechanistic Studies of Transition Metal Catalysis

The highly elaborate equipment associated with some time-resolved photochemical studies makes it possible to observe directly and study quantitatively the reactivity of transient species involved in a catalytic cycle. Time-resolved IR spectroscopy has allowed not only direct observations of "nonclassical" dihydrogen complexes of $\eta_5\text{-}(C_5R_5)M(CO)_4$, M being a Group 5 metal, but also the kinetic study of $\eta_5\text{-}(C_5R_5)M(CO)_3$ intermediates. Photoacoustic calorimetry has provided almost direct evidence that 16-electron species formed by photoejection of CO from a metal carbonyl compound are solvated even in hydrocarbon solvents [123]. Other examples are given in [6]. These types of in-depth studies will certainly increase the understanding of thermal transition metal complex-induced catalysis.

References

[1] G. G. Wubbels, *Acc. Chem. Res.* **1983**, *16*, 285.

[2] (a) A. Albini, *Acc. Chem. Res.* **1984**, *17*, 234; (b) A. Albini, *J. Chem. Educ.* **1986**, *63*, 383.

[3] H. Kisch in *Photocatalysis, Fundamentals and Applications* (Eds.: N. Serpone, E. Pelizzetti), Wiley New York, **1989**, pp. 1–8.

[4] (a) S. Dumas, E. Lastra, L. S. Hegedus, *J. Am. Chem. Soc.* **1995**, *117*, 3368; (b) C. Dubuisson, Y. Fukumoto, L. S. Hegedus, *J. Am. Chem. Soc.* **1995**, *117*, 3697.

[5] J. Plotnikov, *Allgemeine Photochemie*, Walter de Gruyter, Berlin, **1936**, pp. 362–375.

[6] F. Chanon, M. Chanon in *Photocatalysis, Fundamentals and Applications* (Eds.: N. Serpone, E. Pelizzetti), Wiley, New York, **1989**, pp. 489–540.

[7] M. Chanon, L. Eberson in *Photoinduced Electron Transfer*, Part A (Eds.: M. A. Fox, M. Chanon), Elsevier, Amsterdam, **1988**, pp. 409–797.

[8] H. Hennig, D. Rehorek, R. D. Archer, *Coord. Chem. Rev.* **1985**, *61*, 1.

[9] R. G. Salomon, *Tetrahedron* **1983**, *39*, 485.

[10] D. P. Summers, J. C. Luong, M. S. Wrighton, *J. Am. Chem. Soc.* **1981**, *103*, 5238.

[11] S. Oishi, *J. Mol. Catal.* **1987**, *40*, 289.

[12] J. P. Collman, L. S. Hegedus, J. R. Norton, R. G. Finke, *Principles and Applications of Organotransition Metal Chemistry*, University Science Books, Mill Valley, CA, **1987**.

[13] M. Chanon, M. Rajzmann, F. Chanon, *Tetrahedron* **1990**, *46*, 6193.

[14] L. Moggi, A. Juris, D. Sandini, M. F. Manfrin, *Rev. Chem. Intermed.* **1981**, *5*, 107.

[15] A. W. Adamson, *Comments Inorg. Chem.* **1981**, *1*, 33.

[16] W. Strohmeier, L. Weigelt, *J. Organomet. Chem.* **1977**, *133*, C43.

[17] B. H. Byers, T. L. Brown, *J. Am. Chem. Soc.* **1977**, *99*, 2527.

[18] D. Astruc, *Chem. Rev.*, **1988**, *88*, 1189.

[19] (a) D. R. Tyler in *Paramagnetic Organometallic Species in Activation, Selectivity, Catalysis* (Eds.: M. Chanon, M. Julliard, J. C. Poite), Reidel, Dordrecht, **1988**, pp. 201–211; (b) D. R. Tyler, *Acc. Chem. Res.* **1991**, *24*, 325.

[20] R. G. Austin, R. S. Paonessa, P. J. Giordano, M. S. Wrighton, *ACS Adv. Chem. Ser.* **1978**, *168*, 189.

[21] M. A. Schroeder, M. S. Wrighton, *J. Am. Chem. Soc.* **1976**, *98*, 551.

[22] M. I. Bruce, *Coord. Chem. Rev.* **1987**, *76*, 1.

[23] N. G. Connelly, W. E. Geiger, *Adv. Organomet. Chem.* **1984**, *23*, 1; N. G. Connelly, W. E. Geiger, *ibid.* **1985**, *24*, 87.

[24] J. C. Kotz in *Topics in Organic Electrochemistry* (Eds.: A. J. Fry, W. E. Britton), Plenum, New York, **1986**, p. 142.

[25] T. M. Bockman, J. K. Kochi, *J. Am. Chem. Soc.* **1987**, *109*, 7725.

[26] D. Astruc, *Electron Transfer and Radical Processes in Transition Metal Chemistry*, VCH, Weinheim, **1995**.

[27] (a) A. Davison, N. Martinez, *J. Organomet. Chem.* **1974**, *74*, C17; (b) T. G. Attig, R. G. Teller, S.-M. Wu, R. Bau, A. Wojcicki, *J. Am. Chem. Soc.* **1979**, *101*, 619.

[28] C. Kutal, *Coord. Chem. Rev.* **1985**, *64*, 191.

[29] M. J. Mirbach, *EPA Newl* **1984**, *20*, 16.

[30] See also [7], p. 421.

[31] M. Koizumi, S. Kato, N. Mataga, T. Matsuura, Y. Usui, *Photosensitized Reactions*, Kagakudojin Publishing, Kyoto, Japan, **1978**.

[32] A. Fox, A. Poe, R. Ruminski, *J. Am. Chem. Soc.* **1982**, *104*, 7327.

[33] M. Grätzel in *Photoinduced Electron Transfer*, Vol. D (Eds.: M. A. Fox, M. Chanon), Elsevier, Amsterdam, **1988**, p. 394.

[34] *New J. Chem.* **1987**, *11*, (2) special issue devoted to the photochemical conversion and storage of solar energy.
[35] A. Harriman, M. West (Eds.), *Photogeneration of Hydrogen*, Academic Press, London, **1982**.
[36] Heterogeneous photocatalysis: M. A. Fox, M. T. Dulay, *Chem. Rev.* **1993**, *93*, 341.
[37] G. L. Geoffroy, M. S. Wrighton, *Organometallic Photochemistry*, Academic Press, New York, **1970**.
[38] G. J. Ferraudi, *Elements in Inorganic Photochemistry*, Wiley, New York, **1987**.
[39] A. W. Adamson, P. D. Fleischauer, *Concepts of Inorganic Chemistry*, Wiley, New York, **1975**.
[40] V. Balzani, V. Carassiti, *Photochemistry of Coordination Compounds*, Academic Press, New York, **1970**.
[41] K. Kalyanasundaram, M. Grätzel (Eds.), *Photosensitization and Photocatalysis Using Inorganic and Organometallic Compounds*, Kluwer Academic, Amsterdam, **1993**.
[42] G. J. Kavarnos, *Fundamentals of Photoinduced Electron Transfer*, VCH, Weinheim, **1993**.
[43] *J. Photochem. Photobiol. A. Chemistry*, **1994**, *82* (August), special issue.
[44] *Chem. Rev.*, **1992**, *92* (3), special issue devoted to electron transfer reactions.
[45] I. V. Spirina, V. P. Maslennikov, *Russ. Chem. Rev.* **1994**, *63*, 41.
[46] O. V. Gerasimov, V. N. Parmon, *Russ. Chem. Rev.* **1992**, *61*, 154.
[47] *Tetrahedron Symposia in Print No. 39*, **1989**, *45*, special issue devoted to covalently linked donor-acceptor species for mimicry of photosynthetic electron and energy transfer.
[48] J. Sykora, J. Sima, *Coord. Chem. Rev.* **1990**, *107*, special issue devoted to the photochemistry of coordination compounds.
[49] *Coord. Chem. Rev.* **1991**, *111*, special issue devoted to the photochemistry and photophysics of coordination compounds.
[50] *Coord. Chem. Rev.* **1994**, *132*, special issue devoted to the photochemistry and photophysics of coordination compounds.
[51] *Coord. Chem. Rev.* **1993**, *125*, special issue devoted to perspectives in photochemistry.
[52] *Chem. Rev.* **1993**, *93* (1), special issue devoted to photochemistry.
[53] J. Mattay, A. Griesbeck, (Eds.), *Photochemical Key Steps in Organic Synthesis*, VCH, Weinheim, **1994**.
[54] See [6]. The following reaction types have been listed: (a) Geometric isomerization of alkenes; (b) Allylic [1,3] hydrogen shift; (c) Cycloaddition of alkenes, Dimerization, Trimerization, Polymerization; (d) Skeletal rearrangments of alkenes and methathesis; (e) Hydrogenation of alkenes; (f) Additions to alkenes; (g) Additions to C = X; (h) Aliphatic substitutions; (i) Aromatic substitution; (j) Vinyl substitution; (k) Oxidation of alkenes; (l) Oxidation of alcohols; (m) Oxidation of arenes; (n) Oxidative decarboxylation; (o) Oxidation of amines; (p) Oxidation of vinylsilanes and sulfides; (q) Oxidation of benzaldehyde; (r) Dehydrogenations.
[55] P. C. Ford, A. F. Friedman in *Photocatalysis, Fundamentals and Applications* (Eds.: N. Serpone, E. Pelizzetti), Wiley, New York, **1989**, pp. 541–565.
[56] A. Yanagisawa, H. Noritake, H. Yamamoto, *Chem. Lett.* **1988**, 1899.
[57] H. Nagano, Y. Kuno, *J. Chem. Soc., Chem. Comm.* **1994**, 987.
[58] B. Giese, G. Thoma, *Helv. Chim. Acta* **1991**, *74*, 1135.
[59] S. Mori, S. Tatsumi, M. Yasuda, K. Kudo, N. Sugita, *Bull. Chem. Soc. Jpn.* **1991**, *64*, 3017–3022.
[60] S. Mori, H. Matsuyoshi, K. Kudo, N. Sugita, *Chem. Lett.* **1991**, 1397.
[61] Y.-T. Tao, T. J. Chow, J.-T. Lin, C.-C. Lin, M.-T. Chien, C.-C. Lin, Y. L. Chow, G. E. Buono-Core, *J. Chem. Soc. Perkin Trans. 1* **1989**, 2509.
[62] W. Leitner, *Angew. Chem., Int. Ed. Engl.* **1994**, *33*, 173, and references therein.

[63] D. P. Curran, C.-T. Chang, *J. Org. Chem.* **1989**, 54, 3140.

[64] D. P. Curran, M.-H. Chen, D. Kim, *J. Am. Chem. Soc.* **1989**, *111*, 6265.

[65] L. F. Tietze, U. Beifuss, *Angew. Chem., Int. Ed. Engl.* **1993**, *32*, 131.

[66] A. de Meijere, F. E. Meyer, *Angew. Chem., Int. Ed. Engl.* **1994**, *33*, 2379.

[67] C. Destabel, J. D. Kilburn, J. Knight, *Tetrahedron* **1994**, *50*, 11289.

[68] G. Haaima, L. R. Hanton, M.-J. Lynch, S. D. Mawson, A. Routledge, R. T. Weavers, *Tetrahedron* **1994**, *50*, 2161.

[69] J.-P. Dulcère, E. Dumez, R. Faure, *J. Chem. Soc., Chem. Commun.* **1995**, 897.

[70] See [7], p. 415, for a discussion of this matter.

[71] S. Kar, S. Lahiri, *J. Chem. Soc., Chem. Commun.* **1995**, 957.

[72] T. Ishiyama, N. Miyaura, A. Suzuki, *Tetrahedron Lett.* **1991**, *32*, 6923.

[73] T. Ishiyama, M. Murata, A. Suzuki, N. Miyaura, *J. Chem. Soc., Chem. Commun.* **1995**, 295.

[74] F. D. Lewis, S. V. Barancyk, *J. Am. Chem. Soc.* **1989**, *111*, 8653.

[75] D. I. Schuster, G. Lem, N. A. Kaprinidis, *Chem. Rev.* **1993**, *93*, 3.

[76] K. Kakiuchi, B. Yamaguchi, M. Kinugawa, Y. Ue, Y. Tobe, Y. Odaira, *J. Org. Chem.* **1993**, *58*, 2797.

[77] K. Kakiuchi, B. Yamaguchi, Y. Tobe, *J. Org. Chem.* **1991**, *56*, 5745.

[78] C. C. Huval, D. A. Singleton, *J. Org. Chem.* **1994**, *59*, 2020.

[79] W. Zhang, Y. Hua, G. Hoge, P. Dowd, *Tetrahedron Lett.* **1994**, *35*, 3865.

[80] D. P. Curran, H. Liu, *J. Am. Chem. Soc.* **1992**, *114*, 5863.

[81] Asymmetric approach: D. P. Curran, S.-B. Ko, *J. Org. Chem.* **1994**, *59*, 6139.

[82] W. Adam, M. Braun, A. Griesbeck, V. Luccini, E. Staab, B. Will, *J. Am. Chem. Soc.* **1989**, *111*, 203.

[83] W. Adam, M. Richter, *Tetrahedron Lett.* **1992**, *33*, 3461; W. Adam, M. J. Richter, *J. Org. Chem.* **1994**, *59*, 3341.

[84] W. Adam, S. Kömmerling, E.-M. Peters, K. Peters, H. G. von Schnering, M. Schwarm, E. Staab, A. Zahn, *Chem. Ber.* **1988**, *121*, 2151.

[85] W. Adam, O. Gevert, P. Klug, *Tetrahedron Lett.* **1994**, *35*, 1981.

[86] W. Adam, B. Nestler, *J. Am. Chem. Soc.* **1992**, *114*, 6549; W. Adam, B. Nestler, *ibid.* **1993**, *115*, 5041.

[87] W. Adam, B. Nestler, *J. Am. Chem. Soc.* **1993**, *115*, 7226.

[88] M. K. Trost, R. G. Bergman, *Organometallics* **1991**, *10*, 1172.

[89] L. Weber, R. Hommel, J. Behling, G. Haufe, H. Hennig, *J. Am. Chem. Soc.* **1994**, *116*, 2400.

[90] H. Hennig, J. Behling, R. Meusinger, L. Weber, *Chem. Ber.* **1995**, *128*, 229.

[91] T. Sato, S. Yonemochi, *Tetrahedron* **1994**, *50*, 7375.

[92] The biomimetic approach: D. Mansuy, *Coord. Chem. Rev.* **1993**, *125*, 129.

[93] V. W.-W. Yam, C.-M. Che, *New J. Chem.* **1989**, *13*, 707.

[94] C. L. Hill, R. F. Renneke, L. A. Combs, *New J. Chem.* **1989**, *13*, 701.

[95] Review: C. L. Hill, *Synlett* **1995**, 127.

[96] T. Sakakura, T. Sodeyama, M. Tanaka, *New. J. Chem.* **1989**, *13*, 737.

[97] J. A. Maguire, W. T. Boese, A. S. Goldman, *J. Am. Chem. Soc.* **1989**, *111*, 7088.

[98] W. T. Boese, A. S. Goldman, *Organometalllics* **1991**, *10*, 782.

[99] W. D. Jones, R. P. Duttweiler Jr., F. J. Feher, E. T. Hessell, *New. J. Chem.* **1989**, *13*, 725.

[100] F. Minisci, F. Fontana, T. Caronna, L. Zhao, *Tetrahedron Lett.* **1992**, *33*, 3201.

[101] For a recent example cf. T. Osa, Y. Kashiwagi, Y. Yanagisawa, J. M. Bobbitt, *J. Chem. Soc., Chem. Commun.* **1994**, 2535.

[102] T. Hamada, H. Ishida, S. Usui, Y. Watanabe, K. Tsumura, K. Ohkubo, *J. Chem. Soc., Chem. Commun.* **1993**, 909.

[103] R. H. Crabtree, S. H. Brown, C. A. Muedas, C. Boojamra, R. R. Ferguson, *Chemtech* **1991**, *21*, 634.
[104] S. H. Brown, R. H. Crabtree, *J. Am. Chem. Soc.* **1989**, *111*, 2935.
[105] S. H. Brown, R. H. Crabtree, *J. Am. Chem. Soc.* **1989**, *111*, 2946.
[106] C. G. Boojamra, R. H. Crabtree, R. R. Ferguson, C. A. Muedas, *Tetrahedron Lett.* **1989**, *30*, 5583.
[107] J. C. Lee Jr., C. G. Boojamra, R. H. Crabtree, *J. Org. Chem.* **1993**, *58*, 3895.
[108] C. Bartocci, A. Maldotti, G. Varani, V. Carassiti, P. Battioni, D. Mansuy, *J. Chem. Soc., Chem. Commun.* **1989**, 964.
[109] M. Onishi, M. Matsuda, I. Takaki, K. Hiraki, S. Oishi, *Bull. Chem. Soc. Jpn.* **1989**, *62*, 2963.
[110] T. Sakakura, F. Abe, M. Tanaka, *Chem. Lett.* **1990**, 583.
[111] J.-M. Lehn, R. Ziessel, *J. Organomet. Chem.* **1990**, *382*, 157.
[112] S. Kuwabata, H. Uchida, A. Ogawa, S. Hirao, H. Yoneyama, *J. Chem. Soc., Chem. Commun.* **1995**, 829.
[113] I. Willner, D. Mandler, *J. Am. Chem. Soc.* **1989**, *111*, 1330.
[114] S. A. Jackson, P. M. Hodges, M. Poliakoff, J. J. Turner, F.-W. Grevels, *J. Am. Chem. Soc.* **1990**, *112*, 1221.
[115] P. M. Hodges, S. A. Jackson, J. Jacke, M. Poliakoff, J. J. Turner, F.-W. Grevels, *J. Am. Chem. Soc.* **1990**, *112*, 1234.
[116] D. Chmielewski, F.-W. Grevels, J. Jacke, K. Schaffner, *Angew. Chem., Int. Ed. Engl.* **1991**, *30*, 1343.
[117] Y. Ito, K. Kunimoto, S. Miyachi, T. Kako, *Tetrahedron Lett.* **1991**, *32*, 4007.
[118] Y. Ito, *J. Chem. Soc., Chem. Commun.* **1991**, 622.
[119] S. Igueld, M. Baboulène, A. Dicko, M. Montury, *Synthesis* **1989**, 200.
[120] N. Iranpoor, E. Mottaghinejad, *J. Organomet. Chem.* **1992**, *423*, 399.
[121] K. Meier, *Coord. Chem. Rev.* **1991**, *111*, 97.
[122] B. Gita, G. Sundarajan, *Tetrahedron Lett.* **1993**, *34*, 6123.
[123] M. W. George, M. T. Haward, P. A. Hamley, C. Hughes, F. P. A. Johnson, V. K. Popov, M. Poliakoff, *J. Am. Chem. Soc.* **1993**, *115*, 2286.

3.2.10   Olefins from Aldehydes

Wolfgang A. Herrmann

3.2.10.1   Introduction

It was the pioneering work of Georg Wittig [1] that yielded an industrially applicable olefin synthesis by C–C coupling [2]: phosphorus ylides affect a nucleophilic attack at aldehydes and certain other organic keto compounds, resulting in a methylene (alkylidene) group transfer with concomitant formation of the desired olefin and a phosphine oxide. The latter type of compounds represents the thermodynamic driving force of this reaction. It is evident from eq. (1) that the Wittig olefination is a *stoichiometric* process. The phosphine oxide can be recycled by means of reducing silanes, e. g., chlorodimethylsilane or hexachlorodisilane, although the procedures are cumbersome and the yields often low. A recent alternative comprises transformation of the phosphine oxide into the dichloride ($\equiv$ P=O $\rightarrow$ $\equiv$ PCl$_2$), followed by reductive dechlorination with the help of aluminium granulate [3].

$$\underset{H}{\overset{R}{>}}C=O \;+\; R_3'P=CH_2 \;\longrightarrow\; \underset{H}{\overset{R}{>}}C=CH_2 \;+\; R_3'P=O \tag{1}$$

Numerous monographs [4] and reviews [5] on the famous Wittig reaction have been written since its discovery in 1953. The BASF vitamin-A synthesis depends in the final step on a Wittig coupling between vinyl-β-ionol (C$_{15}$) and γ-formyl-crotyl acetate (C$_5$). This application was developed by Pommer et al. [6] of BASF in the 1960s.

Again noncatalytic, the organotitanium-mediated olefination of aldehydes, ketones, and carboxylic esters has been developed by Grubbs et al. [7]. They used the (commercial) "Tebbe reagent" (Structure **1**) as a source of methylene (CH$_2$) groups to be transferred to the keto component (eq. (2a)). A second route with the same overall result implies previous transformation of **1** into a titanacyclobutane **2** which again acts as a methylene transfer reagent. In spite of significant advantages over the Wittig reaction (high selectivities and yields, mild conditions, broad spectrum of keto precursors, e. g., carboxylic esters and cyclic lactones), there remain several drawbacks of this type of olefination: (1) constitutional restriction to titanium–methylene reagents (no higher titanium-alkylidene homologs are available); (2) no perspective of a *catalytic* performance. The latter problem is due to the considerable strength of the titanium–oxo bond. As a matter of fact, the analogy between the P=O and Ti=O products is obvious when considering the formation of the (trimeric) organometallic oxide (Structure **3**).

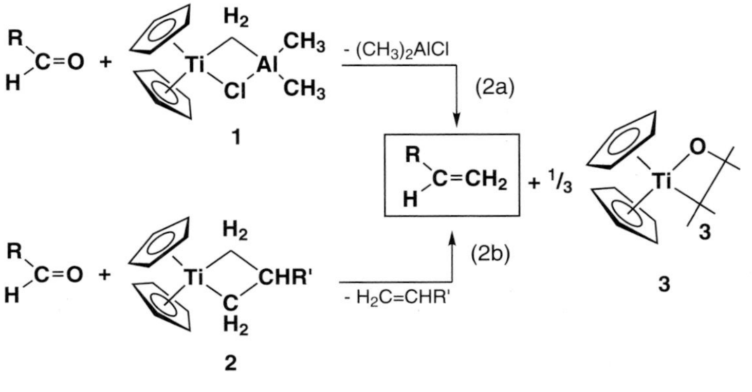

3.2.10.2 The Catalytic Approach

An approach to the problem is summarized in eq. (3). Aldehydes are subject to catalytic olefination when certain aliphatic diazoalkanes are used as alkylidene group transfer reagents; phosphines are necessary to carry off the oxo group, once again reminiscent of the above Wittig reaction. However, the olefination of eq. (3) is catalytic in terms of the olefination components and the deoxygenation reagent [8–10].

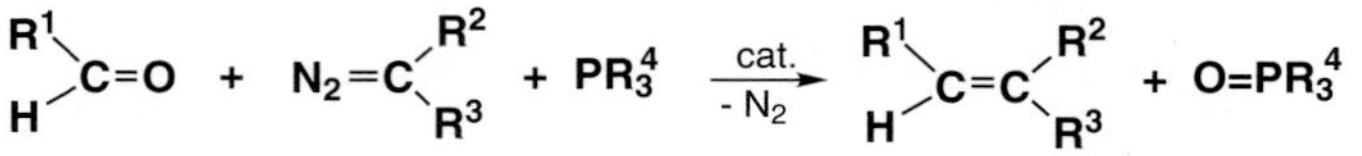

$$\underset{H}{\overset{R^1}{\diagdown}}C=O \ + \ N_2=C\underset{R^3}{\overset{R^2}{\diagup}} \ + \ PR_3^4 \quad \xrightarrow[-\,N_2]{cat.} \quad \underset{H}{\overset{R^1}{\diagdown}}C=C\underset{R^3}{\overset{R^2}{\diagup}} \ + \ O=PR_3^4 \qquad (3)$$

3.2.10.3   Catalysts

It is obvious from the overall equation that a catalyst must generate an intermediate susceptible to C–C coupling, e. g., a metal–carbene species. Three catalytic systems based on coordination compounds have been described: the molybdenocyl (Mo^{VI}) dithiocarbamate **4** [8], the organorhenium (Re^{VII}) oxide **5** [9–11], and the phosphane–rhenium (Re^{V}) chloride **6** [11], all representing high oxidation-state metal–oxo complexes (all three compounds are commercially available, e. g., from Aldrich and Fluka). Methyltrioxorhenium (MTO) **5** is easily synthesized by methylation of dirhenium heptoxide (Re_2O_7) or its carboxylic esters $O_3ReOC(=O)R$ (e. g., R = CF_3) with $Sn(CH_3)_4$ [12, 13]. Binary rhenium oxides (ReO_2, ReO_3, Re_2O_7) and perrhenates are totally inactive, neither is there a reaction in the presence of rhenium–carbonyl complexes [10]. CH_3ReO_3 (**5**) has also been applied in the presence of polymer-bound triphenylphosphine [9–11]. A high-yield synthesis of this catalyst is described in Ref. [13].

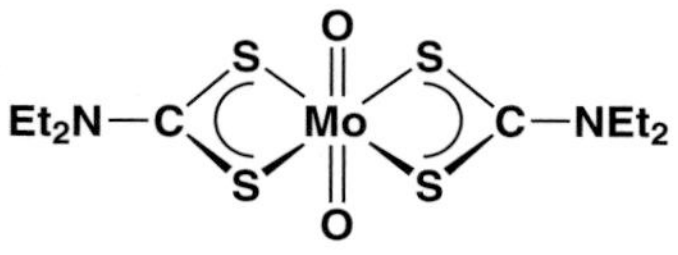

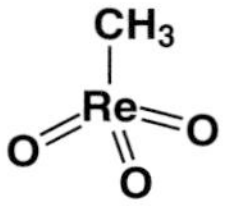

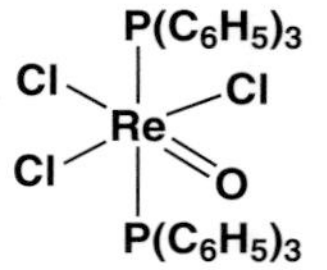

The catalysts **4–6** are employed at concentrations of 10 mol % (Mo) and 1–10 mol % (Re), with respect to the aldehyde. Too low a rhenium concentration favors the formation of ketazines. The reactions are normally conducted at room temperature (see below).

Methyltrioxorhenium(VII) **5** is active even at temperatures as low as –35 °C but is usually administered at 25 °C. The molybdenum(VI) catalyst requires temperatures of 80 °C (boiling benzene) to convert benzaldehyde and ethyl diazoacetate into the corresponding cinnamic ester [8]. Turnovers are not yet sufficient for technical applications but the yields range between 75 and 98 % for the Re-catalyzed process [9–11].

3.2.10.4   Scope of Reaction, Reagents, and Side Reactions

3.2.10.4.1 Diazoalkanes

It is clear from the known diazoalkane reactivity pattern [14] that only derivatives $R^2R^3C=N_2$ can be employed that do not react with phosphines (cf. Section 3.2.10.5). Transformations can be observed for the easy-to-handle diazoacetates and diazomalonates as well as for a number of other compounds, e. g., trimethylsilyl diazomethane. Aryldiazoalkanes, however, form phosphazenes that do not release the alkylidene group. Another problem can arise from the (catalytic) formation of ketazines – again, unreactive byproducts.

3.2.10.4.2 Keto Compounds

So far, arylaldehydes have been the preferred substrates. Normal ketones are of little or no reactivity while strained cycloketones, particularly cyclobutanone, undergo olefination (eq. (4)). Within the series of benzaldehydes, electron-withdrawing substituents (e. g., *p*-NO$_2$) favour the C–C coupling.

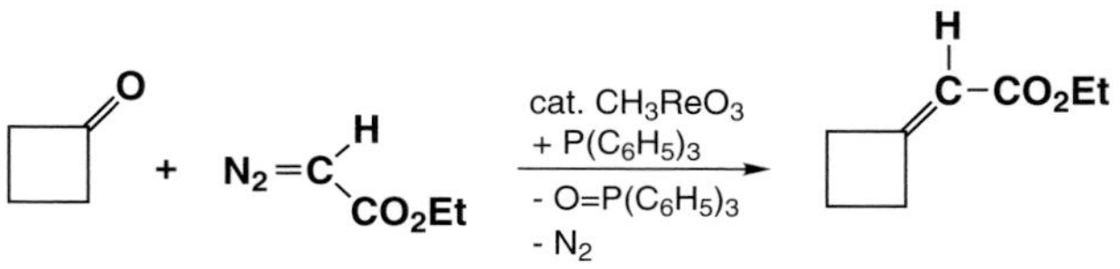

$$\text{(4)}$$

3.2.10.4.3 Phosphines

No serious restrictions with regard to the oxygen-accepting organophosphanes PR^4_3 seem to apply. While triphenylphosphine is the only well-studied substrate, trialkylphosphines and the water-soluble TPPTS also work (TPPTS = tris(sodium-*m*-sulfonatophenyl)phosphine; cf. Section 3.1.1.1). Other deoxygenation reagents have not yet been employed (e. g., silyl and thio compounds).

3.2.10.4.4 Conditions

Depending somewhat on the reactivity of the diazo component, the catalytic C–C coupling of eq. (3) is normally conducted at room temperature. To avoid early N_2 elimination and unwanted side products, the reaction should be finished within several hours.

3.2.10.5 Mechanism

Up to now, there is only slight knowledge of mechanistic details, and no kinetic data are yet available. Both catalytic systems depend on high oxidation-state metal centers, indicating that (reversible) redox steps $Mo^{VI} \rightleftharpoons Mo^{IV}$ and $Re^{VII} \rightleftharpoons Re^{V}$, respectively, are in operation to move the oxo ligand(s) of the catalyst in and out. The following proposal for the mechanism related to methyltrioxorhenium (VII) **5** may apply in principle for the molybdenyl catalyst **4**, too.

3.2.10.5.1 Catalyst Activation

It is often observed that metal oxo complexes undergo reduction by organophosphines. With methyltrioxorhenium(VII), a stoichiometric 1:1 reaction occurs to yield the phosphine oxide complex **7a** of Scheme 1, probably via a primary adduct of type $CH_3ReO_3 \cdot P(C_6H_5)_3$. Careful workup also yielded crystalline **7b**, the structure of which species is composed of CH_3ReO_3 (Lewis acid, Re^{VII}) and $CH_3Re-[P(C_6H_5)_3]_2O_2$ (Lewis base via oxygen, Re^{V}). Irrespectiveof the exact nature of the intermediates, one out of three oxo ligands of catalyst **5** is selectively labilized by the phosphine, as is evident from the facile formation of derivatives CH_3ReO_2L upon addition of π-Ligands L. A prominent example is acetylene (Scheme 1) [15,16], which forms the stable complex $CH_3ReO_2(HC\equiv CH)$ **8** in high yields, while CH_3ReO_3 (**5**) does not react with alkynes in the absence of the "labilizing" (= deoxygenating) phosphine.

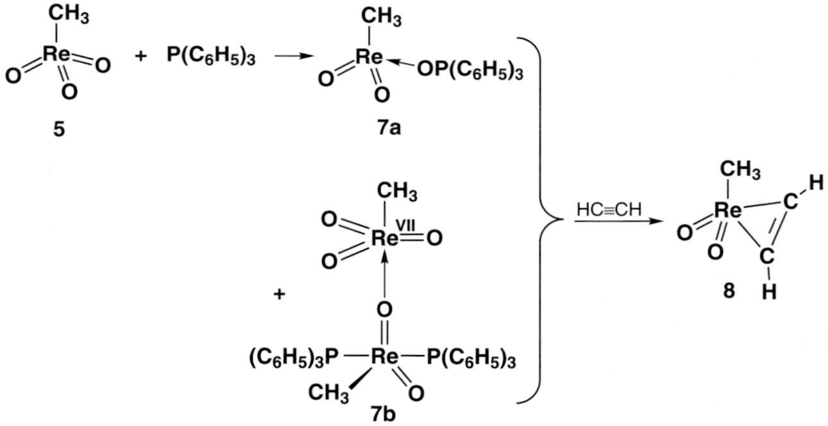

Scheme 1. Deoxygenation of methyltrioxorhenium(VII) by triphenylphosphine.

3.2.10.5.2 Diazoalkane Activation

It is known from previous work that diazoalkanes can form carbene(alkylidene)-metal complexes [17, 18], cf. eq. (5). It is thus reasonable to assume that a metal–carbene **9** is formed from the (phosphine oxide-stabilized) species $\{CH_3Re^VO_2\}$ (eq. (5)). High oxidation-state metal carbene complexes have ample precedent, especially through the work of Schrock et al. [19]. Isolation of type-**8** species may be facilitated by sterically more-demanding auxiliary groups (e. g., C_5H_5 in place of CH_3) or, by using heterocyclic carbenes of pronounced Lewis basicity (e. g., 1.3-imidazolin-2-ylidene [20]).

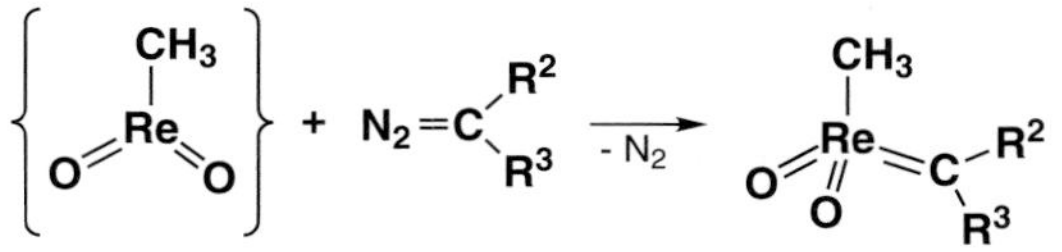

$$\left\{ \begin{array}{c} CH_3 \\ | \\ O{=}Re{=}O \end{array} \right\} \; + \; N_2{=}C\begin{array}{c} R^2 \\ R^3 \end{array} \quad \xrightarrow{-N_2} \quad \begin{array}{c} CH_3 \\ | \\ O{=}Re{=}C{-}R^2 \\ \| \quad \quad R^3 \\ O \end{array} \qquad (5)$$

$$\mathbf{9}$$

3.2.10.5.3 Aldehyde Activation

Since the keto component reacts neither with the catalyst **5** (with or without phosphine) nor with the diazoalkane (exceptions below), it is likely to enter the catalytic cycle at the metal–carbene stage **8**. If this proposal withstands future mechanistic studies, then the regioselectivity – oxygen to the oxophilic rhenium center – is reasonable by considering a metallacyclic intermediate **10** (eq. (6)). This hypothetical "rhena-oxetane" is assumed to eliminate the olefin. As a matter of fact, catalyst **5** has been detected from such precursors (e. g., **7a, b** + diazoalkane + aldehyde) by gas chromatography [9, 10].

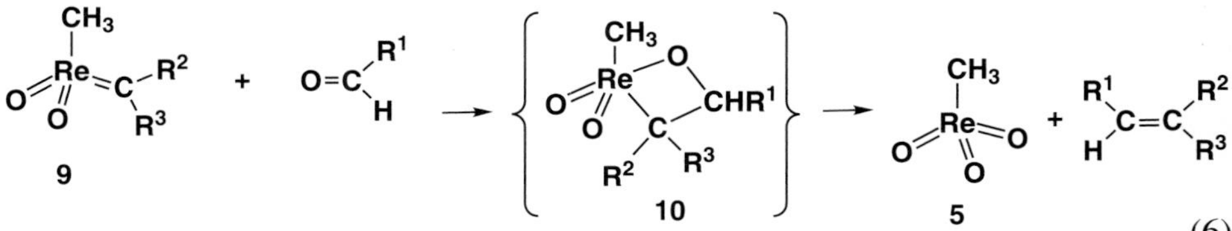

$$\begin{array}{c} CH_3 \\ | \\ O{=}Re{=}C{-}R^2 \\ \| \quad \quad R^3 \\ O \end{array} \; + \; O{=}C\begin{array}{c} R^1 \\ H \end{array} \; \longrightarrow \; \left\{ \begin{array}{c} CH_3 \\ | \quad O \\ O{=}Re \quad \\ \| \quad \quad CHR^1 \\ O \quad C \\ R^2 \quad R^3 \end{array} \right\} \; \longrightarrow \; \begin{array}{c} CH_3 \\ | \\ O{=}Re{=}O \\ \| \\ O \end{array} \; + \; \begin{array}{c} R^1 \quad R^2 \\ C{=}C \\ H \quad R^3 \end{array}$$

$$\mathbf{9} \qquad \qquad \qquad \qquad \mathbf{10} \qquad \qquad \qquad \mathbf{5} \qquad \qquad (6)$$

3.2.10.5.4 The Catalytic Cycle

The mechanistic proposal can be summarized in the catalytic cycle of Scheme 2. The catalyst **5** performs so well because it undergoes reactivation by the phosphine. This is not the case with the organotitanium(IV) oxide **3** (eqs. (2a), (2b)), thus explaining the *non*catalytic reactivity of the Tebbe and Grubbs reagents.

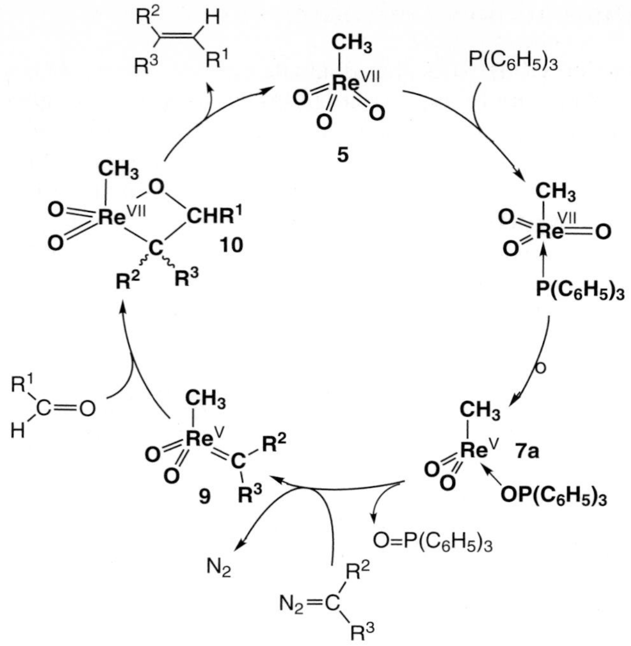

Scheme 2. Proposed mechanism of the oxorhenium-catalyzed olefination of aldehydes.

3.2.10.5.5 Catalyst Assessment

The olefination of aldehydes by diazoalkanes depends on catalysts that accept the keto oxygen from the substrate **S.** and transfer it to an auxiliary reagent **R..** A redox process $M^{n-2} \rightleftharpoons M^n{=}O$ is necessarily included in such a sequence (Scheme 3). The reduced step $O{=}Mo^{IV}(S_2C\text{-}NEt_2)_2$ has in fact been isolated from the phosphine reduction of catalyst **4**, and a carbene complex of structure **11** according to eq. (7) was suggested [8].

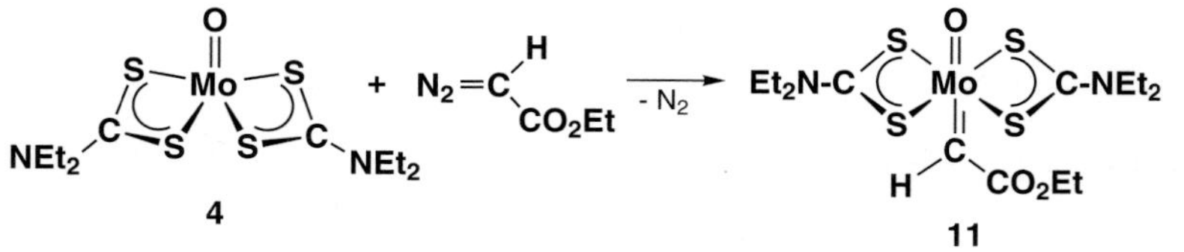

$$(7)$$

From this assessment two requirements concerning the catalyst can be postulated: (1) the oxophilicity has to be high enough to react with the aldehyde in the catalyst's reduced state; (2) the reductive potential has to bei high enough to release the oxo ligand in the catalyst's oxidized state. In addition, the catalyst must have a configuration that allows the C–C coupling of the alkylidene-group carriers (aldehyde, diazoalkane). Low-coordinated species such as CH_3ReO_3 seem specifically to meet these requirements.

Cyclopropanation of the olefinic product does not occur; the catalytic metal–carbene intermediate obviously cannot transfer its CR^2R^3 unit to the olefinic double bond (see, however, metal-catalyzed cyclopropanation, Section 3.1.7).

3.2.10.6   Perspectives

At first sight, the olefination reaction of eq. (3) looks rather special with regard to the diazoalkane substrates and the phosphine which is required to bind the oxygen. However, the most successful diazoalkanes are either commercially available (e. g., ethyl diazoacetate, trimethyl silyldiazomethane) or can easily be prepared (e. g., diazomalonates). Furthermore, phosphines may in the future be replaced by other oxophilic co-reactants, possibly by silyl compounds. Finally, this type of C–C coupling is *catalytic*, an uncommon feature in olefination chemistry. For these reasons and because of the as yet unexplored potential of stereoselectivity should the little recognized reaction become the subject of intense research. For example, the immobilization of CH_3ReO_3 and related compounds on oxidic supports could yield heterogeneous catalysts if the specific balance of oxygen elimination and addition can be maintained (Scheme 3).

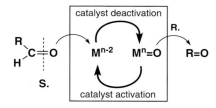

Scheme 3. Schematic mechanism of oxo transfer at high oxidation-state metal complexes (**R.** = reducing agent, e. g. phosphine; **S.** = substrate = oxygen-donating reagent).

References

[1] The German chemist Georg Wittig (1897–1987) was a professor of chemistry at the Technische Hochschule Braunschweig (Germany) and at the universities of Freiburg, Tübingen, and Heidelberg. He received the Nobel Prize for chemistry in 1979, jointly with Herbert C. Brown; cf. W. Tochtermann, *Top. Curr. Chem.* **1988**, *144* (preface).

[2] (a) G. Wittig, G. Geissler, *Liebigs Ann. Chem.* **1953**, *580*, 44; (b) G. Wittig, *Angew. Chem.* **1956**, *68*, 505.

[3] H. G. Hauthal, *Nachr. Chem. Tech. Lab. (Weinheim/Germany)* **1993**, *41*, 1015.

[4] M. Schlosser, *Top. Stereochem.* **1972**, *5*, 1.

[5] Römpp, *Chemie-Lexikon*, Vol. 6, 9th ed., Thieme, Stuttgart, **1992**, p. 5056.

[6] Reviews: (a) H. Pommer, *Angew. Chem.* **1960**, *72*, 811; (b) H. Pommer, *Angew. Chem.* **1977**, *89*, 437; *Angew. Chem., Int. Ed. Engl.* **1977**, *16*, 423; (c) H.-J. Bestman, O. Vos-

trowsky, *Top. Curr. Chem.* **1983**, *109*, 85; (d) H.-J. Bestmann, R. Zimmermann in: *Houben-Weyl, Methoden der Organischen Chemie*, Vol. E1, 4th ed., Thieme, Stuttgart, **1982**, p. 616.

[7] (a) Review: R. H. Grubbs, *Pure Appl. Chem.* **1983**, *55*, 1733; (b) R. H. Grubbs, L. R. Gillom, *J. Am. Chem. Soc.* **1986**, *108*, 733; (c) H.-U. Reissig, *Nachr. Chem. Tech. Lab. (Weinheim/Germany)* **1986**, *34*, 562.

[8] X. Lu, H. Fang, Z. Ni, *J. Organomet. Chem.* **1989**, *373*, 77.

[9] (a) Hoechst AG (W. A. Herrmann), DE 4.101.737 (1991) and DE 4.002.505 (1990); (b) W. A. Herrmann, Mei Wang, *Angew. Chem.* **1991**, *103*, 1709; *Angew. Chem., Int. Ed. Engl.* **1991**, *30*, 1641.

[10] W. A. Herrmann, P. W. Roesky, Mei Wang, W. Scherer, *Organometallics* **1994**, *13*, 4531.

[11] P. W. Roesky, Ph. D. Thesis, Technische Universität München, Germany **1994**.

[12] W. A. Herrmann, W. R. Thiel, F. E. Kühn, R. W. Fischer, M. Kleine, E. Herdtweck, W. Scherer, J. Mink, *Inorg. Chem.* **1993**, *32*, 5188.

[13] (a) W. A. Herrmann, R. W. Fischer, M. Rauch, W. Scherer, *J. Mol. Catal.* **1994**, *86*, 243; (b) W. A. Herrmann (Ed.), *Synthetic Methods in Inorganic and organometallic Chemistry*, Vol. 1, Thieme, Stuttgart, **1996**.

[14] Monograph: M. Regitz, *Diazoalkane: Eigenschaften und Synthesen,* Thieme Verlag, Stuttgart, **1977**.

[15] J. K. Felixberger, J. G. Kuchler, E. Herdtweck, R. A. Paciello, W. A. Herrmann, *Angew. Chem.* **1988**, *100*, 975; *Angew. Chem., Int. Ed. Engl.* **1988**, *27*, 946.

[16] W. A. Herrmann, J. K. Felixberger, J. G. Kuchler, E. Herdtweck, *Z. Naturforsch. Teil B* **1990**, *45*, 876.

[17] Review: W. A. Herrmann, *Angew. Chem.* **1978**, *90*, 855; *Angew. Chem., Int. Ed. Engl.* **1978**, *17*, 800.

[18] Review: W. A. Herrmann, *Adv. Organomet. Chem.* **1982**, *20*, 160.

[19] (a) R. R. Schrock, *J. Organomet. Chem.* **1986**, *300*, 249; (b) R. R. Schrock in *Carbyne Complexes* (Eds: H. Fischer, P. Hofmann, F. R. Kreissl, R. R. Schrock, U. Schubert, K. Weiss), Verlag Chemie, Weinheim, **1988**, pp. 147–204.

[20] W. A. Herrmann, M. Elison, J. Fischer, C. Köcher, G. R. J. Artus, *Angew. Chem.* **1995**, *107*, 2602; *Angew. Chem., Int. Ed. Engl.* **1995**, *34*, 2371, and references cited therein.

3.2.11   Water-Gas Shift Reaction

Wolfgang A. Herrmann, Michael Muehlhofer

3.2.11.1   Introduction

Many catalytic reactions described in this book depend on carbon monoxide and hydrogen as feedstock chemicals. *Hydroformylation* ($CO + H_2$) and simple *hydrogenation* (H_2) are typical examples. In many cases carbon monoxide undergoes side reactions, among which the "water-gas shift reaction" is well studied in terms of the mechanism. This explains why carbon monoxide in the presence of water (e. g., aqueous media) can be used to hydrogenate substrates such as olefins, nitroaromatics, and other unsaturated organic compounds. In a number of industrial processes (e. g., the hydrocarboxylation of ethylene), however, this is an unwanted side reaction.

3.2.11.2   Definition

Steam reforming [1] following eq. (1) is the technology commonly used to pro-duce carbon monoxide and hydrogen in an endothermic reaction (130 kJ/mol).

$$C + H_2O \longrightarrow CO + H_2 \qquad (1)$$

The product is called "water-gas" (German "*Wassergas*"). To optimize the yield of hydrogen, the equilibrium to eq. (2) is employed. This particular process is known as the "water-gas shift reaction", albeit *Konvertierungsgleichgewicht* (as the German *terminus technicus*) seems more appropriate.

$$CO + H_2O \rightleftharpoons CO_2 + H_2 \qquad (2)$$

The hydrogen required for the production of ammonia is generated thus. The industrially most important equilibrium (2) is slightly exothermic (42 kJ/ mol) and is catalyzed by a large number of soluble transition metal complexes, but the commercial plants work with heterogeneous catalysts. In the typical two-stage process, chromium(III) oxides (at 350–450 °C) and copper/zinc oxides (at 200–300 °C) are employed as catalysts [1]. Soluble catalysts are complexes such as $Fe(CO)_5$ and its hydrido derivative $[HFe(CO)_4]^-$, $Ru_3(CO)_{12}$, $[Rh(CO)_2I_2]^-$, and $Pt[P(Pr^i)_3]_3$ [2]. The rhodium complex $[Rh(CO)_2I_2]^-$ – the key catalytic feature in the Monsanto acetic acid process (Section 2.1.2.1) – is particularly efficient: in a solution of acetic acid, CO is converted at 80–90 °C and 0.5 atm into an equimolar amount of CO_2 and an equivalent volume of H_2 [2b]. Although the catalytic turnover is low (five to nine cycles *per diem*), the catalytically active species is quite robust.

3.2.11.3   Mechanism

The water-gas shift mechanism was explored mainly by P. C. Ford et al. in the 1970s [2–4], showing that a number of consecutive organometallic reactions are involved (Scheme 1). All of them have stoichiometric precedents.

The first step is likely to be a so-called "Hieber base reaction" (ⓐ in Scheme 1), describing a nucleophilic attack at metal-coordinated carbon monoxide by hydro-xide. While eq. (3) is the prototypical example as first reported by Walter Hieber in 1932 [5], several other variants have since become known. For example, azides and alkyl/aryl anions attack metal carbonyls according to eqs. (4) and (5), yielding metal isocyanates (via a Curtius-type azide degradation step) and metal carbenes, respectively [6, 7].

$$Fe(CO)_5 + [OH]^- \longrightarrow \left\{ \left[(CO)_4Fe-C\underset{O}{\overset{OH}{\diagup}} \right]^- \right\} \xrightarrow{-CO_2} [HFe(CO)_4]^- \qquad (3)$$

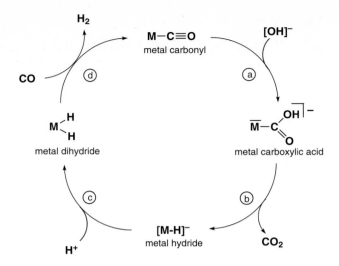

Scheme 1. Mechanistic proposal for the metal-catalyzed water-gas shift reaction; equilibria disregarded. M: metal complex, e. g., $Fe(CO)_4$ (see text).

$$W(CO)_6 + [N_3]^- \longrightarrow \left\{ \left[(CO)_5W-C \begin{smallmatrix} N_3 \\ O \end{smallmatrix} \right]^- \right\} \xrightarrow[-N_2]{} [(CO)_5W-N=C=O]^- \qquad (4)$$

$$W(CO)_6 + [C_6H_5]^- \longrightarrow \left[(CO)_5W-C \begin{smallmatrix} C_6H_5 \\ O \end{smallmatrix} \right]^- \xrightarrow{+ [CH_3]^+} (CO)_5W=C \begin{smallmatrix} C_6H_5 \\ OCH_3 \end{smallmatrix} \qquad (5)$$

A synthetically useful technique for substituting (*C*-electrophilic) carbon monoxide ligands from metal carbonyls is a base reaction with the sufficiently nucleophilic trimethylamine oxide [8]. As represented by eq. (6), the same mechanistic scheme applies, with the zwitterionic intermediate breaking down into the (volatile, weakly coordinating) fragments CO_2 and $N(CH_3)_3$; at the same time, a reactive fragment (e. g., the 16e-system "$Fe(CO)_4$") is being formed.

$$Fe(CO)_5 + (CH_3)_3NO \longrightarrow \left\{ (CO)_4Fe-C \begin{smallmatrix} \overset{+}{O}N(CH_3)_3 \\ O \end{smallmatrix} \right\} \longrightarrow \begin{smallmatrix} "(CO)_4Fe" \\ + CO_2 + N(CH_3)_3 \end{smallmatrix} \qquad (6)$$

This "base reaction" does not depend on hydroxide and proceeds under mild conditions, thus replacing the often less-clean photoelimination of carbon monoxide. Certain cationic metal carbonyls are attacked even by water, a reaction that

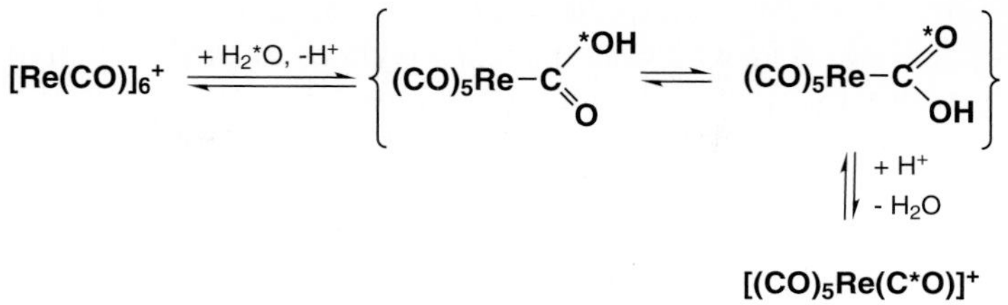

$$[\text{Re(CO)}]_6{}^+ \xrightleftharpoons{+ \text{H}_2{}^*\text{O}, -\text{H}^+} \left\{ (\text{CO})_5\text{Re}-\overset{*\text{OH}}{\underset{\text{O}}{\text{C}}} \rightleftharpoons (\text{CO})_5\text{Re}-\overset{*\text{O}}{\underset{\text{OH}}{\text{C}}} \right\}$$

$$\Big\updownarrow \begin{matrix} + \text{H}^+ \\ - \text{H}_2\text{O} \end{matrix}$$

$$[(\text{CO})_5\text{Re}(\text{C}^*\text{O})]^+$$

Scheme 2. Tautomerization of a rheniumcarboxylic acid as an example of the synthesis of oxygen-labeled metal carbonyls.

can be utilized to from oxygen-labeled derivatives (^{17}O, ^{18}O) from normal CO complexes through a tautomerization step [9]; the cationic complex $[\text{Re(CO)}_6]^+$ is the prototypical example (Scheme 2).

Transition metal carboxylic acids resulting from the above base reaction are normally unstable with respect to decarboxylation, thus undergoing degradation to hydrides (reaction ⓑ in Scheme 1) [10]. Only a few cases of barely stable derivatives are known, such as (η^5-C$_5$H$_5$)Fe(CO)$_2$CO$_2$H [11]. As seen in the historical reaction of eq. (3), the carboxylate system undergoes fast elimination of CO$_2$. (Oxidation of the resulting anionic hydrido complex [HFe(CO)$_4$]$^-$ by MnO$_2$ is the standard procedure for synthesis of Fe$_3$(CO)$_{12}$.)

The anionic metal hydride takes up a proton from the aqueous medium to yield a dihydride (step ⓒ); a known example is H$_2$Fe(CO)$_4$ [5 a, 12]. In the final step ⓓ, hydrogen elimination through replacement by CO closes the catalytic cycle, thus furnishing the water-gas shift reaction of eq. (2).

3.2.11.4   Applications

The catalytic formation of hydrogen has been exploited as a means to reduce substrates by carbon monoxide/water. For example, selenium is converted into selenium hydride, SeH$_2$, in the sequence of eqs. (7a–d) under high-pressure autoclave conditions [13].

$$\text{Se} + \text{CO} \rightleftharpoons \text{Se}{=}\text{C}{=}\text{O} \tag{7a}$$

$$\text{Se}{=}\text{C}{=}\text{O} + \text{H}_2\text{O} \rightleftharpoons \text{H}{-}\text{Se}{-}\overset{\text{OH}}{\underset{\text{O}}{\text{C}}} \tag{7b}$$

$$\text{H}{-}\text{Se}{-}\overset{\text{OH}}{\underset{\text{O}}{\text{C}}} \rightleftharpoons \text{SeH}_2 + \text{CO}_2 \tag{7c}$$

$$\text{SeH}_2 \rightleftharpoons \text{Se} + \text{H}_2 \tag{7d}$$

$$\overline{\text{CO} + \text{H}_2\text{O} \rightleftharpoons \text{CO}_2 + \text{H}_2} \tag{7}$$

Once again, a "carbonyl complex" (SeCO) is initially formed, with the latter forming a carboxylic acid which then decarboxylates to SeH_2. Catalytic amounts of selenium suffice to effect reduction of nitroaromatics according to eq. (8). The reduction equivalent (H_2) originates from the water-gas shift reaction, although SeH_2 also acts as a (strong) reductant! The optimum solvent was found to be *N*-methylpyrrolidone. This sequence shows that not only transition metals are capable of catalyzing the reaction of water with carbon monoxide.

$$\text{—NO}_2 \;+\; 3\,\text{CO} \;+\; \text{H}_2\text{O} \;\xrightarrow{\text{cat. Se}}\; \text{—NH}_2 \;+\; 3\,\text{CO}_2 \qquad (8)$$

A base reaction is also made responsible for the palladium(II)-catalyzed reductive carbonylation of nitroaromatics to isocyanates. Carbon monoxide affects the reduction step $Pd^{2+} \longrightarrow Pd^0$ in protic media [14] according to eq. (9). The consecutive sequence $-NO_2 \longrightarrow -NO \longrightarrow -N \longrightarrow -N{=}C{=}O$ including the two oxygen abstraction steps are possible with the Pd^0 thus generated (cf. Section 3.3.5) [15].

$$\text{Pd}^{2+} \; \underset{-\,\text{CO}}{\overset{+\,\text{CO}}{\rightleftharpoons}} \; \{[\text{Pd-C}{\equiv}\text{O}]^{2+}\} \; \xrightarrow[-\,\text{H}^+]{+\,\text{H}_2\text{O}} \; \left[\text{Pd}^{2+}\!-\!\text{C}\overset{\text{O}}{\underset{\text{OH}}{\diagdown}}\right]^{+} \longrightarrow \; \text{Pd}^0 \;+\; \text{H}^+ \;+\; \text{CO}_2 \qquad (9)$$

Another interesting application is the selective reduction of aldehyde functions under water-gas shift reaction conditions [21]. Starting from $Rh_6(CO)_{16}$ in the presence of an amine in aqueous media, the anionic cluster anion $[Rh_6(CO)_{15}H]^-$ forms via nucleophilic attack of hydroxide, followed by elimination of carbon dioxide. The anionic hydride cluster is thought to be the active species in the reduction of a number of aldehydes according to eq. (10).

$$\underset{R_1}{\overset{O}{\|}}\!\!\!\diagup\!\!\!_H \;+\; \text{CO} \;+\; \text{H}_2\text{O} \;\xrightarrow[\text{Rh}_6(\text{CO})_{16}\ \ \text{benzene}]{10\text{-}15\ \text{atm},\ 80^\circ\text{C},\ \text{amine}}\; \underset{R_1}{\overset{H\quad OH}{\diagup\!\!\diagdown}}_H \;+\; \text{CO}_2 \qquad (10)$$

The reduction proved to be chemoselective for aldehyde functions even in the presence of keto groups. The yield in homogeneous reactions varies between 9 % and 30 %. When aminated polystyrene is used as the basic component, the active anionic species $[Rh_6(CO)_{15}H]^-$ is immobilized via interaction with the supported ammonium cations and leads to an increase in the yields up to 91–99 % [22, 23], while the catalyst can be recovered by simple filtration from the product.

3.2.11.5   The Arco Ethylurethane Process

The selenium-catalyzed reduction of nitrobenzene via the water-gas shift reaction (see eqs. (7a–d)) was exploited by Arco Chemicals to make the ethylurethane **1** according to eq. (11). Here, ethanol is being used in place of water [16].

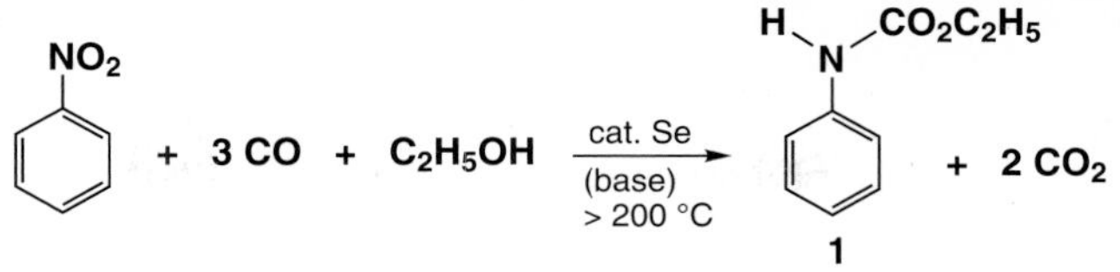

$$\text{(NO}_2\text{-benzene)} + 3\,CO + C_2H_5OH \xrightarrow[\substack{(base) \\ > 200\,°C}]{cat.\ Se} \text{(1)} + 2\,CO_2 \qquad (11)$$

The primary product is aniline (eq. (8)), which then undergoes addition of carbonyl selenide, Se=C=O, in the presence of a strong base [17]. The resulting urethane can further be converted into the methylene diurethane, which is then cracked to the diisocyanate MDI **2**, a key industrial intermediate for the production of polyurethane foams and elastomers (cf. Section 3.3.5). It was probably for toxicity reasons that completion of a technical plant at one of the Arco sites [18] was hampered.

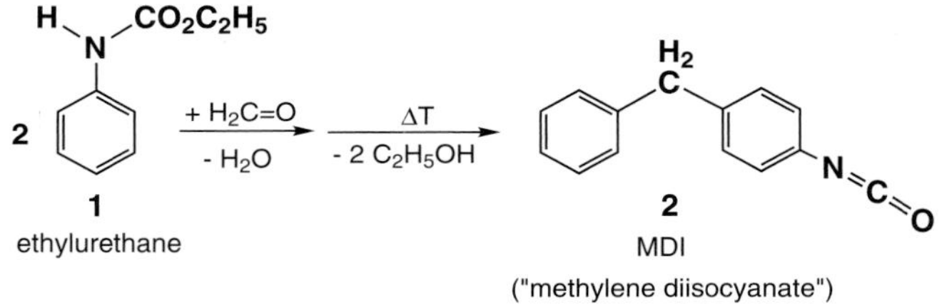

$$\mathbf{2}\ \text{(1, ethylurethane)} \xrightarrow[-\,H_2O]{+\,H_2C=O} \xrightarrow[-\,2\,C_2H_5OH]{\Delta T} \mathbf{2}\ \text{(MDI)} \qquad (12)$$

1
ethylurethane

2
MDI
("methylene diisocyanate")

The ideal method of reductive carbonylation of nitroaromatics would employ synthesis gas according to eq. (13): CO as a *carbonylation* reagent is cheaper in the form of syngas than pure CO, but it is more expensive as a *reducing* agent (eq. (8)) than hydrogen. Unfortunately, there is as yet no catalyst for the overall conversion of eq. (13); only the stepwise reduction works catalytically.

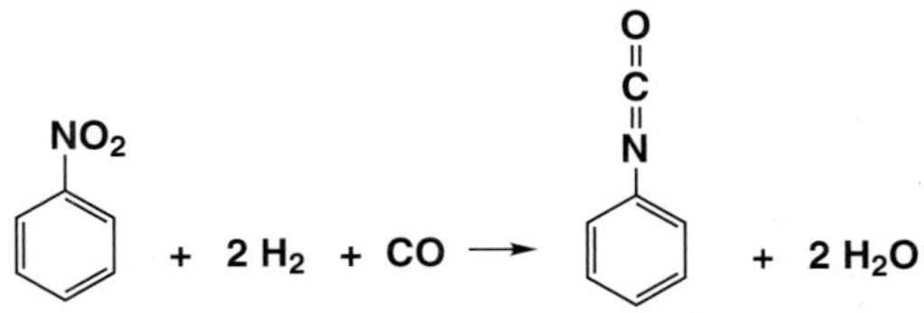

$$\text{(NO}_2\text{-benzene)} + 2\,H_2 + CO \longrightarrow \text{(N=C=O-benzene)} + 2\,H_2O \qquad (13)$$

3.2.11.6 Catalytic Implications and Perspectives

The water-gas shift reaction is normally an unwanted side reaction of homogeneous catalysis when carbon monoxide is engaged as a substrate and if water is present as the medium or as a product. Both a pH-basic medium (formation of the nucleophilic [OH]$^-$) and metals or metal complexes that deprotonate the water favor the shift reaction. For example, in the hydrocarboxylation process to make propionic acid directly from C_2H_4, CO, and H_2O (eq. (14)), the formation of hydrogen via the water-gas shift reaction leads to (minor) hydrogenation and hydroformylation products (cf. Section 2.1.2.2).

$$CH_2{=}CH_2 \ + \ CO \ + \ H_2O \ \longrightarrow \ CH_3CH_2C\overset{\displaystyle O}{\underset{\displaystyle OH}{\Vert}} \tag{14}$$

As a matter of fact, olefin-consuming reactions (by H_2) may be a serious problem in some technical reactions. Palladium complexes and $Co_2(CO)_8$ (commercial products) are typical catalysts. Problems may also arise in the Fischer–Tropsch reaction [19, 20] where iron oxides of a certain basicity (alkaline-metal doping) are being used to catalyze the formation of hydrocarbons according to (the simplified) eq. (15). More details are provided in Section 3.1.8. Since water is inevitably formed, carbon dioxide can also occur. On the other hand, it is doubtful whether the CO/H_2O system will be used for directed reductions of organic compounds, since hydrogen is an extremely abundant industrial chemical. The water-gas shift reaction is thus to be avoided in the vast majority of cases.

$$CO \ + \ 2\,H_2 \ \longrightarrow \ {}^{1}/_{n} \ {\left(\!CH_2\!\right)}_{n} \ + \ H_2O \tag{15}$$

References

[1] (a) Mechanism: K. Tamaru in *Catalysis – Science and Technology* (Eds.: J. R. Anderson, M. Boudart), Springer, Berlin, **1991**, Vol. 9 pp. 93–94; (b) J. R. Rostrup- Nielsen, *ibid.* **1984**, Vol. 5, pp. 57–58; (c) M. A. Vannice, *ibid.* **1982**, Vol. 3, pp. 190–193.

[2] (a) R. M. Laine, R. G. Rinker, P. C. Ford, *J. Am. Chem. Soc.* **1977**, *99*, 252; (b) C. H. Cheng, D. E. Hendrikson, R. Eisenberg, *ibid.* **1977**, *99*, 2791; (c) R. B. King, C. C. Frazier, R. M. Hanes, A. D. King, *ibid.* **1978**, *100*, 2925; (d) T. Yoshida, T. Okano, Y. Ueda, S. Otsuka, *ibid.* **1981**, *103*, 3411; (e) T. Yoshida, Y. Ueda, S. Otsuka, *ibid.* **1978**, *100*, 3941.

[3] Reviews: (a) P. C. Ford, *Acc. Chem. Res.* **1981**, *14*, 31; (b) P. C. Ford, A. Rokocki, *Adv. Organomet. Chem.* **1988**, *28*, 139.

[4] R. M. Laine, R. B. Wilson, "Recent developments in the homogeneous catalysis of the water-gas shift reaction", in *Aspects of Homogeneous Catalysis* (Ed.: R. Ugo), Vol. 5, Reidel (Kluwer), Dordrecht **1984**.

[5] W. A. Herrmann, *J. Organomet. Chem.* **1990**, *382*, 21; (b) Historical review: B. Cornils, W. A. Herrmann, M. Rasch, *Angew. Chem.* **1994**, *106*, 2219; *Angew. Chem., Int. Ed. Engl.* **1994**, *33*, 2144, and references cited therein; (c) Comprehensive text: W. Hieber, *Adv. Organomet. Chem.* **1970**, *8*, 1; (d) M. Catellani, J. Halpern, *Inorg. Chem.* **1980**, *19*, 566.

[6] W. Beck, *J. Organomet. Chem.* **1990**, *383*, 143.

[7] (a) E. O. Fischer, *Adv. Organomet. Chem.* **1976**, *14*, 1; (b) K. H. Dötz, H. Fischer, P. Hoffmann, F. R. Kreissl, U. Schubert, K. Weiss, *Transition Metal Carbene Complexes*, Verlag Chemie, Weinheim, **1983**.

[8] Review: M. J. Albers, N. J. Coville, *Coord. Chem. Rev.* **1984**, *53*, 227.

[9] Other examples: R. H. Crabtree, *The Organometallic Chemistry of the Transition Metals*, Wiley Interscience, New York, **1988**.

[10] N. Grice, S. C. Kao, R. Pettit, *J. Am. Chem. Soc.* **1979**, *101*, 1697.

[11] J. R. Sweet, W. A. G. Graham, *Organometallics* **1982**, *1*, 982.

[12] (a) W. Hieber, F. Leutert, *Z. Anorg. Allgem. Chem.* **1932**, *204*, 145; (b) W. Hieber, H. Vetter, *ibid.* **1933**, *212*, 145.

[13] T. Miyata, K. Kondo, S. Murai, T. Hirashama, N. Sonoda, *Angew. Chem.* **1980**, *92*, 1040; *Angew. Chem., Int. Ed. Engl.* **1980**, *19*, 1008.

[14] V. A. Golodov, Yu. L. Sheludyakov, R. I. Di, V. K. Kokanov, *Kinet. Katal.* **1977**, *18*, 234.

[15] (a) A. L. Balch, D. Petrides, *Inorg. Chem.* **1969**, *8*, 2245; (b) R. G. Little, R. J. Doedens, *ibid.* **1973**, *12*, 536; (c) S. Otsuka, Y. Aotani, Y. Tatsuno, T. Yoshida, *ibid.* **1976**, *15*, 656.

[16] ARCO Chemicals (J. G. Zajacek, J. J. McCoy, K. E. Fuger), US 3.919.279 (1975) and 3.956.360 (1976).

[17] Mitsui Toatsu (H. Seiji, H. Yutaka, M. Katsuhara), US 4.170.708 (1979).

[18] (a) Anon., *Chem. Eng. News*, Oct. 10, **1977**, p. 12; (b) Anon., *Chem. Week*, July 26, **1978**, p. 28.

[19] Reviews: (a) W. A. Herrmann, *Angew. Chem.* **1982**, *94*, 118; *Angew. Chem., Int. Ed. Engl.* **1982**, *21*, 117; (b) C. K. Rofer-DePoorter, *Chem. Rev.* **1981**, *81*, 447.

[20] Monograph: H. H. Storch, N. Golumbic, R. B. Anderson, *The Fischer–Tropsch and Related Syntheses*, Wiley, New York, **1951**.

[21] K. Kaneda, M. Hiraki, T. Imanaka, S. Teranishi, *J. Mol. Catal.* **1980**, *9*, 227; K. Kaneda, Y. Yasumura, T. Imanaka, S. Teranishi, *Chem. Commun.* **1982**, 93.

[22] K. Kaneda, T. Mizugaki, K. Ebitani, *Tetrahedron Lett.* **1997**, *38*, 3005.

[23] J. M. Basset, P. Dufour, L. Huang, A. Choplin, S. G. Sanchez-Delgado, A. Tholier, *J. Orgnomet. Chem.* **1988**, *354*, 354; B. T. Heaton, L. Strona, S. Martinengo, D. Strumolo, R. J. Goodfellow, I. H. Sadler, *J. Chem. Soc., Dalton Trans.* **1982**, 1499.

3.2.12   Catalytic McMurry Coupling: Olefins from Keto Compounds

Wolfgang A. Herrmann, Horst Schneider

3.2.12.1   Introduction

In 1973 Mukaiyama, Tyrlik, and McMurry discovered a remarkably simple reaction that couples aldehydes or keto compounds reductively to olefins [1, 2]. This methodology following eq. (1) differs from that of Section 3.2.10 in that no extra methylene or alkylidene transfer reagent is required. The stereochemistry of the product depends on the nature of the substituents R and whether an open-chain or a cyclic olefin results.

$$\underset{H}{\overset{R}{\diagdown}}C=O \; + \; O=C\underset{R}{\overset{H}{\diagup}} \; + \; "Ti" \; \longrightarrow \; \underset{H}{\overset{R}{\diagdown}}C=C\underset{H}{\overset{R}{\diagup}} \; + \; "TiO_2" \tag{1}$$

The driving force of the reaction is the formation of the strong titanium–oxygen bonds. Low-valent titanium is oxophilic enough to extrude all the oxygen from the substrate. This C–C coupling process, albeit stoichiometric with regard to the (inorganic) coupling reagent(s), has become extraordinarily useful in the synthesis of olefinic compounds, be it either *intra-* or *inter*molecularly. The reaction is compatible with quite a large number of functional groups, e. g., hydroxyl, amide, sulfide, ether, and C–C double bonds. Even phenanthrenes

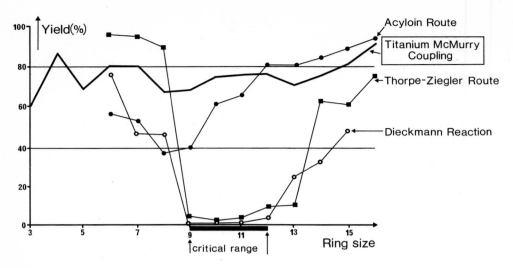

Figure 1. Yields from different synthetic routes to C_3-C_{16} carbocycles.

and heterocycles such as arsole, indole, and pyrazole derivatives are thus accessible.

It is specifically noted that this active-metal supported coupling makes cyclic olefins of otherwise unfavorable ring sizes ($n = 8-12$) available without problems (cf. Figure 1). Reactions following eq. (1) thus belong to the modern synthetic methodology. From a mechanistic point of view, electron transfer (metal to substrate) is of key importance (e. g., pinacolate–titanium intermediates).

3.2.12.2 Stoichiometric Titanium Compounds, Other Reagents, and Mechanistic Aspects

A fair number of coupling reagents has been reported for the McMurry reaction. They mostly contain titanium in an oxidation state smaller than 4. For example, $TiCl_3 \cdot {}^1/_2\ CH_3OCH_2CH_2OCH_3$ [3], and a system described as "$TiCl_3$–$LiAlH_4$–THF," known as the "McMurry reagent". Improvements of the coupling reagent were reported by Bogdanovic' and co-workers, who described a hydridic Ti^{II} species resulting from treatment of $TiCl_3$ with activated MgH_2 [4] according to eq. (2).

$$TiCl_3 \bullet (THF)_3 + MgH_2 \xrightarrow[- 80 \rightarrow 25\ °C]{(THF)} HTiCl(THF)_{-0.5} + MgCl_2 + {}^1/_2\ H_2 \qquad (2)$$

$$\mathbf{1}$$

The same active species (**1**, eq. (2)) could be found in the "McMurry reagent" system where $TiCl_3$ becomes reduced by 0.5 equivalent of $LiAlH_4$ in THF [2f]. Ti^0

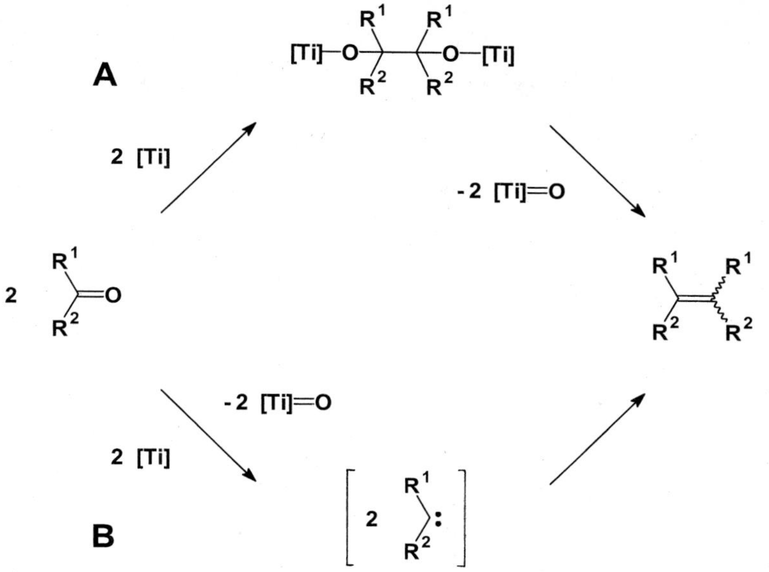

Scheme 1

could be excluded as active species according to EXAFS measurements [5, 6], and more generally, Ti^0 is not presupposed for the McMurry reaction [7]. The mechanism of McMurry type reactions has a dual nature [7, 8], as shown in Scheme 1 [1 a, b].

Dependent of the carbonyl substrate, the Ti compound, the reducing reagents, and other available compounds, the dual nature of the mechanism can be observed. Less hindered ketones follow pathway (**A**) with pinacolate intermediates. Heavily hindered ketones would follow the carbenoid route (**B**) [2 f, 7, 8].

Other efficient reagents include TiCl$_4$/Zn [1 a], TiCl$_3$/Mg [1 b], TiCl$_3$/Zn-Cu [1 e], and TiCl$_3$ · (DME)$_{1.5}$/Zn–Cu [1 d] (DME = 1.2-dimethoxyethane), with the latter resulting as blue crystals from boiling TiCl$_3$ in DME, and M^{1}Cl$_4$/ M^2(Hg) (M^1 = U, Ti, M^2 = alkali metal) [8, 9]. However, the McMurry reaction is stoichiometric, since the titanium–oxo bonds (e. g., TiO$_2$) resist (catalytic) reactivation. Although McMurry coupling reactions play a key role in the synthesis of numerous natural products (e. g., crassin [10], taxol [11], cembren C [12], and mevinolen [13]), textbooks on catalysis hardly mention this otherwise interesting (from a mechanistic point of view) and useful synthetic procedure [14] for the reason mentioned above, and because of problems of reproducibility due to the manifold combination possibilities of the reagents [7]. Iodine also can activate low-valent Ti reagents [15], and external ligands, e. g., *t*-BuOH [16] or pyridine [17], have been used to direct the reaction.

Other aldehyde-to-olefin coupling reagents of low-valent metals have also been described; however, they suffer from being less efficient. The tungsten(III) alkox-

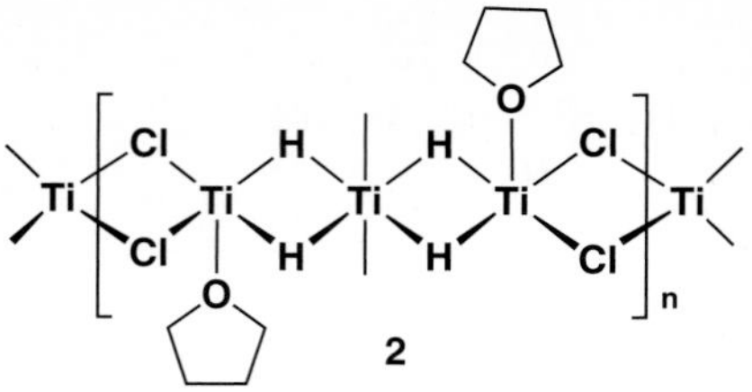

2

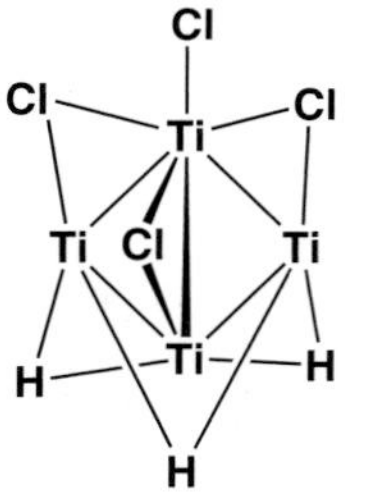

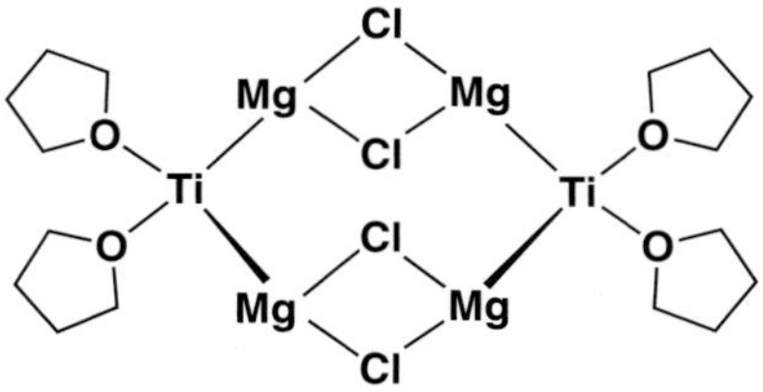

3   **4**

ide $W_2(OCH_2But)_6 \cdot 2py$, for example, gives yields up to 66 % [18], while titanium-mediated couplings are often beyond 90 % yield.

A variety of low-valent species effects the deoxygenation of eq. (1). The exact
nature of Ti_xO_y formed during a McMurry coupling (simplified in eq. (1) as TiO_2)
now becomes better understood. The nature of the active species is often speculative, although firm evidence for structures **2–4** resulting from α-TiCl$_3$/MgH$_2$,
TiCl$_4$/MgH$_2$, or TiCl$_3$/LiAlH$_4$/THF has been presented [2 f, 4].

3.2.12.3 Catalytic Deoxygenation

Fürstner et al. discovered [19] that the combination of TiCl$_3$/Zn dust and
$(CH_3)_3SiCl$ works catalytically with regard to the titanium in reductive aldehyde
coupling. The so-called "instant method" works also in non-etherlike solvents
(e. g., acetonitrile, DMF), and a couple of functional groups are tolerated [2 f].
Equation (3) can be written for the overall reaction, showing that the chlorosilane
reagent is necessary in a twofold stoichiometric amount based on the aldehyde
(ketone) to be coupled. Acetonitrile or DME are the preferred (coordinating) solvents. While $(CH_3)_3SiCl$ as the cheapest of all chlorosilanes requires 5–10 mol %
TiCl$_3$, the bis(chlorosilane) $Cl(CH_3)_2Si–CH_2CH_2–Si(CH_3)_2Cl$ works better
(2 mol % TiCl$_3$). A number of heterocycles, e. g., substituted indoles, is thus available.

$$2\ \overset{R}{\underset{H}{\diagdown}}C{=}O + 4\ (CH_3)_3SiCl \xrightarrow{\text{cat. Ti}} \overset{R}{\underset{H}{\diagdown}}C{=}C\overset{R}{\underset{H}{\diagup}} + 2\ (CH_3)_3Si\text{-}O\text{-}Si(CH_3)_3 \qquad (3)$$

Chlorosilanes are also capable of activating commercial titanium powder (e. g., $(CH_3)_3SiCl$ in boiling DME [20]), which then couples oxoamides such as **5** in isolated yields of $> 92\%$ to the indoles, e. g., **6**, according to eq. (4) [21].

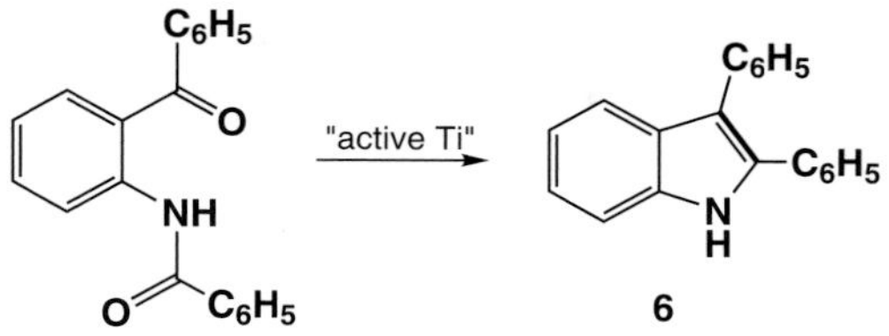

$$\tag{4}$$

Above-stoichiometric amounts of chlorosilanes are necessary for high yields. The reagent exhibits a strong template effect for *intra*molecular coupling processes. Even 36-membered carbocycles and numerous unsaturated crown-ether derivatives can be made in good yields. Beyond that, multiply C–C-unsaturated compounds such as retinal **7** undergo reductive coupling: in case of eq. (5) β-carotin **8** is formed in good yields.

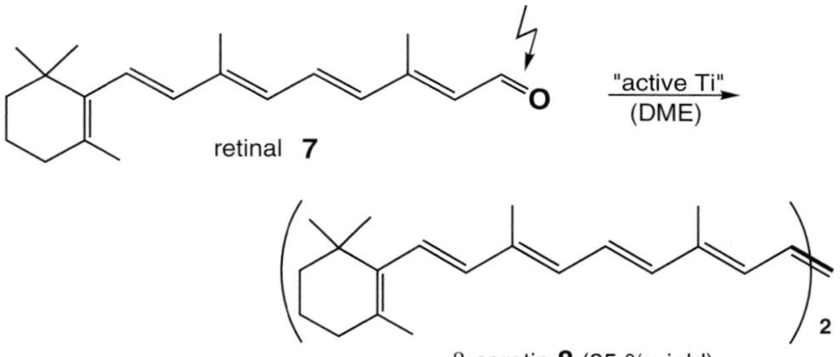

retinal **7**

$$\tag{5}$$

β-carotin **8** (85 % yield)

The proposal of Scheme 2 appears to be a reasonable approach to the problem. The chlorosilane obviously does not only destroy the oxidic layer of titanium powder (activation effect) but also seems to facilitate the electron transfer from the metal to the substrate [21].

3.2.12.4   Perspectives

The catalytic system of Fürstner et al. is to be considered as a breakthrough in reductive deoxygenating C–C coupling with highly oxophilic low-valent titanium [22], albeit not too much has been published since the time of this discovery [19]. The auxiliary oxygen traps (chlorosilanes) are cheap, easy to handle and to remove (e. g., they have low boiling points), and relatively unreactive toward the substrates to be coupled. Beyond that, the catalytic "titanium instant" is an insoluble and thus easy-to-remove reagent. Considering the vast number of bio-logically relevant C–C-unsaturated carbo- and heterocycles, the *catalytic* route is expected to become a major synthetic approach. It is well known that inter-

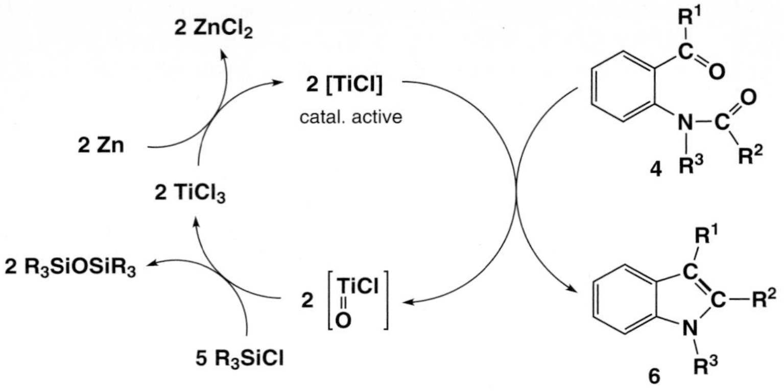

Scheme 2. Catalytic McMurry coupling of oxoamides to indoles according to Fürstner [21]. Preferred substituents: $R^1 = C_6H_5$, $R^2 = CF_3$, CO_2Et, $R^3 = H$. The [TiCl] species is structurally undefined and differs in nature depending upon the conditions of generation.

mediate-size cyclic compounds (e. g., carbocycles of ring size 8–12; cf. Fig. 1) are only accessible in reasonable yields by the McMurry coupling [2]. The catalytic efficiency of titanium (or another cheap metals?) is certainly subject to further improvement. Suffice it to say that an electrochemical reactivation of the metal oxide would be the most elegant solution to the intrinsic problem of the McMurry coupling.

References

[1] (a) T. Mukaiyama, T. Sato, J. Hanna, *Chem. Lett.* **1973**, 1041; (b) S. Tyrlik, I. Wolochowicz, *Bull. Soc. Chim. Fr.* **1973**, 2147; (c) J. E. McMurry, M. P. Fleming, *J. Am. Chem. Soc.* **1974**, *96*, 4708; (d) J. E. McMurry, *J. Org. Chem.* **1978**, *43*, 3255.

[2] Reviews: (a) J. E. McMurry, *Acc. Chem. Res.* **1983**, *16*, 405; (b) B. E. Kahn, R. T. Riecke, *Chem. Rev.* **1988**, *88*, 733; J. E. McMurry, *Chem. Rev.* **1989**, *89*, 1514; (d) C. Betschart, D. Seebach, *Chimia* **1989**, *43*, 39; (e) D. Lenoir, *Synthesis* **1989**, 8830; (f) A. Fürstner, B. Bogdanovic, *Angew. Chem.* **1996**, *108*, 2582; *Angew. Chem. Int. Ed. Engl.* **1996**, *35*, 2442.

[3] J. E. McMurry, T. Lectka, J. G. Rico, *J. Org. Chem.* **1989**, 54, 3748.

[4] (a) L. E. Aleandri, B. Bogdanovic, A. Gaidies, D. J. Jones, S. Liao, A. Michalowicz, J. Rozière, A. Schott, *J. Organomet. Chem.* **1993**, *459*, 87; (b) L. E. Aleandri, S. Becker, B. Bogdanovic, D. J. Jones, J. Rozière, *J. Organomet. Chem.* **1994**, *472*, 97.

[5] H. Bertagnolli, T. S. Ertel, *Angew. Chem.* **1994**, *106*, 15; *Angew. Chem. Int. Ed. Engl.* **1994**, *33*, 45.

[6] (a) R. Dams, M. Malinowski, I. Westdorp, H. J. Geise, *J. Org. Chem.* **1982**, *47*, 248; (b) R. Dams, M. Malinowski, H. J. Geise, *Transition Met. Chem. (London),* **1982**, *7*, 37; (c) *Bull. Soc. Chim. Belg.* **1981**, *90*, 1141.

[7] M. Ephritikhine, *Chem. Commun.* **1998**, 2549.

[8] C. Villiers, M. Ephritikhine, *Angew. Chem.* **1997**, *109*, 2477; *Angew. Chem. Int. Ed. Engl.* **1997**, *36*, 2380.

[9] (a) D. Maury, C. Villiers, M. Ephritikhine, *New J. Chem.* **1997**, *21*, 137; (b) *Angew. Chem.* **1996**, *108*, 1215; *Angew. Chem. Int. Ed. Engl.* **1996**, *35*, 1129; (c) C. Villiers, R. Adam, M. Lance, M. Nierlich, J. Vigner, M. Ephritikhine, *J. Chem. Soc., Chem. Commun.* **1991**, 1144.

[10] (a) W. G. Dauben, T. Z. Wang, R. W. Stephens, *Tetrahedron Lett.* **1990**, 2393; (b) J. E. McMurry, R. G. Dushin, *J. Am. Chem. Soc.* **1990**, *112*, 6942.

[11] (a) K. C. Nicolaou, J. J. Liu, Z. Yang, H. Ueno, E. J. Sorensen, C. F. Claiborne, R. K. Guy, C. K. Hwang, M. Nakada, P. G. Nantermet, *J. Am. Chem. Soc.* **1995**, *117*, 634; (b) K. C. Nicolaou, Z. Yang, J. J. Liu, P. G. Nantermet, C. F. Claiborne, J. Renaud, R. K. Guy, K. Shibayama, *J. Am. Chem. Soc.* **1995**, *117*, 645.

[12] Y. Li, W. Li, Y. Li, *Synth. Commun.* **1994**, *24*, 721.

[13] D. L. J. Clive, K. S. K. Murthy, A. G. H. Wee, J. S. Prasad, G. V. J. da Silva, M. Majewski, P. C. Anderson, C. F. Evans, R. D. Haugen, L. D. Heerze, J. R. Barrie, *J. Am. Chem. Soc.* **1990**, *112*, 3018.

[14] Textbook and review on organic synthesis: (a) K. P. C. Vollhardt, *Organic Chemistry, Structure and Function*, W. H. Freeman, New York, **1999**; (b) D. Seebach, *Angew. Chem.* **1990**, *102*, 1363; *Angew. Chem. Int. Ed. Engl.* **1990**, *29*, 1320.

[15] S. Talukdar, S. K. Nayak, A. Banerji, *J. Org. Chem.* **1998**, *63*, 4925.

[16] T. A. Lipski, M. A. Hilfiker, S. G. Nelson, *J. Org. Chem.* **1997**, *62*, 4566.

[17] N. Balu, S. K. Nayak, A. Banerji, *J. Am. Chem. Soc.* **1996**, *118*, 5932.

[18] M. H. Chisholm, J. A. Klang, *J. Am. Chem. Soc.* **1989**, *111*, 2324.

[19] A. Fürstner, A. Hupperts, A. Ptock, E. Janssen, *J. Org. Chem.* **1994**, *59*, 5215.

[20] A. Fürstner, B. Tesche, *Chem. Mater.* **1998**, *10*, 1968.

[21] A. Fürstner, A. Hupperts, *J. Am. Chem. Soc.* **1995**, *117*, 4468.

[22] Review on "active titanium": A. Fürstner, *Angew. Chem.* **1993**, *105*, 171; *Angew. Chem. Int. Ed. Engl.* **1993**, *32*, 164.

3.2.13 Catalytic Hydrogenation of Heterocyclic Sulfur and Nitrogen Compounds in Raw Oils

Claudio Bianchini, Andrea Meli, Francesco Vizza

3.2.13.1 Introduction

The use of single-site transition metal catalysts to effect the hydrogenation of heteroaromatic sulfur and nitrogen compounds finds its primary impetus in the need for improved understanding of the mechanisms of the hydrodesulfurization (HDS, eq. (1)) and hydrodenitrogenation (HDN, eq. (2)) processes [1]. Indeed, with stringent environmental regulations concerning the amount of sulfur and nitrogen permitted in gasoline and city diesel, the development of new HDS and HDN catalysts is a priority in the petrochemical industry.

$$\text{Hydrodesulfurization} \quad C_aH_bS + c\,H_2 \longrightarrow H_2S + C_aH_d \tag{1}$$

$$\text{Hydrodenitrogenation} \quad C_aH_bN + c\,H_2 \longrightarrow NH_3 + C_aH_d \tag{2}$$

It is difficult to know whether, and if so which, HDS and HDN modeling studies have led to the development of new catalysts with improved performance under actual refinery conditions. There is little doubt, however, that the homogeneous studies have contributed greatly to the elucidation of the binding of the *S*- and *N*-heterocycles to metal centers as well as to the mechanisms of fundamental steps such as the hydrogen transfer from metal to coordinated substrate and the C–S or C–N bond scissions [2]. Reference to this important work is provided in this Section, which, however, is almost exclusively concerned with catalytic hydrogenation reactions of sulfur and nitrogen heteroaromatics. Recently, some homogeneous and heterogeneous single-site metal catalysts have been found to assist effectively the hydrogenation and/or hydrogenolysis of *N*- and *S*-heteroaromatics, and even the desulfurization of thiophenic substrates. These reactions, especially those that have provided data on product distribution, kinetics, and selectivity, are the subject of this Section. For the sake of completeness, it is worth mentioning that both the hydrogenation of *S*- and *N*-heterocycles and their reductive opening by transition metal complexes are also efficient synthetic routes to a variety of fine chemicals and materials containing sulfur or nitrogen [3].

Compounds mentioned in this chapter are: pyrrole (abbreviated PYR), indole (IN), carbazole, pyridine (Py), quinoline (Q), isoquinoline (IQ), acridine (AC), 5,6- and 7,8-benzoquinoline (5,6-BQ; 7,8-BQ), thiophene (T), benzo[*b*]thiophene (BT), dibenzo[*b,d*]thiophene (DBT), etc.

3.2.13.2   Hydrogenation of Sulfur Heterocycles

The principal mechanisms proposed for the heterogeneous HDS of a prototypical *S*-heterocycle, namely benzo[*b*]thiophene (BT), are illustrated in Scheme 1. The plain hydrogenation of C–C double bonds occurs in step **a**, involving the regioselective reduction of BT to dihydrobenzo[*b*]thiophene (DHBT), as well as in step **e**, where styrene is reduced to ethylbenzene.

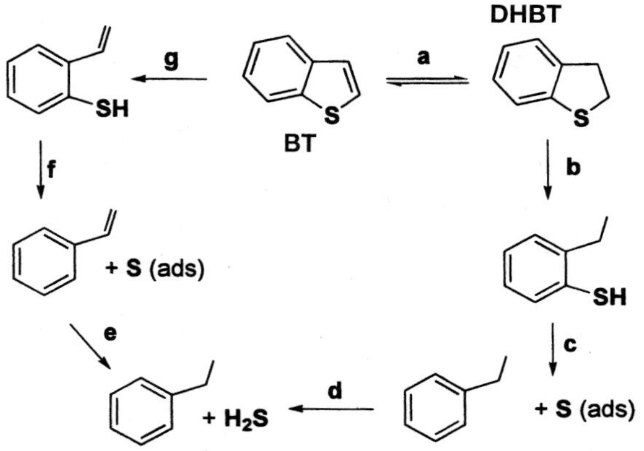

Scheme 1

3.2.13.2.1 Homogeneous Systems

In fluid-solution systems, the plain hydrogenation of thiophenes to thioether products has been found to be catalyzed by various transition metal complexes; surprisingly, none of these catalysts contains either molybdenum or tungsten, which are essential components of heterogeneous HDS catalysts [1 b]. Unpromoted MoS_2, on the other hand, is quite active for the HDS of thioethers with no need of assistance by a late transition metal promoter [1 b].

The pseudo-olefinic character of the C_2–C_3 bond makes BT the easiest thiophene to hydrogenate to the corresponding thioether. Indeed, no example of homogeneous hydrogenation of dibenzo[*b,d*]thiophene (DBT) is known, whereas a case of catalytic reduction of thiophene (T) to tetrahydrothiophene (THT) in 1,2-dichloroethane has been reported to occur with the catalyst precursor $[IrH_2(\eta^1\text{-}S\text{-}T)_2(PPh_2)_2]PF_6$ [5]. BT is actually a better ligand than T or DBT and can form both $\eta^2\text{-}C,C$ and $\eta^1\text{-}S$ complexes as single species or in equilibrium with each other [6].

As a general trend, the homogeneous hydrogenation of thiophenes to thioethers is catalyzed by complexes that are not sterically demanding with relatively electrophilic metal centers (e. g., d^6 metal ions such as Rh^{III}, Ir^{III}, Ru^{II}, or Os^{II}). This is because moderate electron density and low steric hindrance at the metal favor the $\eta^2\text{-}C,C$ coordination mode of the thiophene (precursor to hydride migration) over $\eta^1\text{-}S$ binding (precursor to C–S bond cleavage) [7–10].

In the known hydrogenation catalysts, the metals are stabilized by either cyclopentadienyl or phosphine ligands [2 b, 7–10]. Solvents with good ligating properties generally slow down the hydrogenation rate as they may compete with the substrates for coordination. Relatively drastic reaction conditions are generally employed (20–110 bar; 40–170 °C), but there are hydrogenation reactions that take place even at ambient temperature and pressure with fairly good rates [2 b, 7–10]. In terms of catalytic efficiency, Ru^{II} forms the most active systems with TOFs as high as 500 [9 a].

Commonly accepted hydrogenation mechanisms of BT catalyzed by metal precursors devoid of hydride ligands comprise the usual steps of H_2 oxidative addition, $\eta^2\text{-}C,C$ coordination of the substrate, hydride transfer to form dihydrobenzothienyl, and elimination of DHBT by hydride/dihydrobenzothienyl reductive coupling (Scheme 2). A similar sequence of steps is proposed for catalysts bearing a hydride ligand (Scheme 3). In this case, the reaction with H_2 follows the addition of the substrate and the hydride migration step. Irrespective of the structure of the catalyst, the regioselectivity of the first hydride migration step is still unknown as no hydride (dihydrobenzothienyl) intermediate has ever been intercepted.

The mechanism of hydrogenation of T to THT does not differ significantly from that reported above for the hydrogenation of BT. The only peculiar feature concerns the first hydride migration step (*endo* migration), which generally occurs with regio- and stereospecificity to give a thioallyl intermediate. This is converted to a 2,3-dihydrothiophene ligand, which is hydrogenated like any other alkene.

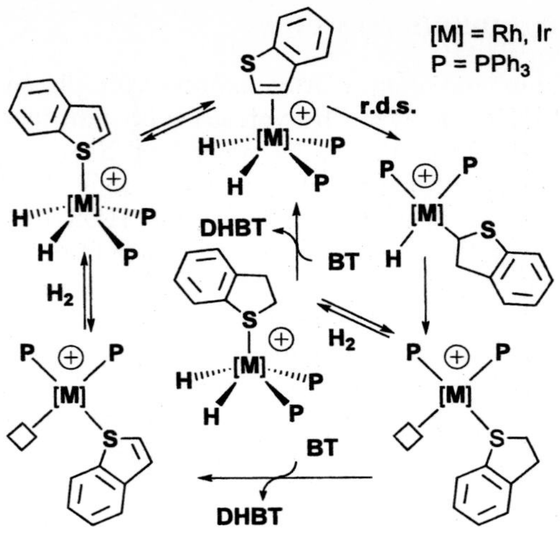

Scheme 2

Kinetic studies of the regioselective hydrogenation of BT to DHBT have been reported for various catalysts. Using the precursors [Rh(PPh$_3$)$_2$(COD)]PF$_6$ (COD = cycloocta-1,5-diene) [8 b] and [Ir(PPh$_3$)$_2$(COD)]PF$_6$ [5, 8 a] in THF or 1,2-dichloroethane, first-order dependence on both catalyst and H$_2$ concentrations and zero-order dependence on BT concentration have been observed, while the hydride migration yielding the dihydrobenzothienyl intermediate has been proposed to be the rate-determining step (rds) (Scheme 2). In contrast, a first-order dependence on catalyst, H$_2$, and substrate concentrations and an rds involving the reversible dissociation of the thioether product from the metal have been reported for

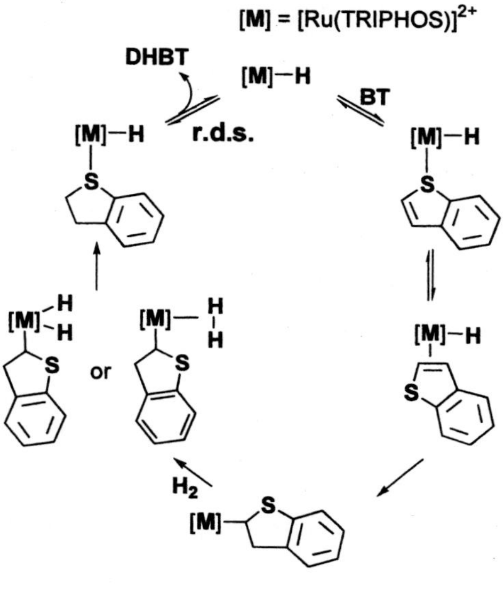

Scheme 3

the hydrogenation of BT promoted by [Ru(MeCN)$_3$(TRIPHOS)]BPh$_4$ in THF [TRIPHOS = MeC(CH$_2$PPh$_2$)$_3$] [9 a] (Scheme 3).

Valuable information on the mechanism of the regioselective BT hydrogenation by soluble metal complexes has been obtained by substituting deuterium for hydrogen gas in the reduction reactions catalyzed by the precursors [Rh(MeCN)$_3$(Cp*)](BF$_4$)$_2$ [7 a] and [Ru(MeCN)$_3$(TRIPHOS)]BPh$_4$ [7 a, 9 a].

In situ high-pressure ^{31}P{^{1}H} and ^{1}H NMR experiments have shown that the hydrogenation of BT with the catalyst precursor [Ru(MeCN)$_3$(TRIPHOS)](BPh$_4$)$_2$ in THF involves the preliminary conversion of MeCN to various nitrogen bases: NH$_2$Et, NHEt$_2$, NEt$_3$, and NH$_3$ [9 a]. The effective catalyst for the hydrogenation of the thiophene was suggested to be the 14e$^-$ fragment [RuH(TRIPHOS)]$^+$, formed via base-assisted heterolytic splitting of H$_2$. At the beginning of the catalytic reaction, the Ru^{II} fragment [RuH(TRIPHOS)]$^+$ was intercepted primarily as the bisacetonitrile complex [RuH(MeCN)$_2$(TRIPHOS)]$^+$, while, at the end of the catalysis, three monohydride complexes stabilized by NH$_3$ or NHEt$_2$ ligands were observed. The formation of Ru–NH$_3$ bonds that are stronger than those with BT or DHBT is in line with the hypothesis according to which the deactivation of the Ru-based HDS catalysts by nitrogen bases is due to the formation of very strong (Ru)$_3$–N or (Ru)$_{3-x}$–NH$_x$ bonds derived from the degradation of the ammonia produced in the concomitant HDN process [1, 9].

Rh and Ir generally go through the hydrogenation catalysis with the III → I → III reduction/oxidation cycle, provided the activation of H$_2$ occurs *via* oxidative addition. In turn, Ru and Os follow the IV → II → IV reduction/oxidation cycle. A constant oxidation state along the whole catalysis cycle should feature the metal center if the activation of dihydrogen occurs via the η^2-H$_2$ pathway and the catalyst contains a hydride ligand [11]. Indeed, the possibility of intermediates containing intact H$_2$ ligands cannot be disregarded in hydrogenation reactions catalyzed by d^6 metal ions. The ability of the η^2-H$_2$ complex [RuH$_2$(H$_2$)$_2$(PCy$_3$)$_2$] to reduce various *S*- and *N*-heterocycles to the corresponding cyclic thioethers and amines has been reported, in fact [12].

An η^2-H$_2$ complex has also been detected along the pathway of hydrogenation of C–S-inserted BT to ethylthiophenol with the [Ir(TRIPHOS)] fragment [13 c]. On the other hand, DFT calculations have suggested that the hydrogenation of thiophenic substrates over Ni$_x$S$_y$ clusters ($x = 3, 4$) may involve adsorbed molecular hydrogen that subsequently undergoes heterolytic activation to give both Ni–SH and Ni–H species [13, 14].

3.2.13.2.2 Aqueous-Biphasic Systems

Although still confined to laboratory scale, aqueous-biphase catalysis (cf. Section 3.1.1.1) and related variations such as supported liquid-phase catalysis (cf. Section 3.1.1.3.5) are emerging as viable techniques for the deep HDS of refined fuels [15].

The selective hydrogenation of *S*- and *N*-heterocycles has been achieved by researchers at PDVSA–INTEVEP with the use of water-soluble Ru^{II} catalysts stabilized by either triphenylphosphine trisulfonate (TPPTS) or triphenylphosphine monosulfonate (TPPMS) ligands [16]. The biphasic reactions were performed under relatively harsh experimental conditions (130–170 °C, 70–110 bar H_2) and gave the selective reduction of the heterocyclic ring irrespective of the heterocycle. It was generally observed that nitrogen compounds did not inhibit the hydrogenation of either T or BT. In some cases, indeed, a promoting effect was observed. For example, the rate of hydrogenation of BT to DHBT catalyzed by various Ru^{II} complexes with either TPPMS or TPPTS in water/decalin quadrupled when quinoline or aniline was used as co-catalyst.

Selective Ru and Rh catalysts for the aqueous-biphasic hydrogenation of BT to DHBT have been obtained using the polydentate phosphines $NaO_3S(C_6H_4)$-$CH_2)_2C(CH_2PPh_2)_2$ ($Na_2DPPPDS$) [17] and $NaO_3S(C_6H_4)CH_2C(CH_2PPh_2)_3$ (NaSULPHOS) [18] (**1, 2**).

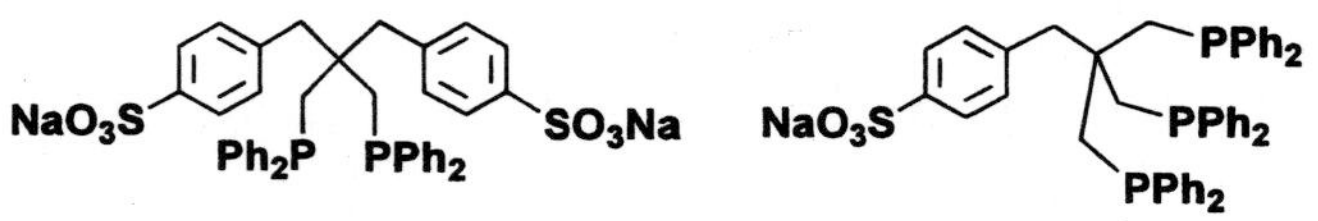

Na₂DPPPDS
1

NaSULPHOS
2

In general, Ru-based catalysts are more efficient than Rh-based catalysts for the selective hydrogenation of BT to DHBT in water/hydrocarbon mixtures [19]. Rhodium forms much better catalysts for the hydrogenolysis of thiophenes to thiols (*vide infra*).

The binuclear complex $Na[\{Ru(SULPHOS)\}_2(\mu\text{-}Cl)_3]$ [20] and the monomeric derivative $[Ru(MeCN)_3(SULPHOS)](SO_3CF_3)$ [9 a] have been employed as precatalysts for the hydrogenation of BT to DHBT in water/decalin or water/*n*-heptane showing a very similar rate (TOF 30) in comparable experimental conditions (100–140 °C, 3 MPa H_2) [19]. It was therefore suggested that the disruption of the dimeric structure of the μ-Cl_3 complex may occur under catalytic conditions.

In aqueous biphasic conditions, the zwitterionic Rh^I complex Rh(COD)(SULPHOS) has been shown to be a modest catalyst for the hydrogenation of BT to DHBT (TOF 5) [10 a].

3.2.13.2.3 Heterogenized Single-Site Systems

The increased selectivity and much milder experimental conditions required for high conversions make molecular catalysts compete with heterogeneous ones in many chemical processes. The application of traditional molecular catalysis in large-scale reactions such as HDS and HDN, however, is not possible. In order to overcome this drawback, many research efforts are being directed toward the heterogenization of molecular catalysts [21]. Successful applications of heterogenized molecular catalysts in several large-volume reactions have already been obtained [21].

The first attempt to hydrogenate sulfur heterocycles with a supported metal catalyst was reported by Fish in 1985; cf. Structure **3** [22]. Interestingly, the initial hydrogenation rate of BT was three times faster for the single-site heterogeneous catalyst than for the homogeneous derivative Rh(PPh$_3$)$_3$Cl. This rate enhancement was also observed for *N*-heterocycles and was attributed to steric requirements for the surroundings of the active metal center in the tethered complex, which would favor the coordination of the heterocycles by disfavoring that of PPh$_3$ [22].

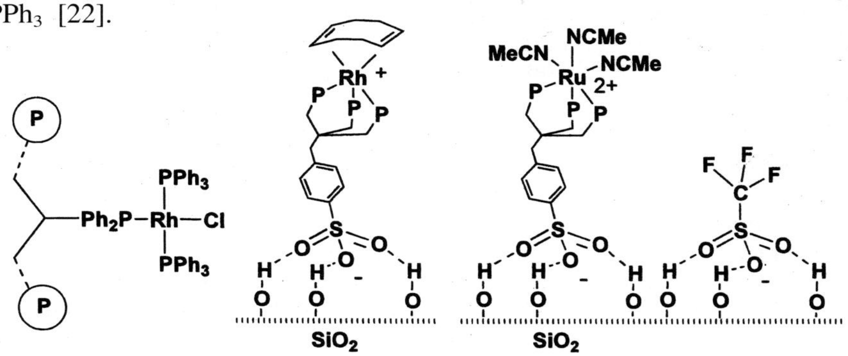

3

The use of polymer-supported metal catalysts for the hydrogenation of thiophenic substrates has recently been extended to Ru and Rh complexes anchored to silica via hydrogen bonding [23, 24].

Inspired by previous work from Angelici and co-workers [25], Bianchini, Psaro, and co-workers have recently anchored the complex Rh(COD)(SULPHOS) via hydrogen bonding to silica-containing Pd nanoparticles (**4**). A sample of Rh(COD)(SULPHOS)/Pd/SiO$_2$ containing 0.5 wt.% Rh^I and 10 wt.% Pd^0 was employed to hydrogenate BT in *n*-octane (3 MPa H$_2$, 100 °C). A 12-fold increase in the hydrogenation rate with no loss of selectivity was observed for the mixed molecular-metal particle catalyst as compared to Rh(COD)(SULPHOS)/SiO$_2$, while the catalyst containing exclusively Pd was inactive. The factors which are responsible for this remarkable synergic effect are not yet understood completely.

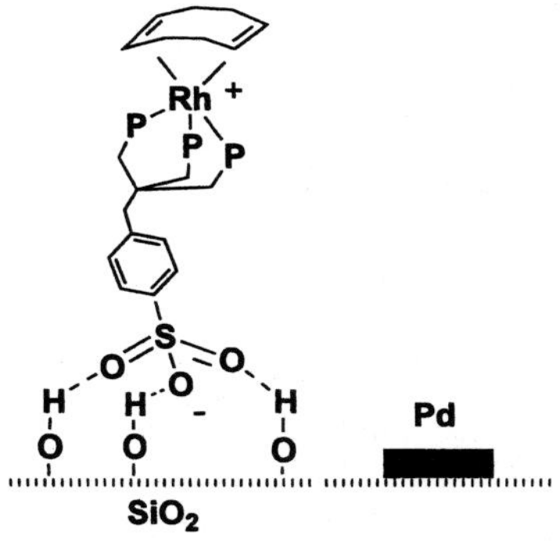

4

3.2.13.3   Hydrogenolysis of Sulfur Heterocycles

The reaction which transforms a thiophenic substrate into the corresponding unsaturated thiol is referred to as hydrogenolysis (eq. (3)). Because of the facile hydrogenation of the unsaturated thiols derived from T or BT, the actual hydrogenolysis products obtained with molecular catalysts are generally the saturated thiols.

$$\text{(3)}$$

The hydrogenolysis of thiophenes to thiols is a reaction that only a few metal complexes catalyze efficiently. Indeed, whereas the metal complexes which are capable of cleaving and then hydrogenating C–S bonds in thiophenes are relatively numerous [2 b], those which do this in catalytic fashion are very few and all are characterized by a well-defined molecular architecture as well as remarkable thermal and chemical stability. C–S bond scission is best accomplished, in fact, by electron-rich, coordinatively unsaturated systems, commonly $16e^-$ tetra- or tricoordinate species, with ligand sets that do not generally allow the fragment to attain the too stable square-planar geometry, e. g., [MH(TRIPHOS)] (M = Rh, Ir) [3, 10, 13], [IrCp*] [26], or [Rh(PMe$_3$)Cp*] [27]. Highly energetic metal fragments with filled orbitals of appropriate symmetry are necessary to lower the barrier to C–S insertion which occurs via $d\pi$(metal) $\rightarrow \pi^*$(C–S) transfer [27]. Moreover, the steric crowding at the metal center must be great enough to disfavor the η^2-C,C bonding mode of the substrate, but not so great to impede the coordination of the substrate via the sulfur atom [27]. In actuality, the insertion of metal fragments into C–S bonds in T, BT, and DBT is a relatively high-energy process, only slightly disfavored over C(sp^2)–H insertion [28], but much easier than C–N insertion [1, 2].

3.2.13.3.1 Homogeneous Systems

Tailoring the electronic and steric characteristics of a metal complex is not sufficient *per se* to form a catalyst for the hydrogenolysis of thiophenes as this reaction generally requires high temperature and H$_2$ pressure to take even in homogeneous phase. So far, only the highly chelating tripodal triphosphine TRIPHOS in combination with Ru^{II}, Rh^I and Ir^I has been found capable of forming catalysts that tolerate the thermal and chemical stress of the hydrogenolysis reactions of thiophenes (3 MPa H$_2$, 100–160 °C, presence of a strong base).

Irrespective of the metal catalyst, the hydrogenolysis rate of any thiophenic substrate is significantly accelerated when a strong Brønsted base, generally KOBu^t in THF, is added to the catalytic mixture in the same concentration as the substrate [9, 10]. The main role of the base is that of speeding up the removal of the thiol product from M(H)(SR) intermediates, which constitutes the rds of all hydrogenolysis reactions reported so far. In some cases, strong bases have been used as co-catalysts to generate M–H bonds by heterolytic splitting of H_2 [9, 10].

The catalytically active species for the homogeneous hydrogenolysis of T (Rh [10 b]), BT (Rh [10], Ru [9 b]), DBT (Ir [29]) and dinaphtho[2,1-*b*:1′,2′-*d*]thiophene (DNT) (Rh [30], Ir [30]) have the general formula $[MH(TRIPHOS)]^n$ (M = Rh, Ir, $n = 0$; M = Ru, $n = -1$). In comparable experimental conditions using the catalyst [RhH(TRIPHOS)], the hydrogenolysis rate was found to decrease in the order

$$BT > T > \text{fused-ring thiophenes higher than DBT} > DBT$$

which reflects the propensity to undergo C–S insertion [1 b, 9, 10, 13].

The proposed mechanisms for the base-assisted hydrogenolysis of prototypical thiophenes to the corresponding thiols catalyzed by $[MH(TRIPHOS)]^n$ catalysts (M = Ru, $n = -1$; Rh, Ir, $n = 0$) are illustrated in Scheme 4.

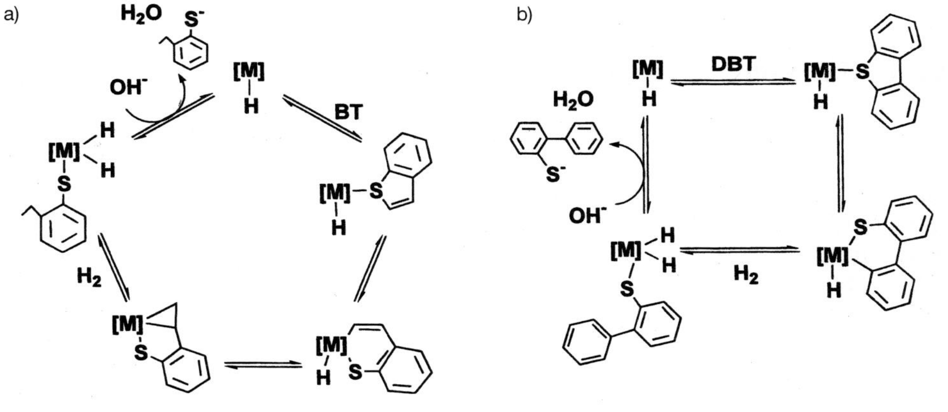

Scheme 4

The mechanisms for the model substrates BT (**a**) and DBT (**b**) involve the steps of C–S insertion, hydrogenation of the C–S inserted thiophene to the corresponding thiolate, base-assisted reductive elimination of the thiol (rds) to complete the cycle (in the catalytic reactions carried out in the absence of base, the displacement of the thiol by the substrate occurs thermally [10 b, c]). The addition of a strong base to the catalytic mixtures results in a remarkable rate enhancement; for example, the TOF relative to the hydrogenolysis of BT to 2-ethylthiophenol catalyzed by [RhH(TRIPHOS)] increases from 12 to 40 by simply adding an excess of KOBu^t to the catalytic mixture [10 b, c].

The importance of the metal oxidation state in controlling the chemoselectivity of hydrogenation of thiophenes is highlighted by the Ru–TRIPHOS case (cf. [9 a, b]).

3.2.13.3.2 Aqueous-Biphasic Systems

The aqueous-biphasic hydrogenolysis of BT has been accomplished in either water *n*-decalin or water naphtha mixtures by simply substituting NaSULPHOS for TRIPHOS in the preparation of the rhodium precursor [Rh(COD)(SUL-PHOS)]. Rather harsh reaction conditions (160 °C, 30 bar H_2) and an equivalent amount of NaOH were required for high conversions of the BT to 2-ethylthiophe-nolate (TOF 16). In these conditions, the thiolate product was totally recovered in the aqueous phase, leaving the hydrocarbon phase formally "desulfurized" [10 a].

It is generally agreed that the mechanisms of the biphasic reactions are quite similar to those proposed in single-phase systems (Scheme 4). Experimental evidence supporting mechanistic analogies in single-phase and biphasic systems has been provided for the hydrogenolysis of BT to 2-ethylthiophenol catalyzed by the 16e$^-$ fragments [RhH(TRIPHOS)] and [RhH(SULPHOS)]$^-$ in THF or H_2O–MeOH/*n*-heptane, respectively [10]. For Ru catalysts cf. [19].

3.2.13.3.3 Heterogeneous Single-Site Systems

A modified version of TRIPHOS has recently been anchored to a crosslinked styrene/divinylbenzene polymer yielding a polymeric material, named POLYTRI-PHOS, containing pendant tripodal triphosphine moieties –C(CH$_2$PPh$_2$)$_3$ [31 a]. The simple reaction of POLYTRIPHOS with a CH$_2$Cl$_2$ solution of [RhCl(COD)]$_2$ in the presence of AgPF$_6$ gives the polystyrene-supported complex [Rh(COD)-(POLYTRIPHOS)]PF$_6$ (Rh 0.94 wt. %) (eq. (4) and Structure **5**).

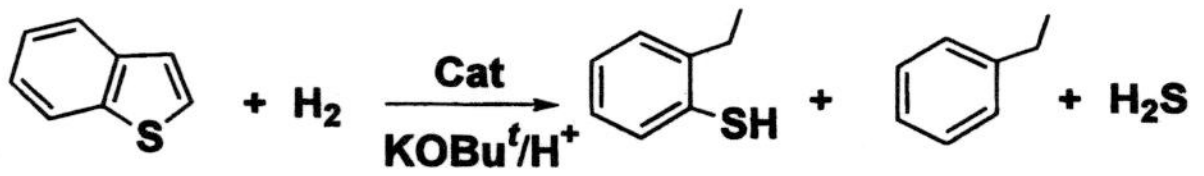

$$ \tag{4} $$

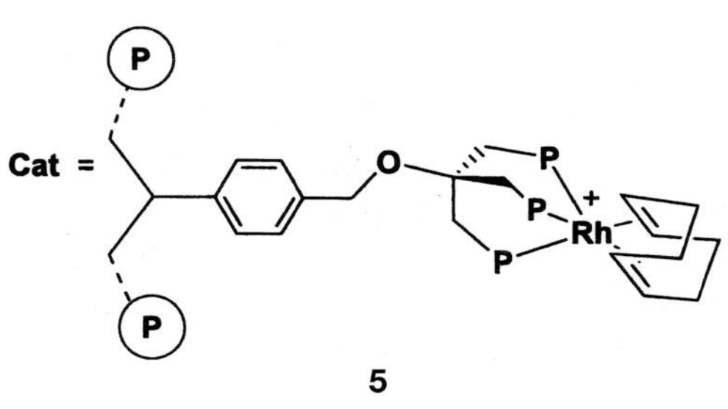

5

The supported Rh complex has been shown to be a powerful catalyst for the hydrogenolysis of BT to 2-ethylthiophenol (TOF 48) and ethylbenzene (TOF 2); this represents the first evidence of a successful single-site catalyst in the heterogeneous HDS of a thiophenic substrate.

3.2.13.4 Hydrodesulfurization in Different Phase Variation Systems

While metal complexes capable of desulfurizing or hydrodesulfurizing thiophenes are relatively numerous, homogeneous catalysts are very rare, being limited to rhodium and iridium TRIPHOS precursors that are unique in tolerating the harsh experimental conditions required for the second C–S bond cleavage of thiophenes [2 b].

The first example of HDS of DBT was obtained using the C–S insertion product [IrH(η^2-C,S-DBT)(TRIPHOS)] in THF under 30 bar of H_2 at 160 °C [29]. 2-Phenylthiophenol, biphenyl, and H_2S were produced in excess of the stoichiometric amounts. Under similar reaction conditions, catalytic production of butane, butenes, 1- butanethiol, and H_2S was observed upon hydrogenation of T in the presence of the [RhH(TRIPHOS)] catalyst and of a strong Brønsted base [10 b]. Both the Ir and Rh systems exhibited very low desulfurization activity, the hydrogenolysis to thiols being the predominant pathway. In the case of the iridium complex, the elimination of H_2S was suggested to proceed via an $M(H)_2(SH)$ intermediate which was not detected during the catalysis. However, the complex [Ir(H)$_2$(SH)-(TRIPHOS)] was prepared independently and its reaction with H_2 under catalytic conditions gave H_2S [29].

The most efficient single-site HDS catalyst remains the polystyrene-supported complex [Rh(COD)(POLYTRIPHOS)]PF$_6$ (**5**), which has been shown to catalyze the HDS of BT yielding ethylbenzene with a TOF of 2 [31 a].

In conclusion, unlike heterogeneous processes with commercial HDS catalysts, single-site catalysts have been found to desulfurize thiophenes (T, BT, DBT) exclusively after these have been converted to saturated thiols or thiolates. No example of catalytic desulfurization of THT or DHBT by a single-site catalyst has ever been reported, although stoichiometric reactions assisted by both mononuclear and polynuclear complexes are known for THT and other cyclic thioethers [2 b, 32].

As previously mentioned, the stoichiometric desulfurization of thiophenes has been achieved with a relatively large number and variety of metal complexes. In general, polynuclear complexes containing both component (Mo or W) and promoter (Ni, Co, Ir, Ru) metals turn out to be more active than mononuclear complexes containing promoter metals [1, 2]. A paradigmatic case has been reported in which the hydrogenolysis of BT to either 2-vinylthiophenol or 2-ethylthiophenol is a facile process for the promoter (Rh), but the desulfurization step to ethylbenzene requires the compulsory assistance of a component metal (W) to take place [33].

3.2.13.5 Hydrogenation of Nitrogen Heterocycles

The principal reaction pathways proposed for the heterogeneous HDN of the prototypical substrate quinoline (Q) are shown in Scheme 5. Unlike HDS, the hydrogenation of both the heterocycle and the carbocycle are preliminary to

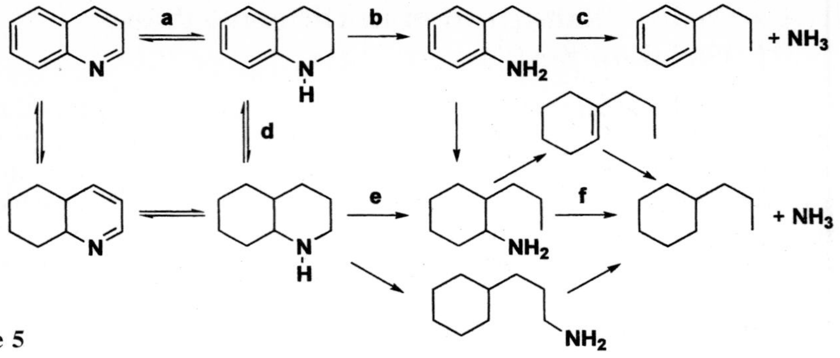

Scheme 5

C–N bond scission. Understanding the hydrogenation mechanism is thus of utmost importance for designing improved HDN catalysts.

On the basis of homogeneous modeling studies, it is now agreed that the η^1-N and η^2-N,C coordination modes of the N-heterocycle are crucial for its hydrogenation and hydrogenolysis, respectively [1, 2 a]. In particular, the regioselective hydrogenation of the heteroaromatic rings is best accomplished by late transition metals in their high oxidation states, preferentially in the presence of protic acids. The C–N insertion is a much more difficult task that proceeds in hydrolytic fashion and apparently requires action on an η^2-N,C-heterocycle by an early transition metal in relatively low oxidation state [1, 2 a].

3.2.13.5.1 Homogeneous Systems

The selective hydrogenation of pyridine (Py) to piperidine was first accomplished in dimethylformamide at ambient pressure with the catalyst system $Rh(Py)_3Cl_3/$ $NaBH_4$, which proved to be active also for the reduction of Q to 1,2,3,4-tetrahydroquinoline (THQ) [34]. Later, various metal carbonyls, $Rh_6(CO)_{16}$, $Fe(CO)_5$, $Mn_2(CO)_8(PBu_3)_2$, and $Co_2(CO)_6(PBu_3)_2$, were employed to reduce Q selectively as well as several polyaromatic heterocycles (5,6-benzoquinoline (5,6-BQ), 7,8-benzoquinoline (7,8-BQ), acridine (AC) and isoquinoline (IQ)), applying either water-gas-shift (WGS) or synthesis-gas (SG) conditions [35]. In all cases, high temperatures (180–200 °C) were required even to give very low TOFs. A remarkable rate enhancement effect was observed by adding a base and a phase-transfer agent along with the catalyst $Fe(CO)_5$ [35 b]. Under WGS conditions, $RuCl_2$-$(CO)_2(PPh_3)_2$ and $Ru_4H_4(CO)_{12}$ were inactive, however, due to competitive coordination of CO. With these precursors, the hydrogenation of the substrates was achieved using only H_2 gas [35 c]. The selective hydrogenation of Q has also been achieved with the Os clusters $H_2Os_3(CO)_{10}$ and $Os_3(CO)_{12}$, which gave a deuteration pattern of THQ with more deuterium in the 4-position and less in the 2-position, suggesting the occurrence of oxidative addition of the Os cluster to C–H bonds in Q and 1,4-hydrogenation as well [36].

The first kinetic and mechanistic studies were reported by Fish and co-workers for the hydrogenation of 2-methylpyridine (2-MePy) to 2-methylpiperidine and of Q to THQ in the presence of the Rh^{III} precursor $[Rh(MeCN)_3Cp^*]^{2+}$ [7 a, 37].

Deuterium gas experiments and *in situ* high-pressure NMR reactions allowed Fish to propose his own mechanism for the hydrogenation of Q to THQ (40 °C, 500 psi H_2, CH_2Cl_2) [7 a, 37].

The catalyst precursor $[Rh(MeCN)_3Cp^*]^{2+}$ was successfully employed to catalyze the selective reduction of various *N*-heterocycles with rates that were found to decrease in the order

$$AC > Q > 5,6\text{-}BQ > 2\text{-}MeQ > 2\text{-}MePy$$

In particular, it was reported that the rate decreases with increasing basicity and steric hindrance at the nitrogen atom. An exception to this rule was 7,8-BQ, which showed the highest relative rate and, in competitive reactions, was found to enhance the rate of hydrogenation of Q and other substrates as well. It was proposed that the rate enhancement effect is occasioned by a concomitant hydrogen transfer mechanism [7 a].

The Fish mechanism was demonstrated to be substantially valid also for the hydrogenation of polyaromatic substrates catalyzed by the Rh^I and Ru^{II} complexes $RhCl(PPh_3)_3$ [7 c] and $RuHCl(PPh_3)_3$ (85 °C, ca. 2 MPa, benzene) [7 b]. For the hydrogenation of Q, it was proposed that the activation of the C_8–H bond in the carbocyclic ring occurs via cyclometallation, while the relative hydrogenation rates decreased in the order

$$\text{phenanthridine (PHT)} > AC > Q > 5,6\text{-}BQ > 7,8\text{-}BQ$$

which again reflects the influence of both steric and electronic effects. All substrates were regioselectively hydrogenated at the heteroaromatic ring; only AC was converted to a mixture of 9,10-dihydroacridine and 1,2,3,4-tetrahydroacridine. The hydrogenation of Q was inhibited by the presence of pyridines and of THQ in the reaction mixture, due to competing coordination to the metal center, while all the other substrates had no appreciable effect on the rate of Q hydrogenation. In the case of the Rh catalyst, a promoting effect on Q reduction was observed in the presence of IN (indole), PYR (pyrrole), carbazole and even of sulfur heterocycles such as BT, BT, and DBT [7 c].

The selective hydrogenation of Q to THQ in relatively harsh experimental conditions (150 °C, 30 bar H_2, toluene) has been investigated by Sánchez-Delgado and Gonzáles with the use of various Ru, Rh, Os, and Ir metal catalysts [38]. The Rh complex was found to be the most active (initial rate ca. 200 mol Q (mol cat)$^{-1}$ h^{-1}), while the Os complex was the least efficient (initial rate ca. 5 mol Q (mol cat)$^{-1}$ h^{-1}). Coordinating solvents such as MeCN or MeOH or added ligands such as CO quenched the catalysis with $(PPh_3)_3RuHCl(CO)$. The addition of Brønsted acids or bases gave adverse effects, the base acting as an inhibitor, while water did not apparently affect the catalytic rate. Similar observations were made by López-Linares and co-workers using Rh, Ir, and Ru catalysts containing Tp or Tp* ligands (Tp = tris[pyrazolyl]borate; Tp* = tris[3,5-dimethylpyrazolyl]borate) [40]. Rhodium formed the most active catalysts, while the presence of ligands capable of competing with Q for coordination, e. g., COE (cyclooctene) and ethylene, decreased the hydrogenation rate due to competitive metal insertion into sp^2 C–H bonds.

The mechanism of Q hydrogenation assisted by [Rh(DOD)(PPh$_3$)$_2$]PF$_6$ has been studied by Sánchez-Delgado with gas-absorption techniques [40]. They reported an experimental rate law of the type $r_i = k_{cat}$ [Rh][H$_2$]^2, the isolation of [Rh(Q)$_2$(COD)]PF$_6$ at the end of the catalysis (370 K, $\leq$ 0.1 MPa H$_2$, toluene), and the observation that the rate of hydrogenation of the partially reduced substrate dihydroquinoline (DHQ) was comparable with that of Q.

Kinetic studies of Q reduction to THQ have been reported also by Rosales and co-workers for the ruthenium complex [RuH(CO)(MeCN)(PPh$_3$)$_2$]BF$_4$ [41]. At low hydrogen pressure, the experimental rate law $r_i = k_{cat}$ [Ru$_0$][H$_2$]^2 is quite similar to that found by Sánchez-Delgado. In contrast, at high H$_2$ pressure, a first-order dependence of the reaction rate with respect to the hydrogen concentration was observed. The proposed mechanism involves a rapid and reversible partial hydrogenation of bonded Q, followed by a rate-determining second hydrogenation of DHQ. The catalyst precursor [RuH(CO)(MeCN)$_2$(PPh$_3$)$_2$]BF$_4$ was also employed to catalyze the hydrogenation of various polyaromatic N-heterocycles under relatively mild conditions (125 °C, 4 bar H$_2$, xylene or toluene) [41]. The reactivity order

$$AC > Q >> 5,6\text{-}BQ > 7,8\text{-}BQ > IN > IQ$$

was in line with previous trends and reflects steric and electronic effects. A kinetic study was carried out of the reduction of AC to 9,10-dihydroacridine. Unlike Q, the experimental rate law was $r = k_{cat}$ [Ru][H$_2$] and the postulated mechanism involves as rds the hydrogenation of coordinated AC in [RuH(CO)(η^1(N)-AC)-(MeCN)(PPh$_3$)$_2$]$^+$ to yield 9,10-dihydroacridine and the coordinatively unsaturated complex [RuH(CO)(MeCN)(PPh$_3$)$_2$]$^+$.

A much more complex kinetic law has been proposed by Macchi, Bianchini, and co-workers for the reaction catalyzed by the complex [Rh(DMAD)(TRIPHOS)]PF$_6$ (DMAD = dimethyl acetylenedicarboxylate) [42]. At 60 °C in the H$_2$ pressure range from 0.4 to 3 MPa and in the range of catalyst concentration from 36 to 110 mM, the rate showed a first-order dependence on both [H$_2$] and [Rh], while the hydrogenation rate was found to be inversely dependent on [Q]. An empiric rate law of the type $r = k``$[Rh][H$_2$][Q]$^2$, where $k`` = k$ $(a + b$ [Q] $+ c$ [Q] $^2)^{-1}$ was proposed to account for the inhibiting effect of high Q concentration and the experimental observation that the rate tends to be second order for very low Q concentrations and zero order for very high Q concentrations. Incorporation of kinetic, deuterium labeling, and high-pressure NMR experiments, and the identification of catalytically relevant intermediates led to a mechanism which differs essentially from that reported by Sánchez-Delgado for the rate-limiting step, i. e. the reversible reduction of the C=N bond instead of that of the C$_3$=C$_4$ bond (irreversible) (Scheme 6). The overall hydrogenation of the C=N bond, which actually disrupts the aromaticity of Q, was proposed as rds also in the light of the independent reduction of isolated 2,3-dihydroquinoline, which, under comparable experimental conditions, was reduced

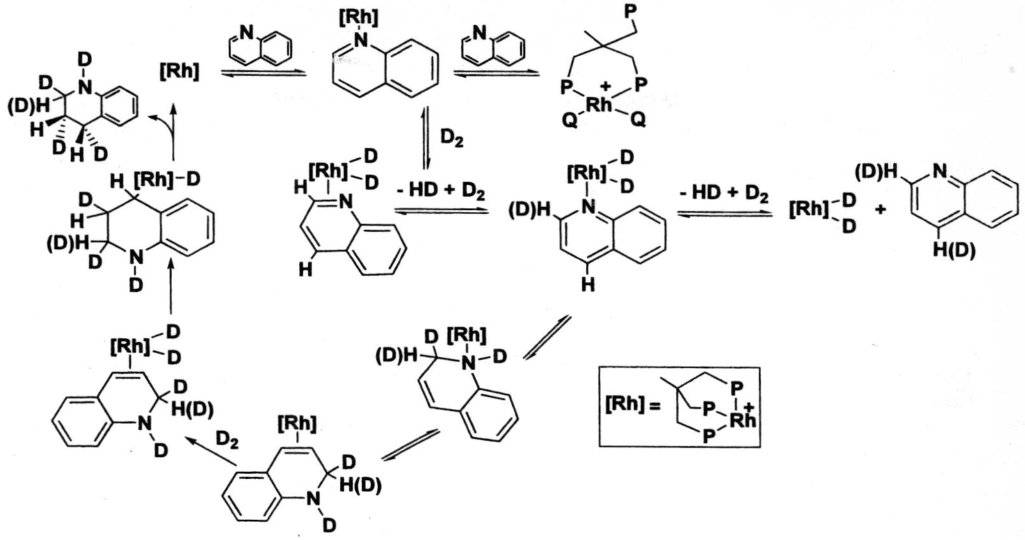

Scheme 6

faster than Q. Moreover, the lack of deuterium incorporation into the carbocyclic ring of both THQ and Q ruled out the intermediacy of η^6-Q or η^6-THQ complexes.

A remarkable rate enhancement was observed by addition of an excess of CF_3SO_3H to the catalytic mixtures (at 40 °C and 3 MPa H_2, the TOF was increased from 40 to 95 by addition of a 20-fold excess of acid). The role of triflic acid was that of aiding the conversion of inactive Rh^I (formed in the basic environment of the reaction) to active Rh^{III}. The addition of strong protic acids was found to be of mandatory importance for generating a catalytically active system from [Ru (MeCN)$_3$(TRIPHOS)](O$_3$SCF$_3$)$_2$, which, under neutral conditions, is almost inactive for Q reduction. Treatment of the Ru compound with H_2 produces ammonia, in fact, which prevails over Q for coordination to the metal center; moreover, in the basic environment of the reaction, traces of water were found to transform catalytically active [RuH(TRIPHOS)]$^+$ into inactive binuclear [Ru$_2$(μ-OH)$_3$(TRIPHOS)$_2$]$^+$ [9 a, b]. In this case, the protic acid inhibits the formation of both NH$_3$ adducts and the (μ-OH)$_3$ binuclear complex, thus allowing the hydrogenation of Q to THQ to proceed smoothly with a TOF of 65 [9 a, b].

In homogeneous phase, IN is much more difficult to reduce than Q, as shown by the limited number of known catalysts (e. g. RuHCl(PPh$_3$)$_3$ [7b] and [RuH-(CO)(MeCN)(PPh$_3$)$_2$]BF$_4$ [43]), which, by the way, are scarcely efficient. The inertness to hydrogenation exhibited by IN has been attributed to its incapability of using the nitrogen atom for coordination. In fact, IN binds metal via the η^6–πC coordination which does not activate the C=N bond and also occupies too many coordination sites, disfavoring the oxidative addition of H_2 to the metal center. To our knowledge, the only catalyst that is able to hydrogenate IN regioselectively to indoline with an acceptable TOF is [Rh(DMAD)(TRIPHOS)]PF$_6$ in the presence of a protic acid [42]. By using equivalent IN and triflic acid concentrations, TOFs as high as 100 were obtained at only 60 °C and 3 MPa H$_2$. It was shown experimentally that indoline was formed by reduction of the *3H*-indolium cation, which possesses a localized C=N bond [42 a].

3.2.13.5.2 Aqueous-Biphasic Systems

The catalysts that effect the hydrogenation of thiophenes in aqueous-biphase systems (see Section 3.2.13.3.2) are also active for the selective reduction of aromatic heterocycles to cyclic amines using aqueous-biphase catalysts (Structures **6–10**).

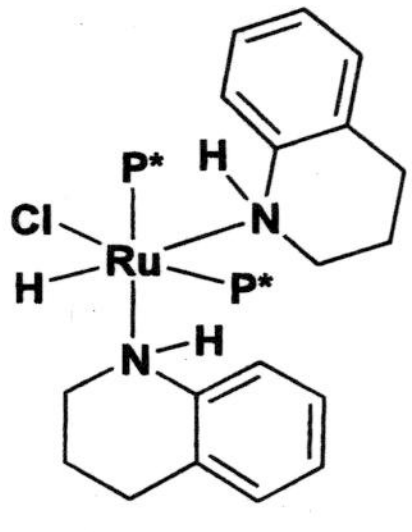

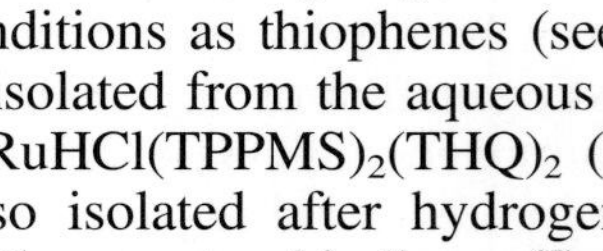

The heterocyclic ring in Q, AC, and IQ has been selectively hydrogenated in water/decalin with Ru^{II} catalysts prepared *in situ* from RuCl$_3$·3H$_2$O and excess TPPMS or TPPTS in the same experimental conditions as thiophenes (see Section 3.2.13.2.2) [44, 45]. The major Ru product isolated from the aqueous phase after hydrogenation of Q in water/decalin was RuHCl(TPPMS)$_2$(THQ)$_2$ (Structure **6**). The same termination product was also isolated after hydrogenation of a Q/BT mixture, which is consistent with the greater binding affinity of amines to Ru^{II} in comparison with thioethers. The regioselective reduction of Q to THQ in water/hydrocarbon mixtures has been achieved with catalyst systems comprising either Na$_2$DPPPDS or NaSULPHOS in combination with Rh or Ru [17, 18, 46].

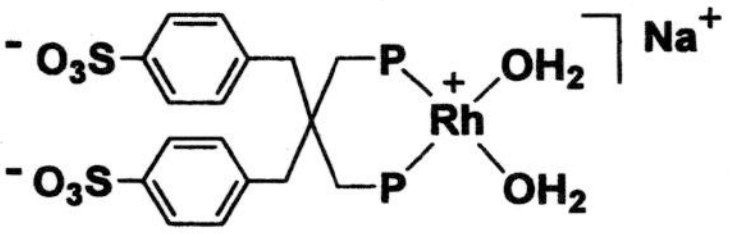

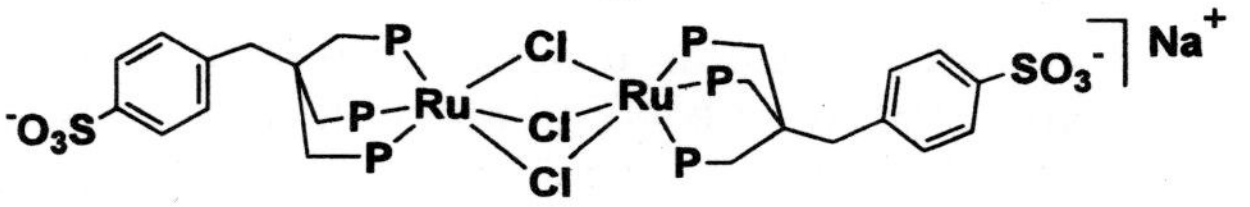

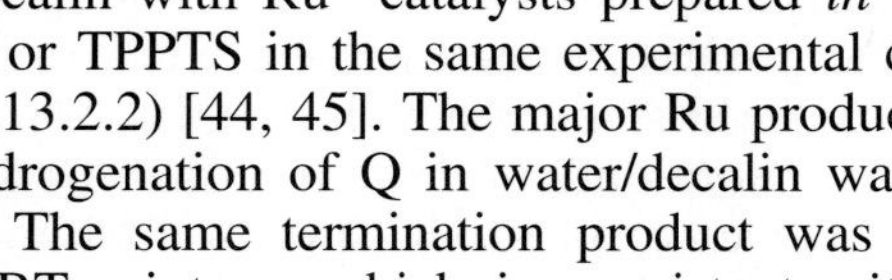

The Rh^I complex [Rh(H$_2$O)$_2$(DPPPDS)]Na was isolated and employed in water/*n*-octane to hydrogenate 1:1 mixtures of Q and BT at 160 °C yielding almost exclusively THQ with a TOF of 50, BT hydrogenation to DHBT being only marginal (TOF 2). A similar selectivity was shown also by the catalytic system RuCl$_3$ · H$_2$O/2Na$_2$DPPPDS prepared *in situ*. In contrast, the binuclear complex Na[{Ru(SULPHOS)}$_2$(μ-Cl)$_3$] was found to hydrogenate Q and BT at comparable rates (TOF = 30 at 140 °C, 3 MPa, water/*n*-heptane).

The mononuclear complex [Ru(MeCN)$_3$(SULPHOS)]$^+$ has been found to catalyze the hydrogenation of Q to THQ in water/*n*-heptane or water/methanol/*n*-heptane at fairly fast initial rates, that, however, decreased remarkably with time due to the formation of the catalytically inactive binuclear μ-hydroxy species [Ru$_2$(μ- OH)$_3$(SULPHOS)$_2$]$^-$ [46]. As in homogeneous phase with the precursor [Ru (MeCN)$_3$(TRIPHOS)](O$_3$SCF$_3$)$_2$ (see Section 3.2.13.5.1), the addition of a strong protic acid in excess enhanced hydrogenation TOF from 7 in the absence of added acid to 37 with 20 equivalents of triflic acid (MeOH/water/*n*-heptane, 100 °C, 3 MPa H$_2$) [46]. Although not essential for good conversions, the use of an acid co-reagent also improves the catalytic performance of the zwitterionic Rh^I complex Rh(COD)(SULPHOS) in the aqueous-biphasic hydrogenation of both Q and IN [46].

3.2.13.5.3 Heterogeneous Single-Site Systems

Various *N*-heteroaromatics have been successfully hydrogenated in the presence of the single-site catalyst ⓟ–Rh(PPh$_2$)$_2$Cl obtained by tethering the soluble precursor Rh(PPh$_3$)$_3$Cl to 2 % crosslinked phosphinated polystyrene-divinylbenzene (benzene, 85 °C, ca. 2.1 MPa H$_2$ [22]). The order of activity was identical to that in homogeneous phase [7 c], AC > Q > 5,6-BQ > 7,8-BQ, but the initial rates of the heterogeneous hydrogenations were from 10 to 20 times faster due to the increased steric hindrance at the tethered Ru (see Section 3.2.13.2.3). The regioselectivity of hydrogenation was even higher than that in homogeneous phase as no formation of 1,2,3,4-tetrahydroacridine was observed. The heterogenized Ru catalyst was also employed to hydrogenate *N*-heterocycles in model coal liquid containing pyrene, tetralin, *p*-cresol, 2-methylpyridine, and methylnaphthalene. A rate enhancement effect was observed which was attributed to the ability of some constituents, especially of *p*-cresol, to stabilize unsaturated Rh species formed in the course of the catalysis.

It has been discovered recently that a single metal site belonging to the HDS/HDN promoter class can hydrogenate both the heterocyclic and carbocyclic rings of Q, although at different rates [31 b]. Under relatively mild experimental conditions (80 °C, 3 MPa H$_2$), the polystyrene-supported complex [Rh(COD)(POLYDIPHOS)]PF$_6$ has been found to hydrogenate Q in *n*-octane yielding THQ (TOF 63) as well as 5,6,7,8-tetrahydroquinoline (^{5}THQ) (TOF 13) and decahydroquinoline (DeHQ) (TOF 8) (POLYDIPHOS = crosslinked styrene/divinylbenzene resin–(C$_6$H$_4$)CH$_2$OCH$_2$C(CH$_3$)C(CH$_2$PPh$_2$)$_2$) (eq. (5)). Independent reactions with isolated samples of THQ and ^{5}THQ showed that both compounds are further reduced to DeHQ. Most importantly, no metal leaching was observed and the catalyst was recycled several times with no loss of catalytic activity and selectivity.

The heterogeneous hydrogenation of Q to THQ has also been achieved with the silica-supported hydrogen-bonded rhodium catalysts Rh(COD)(SULPHOS)/SiO$_2$ and [Ru(MeCN)$_3$(SULPHOS)](SO$_3$CF$_3$)/SiO$_2$ shown in Structure **3**. In the same experimental conditions employed to hydrogenate BT (*n*-octane, 100 °C, 3 MPa H$_2$) (see Section 3.2.13.2.3), THQ was selectively produced with relatively low TOFs (20–30) [46].

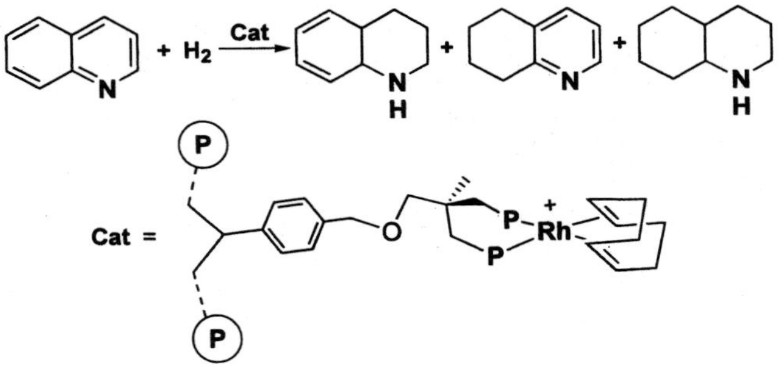

(5)

3.2.13.6   Hydrogenolysis of Nitrogen Heterocycles

While the catalytic hydrogenation of the heterocyclic ring in *N*-heterocycles is relatively facile by molecular catalysis in either homogeneous or heterogenous fashion, the hydrogenolysis of the C–N bonds is considerably more difficult to achieve. Even stoichiometric C–N scissions are very rare [1, 2], which is not surprising as C–N bonds exhibit a higher bond energy than C–S bonds (by 3–9 kcal mol^{-1}). For this reason, the catalysts are less efficient for HDN than for HDS under comparable experimental conditions [1, 2].

To our knowledge, the only example of catalytic hydrogenolysis of a nitrogen heterocycle is the conversion of Py to a mixture of piperidine and various bis-(piperidinyl)alkanes catalyzed by Rh$_6$(CO)$_{16}$ under water-gas shift conditions (150 °C, 800 psi CO) [35 d].

3.2.13.7   Perspectives

Over the last 15 years, the homogeneous studies of HDS and HDN processes have been extremely useful to understand many mechanistic details regarding the coordination of sulfur and nitrogen heterocycles to metal centers, hydrogen transfer from metal to coordinated heterocycle, metal insertion into C–S and C–N bonds, and the desulfurization/denitrogenation paths. Recently, however, there has been a qualitative leap in molecular catalysis so that crossing the borderline between traditional heterogeneous catalysis and molecular catalysis in HDS and HDN is no longer considered a utopia by the experts in the field. Like supported metallocenes in olefin polymerization [47] or phosphine-modified Rh complexes in aqueous hydroformylation of olefins [48], supported metal catalysts, alone or in combination with metal particles or metal sulfides, might indeed be the key to the development of more efficient catalysts for the deep HDS and HDN of fossil fuels.

References

[1] (a) T. Kabe, A. Ishihara, W. Qian, *Hydrodesulfurization and Hydrodenitrogenation*, Wiley-VCH, Tokyo, **1999**; (b) H. Topsøe, B. S. Clausen, F. E. Massoth, *Hydrotreating Catalysis*, Springer-Verlag, Berlin, **1996**.

[2] (a) C. Bianchini, A. Meli, F. Vizza, *Eur. J. Inorg. Chem.* **2000**, 43; (b) C. Bianchini, A. Meli, *Acc. Chem. Res.* **1998**, *31*, 109.

[3] C. Bianchini, A. Meli, *Synlett* **1997**, 643; (b) C. Bianchini, A. Meli, W. Pohl, F. Vizza, G. Barbarella, *Organometallics* **1997**, *16*, 1517.

[4] C. Bianchini, P. Barbaro, G. Scapacci, E. Farnetti, M. Graziani, *Organometallics* **1998**, *17*, 3308.

[5] C. Bianchini, A. Meli, M. Peruzzini, F. Vizza, V. Herrera, R. A. Sánchez-Delgado, *Organometallics* **1994**, *13*, 721.

[6] (a) M. J. Robertson, C. L. Day, R. A. Jacobson, R. J. Angelici, *Organometallics* **1994**, *13*, 179; (b) M.-G. Choi, R. J. Angelici, *Organometallics* **1992**, *11*, 3328.

[7] (a) E. Baralt, S. J. Smith, I. Hurwitz, I. T. Horváth, R. H. Fish, *J. Am. Chem. Soc.* **1992**, *114*, 5187; (b) R. H. Fish, J. L. Tan, A. D. Thormodsen, *Organometallics* **1985**, *4*, 1743; (c) R. H. Fish, J. L. Tan, A. D. Thormodsen, *J. Org. Chem.* **1984**, *49*, 4500.

[8] (a) V. Herrera, A. Fuentes, M. Rosales, R. A. Sánchez-Delgado, C. Bianchini, A. Meli, F. Vizza, *Organometallics* **1997**, *16*, 2465; (b) R. A. Sánchez-Delgado, V. Herrera, L. Rincón, A. Andriollo, G. Martín, *Organometallics* **1994**, *13*, 553; (c) R. A. Sánchez-Delgado, E. González, *Polyhedron* **1989**, *8*, 1431.

[9] (a) C. Bianchini, A. Meli, S. Moneti, W. Oberhauser, F. Vizza, V. Herrera, A. Fuentes, R. A. Sánchez-Delgado, *J. Am. Chem. Soc.* **1999**, *121*, 7071; (b) C. Bianchini, A. Meli, S. Moneti, F. Vizza, , *Organometallics* **1998**, *17*, 2636; (c) C. Bianchini, D. Masi, A. Meli, M. Peruzzini, F. Vizza, F. Zanobini, *Organometallics* **1998**, *17*, 2495.

[10] (a) C. Bianchini, A. Meli, V. Patinec, V. Sernau, F. Vizza, *J. Am. Chem. Soc.* **1997**, *119*, 4945; (b) C. Bianchini, J. Casares, A. Meli, V. Sernau, F. Vizza, R. A. Sánchez-Delgado, *Polyhedron* **1997**, *16*, 3099; (c) C. Bianchini, V. Herrera, M. V. Jiménez, A. Meli, R. A. Sánchez-Delgado, F. Vizza, *J. Am. Chem. Soc.* **1995**, *117*, 8567.

[11] (a) D. M. Heinekey, W. J. J. Oldham, *Chem. Rev.* **1993**, *93*, 913; (b) P. G. Jessop, R. H. Morris, *Coord. Chem. Rev.* **1992**, *121*, 155.

[12] A. F. Borowski, S. Sabo-Etienne, B. Chaudret, *Abstracts ISHC* 12, August 27/September 1, **2000**.

[13] (a) C. Bianchini, P. Frediani, V. Herrera, M. V. Jiménez, A. Meli, L. Rincón, R. A. Sánchez-Delgado, F. Vizza, *J. Am. Chem. Soc.* **1995**, *117*, 4333; (b) C. Bianchini, M. V. Jiménez, A. Meli, F. Vizza, *Organometallics* **1995**, *14*, 3196; (c) C. Bianchini, A. Meli, M. Peruzzini, F. Vizza, S. Moneti, V. Herrera, R. A. Sánchez-Delgado, *J. Am. Chem. Soc.* **1994**, *116*, 4370; (d) C. Bianchini, A. Meli, M. Peruzzini,

F. Vizza, P. Frediani, V. Herrera, R. A. Sánchez-Delgado, *J. Am. Chem. Soc.* **1993**, *115*, 2731.

[14] M. Neurock, R. A. van Santen, *J. Am. Chem. Soc.* **1994**, *116*, 4427.

[15] C. Bianchini, A. Meli, in *Aqueous-Phase Organometallic Catalysis – Concepts and Applications* (Eds.: B. Cornils, W. A. Herrmann), VCH, Weinheim, **1998**, p. 477.

[16] (a) INTEVEP S. A. (D. E. Páez, A. Andriollo, R. A Sánchez-Delgado, N. Valencia, F. López-Linares, R. Galiasso), US 08/657.960 (1996); (b) INTEVEP S. A. (D. E. Páez, A. Andriollo, R. A Sánchez-Delgado, N. Valencia, F. López-Linares, R. Galiasso), Sol. Patente Venezolana 96–1630 (1996).

[17] (a) CNR (C. Bianchini, A. Meli, F. Vizza), (1999), PCT/EP97/06493; (b) CNR (C. Bianchini, A. Meli, F. Vizza), IT FI96A000272 (1996).

[18] C. Bianchini, P. Frediani, V. Sernau, *Organometallics* **1995**, *14*, 5458.

[19] C. Bianchini, A. Meli, S. Moneti F. Vizza, unpublished results.

[20] I. Rojas, F. Lopez Linares, N. Valencia, C. Bianchini, *J. Mol. Catal. A: Chemical* **1999**, *144*, 1.

[21] (a) B. Cornils, W. A. Hermann, in *Applied Homogeneous Catalysis with Organometallic Compounds* (Eds.: B. Cornils, W. A. Herrmann), VCH, New York, **1996**, Vol. 2, p. 575; (b) P. Panster, S. Wieland, in *Applied Homogeneous Catalysis with Organometallic Compounds* (Eds.: B. Cornils, W. A. Herrmann), VCH, New York, **1996**, Vol. 2, p. 605.

[22] R. H. Fish, A. D. Thormondsen, H. Heinemann, *J. Mol. Catal.* **1985**, *31*, 191.

[23] (a) C. Bianchini, V. Dal Santo, A. Meli, W. Oberhauser, R. Psaro, F. Vizza, *Organometallics* **2000**, *19*, 2433; (b) C. Bianchini, D. G. Burnaby, J. Evans, P. Frediani, A. Meli, W. Oberhauser, R. Psaro, L. Sordelli, F. Vizza, *J. Am. Chem. Soc.* **1999**, *121*, 5961.

[24] C. Bianchini, A. Meli, W. Oberhauser, F. Vizza, unpublished results.

[25] H. Gao, R. J. Angelici, *Organometallics* **1999**, *18*, 989 and references therein.

[26] J. Chen, L. M. Daniels, R. J. Angelici, *J. Am. Chem. Soc.* **1990**, *112*, 199.

[27] (a) A. W. Myers, W. D. Jones, *Organometallics* **1996**, *15*, 2905; (b) A. W. Myers, W. D. Jones, S. M. McClements, *J. Am. Chem. Soc.* **1995**, *117*, 11704; (c) L. Dong, S. B. Duckett, K. F. Ohman, W. D. Jones, *J. Am. Chem. Soc.* **1992**, *114*, 151; (d) W. D. Jones, L. Dong, *J. Am. Chem. Soc.* **1991**, *113*, 559.

[28] (a) C. Bianchini, J. A. Casares, D. Masi, A. Meli, W. Pohl, F. Vizza, *J. Organomet. Chem.* **1997**, *541*, 143; (b) C. Bianchini, M. V. Jiménez, A. Meli, S. Moneti, F. Vizza, *J. Organomet. Chem.* **1995**, *504*, 27.

[29] (a) C. Bianchini, M. V. Jiménez, A. Meli, S. Moneti, F. Vizza, V. Herrera, R. A. Sánchez-Delgado, *Organometallics* **1995**, *14*, 2342.

[30] C. Bianchini, D. Fabbri, S. Gladiali, A. Meli, W. Pohl, F. Vizza, *Organometallics* **1996**, *154*, 4604.

[31] (a) C. Bianchini, M. Frediani, F. Vizza, *Chem. Commun.* **2001**, *479*; (b) C. Bianchini, M. Frediani, G. Manlorani, F. Vizza, *Organometallics* **2001**, *20*, 2660.

[32] C. Bianchini, A. Meli, W. Oberhauser, F. Vizza, *Chem. Commun.* **1999**, 671.

[33] C. Bianchini, A. Meli, S. Moneti, F. Vizza, *Organometallics* **1997**, 16, 5696.

[34] I. Jardine, F. J. McQuillin, *J. Chem. Soc. D* **1970**, 626.

[35] (a) S. I. Murahashi, Y. Imada, H. Hirai, *Tetrahedron Lett.* **1987**, *28*, 77; (b) T. J. Lynch, M. Banah, H. D. Kaesz, C. D. Porter, *J. Org. Chem.* **1984**, *49*, 1266; (c) R. H. Fish, A. Thormodsen, G. A. D. Cremer, *J. Am. Chem. Soc.* **1982**, *104*, 5234; (d) R. M. Laine, D. W. Thomas, L. W. Cary, *J. Org. Chem.* **1979**, *44*, 4964.

[36] (a) R. M. Laine, *New J. Chem.* **1987**, *11*, 543; (b) Eisenstadt, C. M. Giandomenico, M. F. Frederick, R. M. Laine, *Organometallics* **1985**, *4*, 2033.

[37] (a) R. H. Fish, H-S. Kim, R. H. Fong, *Organometallics* **1991**, *10*, 770; (b) R. H. Fish, R. H. Fong, A. Than, E. Baralt, *Organometallics* **1991**, *10*, 1209; (c) R. H. Fish,

E. Baralt, H-S. Kim, *Organometallics* **1991**, *10*, 1965–1971; (d) R. H. Fish, H-S. Kim, R. H. Fong, *Organometallics* **1989**, *8*, 1375–1377; (e) R. H. Fish, H-S. Kim, J. E. Babin, R. D. Adams, *Organometallics* **1988**, *7*, 2250.

[38] R. A. Sánchez-Delgado, E. Gonzalez, *Polyhedron* **1989**, *8* 1431.

[39] Y. Alvarado, M. Busolo, F. López-Linares, *J. Mol. Catal. A: Chemical* **1999**, *142*, 163.

[40] R. A. Sánchez-Delgado, D. Rondón, A. Andriollo, V. Herrera, G. Martin, B Chaudret, *Organometallics* **1993**, *12*, 4291.

[41] M. Rosales, Y. Alvarado, M. Boves, R. Rubio, H. Soscun, R. Sánchez-Delgado, *Trans. Met. Chem.* **1995**, *20*, 246.

[42] (a) M. Macchi, Ph. D. Dissertation, Università di Trieste (Italy), **1999**; (b) C. Bianchini, P. Barbaro, M. Macchi, A. Meli, F. Vizza, *Helv. Chim. Acta* **2001**, *84*, 2895.

[43] M. Rosales, J. Navarro, L. Sanchez, A. Gonzales, Y. Alvarado, R. Rubio, C. De la Cruz, T. Rajmankina, *Trans. Met. Chem.* **1996**, *21*, 11.

[44] D. E. Páez, A. Andriollo, F. López-Linares, R. E. Galiasso, J. A. Revete, R. A. Sánchez-Delgado, A. Fuentes, *Am. Chem. Soc. Div. Fuel Chem. Symp. Prepr.* **1998**, *43*, 563.

[45] (a) INTEVEP S. A. (D. E. Páez, A. Andriollo, R. A. Sánchez-Delgado, N. Valencia, R. E. Galiasso, F. López- Linares), US 5.958.223 (1999); (b) INTEVEP S. A. (D. E. Páez, A. Andriollo, R. A. Sánchez-Delgado, N. Valencia, F. López-Linares, R. E. Galiasso), US 5.753.584 (1998).

[46] C. Bianchini, M. Macchi, A. Meli, W. Oberhauser, F. Vizza, manuscript in preparation.

[47] E. Carnahan, G. Jacobsen, *CATTECH* **2000**, *7*, 74.

[48] *Aqueous-Phase Organometallic Catalysis – Concepts and Applications* (Eds.: B. Cornils, W. A. Herrmann), VCH, Weinheim, **1998**, pp. 271–340.

3.2.14   Double-Bond Isomerization of Olefins

Wolfgang A. Herrmann, Martina Prinz

3.2.14.1   Introduction

Olefins display an abundant and versatile coordination chemistry with transition metals. In fact, homogeneous catalysis owes its success mainly to the interactions between olefins and metals: examples include hydroformylation (1938), polymerization (1953), metathesis (1955), and Wacker–Hoechst oxidation (1958). While all these and numerous other reactions involve structural and chemical changes of the olefin, there is yet another, sometimes undesirable, metal-induced phenomenon, olefin isomerization. The C=C double bond may be shifted along the backbone of the olefin to give a mixture of terminal and *cis/trans* internal olefins. This chapter details such double-bond isomerization, without considering skeletal isomerization. Only homogeneously catalysed isomerizations will be outlined [1, 2], although it should be noted that homogeneous and heterogeneous catalysis obey the same mechanistic principles in the isomerization of olefins. Such olefin isomerization is a key step in many industrial processes, among them the Shell higher olefins process (SHOP) (see Section 2.3.1.3 and [1]),

DuPont's butadiene-to-adiponitrile synthesis (see Section 2.5.5.1 and [1]) and the Takasago synthesis of (–)-menthol from α-pinene (see Sections 2.9, 3.2.14.5, and 3.3.1).

3.2.14.2   Catalysts, Scope, and Definition

Olefin isomerization is common in petrochemical refining processes (heterogeneous catalysis) and, of course, follows the thermodynamic driving forces: *trans*-olefins are more stable than their *cis* isomers, and internal olefins more stable than terminal olefins (eq. (1)).

$$\text{R-CH}_2\text{-CH=CH}_2 \quad \rightleftharpoons \quad \text{R-CH=CH-CH}_3 \tag{1}$$

    1-olefin   2-olefin

   A typical example is the near-equilibrium isomerization of 1-octene to a mixture of 2 % 1-octene, 36 % 2-octene, 36 % 3-octene, and 26 % 4-octene (*cis/trans* mixtures) [3]. If 1-butene is allowed to isomerize until it reaches the thermodynamic equilibrium, a mixture of 69 % *trans*-2-butene, 25 % *cis*-2-butene, and 6 % 1-butene is found [4]. Note that the isomerization of olefins is a kinetic phenomenon. In the isomerization of α-olefins, it is the *cis* isomer of the resulting β-olefins that is often formed in kinetic preference and thus these isomers may be isolated as the major product in the early stages of many reactions. The preference for *cis* isomers can be determined by the catalyst used or the presence of certain functional groups in the olefin, e. g., in 1,2-dichloroethylene, 1-chloropropene, and 2-butenecarboxylic nitrile. The β,γ-double bond position dominates in the isomerizations containing carboxylic acids, esters, and nitriles if the β-C atom carries two alkyl groups, e. g., in eq. (2). Furthermore the formation of conjugated di- and oligoolefins is normally favored over isolated double bonds (eq. (3)). This type of isomerization finds application in the synthesis of stereoids, an example of which is illustrated in eq. (4), Here the strong preference of the 14-electron fragment $Fe(CO)_3$ to bind 1,3-dienes is exploited. The $Fe(CO)_3$ group can be oxidatively removed from the isomerized diolefin by means of $FeCl_3$, and in some cases by CrO_3 [5].

$$\begin{array}{cc} \text{CH}_3 & \text{CH}_3 \\ \quad\diagdown & \quad\diagdown \\ \quad\quad\text{C=CH−CH}_2\text{X} & \quad\quad\text{CH−CH=CH−X} \\ \text{CH}_3\diagup & \text{CH}_3\diagup \end{array} \rightleftharpoons \tag{2}$$

X = COOH 94 %	6 %
X = CN 79 %	21 %

$$\begin{array}{cc} \quad\text{H}\quad\text{H}_2 & \quad\text{H}\quad\text{H} \\ \quad\diagup\text{C=}\quad\diagup\quad\diagdown\text{CH}_2 & \quad\diagup\text{C=}\quad\diagup\text{C}\diagdown\text{CH}_3 \\ \text{CH}_3\quad\quad\text{C}\quad\text{C} & \text{CH}_3\quad\quad\text{C}\quad\text{C} \\ \quad\quad\text{H}\quad\text{H} & \quad\quad\text{H}\quad\text{H} \end{array} \tag{3}$$

    1,4-diolefin   2,4-diolefin
   (nonconjugated)   (conjugated)

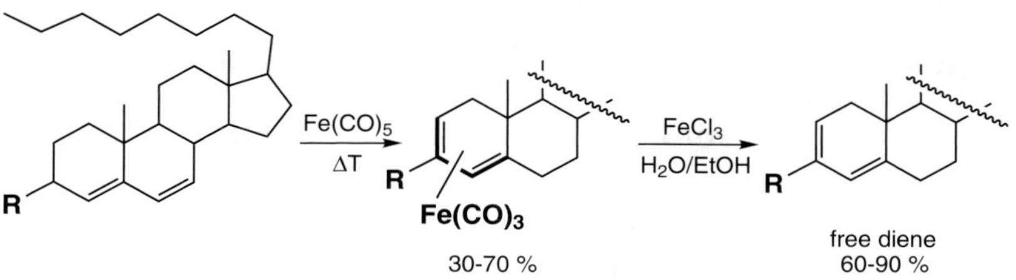

In contrast, metals like Pd and Rh prefer the $1:2,5:6\text{-}\eta^4$-bonding mode (1,5-dienes) of cycloolefinic structures. They rearrange 1,3-dienes in an apparently "contrathermodynamic" way to their 1,5-isomers. The products can be cleaved from the metals by cyanide ions (eq. (5)). Numerous examples are known [6–14].

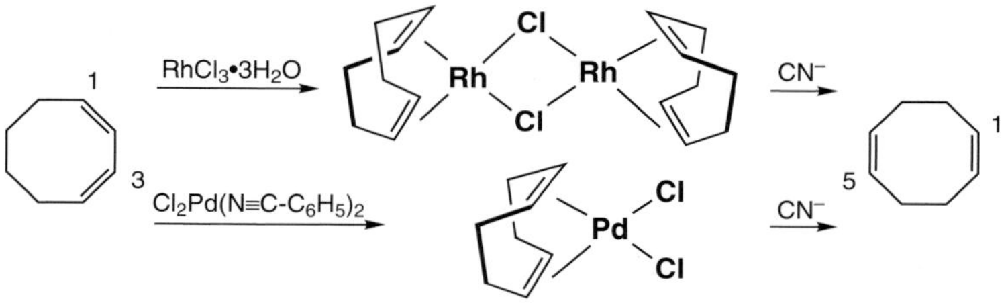

In the case of substituted olefins, the isomers exhibiting the highest degree of branching are thermodynamically favored.

3.2.14.3 Mechanistic Considerations

Depending on the specific nature of the olefin and the metal (complex) in question, two major mechanisms dominate the scene. The coordination chemistry of the metal specifies in many cases the path of olefin isomerization [15].

3.2.14.3.1 The π-Allyl Mechanism (1,3-Hydrogen Shift)

The principle of the π-allyl mechanism is illustrated in Scheme 1. The catalytic process is initiated by coordination of the terminal olefin to the metal followed by activation of the aliphatic $C_\gamma H$-bond, affording the three-carbon arrangement in π-bonding to the metal. The metal-attached hydride has thus two positions to which it may be transferred (α and γ), the α-position being nonproductive and the γ-position leading to the internal olefin. It follows from Scheme 1 that the β-C–H entity is not affected.

Proof for this mechanism is found from the high *cis/trans* ratios of the isomerized olefin formed at an early stage in the reaction [16], and little or no deuterium substitution in the 2-position (β) when deuterated olefins are being used; an example is shown in eq. (6).

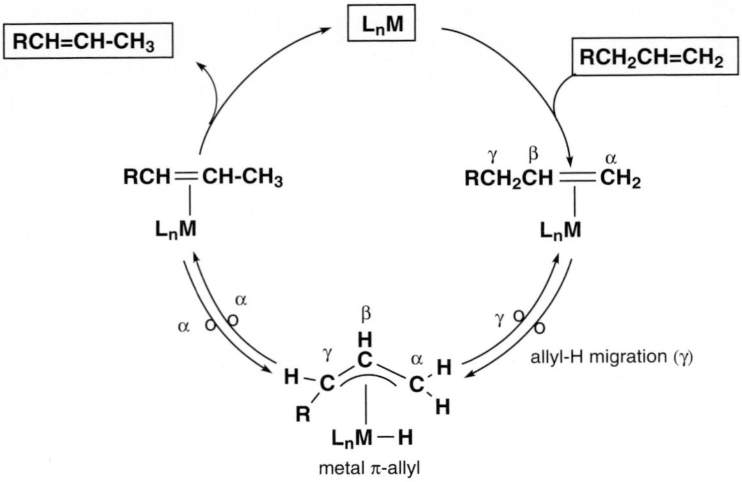

Scheme 1

$$\overset{\alpha \quad \beta \quad \gamma}{CH_2{=}CH{-}CD_2{-}C_2H_5} \;\rightleftharpoons\; \overset{\alpha \quad \beta \quad \gamma}{CH_2D{-}CH{=}CD{-}C_2H_5} \tag{6}$$

The catalysts $Fe_3(CO)_{12}$, $Pd(N{\equiv}CR)_2Cl_2$, $Pd(N{\equiv}CR)_3$, and $ClRh[P(C_6H_5)_3]_3$ (R = alkyl, aryl) are found to follow this mechanism in the olefin isomerization. Also, the isomerization of unconjugated to conjugated double bonds using $Fe(CO)_5$ mentioned previously (see Section 3.2.14.2) follows this mechanism. Note that the ^2D migration in the isomerization of 1,4-cyclohexadiene occurs alongside the metal complexation (FeD intermediates; cf. eq. (7)) [17].

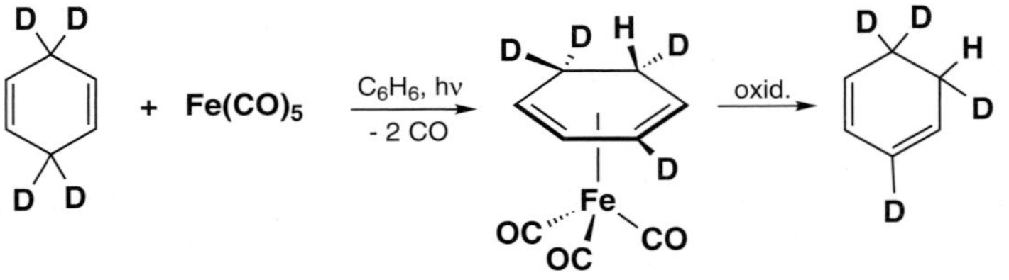

$$\tag{7}$$

A nice model for the π-allyl mechanism has been reported by Bönnemann [18] for the pair of nickel complexes **1** and **2** forming a temperature-dependent equilibrium (eq. (8)).

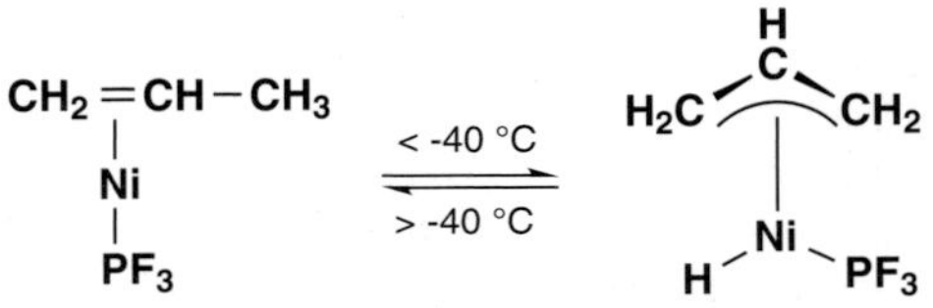

$$\tag{8}$$

3.2.14.3.2 The Alkyl Mechanism (1,2-Hydrogen Shift)

The "alkyl mechanism" is the preferred pathway of isomerization, if the catalytic species contains a metal-bonded hydride as illustrated in Scheme 2. It is reminiscent of the hydroformylation mechanism (see Section 2.1.1).

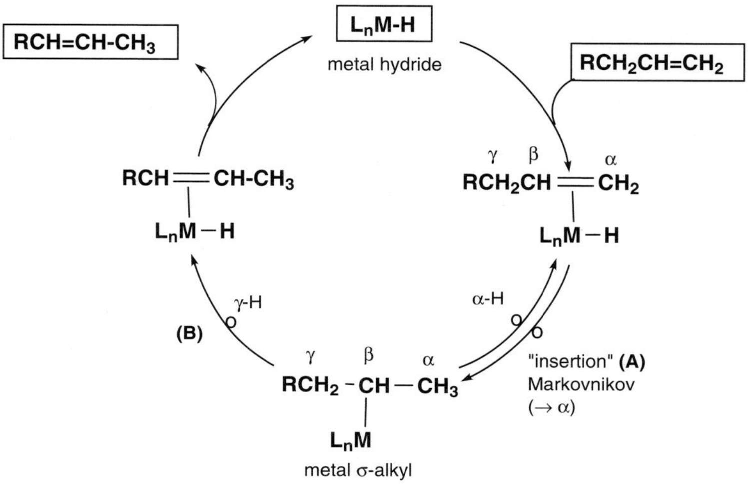

$RCH=CH-CH_3$

L_nM-H
metal hydride

$RCH_2CH=CH_2$

$RCH=CH-CH_3$
L_nM-H

γ β α
$RCH_2CH=CH_2$
L_nM-H

(B)

γ-H

α-H

γ β α
$RCH_2-CH-CH_3$
L_nM
metal σ-alkyl

"insertion" **(A)**
Markovnikov
$(\rightarrow \alpha)$

Scheme 2

Depending on the metal and the specific nature of the ligand sphere (especially its steric bulk), the hydride migration ("insertion", step **A**) in the product-determining step can follow either the anti-Markovnikov or the Markovnikov path (Scheme 2). Only in the latter case and if subsequent γ-H "elimination" takes place (step **B**), does an isomerized olefin result.

Typical catalysts that employ this mechanism are nickel hydrides and Ru hydrides such as $HRhCl[P(C_6H_5)_3]_3$ (which is also a classic olefin hydrogenation catalyst) and $[HNi\{P(C_6H_5)_3\}_3]^+$, which is present in the system $Ni[P(C_6H_5)_3]_4/CF_3COOH$ according to eq. (9) [19, 20].

$$NiL_4 + H^+ \rightleftharpoons [HNiL_4]^+ \xrightarrow[-L]{r.d.} [HNiL_3]^+ \qquad (9)$$

$L = P(C_6H_5)_3$
r.d. = rate-determining step

The distinction between the 1,2- and the 1,3-shift processes is readily demonstrated using D-labeled olefins. In the isomerization of **3** (eqs. (10a) and (10b)) the two pathways are discerned by NMR spectroscopy.

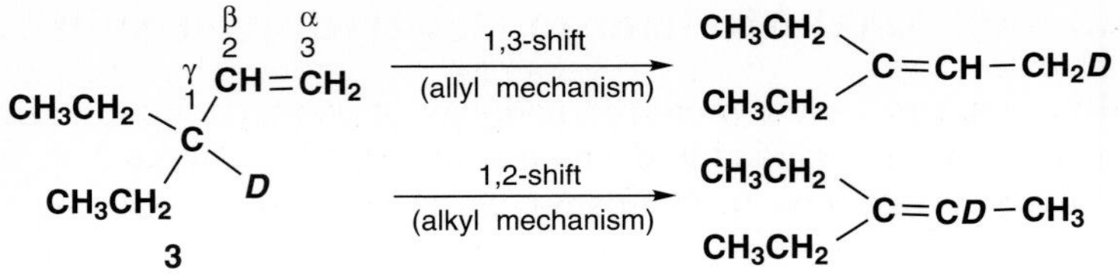

$$CH_3CH_2 \diagdown \underset{\underset{3}{\overset{\overset{\beta}{2}}{CH}}}{\overset{\gamma}{\underset{1}{C}}} = \underset{\overset{\alpha}{3}}{CH_2} \quad \xrightarrow[\text{(allyl mechanism)}]{1,3\text{-shift}} \quad \begin{matrix} CH_3CH_2 \\ CH_3CH_2 \end{matrix} \diagup C = CH - CH_2D \quad (10a)$$

$$CH_3CH_2 \diagup \overset{|}{C} \diagdown D \qquad \xrightarrow[\text{(alkyl mechanism)}]{1,2\text{-shift}} \quad \begin{matrix} CH_3CH_2 \\ CH_3CH_2 \end{matrix} \diagup C = CD - CH_3 \quad (10b)$$

3

3.2.14.4   Applications

As mentioned previously, large-scale olefin isomerization has found application in the SHOP technology (see Section 2.3.1.3 and [1]).

In the BASF synthesis of vitamin A (see Chapter 1) the intermediate β-olefin **5** of eq. (11) is obtained from the isomerization of 6-methyl-6-hepten-2-one **4**, with the latter resulting from condensation of acetone, isobutene, and formaldehyde [21].

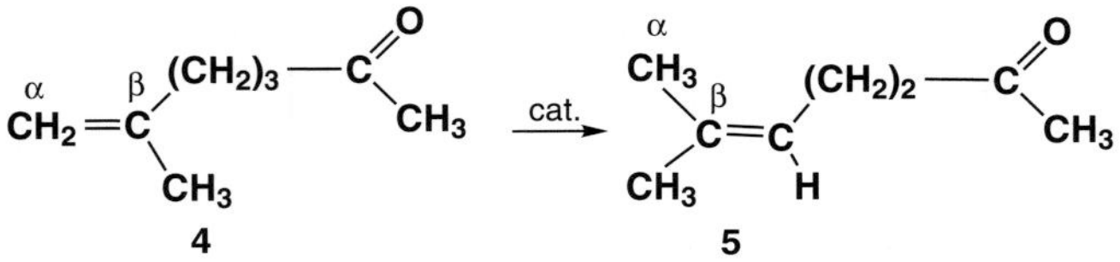

$$\underset{CH_2}{\overset{\alpha}{CH_2}} = \underset{\overset{|}{CH_3}}{\overset{\beta}{C}} - (CH_2)_3 - C \overset{O}{\underset{CH_3}{\diagup}} \quad \xrightarrow{\text{cat.}} \quad \underset{\overset{|}{CH_3}}{\overset{\overset{\alpha}{CH_3}}{C}} = \underset{\overset{|}{H}}{\overset{\beta}{C}} - (CH_2)_2 - C \overset{O}{\underset{CH_3}{\diagup}} \quad (11)$$

4   **5**

A comonomer for the synthesis of ethylene/propene elastomers – 2-ethylidene-norbornene (**7**) – is synthesised via a Diels–Alder cycloaddition of cyclopentadiene and butadiene followed by an isomerization with titanium-based catalysts of the intermediate 2-vinyl derivative **6** in excellent yield (98 %) (eq. (12)) [22].

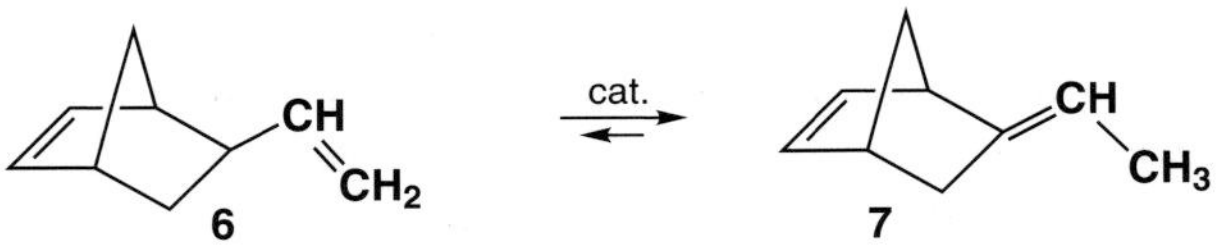

$$(12)$$

6   **7**

The isomerization of functionalized olefins frequently involves a migration process of substituents other than hydrogen.

In the DuPont butadiene-to-adiponitrile synthesis (see Section 2.5.5.1 and [1]), two olefin isomerization steps are employed: rearrangement of **8** via C–C cleavage to the linear isomer **9** (a) is followed by a double-bond shift yielding the terminal olefin **10** (b). The latter is thermodynamically more stable because of the cyano functionality (cf. eq. (13) and Section 2.5).

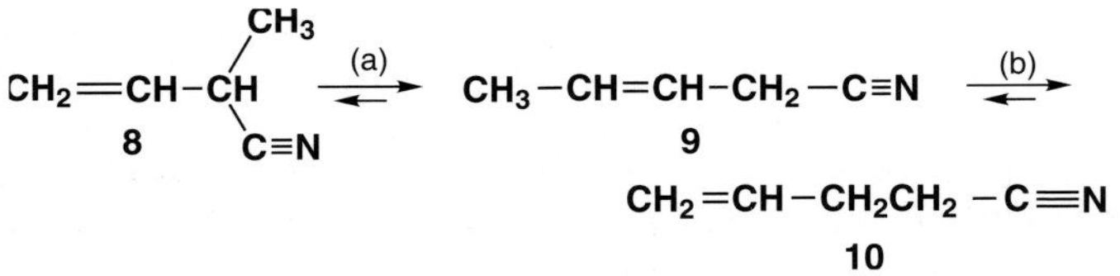

$$CH_2 = CH - \underset{\overset{|}{C \equiv N}}{\overset{\overset{|}{CH_3}}{CH}} \quad \underset{\xleftarrow{}}{\overset{(a)}{\longrightarrow}} \quad CH_3 - CH = CH - CH_2 - C \equiv N \quad \underset{\xleftarrow{}}{\overset{(b)}{\longrightarrow}}$$

8   **9**

$$CH_2 = CH - CH_2CH_2 - C \equiv N \tag{13}$$

10

Another example is the vapor-phase chlorination of butadiene, which gives a mixture of dichlorobutenes of which 3,4-dichloro-1-butene (**12**) is the only desired isomer for the chloroprene synthesis [23, 24] (cf. eq. (14)). It is easily boiled off from the *cis/trans* 1,4-dichloro-2-butenes (123 vs. 155 °C). The migratory isomerization of residual 1,4-isomers **11** is effected by Cu^I complexes and seems to operate through π-olefin/π-allyl intermediates. The chloride probably migrates via the copper center, but no mechanistic details are available.

$$ ClCH_2CH{=}CHCH_2Cl \xrightarrow[]{cat.} ClCH_2-\underset{\underset{\textbf{12}\;\;Cl}{|}}{C}H-CH{=}CH_2 \qquad (14) $$

11

Yet an important application is the analogous isomerization of 1,4-diacetoxy-2-butene (**13**) to the 1,3-isomer **14** (*cis/trans* mixture) with a $Pt^{IV}Cl_4$ catalyst – a key step of the BASF vitamin A synthesis (eq. (15)). The lower-boiling product is enriched to a yield of 95 % and is further hydroformylated to form the vitamin A side chain [25] (see Chapter 1).

$$ AcOCH_2CH{=}CHCH_2OAc \xrightarrow[]{cat.} AcOCH_2-\underset{\underset{\textbf{14}\;\;OAc}{|}}{C}H-CH{=}CH_2 \qquad (15) $$

13

3.2.14.5 Asymmetric Isomerization

Of particular interest is the asymmetric isomerization with chiral catalysts (e. g., eq. (16)), converting allylic alcohols and ethers as well as allylamines into useful synthetic building blocks [26–29].

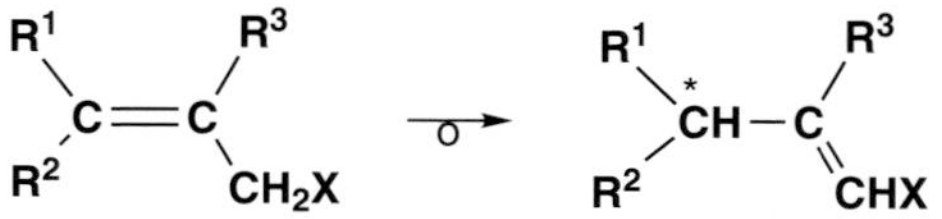

$$ \underset{R^2}{\overset{R^1}{\diagdown}}C{=}C\underset{CH_2X}{\overset{R^3}{\diagup}} \quad \xrightarrow[]{} \quad \underset{R^2}{\overset{R^1}{\diagdown}}\overset{*}{C}H-C\underset{CHX}{\overset{R^3}{\diagup}} \qquad (16) $$

X = OH, OR; NH₂, NR₂

The world's biggest application of asymmetric catalysis is Takasago Perfumery's synthesis of (–)-menthol from myrcene (see Sections 2.9 and 3.3.1) with about 1500 t/a (menthol and other chiral terpenic substances). The key step is the isomerization of geranyldiethylamine with an Rh^I-*S*-BINAP catalyst to citronellal (*E*)-enamine (eqs. (17)) (BINAP = 2,2′-bis(diphenylphosphine)-1,1′-binaphthyl). The geometry of the double bond is 100 % *E*.

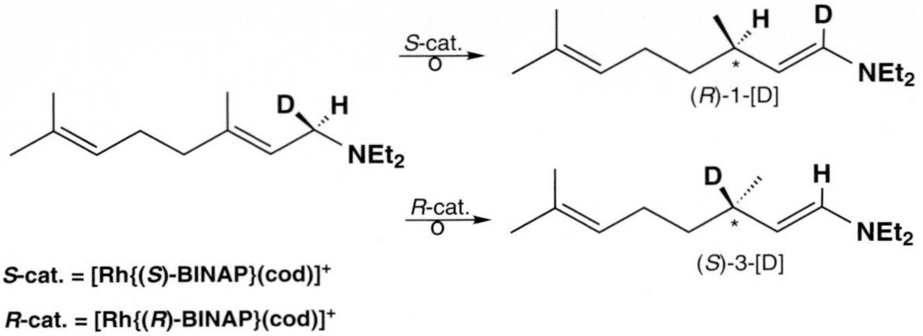
D H
NEt2
S-cat.
R-cat.
H D
NEt2
(R)-1-[D]
D H
NEt2
(S)-3-[D]
S-cat. = [Rh{(S)-BINAP}(cod)]+
R-cat. = [Rh{(R)-BINAP}(cod)]+

(17)

The catalytic process is initiated by coordination of the amino nitrogen atom to the Rh followed by a stereospecific β-hydrogen elimination resulting in an 1,3-hydrogen shift with a suprafacial stereochemistry as determined from D-labeling experiments (eqs. (17)). π-Allyl intermediates account for these unusually clean stereochemical results. The methyl group at the olefinic bond determines the configuration of the transition state. Outstanding enantioselectivity ($\geq 98\,\%$) and high catalyst efficiency (substrate/catalyst ratio ~8000:1) are the remarkable features of this reaction [30–33].

3.2.14.6 Recent Developments

Double-bond isomerization has been exploited as a desired reaction in organic synthesis; examples include the synthesis of steroids. It is also an undesired side reaction of industrially relevant reactions such as hydroformylation (cf. Section 2.1.1), hydrogenation (cf. Section 2.2), and hydrosilyation (cf. Section 2.6), it is a subject of current interest [34–36]. Two promising developments are worth mentioning here because they yielded highly selective catalysts which are, at the same time, easy to handle.

3.2.14.6.1 Organotitanium Catalysts

Special organotitanium cataysts effect regio- and stereoselective isomerizations [37–43]. Titanocene dichloride with various activating reagents (e. g., Grignard compounds, lithium organyls, $LiAlH_4$) has been employed to convert α-olefins into β-olefins with preferred *trans* geometry according to eq. (18) using the immobilized catalyst system **15** in the presence of *t*-butylmagnesium bromide [40].

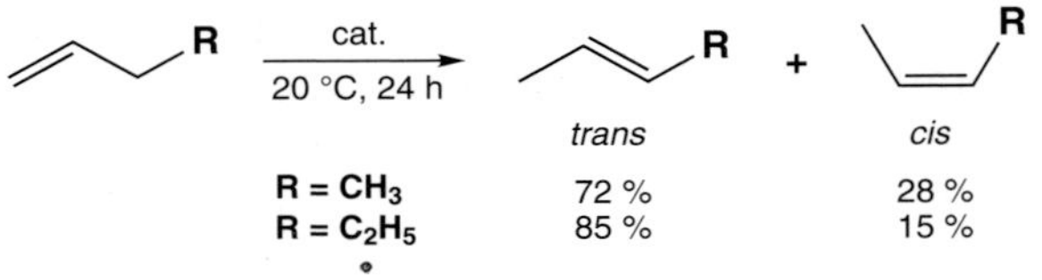
R
cat.
20 °C, 24 h
R
+
R
trans
cis
R = CH3
72 %
28 %
R = C2H5
85 %
15 %

(18)

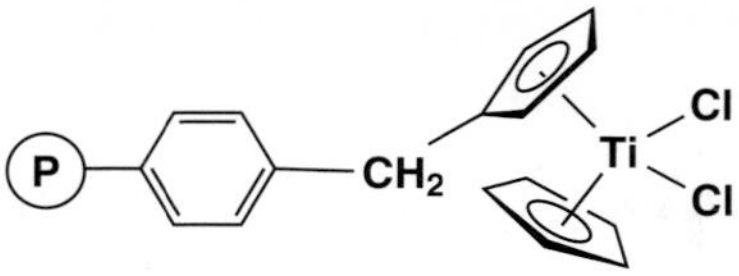

15

Vinylcyclohexane and vinylcyclooctane isomerize quantitatively at 180 °C in the presence of $(\eta^5\text{-}C_5H_5)_2TiCl_2/LiAlH_4$ according to eq. (19) [37].

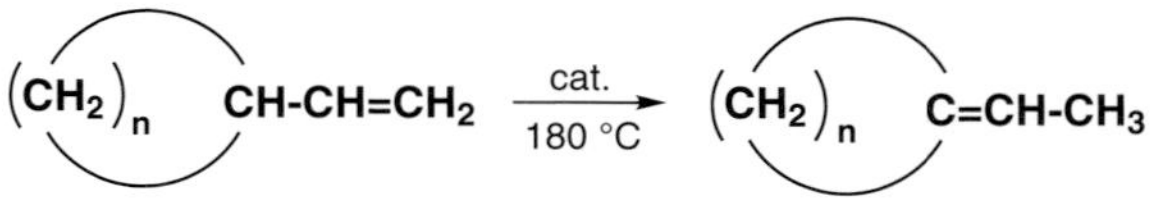

$$\left(CH_2\right)_n \quad CH\text{-}CH\text{=}CH_2 \xrightarrow[180\ °C]{cat.} \left(CH_2\right)_n \quad C\text{=}CH\text{-}CH_3 \tag{19}$$

Nakamura and co-workers discovered outstanding activities and selectivities for the permethylated titanocene $(\eta^5\text{-}C_5Me_5)_2TiCl_2$ (Me = CH_3) in the presence of the reducing agent sodium naphthalide. The olefin isomerization proceeds at ambient termperature, and the preference for the *trans* products seems to depend on the steric bulk of the catalyst [36]. Diolefins yield the conjugated isomers. Examples are given in Table 1. From a mechanistic point of view, Ti^{II} intermediates of type {R$_2$Ti} (R = C$_5$H$_5$, C$_5$Me$_5$) must be invoked, suggesting a π-allyl mechanism. No detailed information is yet available, however, with regard to this question.

Table 1. Isomerization of olefins catalyzed by $(\eta^5\text{-}C_5Me_5)_2TiCl_2/Na$ naphthalide at 20 °C in 60–120 min (olefin/catalyst ratio 100:1): data from [25].

Starting olefin	Product	Yield [%]	*trans* isomer [%]
1-Butene	2-Butene	> 99	99
3-Phenyl-1-propene	1-Phenyl-1-propene	> 99	99
4-Methyl-1-pentene	4-Methyl-2-pentene	25	99
1,4-Pentadiene	1,3-Pentadiene	99	> 99

An asymmetric variant of double-bond isomerization could be achieved with the chiral *ansa*-bis(indenyl)titanium complex [43]. After activation with LiAlH$_4$ it isomerizes *meso,trans*-4-*tert*-butyl-1-vinyl-cyclohexane to the *S*-alkene with remarkable enantioselectivity (80 % *ee*).

3.2.14.6.2 Rhodium Complexes with *N*-Heterocyclic Carbene Ligands

The ligand sphere of an organometallic homogeneous catalyst has in principle two functions: stabilization of the low-valent metal and activation of the metal center by offering vacant coordination sites. Phosphine ligands fulfill these criteria and therefore have played a key role in homogeneous catalysis. As two-electron donors *N*-heterocyclic carbenes resemble phosphines and form remarkably strong metal–carbon bonds with metals from all over the periodic table (see Section 3.1.10 and [44]). In 1994 this ligand class experienced a renaissance with the discovery of the remarkable activity of palladium–*N*-heterocyclic carbene complexes in Heck reactions [45]. The fact that *N*-heterocyclic carbenes are similar to alkylated phosphines resulted in the development of new generations of ruthenium *N*-heterocyclic carbene complexes which are extraordinarily active in ring-opening metathesis reactions (see Section 2.3.3 and [46]).

Rhodium complexes such as **16** with *N*-heterocyclic carbenes can be prepared in one step proceeding from commercially available precursors (e. g. [(η^4-1,5-COD)RhCl]$_2$ (eq. (20)) or Wilkinson's complex [RhCl(PPh$_3$)$_3$] and the free carbene, which is generated from the storable imidazolium salt by deprotonation [47–49]. For more details, see Section 3.1.10.

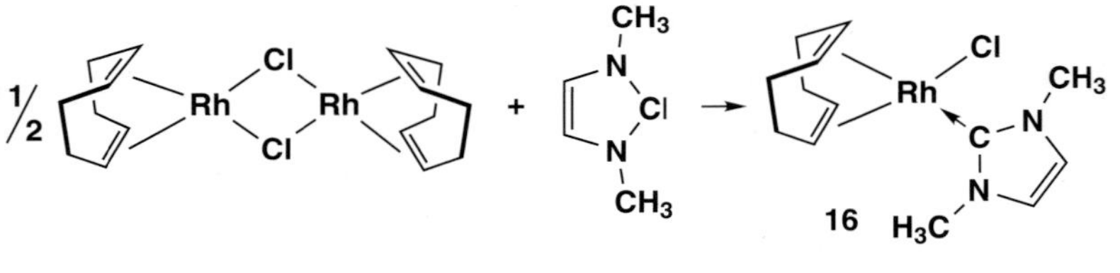

$$(20)$$

3.2.14.7   Perspectives

Double-bond isomerization is one of the major industrial processes in the context of petrochemical oil-refining steps. Selective olefin isomerization under mild conditions is therefore an important goal. New catalysts need to favor a certain isomer kinetically, which means that the speed of rearrangement must be high. As always in homogeneous catalysis, the active species has to maintain its structure for a long time to give reproducible results. The *N*-heterocyclic carbene complexes mentioned above should be borne in mind when further attempts are made at improvement. Of especial charm is stereoselective double-bond isomerization [51], for which new, efficient chiral ligands are warranted.

References

[1] (a) W. A. Herrmann, *Kontakte (Darmstadt),* **1991**, No. 1; (b) W. A. Herrmann, *Kontakte (Darmstadt)*, **1991**, No. 3.

[2] R. A. van Santen, P. W. N. M. van Leeuwen, J. A. Moulijn, B. A. Averill, *Catalysis: an Integrated Approach*, 2nd ed., Elsevier Science, Amsterdam, **1999**, pp. 209–288.

[3] P. A. Verbrugge, G. J. Heisewolf, GB 1.416.317 (1975).

[4] (a) C. A. Tolman, *J. Am. Chem. Soc.* **1992**, *94*, 2999; (b) C. A. Tolman, R. J. McKinney, W. C. Seidel, J. D. Druliner, W. R. Stevens, *Adv. Catal.* **1985**, *33*, 1.

[5] H. Alper, J. T. Edward, *J. Organomet. Chem.* **1968**, *14*, 411.

[6] (a) M. Orchin, *Adv. Catal.* **1966**, *16*, 1; (b) N. R. Davies, *Rev. Pure Appl. Chem.* **1967**, *17*, 83.

[7] C. W. Bird, *Transition Metal Intermediates in Organic Synthesis*, Logos Press/Elek Books, London, **1967**, pp. 69–87.

[8] G. M. Kramer, G. B. McVicker, *Acc. Chem. Res.* **1986**, *19*, 78.

[9] P. N. Rylander, *Organic Synthesis with Noble Metal Catalysts*, Academic Press, New York, **1973**, pp. 145–174.

[10] M. M. T. Khan, A. E. Martell, *Homogeneous Catalysis*, Academic Press, New York, **1974**, pp. 9–37.

[11] G. W. Parshall, S. D. Ittel, *Homogeneous Catalysis*, 2nd ed., John Wiley, New York, **1992**, pp. 9–24.

[12] C. Masters, *Homogeneous Transition Metal Catalysis*, Chapman and Hall, London, **1981**, pp. 70–89.

[13] S. G. Davies, *Organotransition Metal Chemistry: Applications to Organic Synthesis*, Pergamon, Oxford, **1982**, pp. 266–303.

[14] H. M. Colquhoun, J. Holton, D. J. Thompson, M. V. Twigg, *New Pathways for Organic Synthesis*, Plenum, New York, **1984**, pp. 173–193.

[15] J. D. Atwood, *Mechanisms of Inorganic and Organometallic Reactions*, Brooks/Cole, California, **1985**.

[16] M. Turner, J. V. Jouanne, H.-D. Brauer, H. Kelm, *J. Mol. Catal.* **1979**, *5*, 425, 433, 447.

[17] H. Alper, P. C. LePort, *J. Am. Chem. Soc.* **1969**, *91*, 7553.

[18] H. Bönnemann, *Angew. Chem.* **1973**, *85*, 1024; *Angew. Chem., Int. Ed. Engl.* **1973**, *12*, 964.

[19] D. Evans, J. Osborn, G. Wilkinson, *J. Chem. Soc. (London) A* **1968**, 3133.

[20] C. P. Casey, C. R. Cyr, *J. Am. Chem. Soc.* **1973**, *95*, 2248.

[21] H. Pommer, A. Nurrenbach, *Pure Appl. Chem.* **1975**, *43*, 527.

[22] G. Ver Strate, *Encycl. Polym. Sci.* **1986**, *6*, 522.

[23] F. J. Bellringer, C. E. Hollis, *Hydrocarbon Process*, **1968**, *47* (11), 127.

[24] G. W. Parshall, S. D. Ittel, *Homogeneous Catalysis*, 2nd ed., John Wiley, New York, **1992**, pp. 300–302.

[25] BASF AG (J. Hartig, H.-M. Weitz, R. Schnabel), DE 2.747.634 (1979).

[26] M. Beller, C. Bolm, *Transition Metals for Organic Synthesis*, Wiley-VCH, Weinheim, **1998**, p. 147.

[27] L. J. Gazzard, W. B. Motherwell, D. A. Sandham, *J. Chem. Soc., Perkin Trans.* **1999**, *1*, 979.

[28] S. Fuss, J. Harder, *FEMS Microbiol. Lett.* **1997**, *149*, 71.

[29] D. Baudry, M. Ephritikhine, H. Felkin, *J. Chem. Soc., Chem. Commun.* **1978**, 694.

[30] K. Tani, T. Yamagata, S. Otsuka, S. Akutagawa, H. Kumobayashi, T. Taketomi, H. Takaya, A. Miayshita, R. Noyori, *J. Chem. Soc., Chem. Commun.* **1982**, 600.

[31] K. Tani, T. Yamagata, S. Akutagawa, H. Kumobayashi, T. Taketomi, H. Takaya, A. Miayshita, R. Noyori, S. Otsuke, *J. Am. Chem. Soc.* **1984**, *106*, 5208.

[32] K. Tani, T. Yamagata, Y. Tasuno, Y. Yamagata, T. Tomita, S. Akutaga, H. Kumobayashi, S. Otsuka, *Angew. Chem.* **1985**, *97*, 232; *Angew. Chem., Int. Ed. Engl.* **1985**, *24*, 217.

[33] T. Faitig, J. Soulie, J. Y. Collemand, *Tetrahedron* **2000**, *56*, 101.

[34] A. Haynes, J. McNish, J. M. Pearson, *J. Organomet. Chem.* **1998**, *551*, 339.

[35] F. M. Moghaddan, R. Emanj, *Synth. Commun.* **1997**, *27*, 4073.

[36] N. S. Sampson, I. J. Kass, *J. Am. Chem. Soc.* **1997**, *119*, 855.

[37] M. Akita, H. Yasuda, K. Nagasuna, A. Nakamura, *Bull. Chem. Soc. Jpn.* **1983**, *56*, 554.

[38] R. H. Grubbs, C. Gibbons, L. C. Kroll, W. D. Bonds, Jr., C. H. Brubaker Jr., *J. Am. Chem. Soc.* **1973**, *95*, 2373.

[39] W. D. Bonds, Jr., C. H. Brubaker, Jr., E. S. Chandrasekaran, C. Gibbons, R. H. Grubbs, L. C. Kroll, *J. Am. Chem. Soc.* **1975**, *97*, 2128.

[40] D. E. Bergbreiter, G. L. Parson, *J. Organomet. Chem.* **1981**, *208*, 47.

[41] C.-P. Lau, B.-H. Chang, R. H. Grubbs, C. H. Brubaker, Jr., *J. Organomet. Chem.* **1981**, *214*, 325.

[42] K. Mach, F. Turecek, H. Antropiusova, L. Petrusova, V. Hanus, *Synthesis* **1982**, 53.

[43] Z. Chen, R. Halterman, *J. Am. Chem. Soc.* **1992**, *114*, 2276.

[44] (a) Hoechst AG (W. A. Herrmann, M. Elison, J. Fischer, Ch. Köcher), DE 4.447.066 (1994). Reviews: (b) W. A. Herrmann, C. Köcher, *Angew. Chem.* **1997**, *109*, 2256; *Angew. Chem., Int. Ed. Engl.* **1997**, *36*, 2162; (c) D. Bourissou, O. Guerret, F. P. Gabbaï, G. Bertrand, *Chem. Rev.* **2000**, *100*, 39; (d) T. Weskamp, V. P. W. Böhm, W. A. Herrmann, *J. Organomet. Chem.* **2000**, *600*, 12.

[45] (a) W. A. Herrmann, M. Elison, J. Fischer, Ch. Köcher, G. R. J. Artus, *Angew. Chem.* **1995**, *107*, 2602; *Angew. Chem., Int. Ed. Engl.* **1995**, *34*, 2371; (b) J. Fischer, *Ph. D. Thesis*, Technische Universität München, **1996**; (c) W. A. Herrmann, *Angew. Chem., Int. Ed. Engl.* **2002**, in press (review article on *N*-heterocyclic carbenes in catalysis).

[46] (a) T. Weskamp, W. C. Schattenmann, M. Spiegler, W. A. Herrmann, *Angew. Chem.* **1998**, *110*, 2631; *Angew. Chem. Int. Ed.* **1998**, *37*, 2490; (b) T. Weskamp, F. J. Kohl, W. Hieringer, D. Gleich, W. A. Herrmann, *Angew. Chem.* **1999**, *38*, 2416; *Angew. Chem., Int. Ed.* **1999**, *38*, 2416.

[47] W. A. Herrmann, M. Elison, J. Fischer, C. Köcher, *Chem. Eur. J.* **1996**, *2*, 772.

[48] M. Prinz, *Diplomarbeit*, Technische Universität München, **1997**.

[49] A. C. Chen, L. Ren, A. Decken, C. M. Crudden, *Organometallics*, **2000**, *19*, 3459.

[50] W. A. Herrmann, J. Unruh, Ch. Köcher, J. Fischer, unpublished results, **1995/6**.

[51] (a) R. E. Merrill, *CHEMTECH* **1981**, *11*, 118; (b) S. Otsuka, K. Tani, in *Asymmetric Synthesis,* Vol. 5 (Ed.: J. D. Morrison), Academic Press, New York, **1985**, Chapter 6, p. 171.

3.3 Special Products

3.3.1 Enantioselective Synthesis

Hans-Ulrich Blaser, Benoît Pugin, Felix Spindler

3.3.1.1 Introduction and Background

For many applications of chiral compounds, the racemic form will no longer be accepted [1, 2]. As a consequence, the importance of enantioselective synthesis in general and of enantioselective catalysis in particular will undoubtedly increase. There are various methods available to prepare only one enantiomer of a chiral product [3]. The *resolution of racemates* is probably still used most often despite the fact that the yield of the desired enantiomer is at best 50 % [4 a]. If the undesired enantiomer cannot be isomerized and recycled, it must be disposed of. Similar problems occur when applying stoichiometric chiral reagents or chiral auxiliaries. This question does not arise if a starting material from the chiral pool (isolated from natural products or produced by fermentation) can be used, since nature has already produced the desired absolute stereochemistry. However, for larger-scale applications it is not always possible to find the suitable starting material. Therefore, *enantioselective catalysis* with either biocatalysts or chiral chemical catalysts will be applied more frequently in the future because the chiral auxiliary is required only in substoichiometric quantities. However, it must be stressed that every method mentioned above can be the most suitable one for solving a particular problem. There are many factors that influence economical and ecological aspects and no single approach is able to meet all the requirements of an industrial process.

Most of the useful enantioselective homogeneous catalysts consist of a central metal atom and a chiral ligand. Somewhat simplified, the activation of a substrate occurs by binding to the metal center whereas the stereocontrol of the transformation is exerted by the chiral ligand, resulting in the preferential formation of one enantiomer. While most applications are in the field of *asymmetric synthesis* starting from a prochiral substrate, *kinetic resolution*, i. e., the preferential transformation of one enantiomer of a racemic substrate, is of growing technical importance [5]. Up to now, relatively few homogeneous enantioselective catalysts have been used on an industrial scale [6]. One reason is that enantioselective homogeneous catalysis is a relatively recent discipline, but there are many others and these will be discussed below.

In this overview, the opportunities and problems associated with the industrial application of chiral metal complexes will be analyzed in detail. In Section 3.3.1.2, the critical factors are discussed which affect the feasibility of an enantioselective catalyst. In the following Sections, important families of chiral ligands are listed and finally about 40 types of catalytic transformations are described and characterized regarding enantioselectivity, catalyst activity, and productivity, and their potential for technical applications is assessed.

3.3.1.2   Critical Factors for the Technical Application of Homogeneous Enantioselective Catalysts

The application of homogeneous enantioselective catalysts on a technical scale presents some very special challenges and problems [3, 4, 6, 7]. Some of these problems are due to the special manufacturing situation of the products involved, others to the nature of the enantioselective catalytic processes.

3.3.1.2.1   Characteristics of the Manufacture of Enantiomerically Pure Products

Optically pure compounds will be used above all as pharmaceuticals and vitamins [1], as agrochemicals [2], and as flavors and fragrances [8]. Other potential but at present less important applications are as chiral polymers, as materials with nonlinear optical properties, or for ferroelectric liquid crystals [4 b, 9]. The manufacture of pharmaceuticals and agrochemicals can be characterized as follows (typical numbers are given in parentheses):

(1) Multifunctional molecules produced via multistep syntheses (five to ten steps or more for pharmaceuticals, and three to seven for agrochemicals) with short product lives (often less than 20 years).
(2) Relatively small-scale products (1–1000 t/a for pharmaceuticals, 500–10 000 t/a for agrochemicals), usually produced in multipurpose batch equipment.
(3) High purity requirements (usually $> 99\%$ and < 10 ppm metal residue in pharmaceuticals).
(4) High added values and therefore tolerant to higher process costs (especially for very effective, small-scale products).
(5) Short development time for the production process (less than a few months to 1–2 years) since time to market affects the profitability of the product.
(6) Synthetic route often designed around the enantioselective catalysis as key step.

3.3.1.2.2   Characteristics of Enantioselective Catalytic Processes

Homogeneous enantioselective catalysis is a relatively young but rapidly expanding field. Up to 1985, only few catalysts affording enantioselectivities up to 95 % were known [10]. This situation has changed dramatically in recent years and there are now a large number of chiral catalysts known that catalyze a variety of transformation with enantiomeric excesses (*ee*) $> 98\%$ [11, 12]. What still remains a major challenge is the fact that it is difficult to transfer the results obtained for a particular substrate to even a close analog due to the high substrate specificity (low tolerance for structure variation even within a class of substrates). Technical applications of enantioselective catalysts are also hampered because there is little information on catalyst activity or other aspects available (in the literature *enantioselectivity* is the dominant criterion) and because few applications with "real" substrates exist (usually simple model reactions are studied). Finally, chiral ligands and many metal precursors are expensive and/or not easily available.

3.3.1.2.3 Critical Factors for the Application of Enantioselective Catalysts

In the final analysis, the choice of a specific catalytic step is usually determined by the answer to two questions:

(1) Can the costs for the overall manufacturing process compete with alternative routes?
(2) Can the catalytic step be developed in the given time frame?

As a consequence of the peculiarities of enantioselective catalysis described above, the following critical factors often determine the viability of an enantioselective process:

(1) *Enantioselectivity*, expressed as enantiomeric excess (% *ee*), i. e., % desired – % undesired enantiomer. The *ee* of a catalyst should be > 99 % for pharmaceuticals if no purification is possible (via recrystallization or at a later stage via separation of diastereomeric intermediates). This case is quite rare and *ee* values of > 90 % are often acceptable; for agrochemicals *ee* values of > 80 % can be sufficient.
(2) *Chemoselectivity* (or functional group tolerance) will be very important when multifunctional substrates are involved.
(3) *Catalyst productivity*, given as substrate/catalyst ratio (*s/c*) or turnover number (TON), determines catalyst costs. These *s/c* ratios ought to be > 1000 for small-scale, high-value products and > 50 000 for large-scale or less-expensive products (catalyst re-use increases the productivity).
(4) *Catalyst activity*, given as turnover frequency for > 95 % conversion (TOF$_{av}$, h^{-1}), determines the production capacity. TOF$_{av}$ ought to be > 500 h^{-1} for small-scale and > 10 000 h^{-1} for large-scale products.
(5) *Availability and cost of ligands*. In the majority of cases the ligands of the organometallic catalysts are chiral diphosphines which need special synthetic know-how and can be rather expensive. Typical prices are US$ 100–500/g for laboratory quantities and US$ 5000 to > US$ 20 000/kg on a larger scale. Chiral ligands used for early transition metals are usually cheaper.
(6) *Availability and cost of starting materials*. Starting materials are often expensive and difficult to manufacture on a large scale with the required quality.
(7) *Development time*. This can be crucial if an optimal ligand has to be developed for a particular substrate (substrate specificity) and when not much is known on the catalytic process (technological maturity).

For most other aspects such as catalyst stability and sensitivity, handling problems, catalyst separation, space-time yield, poisoning, chemoselectivity, process sensitivity, toxicity, safety, special equipment, etc., enantioselective catalysts have similar problems and requirements compared to nonchiral homogeneous catalysts.

Which of these criteria will be critical for the development of a specific process will depend on the particular catalyst and transformation. The following factors have to be considered: the field of application and the price of the active compound (added value of the catalytic step), the scale of the process, the technical experience and the production facilities of a company, the maturity of the catalytic process, and last but not least, the chemist who plans the synthesis must be aware of the catalytic route!

3.3.1.3   State-of-the-Art and Evaluation of Catalytic Transformations

3.3.1.3.1   General Comments

In the last few years, information on industrial applications has increased both in quantity and in quality because smaller technology-based companies especially are prepared to publish relevant results (cf. Table 1) [6]. From the values in this

Table 1. Statistics for the industrial application of enantioselective catalytic reactions.

Transformation	Production[a]		Pilot[b]		Bench-scale[c]
	> 5 t/a	< 5 t/a	> 50 kg	< 50 kg	
Hydrogenation of enamides	1	1	2	6	4
Hydrogenation of C=C–COOR and C=C–CH–OH	1	0	3	4	6
Hydrogenation of other C=C	1	0	1	2	2
Hydrogenation of *a*- and *b*-functionalized C=O	2	2	3	6	4
Hydrogenation/reduction of other C=O	0	0	0	1	4
Hydrogenation of C=N	1	0	1	0	0
Dihydroxylation of C=C	0	1	0	0	4
Epoxidation of C=C, oxidation of sulfide	2	1	2	0	2
Isomerization, epoxide opening, addition	2	0	3	0	1
Total	10	5	15	19	27

[a] *Production processes* are operated on a regular basis. [b] *Pilot processes* are technically on a similar level but are not (yet) applied on a regular basis. [c] *Bench-scale processes* have an optimized catalyst system and have been carried out on a kilogram scale.

table, it is evident that hydrogenation (enantioselective h.; cf. Section 2.2) is the transformation with the highest industrial impact, followed by epoxidation and dihydroxylation reactions. The success with epoxidation and dihydroxylation reactions can be attributed essentially to the efforts of Sharpless, Katsuki, and Jacobsen (cf., e. g., Section 3.3.2). Nevertheless, as will be shown in the following sections, other catalytic transformations have the potential for industrial use too.

In the next sections, synthetically useful enantioselective reactions and the corresponding catalysts are reviewed. Critical issues will be discussed and an overall assessment of the technical maturity will be given.

3.3.1.3.2 Chiral Ligands

Chiral ligands are obviously at the heart of every enantioselective organometallic catalyst. Over the years a number of ligand types and families have achieved what Jacobsen once called a "privileged" status. This means that certain ligand types have a very broad scope, in many cases because they can easily be tailored for a specific substrate, often have a modular character, and are available in larger quantities, e. g., **1–16**. Many of these catalysts tolerate a wide range of functional groups and are also chemoselective.

BIAR biaryl and heterobiaryl diphosphines

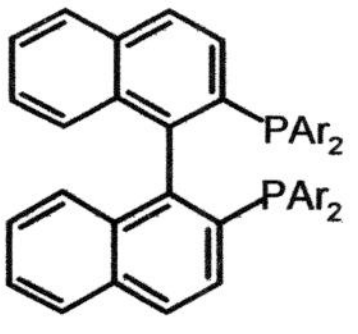

1 BINAP 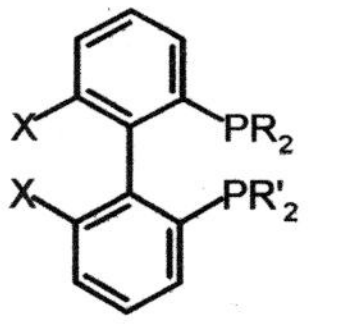 **2** BIPHEP 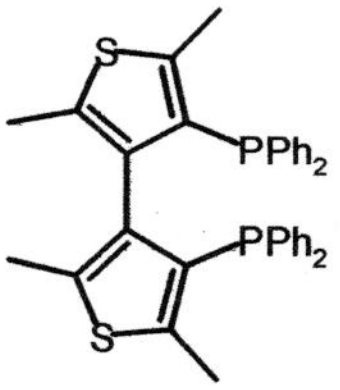 **3** TMBTP

BINO, PAMID binol-based ligands

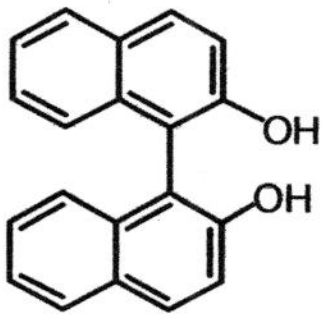

4 BINOL 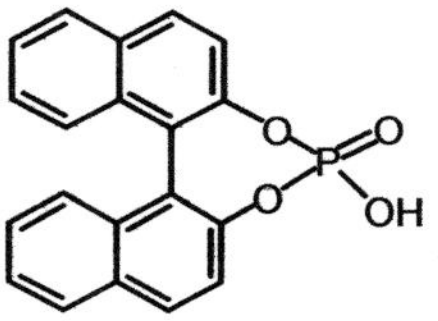 **5** BINOP 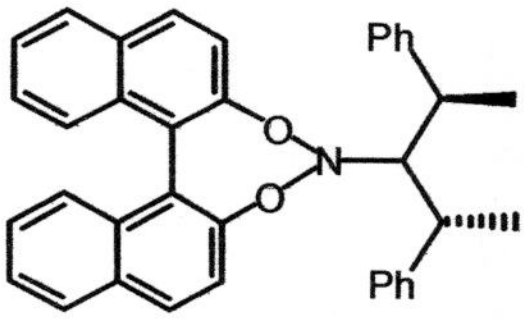 **6** PAMID

FERRO ferrocenyl-based ligands

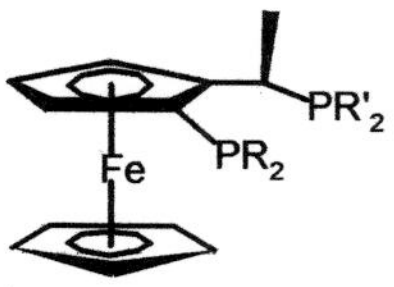

7 JOSIPHOS 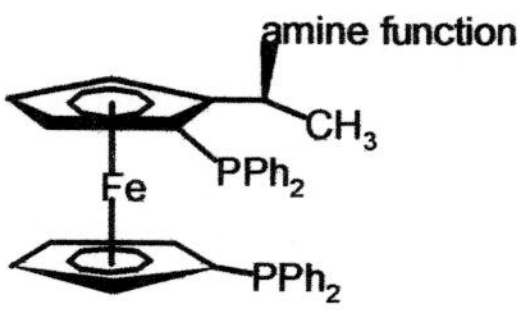 **8** BPPFA 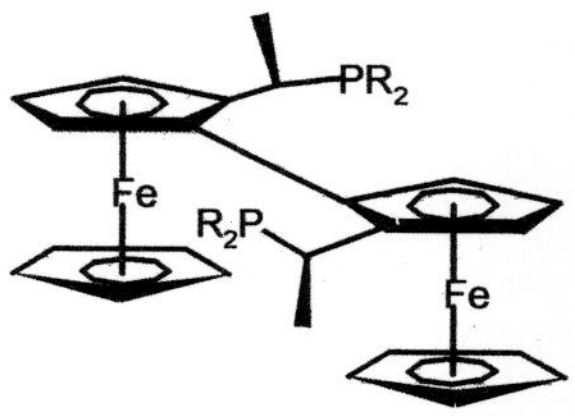 **9** TRAP

PCYCL
ligands with
cyclic phosphine

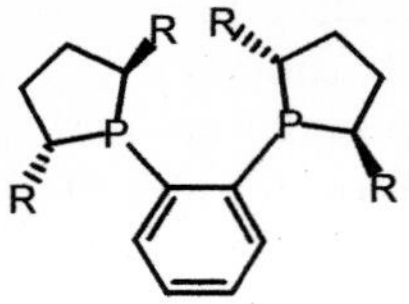

10
DuPHOS

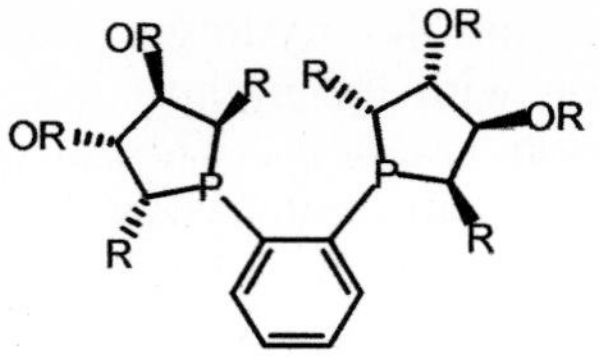

11
ROPHOS

OXAZ, P^OXAZ
oxazoline-derivated
ligands

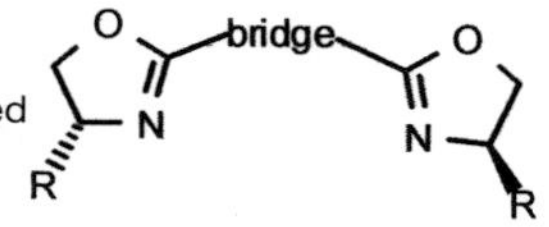

examples
BISOXAZOLINE
PYBOX

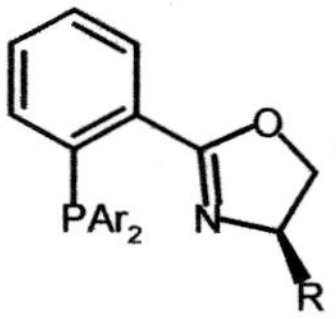

P^OXAZ

12

13

Various types

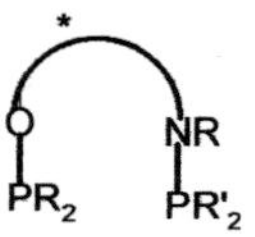

14
NOP
backbone derived
from amino acids

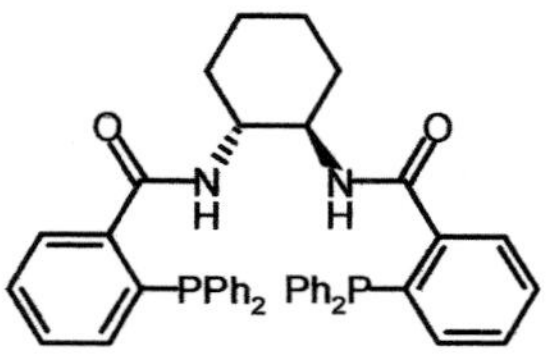

15
PNNP

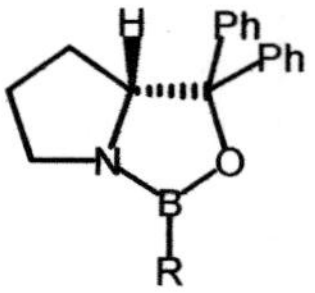

16
OXABOR

In order to give an impression of the structural variety of chiral ligands, representative examples of the preferred ligand types (and abbreviations) mentioned in the following tables are given with these structures. Generally, the abbreviation X^Y is used for bidentate ligands with a chiral backbone and X and Y as coordinating atoms. TART and CINCH are tartaric acid and cinchona derivatives, respectively. All other abbreviations have the meanings used in [11].

3.3.1.3.3 Addition to C=C Groups

Hydrogenation of Olefins

The enantioselective hydrogenation of olefins is the best-studied reaction with the most industrial applications [6, 11 a, 12 a]. Over the most recent decades, a few privileged substitution patterns have evolved that almost guarantee high *ee* values (olefins **17**, **18**) and the state-of-the-art is summarized in Table 2. With few exceptions, Rh and Ru complexes of a limited number of chiral diphosphine families are the preferred catalysts but, in any case, the optimal complex (metal, ligand, anion, etc.) has to be determined for each substrate.

Table 2. State-of-the-art for the hydrogenation of olefins (cf. structures **17–39**).

Substrate	ee [%][a]	TON[a]	TOF [h^{-1}][a]	Preferred catalyst types[b]
Type **17** (enamides, enol acetates, itaconates)	90–98	1000–20 000	200–5000	Rh/PCYCL, Rh/FERRO, Ru/BIAR, Rh/PPM
Type **18** (C=C–C–OH)	80–95	10 000–50 000	1000–5000	Ru/BIAR
Type **18** (C=C–COOH)	85–95	2000–10 000	500–3000	Ru/BIAR, (Rh/PCYCL)
tetrasubstituted C=C	85–95	500–2000	200–500	Ru/BIAR, Ru/PCYCL, Rh/FERRO
C=C without privileged function	80–95	20–100	2–5	Ru/BIAR, Ir/P^OXAZ, Rh/PCYCL

[a] Typical range for suitable substrate and optimized catalyst. [b] For structures see **1–16**.

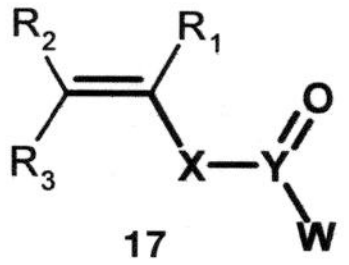

Hydrogenation of enamides **17** (X = NR, Y = C, W = R), especially with R_1 = COOR, is not only the best-known test reaction but also has a very high potential for the production of pesticides or pharmaceuticals. The original motivation for developing this reaction type was the manufacture of α-amino acids but except for small-scale applications such as L-dopa, the preparation of the enamide substrates was too expensive and most amino acids are now produced via biocatalytic methods [13]. Many different substrates of type **17** can be hydrogenated with *ee* values between 95 and 99 % [11a, 12a] but much less is reported on catalyst activity and productivity. In general, more and larger substituents lead to a decrease in catalyst activity either when directly attached to C=C or when bound to X or W.

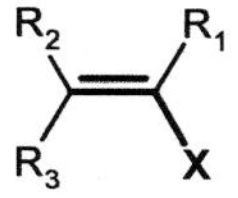

The selected examples (**19–27**) of applications with type **17** substrates show what level of catalyst performance can be achieved by process development (the development status and the relevant company are also shown). In addition, the selection demonstrates the variety of structural motifs found in biologically active compounds and the need for tolerance of functional groups such as pyridyl, cyano, thienyl groups or other C=C or C=O functions. Preferred catalysts and ranges for *ee* values, TONs and TOFs are listed in Table 2. Other type **17** substrates with a similar industrial potential are itaconic and phosphonic acid derivatives.

The hydrogenation of allylic alcohols and α,β-unsaturated acids (type **18**) is another class of transformations with a high industrial success rate. Again, some illustrative examples are Structures **28–33**; preferred catalysts are Ru/BINAP

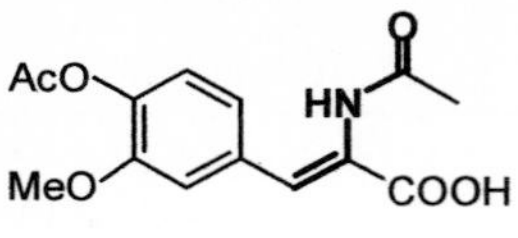

19
Rh/DIPAMP; *ee* 95 %
TON 20 000; TOF 1000
smal-scale production
Monsanto [14]

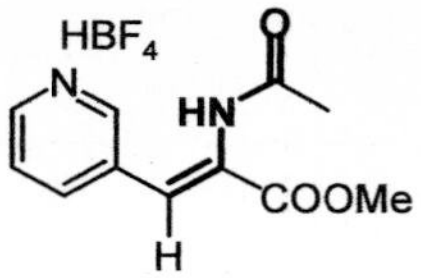

20
Rh/DuPHOS; *ee* 98 %
TON 20 000; TOF n.a.
pilot process, > 200 kg
ChiroTech [15]

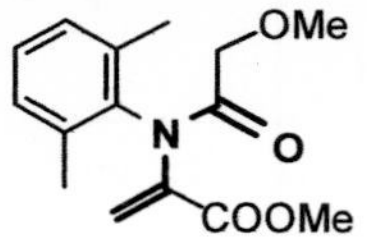

21
Rh/DuPHOS; *ee* 96 %
TON 50 000; TOF 5200
pilot process, kg scale
Ciba–Geigy/Solvias [16]

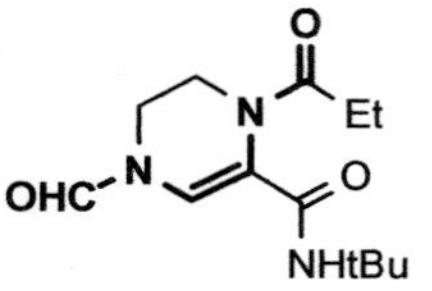

22
Rh/JOSIPHOS; *ee* 97 %
TON 1000; TOF 450
pilot process, > 200 kg
Lonza [17]

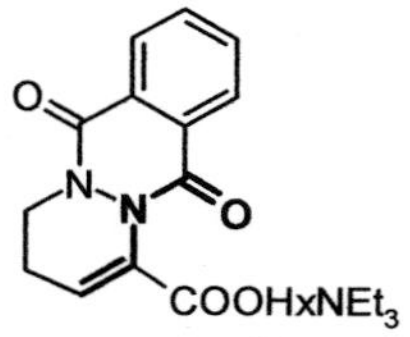

23
Ru/BIPHEP; *ee* > 99 %
TON 20 000; TOF 830
pilot process, > 10 kg
Roche [12 b, 18 a]

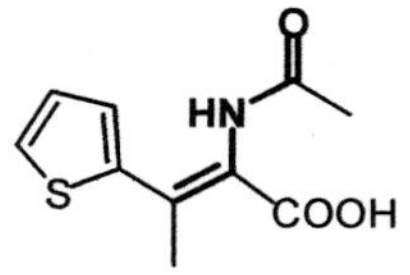

24
Rh/PBM; *ee* 99 %
TON 1600; TOF 1600
bench-scale process
HoechstMarionRoussel
[18 b, 19 a]

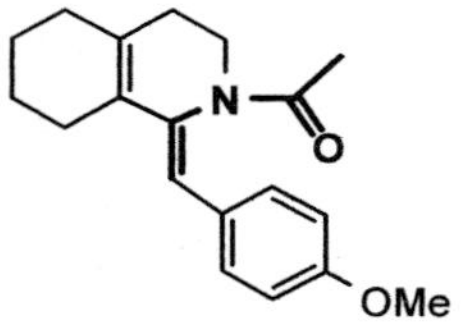

25
Ru/BIPHEP; *ee* 98 %
TON 20 000; TOF 6600
bench scale process
Roche [12 b, 18 a]

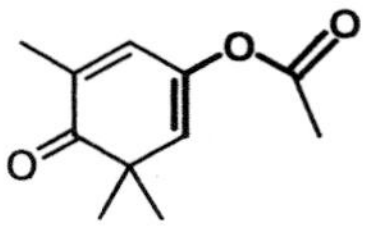

26
Rh/DuPHOS; *ee* 98 %
TON 20 000; TOF 5000
pilot process, multi kg
Roche [18 a]

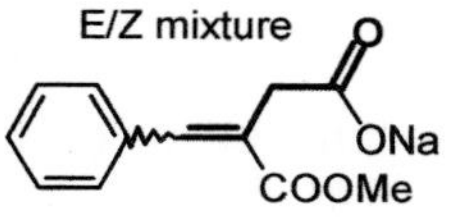

27
Rh/DuPHOS; *ee* 97 %
TON 1000; TOF n.a.
pilot process, > kg scale
ChiroTech [20]

and Ru/BIPHEP but other complexes are also useful. Very high TONs and TOFs have been achieved for simple allylic alcohols, but more complex substrates and especially α,β-unsaturated acids are reduced with lower efficiency.

Homogeneous hydrogenations of tetrasubstituted olefins are still rare, even though high *ee* values and reasonable activities can now be achieved. Three commercial examples are Structures **34–36**; preferred catalysts for this reaction are Rh and Ru/JOSIPHOS and Ru/DuPHOS complexes, Ru/BIPHEP and Rh/TRAP catalysts are also effective [11 a].

The hydrogenation of alkenes without "privileged" functional groups has not been investigated systematically, probably because much more effort is required to achieve good enantioselectivity. Successful examples are Structures **37–39**. Of special interest are the Ir phosphine dihydrooxazole (P^OXAZ) catalysts [27], even though their functional group tolerance is relatively low.

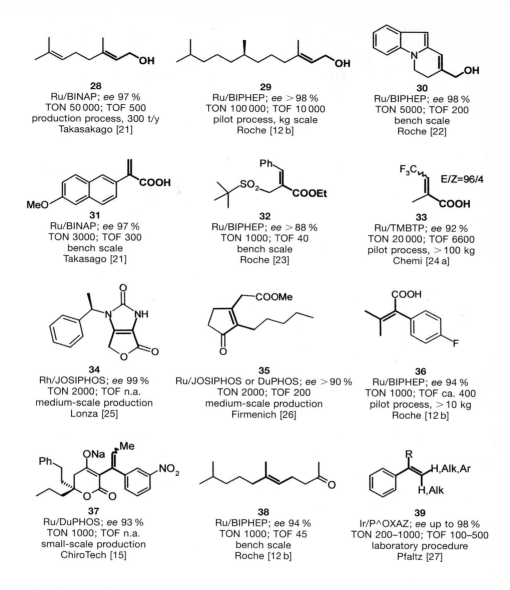

28
Ru/BINAP; *ee* 97 %
TON 50 000; TOF 500
production process, 300 t/y
Takasakago [21]

29
Ru/BIPHEP; *ee* > 98 %
TON 100 000; TOF 10 000
pilot process, kg scale
Roche [12 b]

30
Ru/BIPHEP; *ee* 98 %
TON 5000; TOF 200
bench scale
Roche [22]

31
Ru/BINAP; *ee* 97 %
TON 3000; TOF 300
bench scale
Takasago [21]

32
Ru/BIPHEP; *ee* > 88 %
TON 1000; TOF 40
bench scale
Roche [23]

33
Ru/TMBTP; *ee* 92 %
TON 20 000; TOF 6600
pilot process, > 100 kg
Chemi [24 a]

34
Rh/JOSIPHOS; *ee* 99 %
TON 2000; TOF n.a.
medium-scale production
Lonza [25]

35
Ru/JOSIPHOS or DuPHOS; *ee* > 90 %
TON 2000; TOF 200
medium-scale production
Firmenich [26]

36
Ru/BIPHEP; *ee* 94 %
TON 1000; TOF ca. 400
pilot process, > 10 kg
Roche [12 b]

37
Ru/DuPHOS; *ee* 93 %
TON 1000; TOF n.a.
small-scale production
ChiroTech [15]

38
Ru/BIPHEP; *ee* 94 %
TON 1000; TOF 45
bench scale
Roche [12 b]

39
Ir/P^OXAZ; *ee* up to 98 %
TON 200–1000; TOF 100–500
laboratory procedure
Pfaltz [27]

For the enantioselective reduction of olefins, there are few alternatives to homogeneous hydrogenation because neither transfer hydrogenations with hydrogen donors such as HCOOH/NEt₃ [28] nor chiral heterogeneous catalysts [12 c] are ready for larger-scale applications.

Oxidation of Olefins

Enantioselective oxidation of olefins is a very elegant way of introducing oxygen and in some cases also nitrogen functions into molecules. The catalytic methods with the highest industrial potential are epoxidation and dihydroxylation, and the kinetic resolution of racemic terminal epoxides (Table 3).

Table 3. State-of-the-art for the oxidation of olefins (see structures **40–45**).

Reaction	*ee* [%][a]	TON[a]	TOF [h^{-1}][a]	Preferred catalyst types[b]
Epoxidation of allylic alcohols	85–95	10–40	up to 20	Ti/TART
Epoxidation of C=C	80–95	50–2000	50–200	Mn/SALEN
Dihydroxylation of C=C	85–95	100–500	50–100	Os/CINCH
Kinetic resolution of epoxides	98–99	500–1000	20–40	Co,Cr/SALEN

[a] Typical range for suitable substrate and optimized catalyst. [b] Structures **1–16**.

The epoxidation of allylic alcohols (Structures **40–42**) using Ti/diisopropyl tartrate (Ti/DIPT) or Ti/diethyl tartrate (Ti/DET) catalysts has been applied in numerous multistep syntheses of bioactive compounds [11 b, 12 d]. In presence of molecular sieves, the catalyst is effective for a variety of substituents at the C=C bond and tolerates most functional groups with good to high *ee* values but rather low activity. However, application on a larger scale is still restricted, selected examples of which are given with Structures **40–42** (for details see [6]). The most important is the manufacture of glycidol developed by Arco and now in operation at PPG–Sipsy [4 c]. The reaction has been carefully optimized and is run with cumyl hydroperoxide as oxidant. An interesting new development is a Ta/DIPT attached to silica (*ee* values up to 97 %, TON up to 25 and TOF < 1) [30] but its synthetic potential has not yet been explored.

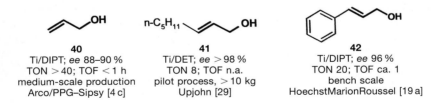

40	**41**	**42**
Ti/DIPT; *ee* 88–90 %	Ti/DET; *ee* > 98 %	Ti/DIPT; *ee* 96 %
TON > 40; TOF < 1 h	TON 8; TOF n.a.	TON 20; TOF ca. 1
medium-scale production	pilot process, > 10 kg	bench scale
Arco/PPG–Sipsy [4 c]	Upjohn [29]	HoechstMarionRoussel [19 a]

In the last few years, the epoxidation of unfunctionalized olefins using cheap NaOCl as oxidizing agent has been developed industrially by Rhodia ChiRex in collaboration with Jacobsen [31] and an example is given with Structure **43**. Mn/SALEN-type catalysts give good results for terminal and *cis*-substituted olefins with *ee* values up to > 97 % with moderate to good catalytic activity [11 c, 12 e]. New developments are the discovery of the beneficial effect of pyridine *N*-oxides [32 a] and of new types of SALEN ligands by Katsuki with

TONs up to 9000 [11 c]. Of potential interest is the use of ionic liquids which allow recycling of the catalyst (cf. Section 3.1.1.2.2) [33]. SALEN complexes are also eminently suitable for the kinetic resolution of epoxides [12 u]; especially promising for commercial applications is hydrolytic ring-opening using Co/SALEN complexes [34] (**42–45**). *α,β*-Unsaturated ketones can be epoxidized with hydrogen peroxide in presence of a polypeptide catalyst with *ee* values up to > 98 % [35].

43
epoxidation
Mn/SALEN; *ee* 88 %
TON > 250; TOF ca. 250
small-scale process
Rhodia ChiRex [31]

44
dihydroxylation
Os/(DHQD)2PHAL; *ee* 95 %
TON ca. 500; TOF 50–100
pilot process, > 10 kg
Rhodia ChiRex [19 b]

45
hydrolytic kinetic resolution
Co/SALEN; *k*(rel) ca. 400, *ee* 98 + 99 %
TON (recycl) > 1500; TOF ca. 40
medium-scale process
Rhodia ChiRex [34]

The asymmetric dihydroxylation (AD) of olefins leads to *cis*-diols with high to very high *ee* values using Os/CINCH complexes [11 d, e, 12 f]). This reaction has also been developed by Rhodia ChiRex and is carried out on commercial scale on request [31]. $K_3Fe(CN)_6$–K_2CO_3, the oxidant used in the commercially available AD mixes is problematic on a larger scale. Recently, it has been shown that oxygen can be used instead, which is more promising for industrial applications [36].

Allylic oxidation [12 g], aminohydroxylation [11 e], and aziridination [12 h] are not yet mature for technical use, even though in specific cases very high *ee* values have been achieved.

Miscellaneous Addition Reactions to C=C Groups

Addition reactions to olefins can be used both for the construction and for the functionalization of molecules. Accordingly, chiral catalysts have been developed for many different types of reactions, often with very high enantioselectivity. Unfortunately, most either have a narrow synthetic scope or are not yet developed for immediate industrial application due to insufficient activities and/or productivities. These reactions include hydrocarbonylation [11 f], hydrosilylation [12 i], hydroboration [12 j], hydrocyanation [12 k], Michael addition [11 g, 12 l, 12 m], Diels–Alder reaction [11 h, 12 n] and the insertion of carbenes in C–H bonds [11 i, 12 p, 12 q, 38]. Cyclopropanation [11 i, 12 p, 12 q] and the isomerization of allylamines [12 s] are already used commercially for the manufacture of Cilastatin (one of the first industrial processes) [12 r], and citronellol and menthol (presently the second largest enantioselective process) [12 t] respectively.

3.3.1.3.4 Addition to C=O Groups

Reduction of Ketones

The hydrogenation of ketones using Rh and Ru diphosphine catalysts is the most versatile and efficient method for the synthesis of a large variety of chiral alcohols (see Structures **46–54** [11 a, 12 v]). While Rh diphosphine catalysts are often substrate-specific, several Ru/BIAR-type catalysts have a fairly broad scope. These catalysts are effective for the hydrogenation of functionalized ketones such as β-keto esters and 2-amino and 2-hydroxy ketones with high *ee* values and often reasonable TONs and TOFs. Due to the low activity of homogeneous catalysts, α-keto esters are still preferentially hydrogenated with heterogeneous cinchona-modified Pt catalysts [12 c]. New Ru/BINAP/chiral diamine catalysts have been developed which effectively hydrogenate aryl ketones (TON up to 2 400 000) and are also suitable for α,β-unsaturated ketones [11 a]. Unfunctionalized alkyl ketones are still a problem: *ee* values > 90 % have been reported for only a few rare cases [11 a]. Structures **46–54** are a selection of ketones for which industrial processes have been developed. Also here, tolerance for functional groups such as pyridines and C–Cl and C=C bonds is important.

Other reducing agents are of interest, especially for small-scale reductions and/ or when no hydrogenation facilities are available. The reduction with BH₃ adducts in presence of catalytic amounts of amino alcohols [12 zc] has already found some industrial applications, especially by PPG–Sipsy and Rhodia ChiRex (see Structures **55–57**). Transfer hydrogenation [11 a] using isopropanol as reducing agent shows some promise for the reduction of aryl ketones because very efficient Rh and Ru transfer hydrogenation catalysts with new bidentate N^N, N^O and P^N ligands have been developed in the last few years. Hydrosilylation [11 q] is of less

Table 4. State-of-the-art for the reduction of functionalized ketones (see Structures **46–57**).

Substrate/ Reducing agent	*ee* [%][a]	TON[a]	TOF [h⁻¹][a]	Preferred catalyst types[b]
RCOCHRCOX (X = OH, OR, R)/H₂	90–95	5000–50 000	2000–10 000	Rh/BIAR
RCOCOOR/H₂	90–95	1000–5000	10–500	Rh/NOP, Ru/BIAR, various
RCOCHRX/H₂ X = NHR, OH	90–95	1000–5000	100–500	Ru/BIAR, Rh/FERRO, Rh/NOP, Rh/DIOP
ArCOR/H₂	90–95	5000–20 000	500–10 000	Ru/BIAR-diamine
ArCOR/R₂CHOH	85–95	1000–5000	100–500	O^N, N^N, P^N
Ketone/BH₃	85–95	20–50	5–10	OXABOR

[a] Typical range for suitable substrate and optimized catalyst. [b] Structures **1–16**.

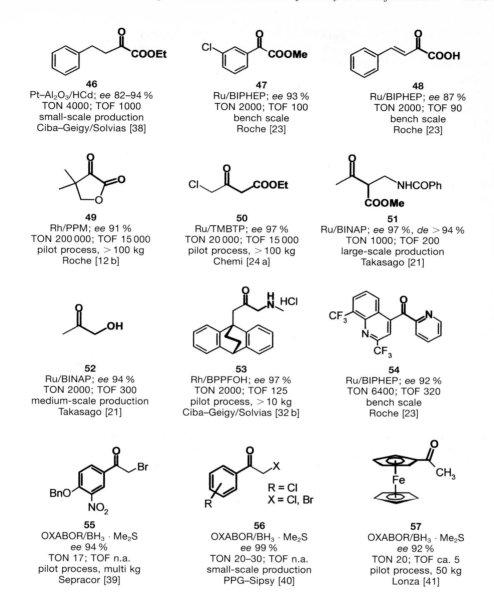

46
Pt–Al$_2$O$_3$/HCd; *ee* 82–94 %
TON 4000; TOF 1000
small-scale production
Ciba–Geigy/Solvias [38]

47
Ru/BIPHEP; *ee* 93 %
TON 2000; TOF 100
bench scale
Roche [23]

48
Ru/BIPHEP; *ee* 87 %
TON 2000; TOF 90
bench scale
Roche [23]

49
Rh/PPM; *ee* 91 %
TON 200 000; TOF 15 000
pilot process, > 100 kg
Roche [12 b]

50
Ru/TMBTP; *ee* 97 %
TON 20 000; TOF 15 000
pilot process, > 100 kg
Chemi [24 a]

51
Ru/BINAP; *ee* 97 %, *de* > 94 %
TON 1000; TOF 200
large-scale production
Takasago [21]

52
Ru/BINAP; *ee* 94 %
TON 2000; TOF 300
medium-scale production
Takasago [21]

53
Rh/BPPFOH; *ee* 97 %
TON 2000; TOF 125
pilot process, > 10 kg
Ciba–Geigy/Solvias [32 b]

54
Ru/BIPHEP; *ee* 92 %
TON 6400; TOF 320
bench scale
Roche [23]

55
OXABOR/BH$_3$ · Me$_2$S
ee 94 %
TON 17; TOF n.a.
pilot process, multi kg
Sepracor [39]

56
OXABOR/BH$_3$ · Me$_2$S
ee 99 %
TON 20–30; TOF n.a.
small-scale production
PPG–Sipsy [40]

57
OXABOR/BH$_3$ · Me$_2$S
ee 92 %
TON 20; TOF ca. 5
pilot process, 50 kg
Lonza [41]

interest since silanes are very expensive. Activities and productivities for some of these methods are often low and for large-scale processes the disposal of wastes from the stoichiometric reducing agent could be problematic.

Miscellaneous Addition Reactions to C=O Groups

Addition reactions to carbonyl groups are very important in synthetic methodology. Even though a wealth of catalysts with high enantioselectivity have been developed in recent years (Table 5), there are only a few commercial applications. Most have low to medium catalytic activity and productivity. The aldol reaction [11 k, 11 l, 12 w], ene reaction [11 m, 12 x] and hetero Diels–Alder reaction [11 h, 12 o] are catalyzed by early transition metal and lanthanide complexes. The addition reaction of ZnR_2 and similar reagents to aldehydes [12 y] in presence of catalytic amounts of amino alcohols or early transition metal complexes has few synthetic applications. Industrial syntheses have been reported for the gold–aldol reaction [11 k, 42] as an interesting approach to β-hydroxy amino acids and for the nitro–aldol reaction [12 z, 43] (eqs. (1) and (2)).

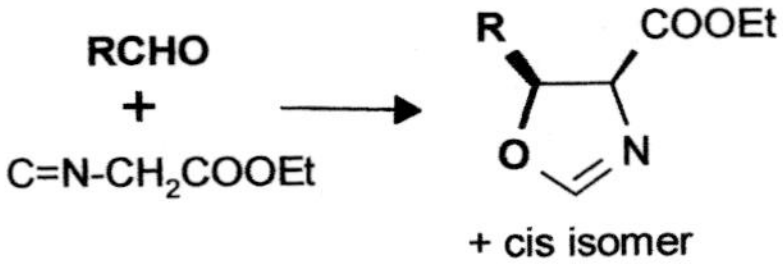

$$\begin{array}{c}\text{RCHO} \\ + \\ \text{C=N-CH}_2\text{COOEt}\end{array} \longrightarrow \begin{array}{c}\text{R} \qquad \text{COOEt}\end{array}$$

(1)

+ cis isomer

gold–aldol; Au/BPPFA
ee 85–90 %, *trans/cis* 20;
TON 100; TOF 5
bench scale, kg scale
Ciba–Geigy [42]

(2)

matched case

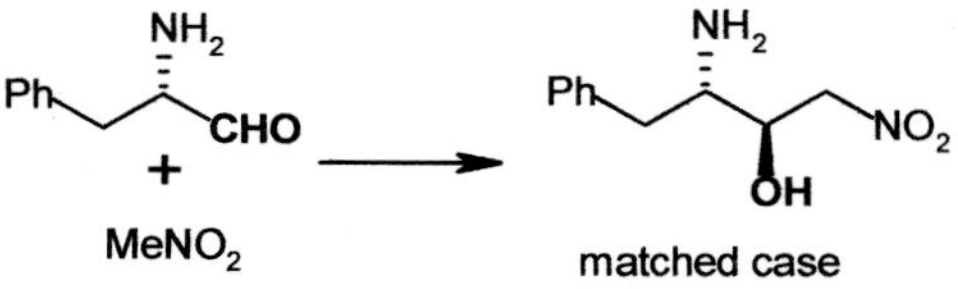

nitro–aldol; LaLi/BINOL complex
ee 96 %, *de* 98 %; TON 30; TOF < 1
small scale process
Kaneka [43]

Table 5. State-of-the-art for addition reactions to C=O (see eqs. (1) and (2)).

Reaction	ee [%][a]	TON[a]	TOF [h^{-1}][a]	Preferred catalyst types[b]
Aldol reaction	90–95	5–20	1–10	Ln/BINOL, Ag/BIAR, Cu/OXAZ
Ene reaction	90–95	5–20	1–10	Ti/BINOL
Addition of MR to RCHO	90–95	5–100	1–20	N^O, O^O, N^N
Hetero Diels–Alder	85.95	10–50	2–10	Cu/OXAZ, N^O, O^O, N^N

[a] Typical range for suitable substrate and optimized catalyst. [b] Structures **1–16**.

3.3.1.3.5 Reduction of and Addition to C=N Groups

Although chiral amines are important intermediates for biologically active compounds, the asymmetric hydrogenation of C=N has been investigated less systematically than that of C=C and C=O groups [11 a, 12 za]. In recent years various Rh and Ir diphosphine complexes were developed with reasonable enantioselectivities. Unfortunately, Rh complexes often have relatively low catalyst activities and productivities and Ir complexes tend to deactivate. The hydrogenation of acyl hydrazones with Rh/DUPHOS with *ee* values up to 95 % and a Ti/EBTHI catalyst for cyclic imines (*ee* > 98 %) have some synthetic potential, but the Ti catalysts unfortunately have a low functional group tolerance and very poor catalytic activity. Good to high enantioselectivities can be achieved with transfer hydrogenation and BH$_3$ reduction with medium to very low catalyst activities. With the exception of Structures **58–60**, the metolachlor process carried out by Ciba–Geigy/Syngenta (with a volume of > 10 000 t/y the largest known production process) [44], and a pilot process developed by Lonza [25], no industrial applications are known. Recently, the first example of a reductive alkylation reaction with high TON and TOF values has been described, an interesting variant from an industrial point of view [45].

58
hydrogenation
Ir/JOSIPHOS; *ee* 80 %
TON 2 000 000; TOF 400 000
very large-scale production
Ciba–Geigy/Syngenta/Solvias [44]

59
hydrogenation
Ir/JOSIPHOS; *ee* 90 %
TON 1500; TOF n.a.
pilot process, > 100 kg
Lonza (Solvias) [25]

60
hydrophosphonylation
YbK/BINOL; *ee* 92–96 %
TON 20; TOF < 1
small-scale production
Hokko Chemical Co. [46]

Several addition reactions to C=N groups have been developed in recent years with a high synthetic potential but with no commercial use so far [12 zb]. The addition reaction of (MeO)$_2$PHO to cyclic imines (**58–60**), an interesting method for the preparation of *α*-amino phosphonic acids, seems to be an exception [46]. While *ee* values of the heterobimetallic catalyst (cf. Section 3.1.5) are very high, TON and TOF values are relatively low.

3.3.1.3.6 Miscellaneous Transformations

Even though most of the reactions in Table 6 form new C–C bonds asymmetrically, none has been developed to really technical maturity, major problems being (as usual) catalyst activities and productivities, and possibly also the synthetic scope. The Ni/P^N-catalyzed cross-coupling reactions [11 r] tolerate only

Table 6. State-of-the-art for miscellaneous transformations (see Structures **61** and **62**).

Reaction	ee [%][a]	TON[a]	TOF [h^{-1}][a]	Preferred catalyst types[b]
Cross-coupling	80–90	500–200	2–20	Ni/P^N
Allylic substitution	85–95	50–1000	20–100	Pd/PNNP, Pd/P^OXAZ, Pd/OXAZOL, various
Heck	80–95	10–100	1–10	Pd/BIAR, Pd/P^OXAZ
Sulfide oxidation	80–95	2–20	1–5	Ti/TART

[a] Typical range for suitable substrate and optimized catalyst. [b] Structures **1–16**.

a few functional groups. Nucleophilic allylic substitution reactions [11 n] with *C*- and *N*-nucleophiles catalyzed by Pd/P^P, Pd/P^N, and Pd/N^N complexes have recently been applied not just in model studies but also in synthetic applications. The asymmetric Heck reaction is still in an exploratory phase even though some syntheses of natural products have been reported [11 o]. The oxidation of aromatic sulfides [11 p] using Ti/TART catalysts exhibits good enantioselectivities but usually very low catalytic activities; nevertheless two industrial applications are on record (Structures **61** and **62**). One of them is being used to make the chiral switch for one of the best selling antiulcer drugs [24 b, 47].

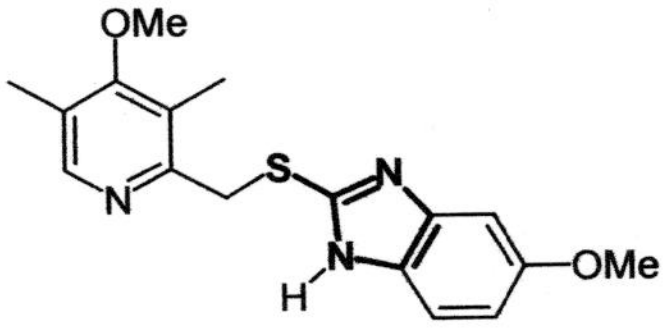

61
Ti/TART; *ee* 92–93 %
TON 3–4; TOF 3–4
medium-scale production
AstraZeneca [24 b, 47]

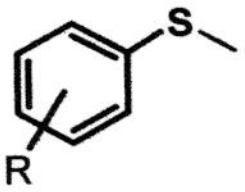

62
Ti/TART; *ee* 98 %
TON n.a.; TOF n.a.
pilot process, < 100 kg
Lonza [17]

3.3.1.4   Conclusions and Prospects

Since the publication of the first edition of this book in 1996, the industrial application of enantioselective homogeneous catalysts has made significant progress. The list of processes suitable for the manufacture of enantiomerically enriched compounds is compiled in [6]. Few have actually been implemented as production processes and run on a regular basis but there is every reason to assume that this technology is here to stay. The number of commercial applications will increase in the near future because development chemists who realize technical processes will be more aware of the potential of enantioselective catalysis. More and more specialized technology companies such as Solvias, ChiRex, or ChiroTech are developing the know-how and experience to use enantioselective catalytic processes and to produce technical quantities of the chiral ligands. Jacobsen [12 zd] predicted the following trends for the next few years: design of new ligands (e. g., Section 3.1.10), catalysts, and transformations with good synthetic potential, in many instances by applying combinatorial approaches (cf. Section 3.1.3); the development of more practical catalyst systems, i. e., with higher activity, productivity, and robustness in part via high-troughput experimentation; and, finally, a deeper understanding of the underlying mechanisms that will help to make catalyst design more rational (cf. Sections 3.1.3 and 3.1.4).

References

[1] For periodic updates on chiral pharmaceuticals, see: S. T. Stinson, *Chem. Eng. News* **1998**, September 21, 83; S. T. Stinson, *Chem. Eng. News* **1999**, November 22, S. T. Stinson, *Chem. Eng. News* **2001**, May 14, 45.

[2] G. M. Ramos Tombo, H. U. Blaser, in *Pesticide Chemistry and Bioscience* (Eds.: G. T. Brooks, T. R. Roberts), Royal Society of Chemistry, Cambridge, **1999**, p. 33 and references cited therein.

[3] J. Crosby, in *Chirality in Industry I* (Eds.: A. N. Collins, G. N. Sheldrake, J. Crosby), John Wiley, Chichester, **1992**, p. 1.

[4] *Chirality in Industry II* (Eds.: A. N. Collins, G. N. Sheldrake, J. Crosby), John Wiley, Chichester, **1997**: (a) for an overview, see A. Bruggink, p. 81; (b) D. Pauluth, A. E. F. Wächter, p. 263; (c) W. P. Shum, M. J. Cannarsa, p. 363; (d) B. A. Astleford, L. O. Weigel, p. 99; (e) J. C. Caille, M. Bulliard, B. Laboue, p. 391.

[5] J. M. Keith, J. F. Larrow, E. N. Jacobsen, *Adv. Synth. Catal.* **2001**, *343*, 5.

[6] For a recent compilation of known industrial processes see: H. U. Blaser, F. Spindler, M. Studer, *Appl. Catal.* A: General **2001**, *221*, 119.

[7] R. A. Sheldon, *Chirotechnology*, Marcel Decker, New York, **1993**.

[8] R. Noyori, *Chemtech* **1992**, *22*, 366.

[9] E. Polastro, in *Chiral Reaction in Heterogeneous Catalysis* (Eds.: G. Jannes, V. Dubois), Plenum Press , New York, **1995**, p. 5.

[10] *Asymmetric Synthesis Vol. 5* (Ed.: J. D. Morrison), Academic Press, New York, **1985**.

[11] *Catalytic Asymmetric Synthesis* (Ed.: I. Ojima), Wiley-VCH, Weinheim, **2000**: (a) T. Ohkuma, M. Kitamura, R. Noyori, p. 1; (b) R. A. Johnson, K. B. Sharpless, p. 231; (c) T. Katsuki, p. 287; (d) R. A. Johnson, K. B. Sharpless, p. 357; (e) C. Bolm, J. P. Hildebrand, K. Muniz, p. 399; (f) K. Nozaki, I. Ojima, p. 429; (g) M. Kanai, M. Shibasaki, p. 569; (h) K. Maruoka, p. 467; (i) M. P. Doyle, p. 191; (k) M. Sawamura, Y. Ito, p. 493; (l) E. M. Carreira, p. 513; (m) M. Ogasawara, T. Hayashi, p. 651; (n) B. M. Trost, C. Lee, p. 593; (o) Y. Donde, L. E. Overmann, p. 675; (p) H. B. Kagan, p. 327; (q) K. Mikami, T. Nakai, p. 543; (r) H. Nishiyama, K. Itoh, p. 111; and references cited in these reviews.

[12] *Comprehensive Asymmetric Catalysis* (Eds.: E. N. Jacobsen, H. Yamamoto, A. Pfaltz), Springer, Berlin, **1999**: (a) J. M. Brown, p. 121; (b) R. Schmid, M. Scalone, p. 1439; (c) H. U. Blaser, M. Studer, p. 1353; (d) T. Katsuki, p. 621; (e) E. N. Jacobsen, M. H. Wu, p. 649; (f) I. E. Marko, J. S. Svendsen, p. 713; (g) T. Katsuki, p. 791; (h) E. N. Jacobsen, p. 607; (i) T. Hayashi, p. 319; (j) T. Hayashi, p. 351; (k) T. V. RajanBabu, A. L. Casalnuovo, p. 367; (l) K. Tomioka, Y. Nagaoka, p. 1105; (m) M. Yamaguchi, p. 1121; (n) D. A. Evans, J. S. Johnson, p. 1177; (o) T.Ooi, K. Maruoka, p. 1237; (p) A. Pfaltz,

p. 513; (q) K. M. Lydon, M. A. McKervey, p. 539; (r) T. Aratani, p. 1451; (s) S. Akutagawa, p. 813; (t) S. Akutagawa, p. 1461; (u) E. Jacobsen, M. H. Wu, p. 1309; (v) T. Ohkuma, R. Noyori, p. 199; w); E. M. Carreira, p. 997; (x) K. Mikami, M. Terada, p. 1143;
(y) K. Soai, T. Shibata, p. 911; (z) M. Shibasaki, H. Gröger, p. 1075; (za) H. U. Blaser, F.
Spindler, p. 247; (zb) S. E. Denmark, O. J.-C. Nicaise, p. 923; (zc) S. Itsuno, p. 289; (zd)
E. N. Jacobsen, p. 1473; and references cited in these reviews.

[13] A. S. Bommarius, M. Schwarm, K. Drauz, *Chimia* **2001**, *55*, 50.

[14] W. S. Knowles, *Chem. Ind. (Dekker)* **1996**, *68*, 141; W. S. Knowles, *Acc. Chem. Res.*
1983, *16*, 106 and *J. Chem. Ed.* **1986**, *63*, 222.

[15] M. J. Burk (ChiroTech), personal communication.

[16] H. U. Blaser, F. Spindler, *Topics Catal.* **1997**, *4*, 275.

[17] W. Brieden, *Proc. ChiraSource '99 Symposium 1999,* The Catalyst Group, Spring
House, USA, **1999**; W. Brieden (Lonza AG), personal communication.

[18] *Proc. ChiraTech '97 Symposium 1997,* The Catalyst Group, Spring House, USA, **1997**:
(a) M. Scalone, R. Schmid, E. A. Broger, W. Burkart, M. Cereghetti, Y. Crameri,
J. Foricher, M. Henning, F. Kienzle, F. Montavon, G. Schoettel, D. Tesauro, S. Wang,
R. Zell, U. Zutter; (b) H. Jendralla.

[19] *Proc. ChiraTech '96 Symposium 1996,* The Catalyst Group, Spring House, USA, **1996**:
(a) G. Beck; (b) A. A. Smith.

[20] M. J. Burk, F. Bienewald, M. Harris, A. Zanotti-Gerosa, *Angew. Chem. Int. Ed.* **1998**,
37, 1931.

[21] S. Akutagawa, *Appl. Catal.* **1995**, *128*, 171; H. Kumobayashi, *Recl. Trav. Chim. Pays-
Bas* **1996**, *115*, 201.

[22] E. A. Broger (Roche), *Book of Abstracts of EuropaCat I 1993,* and personal communication.

[23] R. Schmid, E. A. Broger, *Proc. Chiral Europe '94 Symposium,* Spring Innovations,
Stockport, UK, **1994**, p. 79.

[24] *Proc. ChiraSource 2000 Symposium, 2000,* The Catalyst Group, Spring House, USA,
2000: (a) T. Benincori, S. Rizzo, F. Sannicolo, O. Piccolo; (b) H. J. Federsel.

[25] R. Imwinkelried, *Chimia* **1997**, *51*, 300.

[26] V. Rautenstrauch *Proc. Int. Symposium on Chirality, 1999,* Spring Innovations, Stockport, UK, **1999**, p. 204.

[27] A. Lightfoot, P. Schnider, A. Pfaltz, *Angew. Chem., Int. Ed.* **1998**, *37*, 2897.

[28] W. Leitner, J. M. Brown, H. Brunner, *J. Am. Chem. Soc.* **1993**, *115*, 152; M. Saburi,
M. Ogasawara, T. Takahashi,Y. Uchida, *Tetrahedron Lett.* **1992**, *33*, 5783; and references
therein.

[29] B. K. Sharpless, *Janssen Chem. Acta* **1988**, *6*, 3.

[30] D. Meunier, A. Piechaczyk, A. de Mallmann, J.-M. Basset, Angew. Chem. **1999**, *111*,
3738.

[31] See information given in *www.chirex.com* (technology).

[32] *Process Chemistry in the Pharmaceutical Industry* (Ed.: K. G. Gadamasetti), Marcel
Dekker, New York, **1999**: (a) C. H. Senanayake, E. N. Jacobsen, p. 347; (b) H. U. Blaser,
R. Gamboni, G. Rihs, G. Sedelmeier, E. Schaub, E. Schmidt, B. Schmitz, F. Spindler,
Hj. Wetter, p. 189.

[33] C. E. Song, E. J. Roh, *Chem. Commun.* **2000**, 837.

[34] J. M. Keith, J. F. Larrow, E. N. Jacobsen, *Adv. Synth. Catal.* **2001**, *1*, 5.

[35] M. Porter, J. Skidmore, *Chem. Commun.* **2000**, 1215 and references cited therein.

[36] C. Döbler, G. M. Mehltretter, U. Sundermeier, M. Beller, *J. Am. Chem. Soc.* **2000**, *122*,
10289.

[37] M. P. Doyle, M. N. Protopopova, *Proc. Chiral USA '97 Symposium,* Spring Innovations,
Stockport, UK, **1997**, p. 11.

[38] H. U. Blaser, M. Studer, *Chirality* **1999**, *11*, 459.

[39] R. Hett, Q. K. Fang, Y. Gao, S. A. Wald, C. H. Senanayake, *Org. Proc. Res. Dev.* **1998**, *2*, 96; A. K. Ghosh, S. Fidanze, C. H. Senanayake, *Synthesis* **1998**, 937.

[40] J. C. Caille (PPG–Sipsy), personal communication.

[41] W. Brieden, WO 9616971 (1994) (assigned to Lonza AG) and W. Brieden (Lonza), personal communication.

[42] A. Togni, S. D. Pastor, G. Rihs, *Helv. Chim. Acta* **1989**, *72*, 1471.

[43] H. Sasai, W.-S. Kim, T. Suzuki, M. Shibasaki, *Tetrahedron Lett.* **1994**, *35*, 6123; M. Shibasaki (University of Tokyo), personal communication.

[44] H. U. Blaser, H. P. Buser, K. Coers, R. Hanreich, H. P. Jalett, E. Jelsch, B. Pugin, H. D. Schneider, F. Spindler, A. Wegmann, *Chimia* **1999**, *53*, 275.

[45] H. U Blaser, H. P. Buser, H. P. Jalett, B. Pugin, F. Spindler, *Synlett* **1999**, 867.

[46] H. Gröger, Y. Saida, H. Sasai, K. Yamaguchi, J. Martens, M. Shibasaki, *J. Am. Chem. Soc.* **1998**, *120*, 3089; M. Shibasaki (University of Tokyo), personal communication.

[47] H. Cotton, T. Elebring, M. Larsson, L. Li, H. Sörensen, S. von Unge, *Tetrahedron: Asymmetry* **2000**, *11*, 8319; H. J. Federsel (AstraZeneca), personal communication.

3.3.2   Diols via Catalytic Dihydroxylation

Matthias Beller, K. Barry Sharpless

3.3.2.1   Introduction

The oxidative functionalization of olefins is of major importance for both organic synthesis and the industrial production of bulk and fine chemicals. Among the different oxidation products of olefins, 1,2-diols are used in a wide variety of applications. Ethylene glycol and propylene glycol are produced on a multi-million ton scale per annum, due to their importance as polyester monomers and anti-freeze agents [1]. A number of 1,2-diols such as 2,3-dimethyl-2,3-butanediol, 1,2-octanediol, 1,2-hexanediol, 1,2-pentanediol, and 1,2- and 2,3-butanediol are of interest for the fine chemicals industry. In addition, chiral 1,2-diols are employed as intermediates for pharmaceuticals and agrochemicals. At present 1,2-diols, e. g., 2,3-dimethyl-2,3-butanediol, 1,2-pentanediol and higher nonfunctionalized glycols obtained thanks to the availability of cheap terminal olefins (SHOP process; cf. Section 2.3.1.3), have so far been manufactured industrially by the reaction of alkenes with organic peracids via the corresponding epoxides [1]. Usually performic acid or peracetic acid produced *in situ* by mixing hydrogen peroxide with the carboxylic acid have been employed as oxidants.

Besides stoichiometric epoxidation and subsequent hydrolysis to diols, metal-catalyzed methods for converting olefins to glycols are also known in the literature. The classical method utilizes hydrogen peroxide in the presence of catalytic amounts of acidic metal oxides (Milas reagents) [2]. Typically, strong oxidants such as osmium [3] and ruthenium tetroxides [4], permanganate [5], and chro-

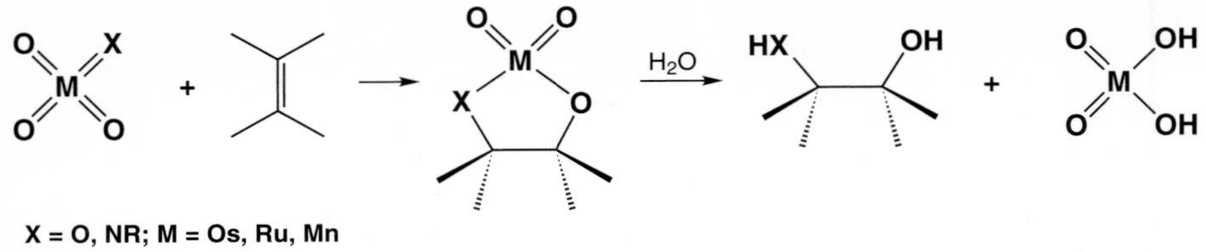

X = O, NR; M = Os, Ru, Mn

Scheme 1. General representation of dihydroxylation and related reactions.

mium(VI) are used as oxometals. The first three reagents are considered to effect directly the addition of two hydroxy groups to double bonds. The intermediate cyclic esters could be either hydrolyzed to glycols or undergo C–C bond cleavage to carbonyl compounds. A simplified representation of dihydroxylation and related oxyamination reactions is shown in Scheme 1.

As an oxometal component, osmium tetroxide is the most reliable reagent on the laboratory scale to produce *cis*-diols. Ruthenium tetroxide in the presence of NaIO$_4$ effects oxidative cleavage of olefins [4], but has been successfully employed for so-called *lightning dihydroxylation* reactions using a two-phase medium [6].

Because most olefins are prochiral starting materials, the dihydroxylation reaction creates one or two new stereogenic centers in the products. Since the discovery of the first stoichiometric asymmetric dihydroxylations [7], catalytic versions with considerable improvements in both scope and enantioselectivity have been developed [8]. From the standpoint of general applicability, scope, and limitations, the osmium-catalyzed asymmetric dihydroxylation (AD) of alkenes has reached a level of effectiveness which is unique among asymmetric catalytic methods. As there are recent reviews in this field [9], this section is primarily oriented toward a summary of aspects of fundamental understanding and interesting practical application of catalytic dihydroxylations.

3.3.2.2   History and General Features of Osmium Catalyzed Dihydroxylation Reactions

The dihydroxylation of olefins with osmium compounds has been known since the first work of Philipps in 1894 [10] and was also pioneered by Criegee in the 1930s using OsO$_4$ stoichiometrically [11]. The chief drawback of using stoichiometric amounts of expensive OsO$_4$ has been overcome by inclusion of a co-oxidant in the reaction which reoxidizes the osmium(VI) species to the osmium tetroxide oxidation level. This allows for the use of the metal in catalytic amounts. Historically, chlorates [12] and hydrogen peroxide in *t*-butanol [13] were first applied as co-oxidants. With hydrogen peroxide the reaction is reported to proceed via formation of peroxoosmic acid, H$_2$OsO$_6$, which causes cleavage of intermediate diols to carbonyl compounds. However, Bäckvall and co-workers were recently able to improve the H_2O_2 reoxidation process significantly by using *N*-methylmorpholine together with flavin as co-catalysts in the presence of hydrogen peroxide [58].

Other reoxidants which minimize overoxidation are *t*-butyl hydroperoxide in the presence of Et_4NOH [4], tertiary amine oxides, and most importantly *N*-methylmorpholine *N*-oxide (NMO) (Upjohn process) [14], although for tri- and particularly tetrasubstituted alkenes as substrates, trimethylaminoxide is superior to NMO [14 c]. The introduction of potassium hexacyanoferrate(III) in the presence of potassium carbonate [15] substantially improved the selectivities in chiral dihydroxylations [16], although it was first reported as a co-oxidant in 1975 [17]. Industrial efforts led to an electrochemical oxidation of potassium ferrocyanide to ferricyanide in order to use electricity as the actual co-oxidant [18].

Oxygen is the most economical as well as the most environmentally friendly oxidation reagent known. However until very recently only a few investigations using O_2 in dihydroxylation reactions had been carried out. Initially it was demonstrated by several groups that in the presence of OsO_4 and oxygen mainly nonselective oxidation reactions take place [19]. Krief et al. successfully designed a reaction system consisting of oxygen with catalytic amounts of OsO_4 and of selenides for the dihydroxylation of *α*-methylstyrene under irradiation with visible light [20]. More recently Beller and co-workers reported that the Os-catalyzed dihydroxylation of aliphatic and aromatic olefins proceeds efficiently in the presence of dioxygen at ambient conditions [59]. The new dihydroxylation procedure constitutes a significant advance compared with other reoxidation procedures. The yield of the diol remains good to very good (87–96 %), independently of the oxidant used. The dihydroxylation process with oxygen is clearly the most ecologically favorable procedure, when the production of waste from a stoichiometric reoxidant is considered. In the presence of $K_3[Fe(CN)_6]$ approximately 8.1 kg of iron salts per kg of product are formed. However, in the case of the Krief or Bäckvall procedure significant amounts of by-products also arise due to the large amounts of co-catalysts and co-oxidants used. It should be noted that only salts and by-products formed from the oxidant have been included in the calculation. Other waste products have not been considered.

With regard to the price and safety issues it is important to note that it is also possible to use air rather than pure oxygen gas as stoichiometric oxidant [60].

Considering the chemoselectivity of the process and that olefins are the starting materials, no other known organic reaction combines such enormous scope with such high selectivity. Although some electron-deficient olefins have long been described as "bad" substrates, Herrmann et al. demonstrated that even perfluorinated olefins could be efficiently dihydroxylated [21].

In general, dihydroxylations are carried out in mixtures of aqueous and organic solvents, although catalytic osmylations have been performed under virtually anhydrous conditions in toluene [21] or dichloromethane [22]. In combination with water, organic solvents such as acetone, *t*-butanol, methyl *t*-butyl ether, and others are employed.

It had already been recognized by Criegee that addition of certain ligands, e. g., amines, greatly accelerates the rate of formation of osmium(VI) ester complexes

[23]. This, together with the finding of Hentges and Sharpless [7] that stoichiometric amounts of chiral ligands derived from cinchona alkaloids can transfer
chirality from the catalyst to olefins, has opened the door for the development
of catalytic asymmetric methods. An important advance regarding the reuse of
the expensive osmium catalyst has been reported by Jacobs et al. [61]. They
immobilized OsO_4 elegantly to a tetrasubstituted olefin which is covalently linked
to a silica support. The Os^{VI} monoglycolate complex is then oxidized to a Os^{VIII}
glycolate complex which is able to react with additional olefins. Due to the
much slower hydrolysis of the tetrasubstituted glycolate, the catalyst can be
recycled.

An interesting offshoot of the work on osmium-catalyzed dihydroxylations is
vicinal hydroxyamination [24]. Here, imido analogs of OsO_4 react with olefins
to produce β-aminoalcohols by a *cis*-addition process. The oxyamination reaction
can be made catalytic in OsO_4 by employing chloramine salts of arylsulfonamides
($ArSO_2NClNa$) or carbamates.

3.3.2.3   Mechanism of Osmium-Catalyzed Dihydroxylations

In the last decade the mechanism of the osmium-catalyzed dihydroxylation was
discussed extensively. Originally, Böseken [25] suggested that the reaction
proceeds by a thermally allowed concerted [3 + 2] cycloaddition leading directly
to the monoglycolate ester, while Sharpless et al. [26] proposed an alternative
reversible [2 + 2] cycloaddition leading to a metallaoxetane intermediate which
undergoes irreversible reductive insertionof the Os–C bond into an Os=O bond
leading to the monoglycolate ester (Scheme 2).

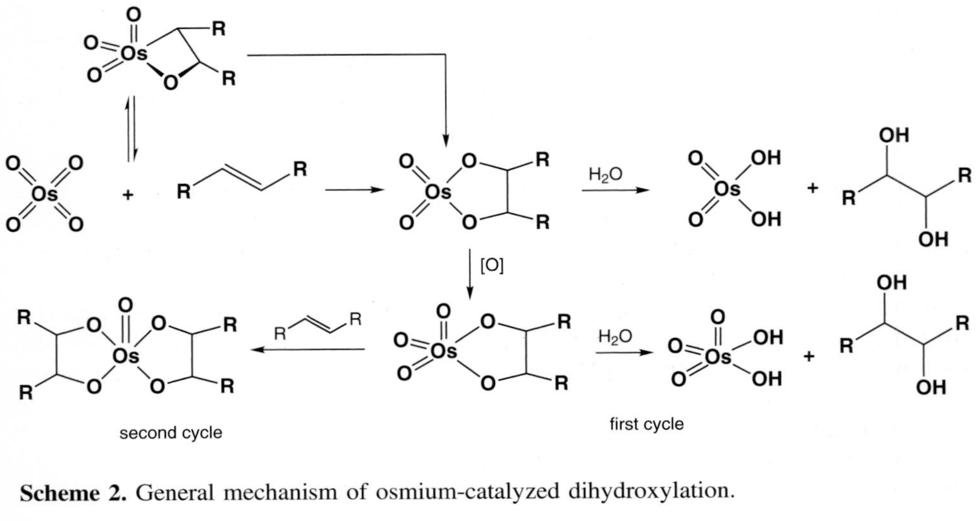

Scheme 2. General mechanism of osmium-catalyzed dihydroxylation.

Recent theoretical investigations clearly favor the [3 + 2] mechanism [27, 62]. The calculation of the respective transition states using DFT methods show significantly lower activation barriers for the [3 + 2] addition compared with the [2 + 2] reaction path. Subsequently, these results were also supported by the theoretical and experimental determination of the kinetic isotope effect of the AD reaction [28].

Depending on the reaction media and the substrates, the rate-determining step in catalytic dihydroxylations can be either the attack of the OsO_4 on the olefin [28], or oxidation of the Os^{VI} glycolate complex to the Os^{VIII} complex [29], or in cases of bulky olefins the hydrolysis of the osmium glycolate complexes. The problem of hydrolysis could be overcome by the addition of methyl sulfonamide [9 b, 30] or sometimes tetraethylammonium acetate. In the presence of one equivalent of $CH_3SO_2NH_2$ dihydroxylations could be as much as 50 times faster. Alternatively, the hydrolysis of sterically hindered osmium glycolates can be performed more efficiently under controlled pH conditions. By using buffered solutions (pH 11–13) or by applying an autotitrator the dihydroxylation can be significantly speeded up [63].

Under homogeneous conditions with the co-oxidant in the same phase as the intermediate osmium(VI) glycolate, two competitive catalytic cycles can operate, involving either direct hydrolysis of the reoxidized osmate(VIII)–glycolate complex or its reaction with a second olefin to give an osmium(VI) bisglycolate ("second cycle") [31]. For enantioselective dihydroxylations the low selectivity-generating second cycle could be completely suppressed by the use of $K_3Fe(CN)_6$ in the presence of K_2CO_3 [16].

Of particular mechanistic interest in AD is the question of how chirality is transmitted from the chiral alkaloid ligand to the Os^{VI} glycolate complex [33].

3.3.2.4 Scope and Limitation of Asymmetric Dihydroxylation

The enormous synthetic utility of AD depends on the one hand on the broad applicability of the osmium-catalyzed dihydroxylation for nearly every class of olefins, and on the other hand on the high selectivities which can be reached with optimized catalyst–ligand systems.

In the past it has been shown that AD is responsive to substantial enantioselectivity improvement through ligand variation. Chiral auxiliaries used for effecting asymmetric dihydroxylation are mainly cinchona alkaloid derivatives [8], some monodentate amine ligands [34], and a variety of bidentate chiral diamines [35] (Structures **1–10**). Complexes derived from osmium tetroxide with diamines do not undergo catalytic turnover because diamines form very stable chelate complexes with the Os^{VI} glycolate products, whereas dihydroquinidine and dihydroquinine derivatives induce very effective catalysis. The Sharpless group has undertaken a systematic ligand optimization study in recent years [8, 9 b]. It soon became clear that the binding constant of the ligand to OsO_4 is important to deliver selectivity. Consequently, quinuclidine derivatives which show much higher affinity to OsO$_4$ (e. g., compared with pyridine) were used. Interestingly, nature provides quinine and quinidine, "pseudoenantiomeric" cinchona alkaloids, as starting materials for ligand variation. So far more than 500 cinchona alkaloid derivatives have been tested. Other groups described minor modifications of the ligands originally discovered, but the corresponding catalyst systems showed no real methodological improvements [36].

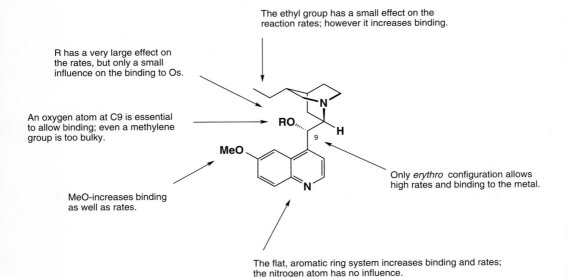

1
Dihydroquinidine (DHQD)

2
Dihydroquinine (DHQ)

Sharpless et al. [9]

3
Diamines [30]

Snyder, Ito, Corey, Fuji, Tomioka, Hanessian, Hirama et al.

X-ray analysis of osmium tetroxide–cinchona alkaloid complexes [37] demonstrated that the chiral center in the alkaloid ligand is quite remote from the oxo ligand. Therefore it is unlikely that the complex itself is responsible for the

The ethyl group has a small effect on the reaction rates; however it increases binding.

R has a very large effect on the rates, but only a small influence on the binding to Os.

An oxygen atom at C9 is essential to allow binding; even a methylene group is too bulky.

Only *erythro* configuration allows high rates and binding to the metal.

MeO-increases binding as well as rates.

The flat, aromatic ring system increases binding and rates; the nitrogen atom has no influence.

Figure 1. Influence of structural features of the cinchona ligands on binding and reaction rates.

high enantioselectivities observed in the addition to alkenes. Nevertheless, the alkaloid core is ideally set up to ensure high rates, binding, and solubility. It soon became evident that the rates and enantioselectivity are influenced considerably by the nature of the O-9 substituent, while binding to OsO_4 is almost independent of that substituent. Variations in the alkaloid backbone have only relatively minor effects. The relationship between ligand structure, binding, and reaction rate is generalized in Figure 1.

Careful ligand screening has led to three different ligand classes based on cinchona alkaloids, which taken together are very effective catalysts for nearly every olefin (*vide infra*) with the six possible substitution patterns. This grouping is shown in Figure 2.

The phthalazine (PHAL) (**4**) [38] and diphenylpyrimidine (PYR) (**5**) [39] ligands contain two independent alkaloid units, attached to a heterocyclic spacer, while the indolinyl carbamyl (IND) (**6**) [40] ligand is attached to only one alkaloid. PHAL ligands are recommended for 1,1- and 1,2-*trans* disubstituted

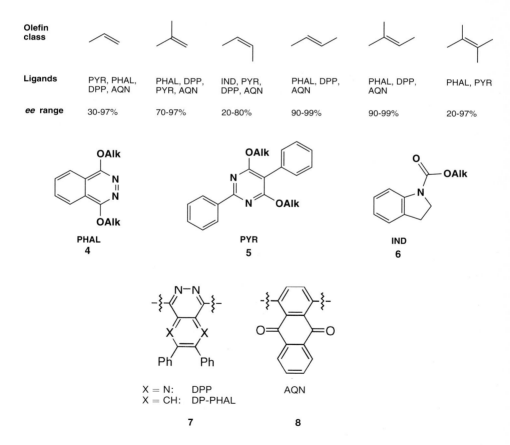

Figure 2. Ligand types (Structures **4-6**) for the different olefin classes.

as well as trisubstituted olefins, but are especially well suited to accommodate olefins with flat aromatic substituents. PYR ligands are the ligands of choice for monosubstituted terminal olefins, especially aliphatic olefins. Although *cis*-olefins were for a long time resistant to selective dihydroxylation, the discovery of the IND ligands [40] made it possible for the first time to dihydroxylate several *cis*-olefins stereoselectively in a practical manner. The resulting stereoselectivity can be predicted in general with remarkable success by reference to a mnemonic device. In Figure 3 the substrate is aligned horizontally in this device in such a way as to minimize steric interactions. The hydroxylation then takes place predominantly from one face, determined by the choice of ligand.

A common feature of all osmylations under the standard catalytic AD conditions in the presence of $K_3Fe(CN)_6$ is the rate enhancement in the presence of the aforementioned ligands [41]. As an example of this ligand accelerated catalysis (LAC), rate improvements up to 10^4 were observed for the reaction of OsO_4 with 2-vinylnaphthalene with $(DHQD)_2PHAL$ as ligand. This ligand acceleration phenomenon supports extremely efficient catalysis, as shown in the homogeneous dihydroxylation (acetone/water, NMO) of 2-vinylnaphthalene in the presence of $(DHQD)_2PHAL$ (**1/4**) as ligand. Here, a turnover number of 3000 has been determined [9 b].

Interestingly, the rate acceleration is significantly greater with the alkaloid derivative than with simple quinuclidine – a hint that the rate accelerations are not directly related to the ground-state binding energy between the ligand and OsO_4, which is higher for quinuclidine. Instead, "enzyme-like" noncovalent binding in a binding pocket of the ligand seems to be the cause of the substantial rate accelerations [42] and the ability to deliver high enantioselectivities without the need for binding tether groups in the substrate. Thus, contrary to most developed catalytic asymmetric methods, even totally unfunctionalized substrates yield prod-

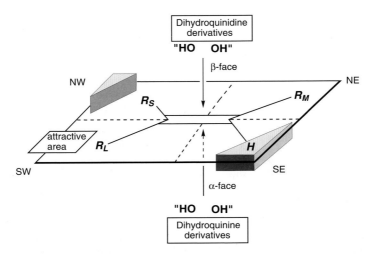

Figure 3. A mnemonic device for predicting enantiofacial selection.

Table 1. AD of nonfunctionalized olefins: selectivities [%] [a]

Olefin	Ligand		
	PHAL (DHQD) [1/4]	PYR (DHQD) [1/5]	IND (DHQD) [1/6]
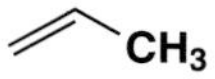 CH₃ olefin	36 (R)	49 (R)	
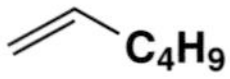 C₄H₉ olefin	80 (R)	89 (R)	
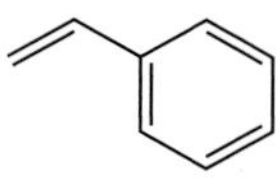 phenyl vinyl olefin	97 (R)	80 (R)	
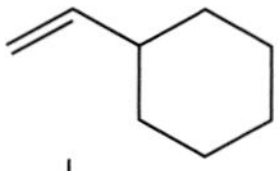 cyclohexyl allyl olefin	88 (R)	96 (R)	
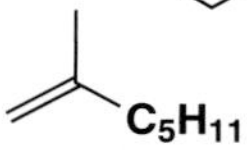 C₅H₁₁ isopropenyl olefin	78 (R)	76 (R)	
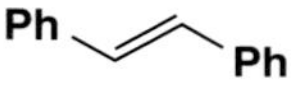 Ph—CH=CH—Ph	> 99 (R)		
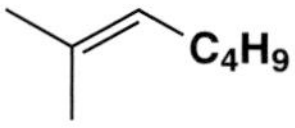 C₄H₉ trisubstituted olefin	98 (R)		
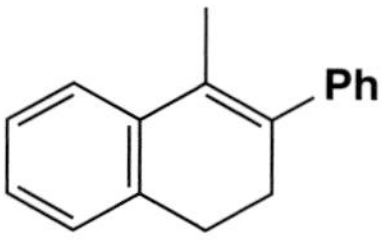 dihydronaphthalene Ph olefin	83 (R,R)	85 (R,R)	
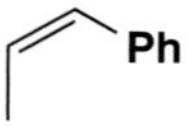 Ph allyl olefin			72 (1R,2S)
cyclohexyl cis olefin			56 (1R,2S)

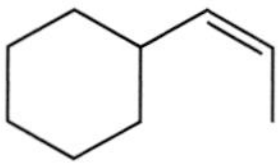

[a] For original references see [9 b].

ucts with reasonable to very high selectivities. For illustration, Table 1 shows selected examples from the large numbers of simple olefins which have been used so far [9 b].

Moreover, and more importantly for practical purposes, functionalized olefins with nearly all kinds of substituents attached to the olefin can be dihydroxylated. Thus, acrylic acid esters, unsaturated amides and ketones, dienes, enynes, vinyl silanes, acrolein acetals, and allylic halides, as well as allylic ethers and sulfur-

Table 2. AD of functionalized olefins: selectivities [%] [a]

Olefin	PHAL (DHQD) [1/4]	PYR (DHQD) [1/5]	IND (DHQD) [1/6]
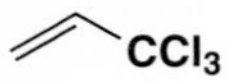 (olefin) CCl₃	70 (*S*)	86 (*S*)	
 (olefin) SiMe₃	46 (*S*)	88 (*R*)	
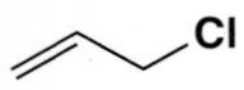 (olefin) Cl	63 (*S*)	53 (*S*)	
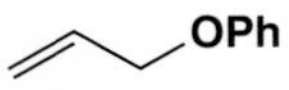 (olefin) OPh	88 (*S*)	43 (*S*)	
(olefin) CO₂Et	67 (*S*)		
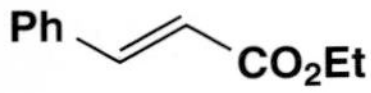 Ph (olefin) CO₂Et	97 (2*S*,3*R*)		
Ph (olefin) Ph, OMe	99 (*R*)		
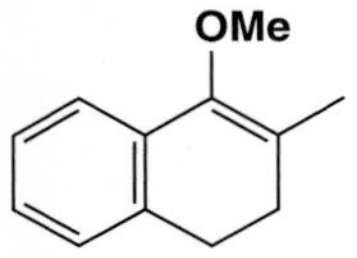 (olefin) OMe, naphthalene	64 (*S*)	41 (*S*)	
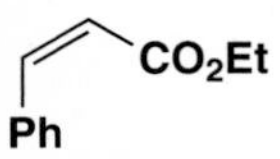 (olefin) CO₂Et, Ph			78 (2*R*,3*R*)

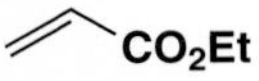

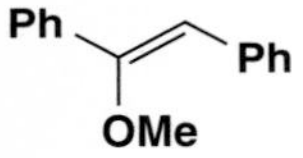

[a] For original references see [9 b].

containing olefins [9], have been successfully dihydroxylated. In Table 2 an attempt is made to summarize representative examples in this area. Due to the fact that OsO₄ reacts as an electrophilic reagent, osmylation of unsaturated carbonyl compounds can be a very slow process. This problem has been solved by increasing the amount of catalyst to 1 mol % and adding one equivalent of methyl sulfonamide [29]. A special case is the asymmetric dihydroxylation of enol ethers and ketone acetals leading directly in a one-pot process to hydroxycarbonyl compounds [9].

It is clear that there is an exception to every rule: even in AD there are a few cases known where other ligands gave improved stereoselectivities compared with PHAL, PYR, or IND. Thus, allylic phosphine oxides undergo AD to yield diols which could be used for the synthesis of optically active allylic alcohols [43].

Warren and co-workers reported best enantiomeric excesses with the original *p*-chlorobenzoate or phenanthryl ether ligands that contain only one quinuclidine unit [44]. A similar trend in enantioselectivity has been reported for the asymmetric dihydroxylation of allylic trimethylsilanes [45].

To explore the possibility of recycling alkaloid–OsO$_4$ complexes, several polymer-bound alkaloid derivatives have been used for heterogeneous catalytic asymmetric dihydroxylations. As chiral ligands, polymerized cinchona alkaloids or copolymers of quinine derivatives with acrylonitrile or styrene were studied [46]. In general, lower selectivities and decreased rates were observed.

Sharpless and co-workers reported the first catalytic asymmetric hydroxyamidation method [47]. Enantioselectivities between 33–81% could be obtained with disubstituted *cis*- or *trans*-olefins in the presence of K$_2$OsO$_2$(OH)$_4$, TsNClNa · 3H$_2$O and (DHQD)$_2$PHAL or (DHQ)$_2$PHAL as ligands. This methodical improvement is another breakthrough in asymmetric catalysis because it offers easy access to chiral β-aminoalcohols which are widely used as pharmaceuticals from cheap olefins. Despite the sometimes moderate enantioselectivities the method is already useful for practical purposes because selectivities could be improved by simple crystallization.

3.3.2.5 Selected Applications of Osmium-Catalyzed Dihydroxylations

From an industrial point of view, olefins are in principle a ubiquitious feedstock for the synthesis of diols. From the standpoint of economically interesting targets, three areas have to be distinguished. In the area of commodity products, ethylene glycol and propylene glycol are valuable targets for osmium-catalyzed air oxidation [59, 60]. Catalyst lifetime and activity in the presence of air as oxidant still have to be improved. The same is true for bulk intermediates and fine chemicals like 1,2-pentanediol, pinacol, and others. On the other hand, chiral diols – as intermediates mainly for pharmaceuticals but also for fungicides, insecticides, and pesticides – will tolerate production costs with terminal oxidants other than air. In this respect the electrocatalytic AD and the new Bäckvall variant especially offer advantages. The use of optically pure diols as valuable materials is promising because there exists a wealth of chemical knowledge for the differentiation and manipulation of the hydroxyl groups of diols, which has recently been reviewed [9 b]. In order to enable further refinement, activation of diols has been pursued by selective arenesulfonylation; reactions to cyclic sulfates, halohydrin esters and epoxides; and formation of cyclic carbamates and lactones.

Synthetic applications of AD which have already appeared and which are of potential industrial interest include the synthesis of propranolol (**9**) [48], diltiazem (**10**) [49], carnitine, and 4-amino-3-hydroxybutyric acid (**11**) [50], azole antifungals (**12**) [51], chloramphenicol (**13**) [52], reticuline intermediates (**14**) [53], camptothecin analogs (**15**) [54], khellactone (**16**) derivatives [55], taxol C-13 side chain (**17**) [56], halosarin [64], dehydro-*exo*-brevicomin [65], and antimalarial active cyclopenteno-1,2,4-trioxanes [57], as summarized in Figure 4.

Structures

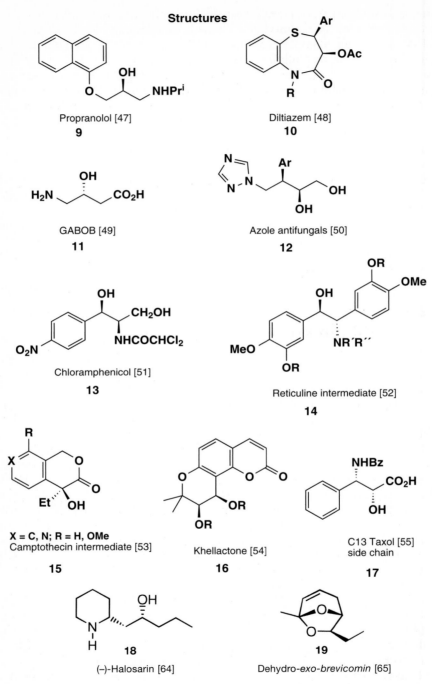

Propranolol [47]
9

Diltiazem [48]
10

GABOB [49]
11

Azole antifungals [50]
12

Chloramphenicol [51]
13

Reticuline intermediate [52]
14

X = C, N; R = H, OMe
Camptothecin intermediate [53]
15

Khellactone [54]
16

C13 Taxol [55]
side chain
17

(–)-Halosarin [64]
18

Dehydro-*exo-brevicomin* [65]
19

Figure 4. Selected examples of chiral diols of potential industrial interest made by AD.

Besides the synthesis of intermediates for pharmacologically active compounds, asymmetric dihydroxylations have been successfully applied to other fields: synthesis of natural products, as well as of a number of chiral auxiliaries for other asymmetric transformations. More detailed information is available in [9b]. In conclusion, the possible synthetic transformations of chiral diols to interesting building blocks and the technically useful characteristics of the osmium-catalyzed process make it very likely that future industrial realizations of this methodology will be seen in the area of "finest chemical synthesis".

References

[1] J. Schlossig, F. Merger, J. Paetsch, H. Gräfje, W. Reiss, F. Heinrich, N. Wilke, P. T. von Bramer, G. B. Bowen, G. Pohl, H. Gaube, P. Werle, L. Scott in *Ullmann's Encycl. Ind. Chem. 5th ed.*, **1985**, Vol. A1, p. 305.

[2] (a) N. A. Milas, S. Sussman, *J. Am. Chem. Soc.* **1936**, *58*, 1302; (b) N. A. Milas, *J. Am. Chem. Soc.* **1937**, *59*, 2342.

[3] (a) M. Schröder, *Chem. Rev.* **1980**, *80*, 187; (b) J. L. Courtney in *Organic Syntheses by Oxidation with Metal Compounds* (Eds.: W. J. Mijs, C. R. H. I. de Jonge), Plenum, New York, **1986**, p. 449.

[4] (a) K. B. Sharpless, K. Akashi, *J. Am. Chem. Soc.* **1976**, *98*, 1986; (b) P. H. J. Carlsen, T. Katsuki, V. S. Martin, K. B. Sharpless, *J. Org. Chem.* **1981**, *46*, 3936; (c) F. X. Webster, J. Rivas-Enterrios, R. M. Silverstein, *J. Org. Chem.* **1987**, *52*, 689; (d) V. S. Martin, M. T. Nunez, C. E. Tonn, *Tetrahedron Lett.* **1988**, *29*, 2701; (e) M. Caron, P. R. Carlier, K. B. Sharpless, *J. Org. Chem.* **1988**, *53*, 5185.

[5] (a) A. J. Fatiadi, *Synthesis* **1987**, 85; (b) D. G. Lee, T. Chen, *J. Am. Chem. Soc.* **1989**, *111*, 7534; (c) T. Ogino, N. Kikuiri, *J. Am. Chem. Soc.* **1989**, *111*, 6175.

[6] T. K. M. Shing, V. W.-F. Tai, E. K. W. Tam, *Angew. Chem.* **1994**, *106*, 2408; *Angew. Chem., Int. Ed. Engl.* **1994**, *33*, 2312.

[7] S. G. Hentges, K. B. Sharpless, *J. Am. Chem. Soc.* **1980**, *102*, 4263.

[8] Selected examples: (a) E. N. Jacobsen, I. Marko, W. S. Mungall, G. Schröder, K. B. Sharpless, *J. Am. Chem. Soc.* **1988**, *110*, 1968; (b) E. N. Jacobsen, I. Marko, M. B. France, J. S. Svendsen, K. B. Sharpless, *J. Am. Chem. Soc.* **1989**, *111*, 737; (c) K. B. Sharpless, W. Amberg, M. Beller, H. Chen, J. Hartung, Y. Kawanami, D. Lübben, E. Manoury, Y. Ogino, T. Shibata, T. Ukita, *J. Org. Chem.* **1990**, *56*, 4585; (d) K. B. Sharpless, W. Amberg, Y. L. Bennani, G. A. Crispino, J. Hartung, K.-S. Jeong, H. L. Kwong, K. Morikawa, Z.-M. Wang, D. Xu, X.-L. Zhang, *J. Org. Chem.* **1992**, *57*, 2768; (e) G. A. Crispino, P. T. Ho, K. B. Sharpless, *Science* **1993**, *259*, 64; (f) G. A. Crispino, K.-Y. Jeoung, H. C. Kolb, Z.-M. Wang, D. Xu, K. B. Sharpless, *J. Org. Chem.* **1993**, *58*, 3785; (g) H. Becker, S. B. King, M. Taniguchi, K. P. M. Vanhessche, K. B. Sharpless, *J. Org. Chem.* **1995**, *60*, 3940; (h) H. Becker, K. B. Sharpless, *Angew. Chem.* **1996**, *108*, 447; *Angew. Chem. Int. Ed.* **1996**, *35*, 448; (i) L. Wang, K. B. Sharpless, *J. Am. Chem. Soc.* **1992**, *114*, 7568.

[9] Reviews: (a) R. A. Johnson, K. B. Sharpless in *Catalytic Asymmetric Synthesis* (Ed.: I. Ojima), VCH, Weinheim, **1993**, p. 227; (b) H. C. Kolb, M. S. Van Nieuwenhze, K. B. Sharpless, *Chem. Rev.* **1994**, *94*, 2483; (c) H. Waldmann, *Nachr. Chem. Tech. Lab.* **1992**, *40*, 702; (d) B. B. Lohray, *Tetrahedron Asymm.* **1992**, *3*, 1317; (e) H. C. Kolb, K. B. Sharpless, in *Transition Metals for Organic Synthesis, Vol. 2* (Eds.: M. Beller, C. Bolm), VCH, Weinheim, **1998**, p. 219; (f) I. E. Markó, J. S. Svendsen, in *Compre-*

hensive Asymmetric Catalysis II (Eds.: E. N. Jacobsen, A. Pfaltz, H. Yamamoto), Springer, Berlin, **1999**, p. 713.

[10] F. C. Philipps, *Z. Anorg. Allg. Chem.* **1894**, *6*, 229.

[11] (a) R. Criegee, *Liebigs Ann. Chem.* **1936**, *522*, 75; (b) R. Criegee, *Angew. Chem.* **1937**, *50*, 153.

[12] K. A. Hofmann, *Chem. Ber.* **1912**, *45*, 3329.

[13] N. A. Milas, J.-H. Trepagnier, J. T. Nolan, M. I. Iliopulos, *J. Am. Chem. Soc.* **1959**, *81*, 4730.

[14] (a) Upjohn (W. P. Schneider, A. V. McIntosh), US 2.769.824 (1956); (b) V. Van Rheenen, R. C. Kelly, D. Y. Cha, *Tetrahedron Lett.* **1976**, *17*, 1973; (c) R. Ray, D. S. Matteson, *Tetrahedron Lett.* **1980**, *21*, 449.

[15] M. Minato, K. Yamamoto, J. Tsuji, *J. Org. Chem.* **1990**, *55*, 766.

[16] Y. Ogino, H. Chen, H.-L. Kwong, K. B. Sharpless, *Tetrahedron Lett.* **1991**, *32*, 3965.

[17] M. P. Singh, H. S. Singh, B. S. Arya, A. K. Singh, A. K. Sisodia, *Indian J. Chem.* **1975**, *13*, 112.

[18] (a) Sepracor Inc. (Y. Gao, C. M. Zepp), PCT Int. Appl. WO 9.317.150 (1994); (b) Anon., *Chem. Eng. News.* **1994**, *72*(24), 41.

[19] (a) J. F. Cairns, H. L. Roberts, *J. Chem. Soc. C* **1968**, 640; (b) Exxon Corp. (R. C. Michaelson, R. G. Austin), EP 0.077.201 (1982); *Chem. Abstr.* **1983**, *99*, 70198p; (c) Exxon Corp. (R. S. Myers, R. C. Michaelson, R. G. Austin), US 4.496.779 (1984); *Chem. Abstr.* **1985**, *102*, 148721f; (d) Exxon Corp. (R. C. Michaelson, R. G. Austin), US 4.533.772 (1985); *Chem. Abstr.* **1986**, *104*, 89183y; (e) R. G. Austin, R. C. Michaelson, R. S. Myers, in *Catalysis of Organic Reactions*, Dekker, New York, **1985**, 269; (f) Celanese Corp., GB 1.028.940 (1966); *Chem. Abstr.* **1966**, *65*, 3064f.

[20] (a) A. Krief, C. Colaux-Castillo, *Tetrahedron Lett.* **1999**, *40*, 4189; (b) A. Krief, C. Delmotte, C. Colaux-Castillo, *Pure Appl. Chem.* **2000**, *72*, 1709.

[21] W. A. Herrmann, S. J. Eder, W. Scherer, *Angew. Chem.* **1992**, *104*, 1371; *Angew. Chem., Int. Ed. Engl.* **1992**, *31*, 1345.

[22] G. Poli, *Tetrahedron Lett.* **1989**, *29*, 7385.

[23] R. Criegee, B. Marchand, H. Wannowius, *Liebigs Ann. Chem.* **1942**, *550*, 99.

[24] (a) K. B. Sharpless, D. W. Pattrick, L. K. Truesdale, S. A. Biller, *J. Am. Chem. Soc.* **1975**, *97*, 2305; (b) D. W. Pattrick, L. K. Truesdale, S. A. Biller, K. B. Sharpless, *J. Org. Chem.* **1978**, *43*, 2628; (c) E. Herranz, K. B. Sharpless, *J. Org. Chem.* **1978**, *43*, 2544; (d) E. Herranz, S. A. Biller, K. B. Sharpless, *J. Am. Chem. Soc.* **1978**, *100*, 3596.

[25] J. Böseken, *Recl. Trav. Chim.* **1922**, *41*, 199.

[26] K. B. Sharpless, A. Y. Teranishi, J.-E. Bäckvall, *J. Am. Chem. Soc.* **1977**, *99*, 3120.

[27] (a) S. Dapprich, G. Ujaque, F. Maseras, A. Lledós, D. G. Musaev, K. Morokuma, *J. Am. Chem. Soc.* **1996**, *118*, 11660; (b) U. Pidun, C. Boehme, G. Frenking, *Angew. Chem.* **1996**, *108*, 3008; *Angew. Chem., Int. Ed. Engl.* **1996**, *35*, 2817; (c) M. Torrent, L. Deng, M. Sola, T. Ziegler, *Organometallics* **1997**, *16*, 13.

[28] E. N. Jacobsen, I. Marko, M. B. France, J. S. Svendsen, K. B. Sharpless, *J. Am. Chem. Soc.* **1989**, *111*, 737.

[29] E. Erdik, D. S. Matteson, *J. Org. Chem.* **1989**, *54*, 2472.

[30] Y. L. Benanni, K. B. Sharpless, *Tetrahedron Lett.* **1993**, *34*, 2079.

[31] (a) J. P. S. Wai, I. Marko, J. S. Svendsen, M. G. Finn, E. N. Jacobsen, K. B. Sharpless, *J. Am. Chem. Soc.* **1989**, *111*, 1123; (b) E. N. Jacobsen, I. Marko, W. S. Mungall, G. Schröder, K. B. Sharpless, *J. Am. Chem. Soc.* **1988**, *110*, 1968.

[32] R. L. Haltermann, M. A. McEvoy, *J. Am. Chem. Soc.* **1992**, *114*, 980.

[33] T. Göbel, K. B. Sharpless, *Angew. Chem.* **1993**, *105*, 1417; *Angew. Chem., Int. Ed. Engl.* **1993**, *32*, 1329.

[34] (a) T. Oishi, M. Hirama, *Tetrahedron Lett.* **1992**, *33*, 639; (b) Y. Imada, T. Saito, T. Kawakami, S.-I. Murahashi, *Tetrahedron Lett.* **1992**, *33*, 5081.

[35] (a) T. Yamada, K. Narasaka, *Chem. Lett.* **1986**, 131; (b) M. Tokles, J. K. Snyder, *Tetrahedron Lett.* **1986**, *27*, 3951; (c) K. Tomioka, M. Nakajima, K. Koga, *J. Am. Chem. Soc.* **1987**, *109*, 6213; (d) E. J. Corey, P. D. Jardin, S. Virgil, P.-W. Yuen, R. D. Connel, *J. Am. Chem. Soc.* **1989**, *111*, 9243; (e) M. Nakajima, K. Tomioka, Y. Itaka, K. Koga, *Tetrahedron* **1993**, *49*, 10793.

[36] (a) G. A. Crispino, A. Makita, Z.-M. Wang, K. B. Sharpless, *Tetrahedron Lett.* **1994**, *35*, 543; (b) E. J. Corey, M. C. Noe, M. J. Grogan, *Tetrahedron Lett.* **1994**, *35*, 6427; (c) E. J. Corey, M. C. Noe, *J. Am. Chem. Soc.* **1993**, *115*, 12579; (d) E. J. Corey, M. C. Noe, S. Sarshar, *J. Am. Chem. Soc.* **1993**, *115*, 3828; (e) B. B. Lohray, V. Bushan, *Tetrahedron Lett.* **1992**, *33*, 5113.

[37] (a) J. S. Svendsen, I. Marko, E. N. Jacobsen, C. P. Rao, S. Bott, K. B. Sharpless, *J. Org. Chem.* **1989**, *54*, 2263; (b) R. M. Pearlstein, B. K. Blackburn, W. M. Davis, K. B. Sharpless, *Angew. Chem.* **1990**, *102*, 710; *Angew. Chem., Int. Ed. Engl.* **1990**, *29*, 639.

[38] K. B. Sharpless, W. Amberg, Y. L. Bennani, G. A. Crispino, J. Hartung, K.-S. Jeong, H.-L. Kwong, K. Morikawa, Z.-M. Wang, D. Xu, X.-L. Zhang, *J. Org. Chem.* **1992**, *57*, 2768.

[39] G. A. Crispino, K.-S. Jeong, H. C. Kolb, Z.-M. Wang, D. Xu, K. B. Sharpless, *J. Org. Chem.* **1993**, *58*, 3785.

[40] (a) L. Wang, K. B. Sharpless, *J. Am. Chem. Soc.* **1992**, *114*, 7568; (b) Z.-M. Wang, K. Kakiuchi, K. B. Sharpless, *J. Org. Chem.* **1994**, *59*, 6895.

[41] D. J. Berrisford, C. Bolm, K. B. Sharpless, *Angew. Chem.* **1995**, *107*, 1159; *Angew. Chem., Int. Ed. Engl.* **1995**, *34*, 1059.

[42] H. C. Kolb, P. G. Andersson, K. B. Sharpless, *J. Am. Chem. Soc.* **1994**, *116*, 1278.

[43] N. J. S. Harmat, S. Warren, *Tetrahedron Lett.* **1990**, *31*, 2473.

[44] A. Nelson, P. O'Brien, S. Warren, *Tetrahedron Lett.* **1995**, *36*, 2685.

[45] S. Okamato, K. Tani, F. Sato, K. B. Sharpless, *Tetrahedron Lett.* **1993**, *34*, 2509.

[46] (a) B. H. Kim, K. B. Sharpless, *Tetrahedron Lett.* **1990**, *31*, 3003; (b) D. Pini, A. Petri, A. Nardi, C. Rosini, P. Salvadori, *Tetrahedron Lett.* **1991**, *32*, 5175; (c) B. B. Lohray, A. Thomas, P. Chittari, J. R. Ahuja, P. K. Dhal, *Tetrahedron Lett.* **1992**, *33*, 5453.

[47] G. Li, H.-T. Chang, K. B. Sharpless, *Angew. Chem.* **1996**, *35*, 451; *Angew. Chem.* **1996**, *108*, 449.

[48] Z.-M. Whang, X.-L. Zhang, K. B. Sharpless, *Tetrahedron Lett.* **1993**, *34*, 2267.

[49] (a) ICI Australia Operations (M. Gredley) PCT Int. Appl. WO 8.902.428 (1989); (b) K. G. Watson, Y. M. Fung, M. Gredley, G. J. Bird, W. R. Jackson, H. Gountzos, B. R. Matthews, *J. Chem. Soc., Chem. Commun.* **1990**, 1018.

[50] H. C. Kolb, Y. L. Bennani, K. B. Sharpless, *Tetrahedron Asymm.* **1993**, *4*, 133.

[51] P. Blundell, A. K. Ganguly, V. M. Girijavallabhan, *Synlett* **1994**, 263.

[52] A. V. R. Rao, S. P. Rao, M. N. Bhanu, *J. Chem. Soc., Chem. Commun.* **1992**, 859.

[53] R. Hirsenkorn, *Tetrahedron Lett.* **1990**, 7591.

[54] (a) D. P. Curran, S.-B. Ko, *J. Org. Chem.* **1994**, *59*, 6139; (b) F. G. Fang, S. Xie, M. W. Lowery, *J. Org. Chem.* **1994**, *59*, 6142; (c) S.-S. Jew, K.-D. Ok, H.-J. Kim, M. G. Kim, J. M. Kim, J. M. Hah, Y.-S. Cho, *Tetrahedron Asymm.* **1995**, *6*, 1245.

[55] L. Xie, M. T. Crimmins, K.-H. Lee, *Tetrahedron Lett.* **1995**, *36*, 4529.

[56] Z.-M. Wang, H. C. Kolb, K. B. Sharpless, *J. Org. Chem.* **1994**, *59*, 5104.

[57] C. W. Jefford, D. Misra, A. P. Dishington, G. Timari, J.-C. Rossier, G. Bernardinelli, *Tetrahedron Lett.* **1994**, *35*, 6275.

[58] (a) K. Bergstad, S. Y. Jonsson, J.-E. Bäckvall, *J. Am. Chem. Soc.* **1999**, *121*, 10424; (b) S. Y. Jonsson, K. Färnegårdh, J.-E. Bäckvall, *J. Am. Chem. Soc.* **2001**, *123*, 1365.

[59] (a) C. Döbler, G. Mehltretter, M. Beller, *Angew. Chem. Int. Ed.* **1999**, *38*, 3026; (b) C. Döbler, G. Mehltretter, U. Sundermeier, M. Beller, *J. Am. Chem. Soc.* **2000**, *122*, 10289.

[60] C. Döbler, G. Mehltretter, U. Sundermeier, M. Beller, *J. Organomet. Chem.* **2001**, *621*, 70.

[61] A. Severeyns, D. E. de Vos, L. Fiermans, F. Verpoort, P. J. Grobet, P. A. Jacobs, *Angew. Chem.* **2001**, *113*, 606.

[62] A. J. DelMonte, J. Haller, K. N. Houk, K. B. Sharpless, D. A. Singleton, T. Strassner, A. A. Thomas, *J. Am. Chem. Soc.* **1997**, *119*, 9907.

[63] G. Mehltretter, C. Döbler, U. Sundermeier, M. Beller, *Tetrahedron Lett.* **2000**, *41*, 8083.

[64] H. Takahata, M. Kobuta, T. Momose, *Tetrahedron Lett.* **1997**, *38*, 3451.

[65] T. Tashiro, K. Mori, *Eur. J. Org. Chem.* **1999**, 2167.

3.3.3 Hydrovinylation

Peter W. Jolly, Günther Wilke

3.3.3.1 Introduction

The hydrovinylation reaction has its origin in the observations made in 1963 that propene dimerizes at a quite remarkable rate in the presence of certain organo-nickel catalysts and that the product distribution can be influenced by introducing auxiliary P-donor ligands [1]. In 1967 it was discovered that in the presence of the chiral ligand P(*trans*-myrtanyl)$_3$, 2-butene can be co-dimerized with propene to give 4-methyl-2-hexene in an enantioselective manner and the extension of this co-dimerization reaction to ethylene has become known as hydrovinylation.

$$\text{(1)}$$

Hydrovinylation is thus the addition of the elements of ethylene (H/CH=CH$_2$) to the neighboring C-atoms of a second alkene molecule (eq. (1)). The term has been coined in analogy to hydroformylation (the addition of H/CHO) and although it does have its merits, it is rather general and in its widest sense would include the whole range of ethylene oligomerization reactions from dimerization to polymerization as well as the co-oligomerization of ethylene with substituted alkenes. For the purpose of this review, we have therefore restricted ourselves to reactions in which ethylene is codimerized with activated alkenes or with cyclic 1,3-dienes; particular attention is given to those reactions in which a new chiral center is generated (eq. (1)). Related reactions involving noncyclic 1,3-dienes have not been included since this topic, and in particular the hydrovinylation of buta-1,3-diene to hexa-1,4-diene, has been adequately reviewed [2].

The historical development of the field and the results obtained by the principal authors have been presented in a series of review articles [3–7] and doctoral theses [8 a–i].

3.3.3.2　The Catalyst

The most active hydrovinylation catalysts contain nickel or palladium. Reactions have been reported which involve ruthenium [9], rhodium [9–11] or cobalt [12] but in these cases the reaction is invariably accompanied by considerable isomerization of the primary product. Isomerization is also the main reaction observed using ligand-free palladium catalysts, such as $PdCl_2$ or $[(PhCH:CH_2)PdCl_2]_2$ [13–15], but this can be suppressed by adding suitable P-donor ligands and active catalysts have been derived from palladium salts, e. g., $Pd(PhCN)_2Cl_2$–$AgBF_4$–PBu_3 [16], aryl–Pd compounds, e. g., $PhPd(PPh_3)_2Br$–$BF_3 \cdot OEt_2$–H_2O [17], alkene–Pd compounds, e. g., $[(PhCH:CH_2)_2PdCl_2]_2$–$BF_3 \cdot OEt_2$–PPh_3 [18], or η^3-allyl-Pd compounds, e. g., $[(\eta^3$-$C_3H_5)Pd(Ph_2PC_2H_4CO_2R)]^+$ SbF_6^- [18–20]. The main interest has, however, concentrated on nickel-containing catalysts and although investigations have been reported involving a nickel salt, e. g., Ni-$(acac)_2$–Et_2AlBr/Et_3Al–PBu_3 [11, 12, 21–27], aryl–Ni compounds, e. g., mesityl-$Ni(PPh_3)_2Br$–$BF_3 \cdot OEt_2$ [28–32], and alkene–Ni compounds, e. g., $(cod)_2$-Ni–Et_2AlCl–$Ph_2PN(Me)R$ [25, 27, 33–36], most attention has been given to catalysts prepared by treating $[(\eta^3$-$C_3H_5)NiCl]_2$ with a Lewis acid and a P-donor ligand [3–8, 37–39].

The most active nickel and palladium catalysts are either ionic or contain a Lewis acid as a co-catalyst. In the case of palladium, activation has been reported in the presence of $BF_3 \cdot OEt_2$ [17, 18] while ionic species have been prepared by reacting $[(\eta^3$-2-$MeC_3H_4)Pd(cod)]^+BF_4^-$ with a donor ligand [20] or by treating the appropriate halide with a silver salt (e. g., eq. (2)) [16, 19, 20].

$$\qquad\qquad\qquad\qquad\qquad\qquad\qquad (2)$$

The active nickel catalysts have been prepared similarly but here the most frequently used Lewis acid is $Et_3Al_2Cl_3$ or a related organoaluminum species, while individual examples have been reported which involve $BF_3 \cdot OEt_2$ or BBr_3 [21, 23, 29–32] or methyl aluminoxane (MAO) [8 h–i, 37]. One active ionic species, namely $[mesitylNi(P(CH_2Ph)_3)_2(MeCN)]^+BF_4^-$ [28], has been reported while others have been prepared *in situ* by reacting $[(\eta^3$-$C_3H_5)NiCl]_2$ with a silver salt in the presence of a donor ligand [4, 6, 7, 8g].

Bearing in mind that in many cases the active species is believed either to be ionic or to contain a strongly polarized metal–halide bond (through interaction with the Lewis acid), it is not surprising that the preferred solvent for the hydrovinylation reaction is CH_2Cl_2 or C_6H_5Cl. However, examples have been reported where the reaction proceeds satisfactorily in acetone [16], THF [28], dioxane [28], toluene [31, 32, 35] or *p*-xylene [25]. In two cases, the effect of varying the solvent has been studied [21, 27].

A P-donor ligand is generally an essential component of the hydrovinylation catalyst. In a few cases it has been demonstrated that the catalyst is inactive in the absence of a suitable ligand (e. g., the hydrovinylation of cyclopentadiene [8 g]). Enantioselective control is invariably associated with the presence of a chiral ligand and particular attention has been given to systems containing Horner phosphines in which the ligand has chiral centers at phosphorus and/or at a P-bonded organic group, e. g., PBu^t(Ph)Me, P(menthyl)(Bu^t)Et. The effect of the donor ligand upon the reaction is discussed in detail in Section 3.3.3.3; here we confine ourselves to a short discussion of the effect upon the activity of the catalyst. Generalizations are, however, not possible since the effects are metal- and alkene-specific and each class of reaction will be treated separately. The structures of the 1-azaphospholene and related ligands are shown in Figure 1.

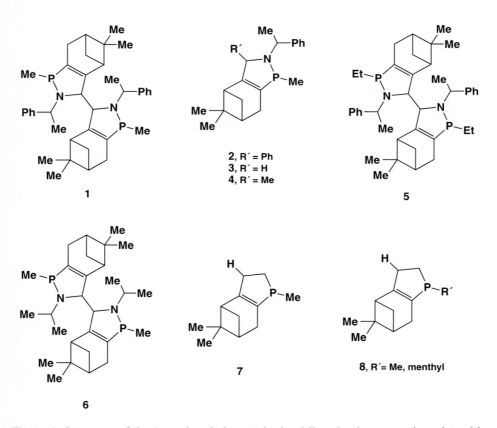

Figure 1. Structures of the 1-azaphospholene and related ligands; the convention adopted in the text refers to the configuration of the starting material used in their preparation, e. g., (*R,R*)-**1** is prepared from (–)-(*R*)-myrtenal and (+)-(*R*)-1-phenylethylamine.

3.3.3.2.1 The Hydrovinylation of Styrene

The activity of the $[(\eta^3\text{-}2\text{-MeC}_3\text{H}_4)\text{Pd(cod)}]^+\text{BF}_4^-\text{-}2\text{PR}_3$ catalysts (based upon styrene conversion) is found to increase in the order [20]:

$$\text{PPr}_3^i < \text{P(O-menthyl)}_2\text{Ph} < \text{PPh}_2\text{(O-menthyl)}$$
$$< \text{P(O-menthyl)}_3 < \text{PPh}_3 < \text{P(OPh)}_3$$

Since the active species is believed to contain only one ligand molecule, it is not surprising that in the presence of bidentate ligands, the conversion is either very low (e. g., diop or $\text{Ph}_2\text{PC}_2\text{H}_4\text{NMe}_2$) or that no reaction occurs (e. g., $\text{Ph}_2\text{PC(Me)HC(Me)HPPh}_2$). In contrast, complexes containing a hemilabile ligand such as $\text{Ph}_2\text{P(CH}_2)_n\text{CO}_2\text{R}$ (n = 1–3, R = Me, Et, menthyl) show a higher activity than systems containing monodentate ligands and this is attributed to the facile displacement of the O-donor atom from the metal by the reacting alkene [19, 20].

Nickel catalysts have been reported which are modified by a range of tertiary phosphines and phosphites, e. g., $\text{P(CH}_2\text{Ph})_3$ [28], $\text{P(menthyl)}_2\text{Me}$ [4, 16], PPh_3 [29, 30] and P(OPh)_3 [21–24] as well as 1-substituted azaphospholenes [5, 8 h–i, 37]. There is some evidence that the activity increases with the steric requirements of the ligand. For example, the $[\text{mesitylNi(PR}_3)_2(\text{MeCN)}]^+\text{BF}_4^-$ catalyst containing $\text{P(CH}_2\text{Ph})_3$ is eight times more active than the analogous PBu_3^i-modified system [28], while an active catalyst is formed in the presence of **2** (R′ = Ph) or (*R,R*)-**1** whereas no activity is observed in the presence of **3** or **4** (R′ = H, Me) [8 i, 37]. Initial results also suggest that catalysts containing the bidentate ligands **9** [8 i] and $(\text{PhCH}_2)_2\text{PCH}_2\text{P(CH}_2\text{Ph})_2$ [28] are inactive.

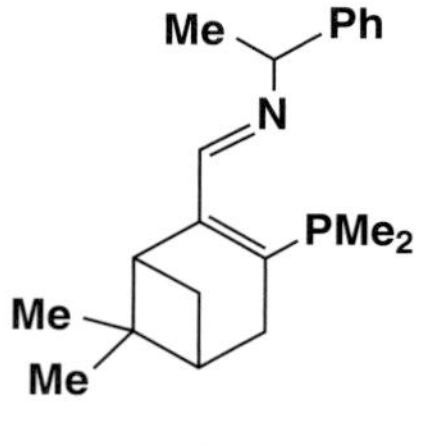

9

3.3.3.2.2 The Hydrovinylation of Bicycloheptene

The only catalysts which have been investigated are nickel-containing systems of the type $[(\eta^3\text{-C}_3\text{H}_5)\text{NiCl}]_2\text{–Et}_3\text{Al}_2\text{Cl}_3\text{–Lig}$. It has been shown that the activity of the catalyst modified by $\text{P(menthyl)}_2\text{Pr}^i$ is independent of the ligand concentration within the range Ni:P = 1:1 to 1:3 [3, 8 c]. The activity of the catalyst modified by the 1-substituted azaphospholene ligands is very sensitive to changes in the geometry of the ligand: the species having Et substituents at phosphorus (**5**) or an iso-propyl group at nitrogen (**6**) as well as the *P*-Me-substituted dimer **1** having an

(*R,S*)-configuration (see caption to Figure 1) show similar poor to moderate activity, but the (*R,R*)-isomer of **1** (or the related (*S,S*)-isomer) is remarkably active (TON 20 000 cycles/Ni-atom h at –65 °C). In contrast, the catalyst involving the monomeric azaphospholene **3** (R' = H) shows very low activity, as do catalysts modified by the phospholene and phospholane derivatives **7** and **8** (R' = Me, menthyl). The phosphaimidazoline derivative **10** (R' = menthyl) is reported to show some activity whereas the analogous species where R' is Me or Ph are inactive, as are the related systems R'P(N(Me)CH(Ph)Me)$_2$ [5, 8 e, 40].

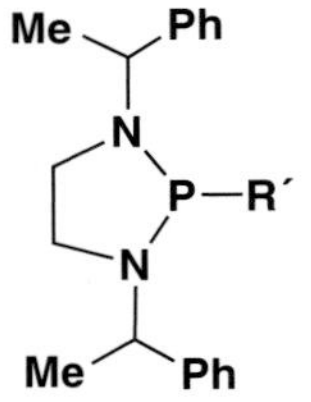

10

3.3.3.2.3 The Hydrovinylation of Cyclic 1,3-Dienes

The reaction has only been reported using nickel catalysts in the presence of tertiary phosphines (e. g., PBu$_3$, P(myrtanyl)$_3$ [3, 25, 27]), aminophosphines (e. g., **11** and **12** [27, 34]), aminophosphine phosphinites (e. g., **13** [33]) and 1-substituted azaphospholenes (e. g., (*R,R*)-**1** [6, 7, 8 g]). In addition, a catalyst has been prepared by grafting an aminophosphine onto a styrene/2 % divinylbenzene copolymer and reacting the product (**14**) with (cod)$_2$Ni-Et$_2$AlCl [35]. In contrast to **12** and **13**, bidentate ligands such as Ph$_2$PC$_2$H$_4$PPh$_2$ and dipyridyl are reported to deactivate the catalyst [25], as do Ph$_2$PCl, P(SR)$_3$, R$_3$As, Ph$_3$Sb and NEt$_3$ [25, 27].

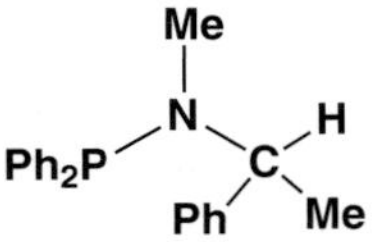

11 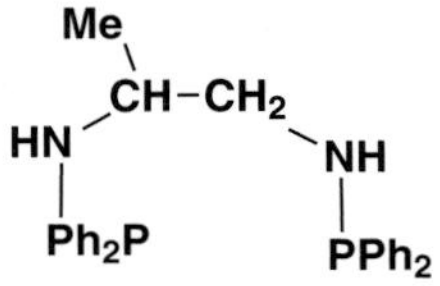   **12** 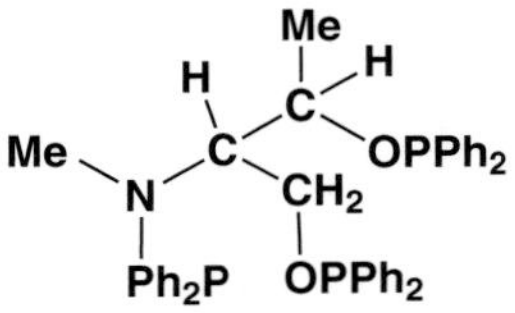   **13**

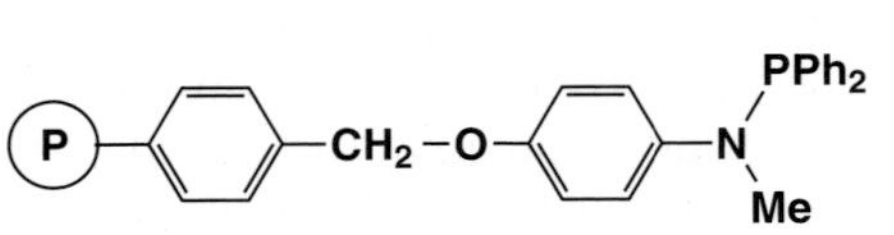

14

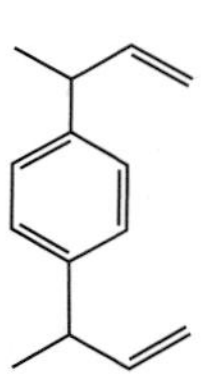

15

The effect of varying the P-donor ligand upon the activity of the catalyst has been studied in detail for the hydrovinylation of cyclopentadiene [8 g]: whereas catalysts involving PPr_3^i, PCy_3, $PPr_2^iBu^t$ and PMe_3 are inactive, that involving PPh_3 is of comparable activity to the chiral ligands which are shown below in the order of decreasing activity:

$$PBu_2^i(menthyl) > (R,R)\text{-}\mathbf{1} > PPh_2(menthyl) > P(menthyl)(Bu)Me$$

The optimal Ni:ligand ratio appears to be ligand- and substrate-dependent. In the case of the hydrovinylation of cyclohexa-1,3-diene in the presence of **11**, changing the ratio from 1:1 to 1:10 has little effect upon the activity [34], while for PBu_3 it is claimed that 1:2 is optimal [25], whereas with PPh_3 the best results are obtained with a 1:1 ratio. In this last example, a 1:5 ratio leads to catalyst deactivation [27]. For many of the reactions involving chiral ligands a ratio of 1:1.2 has been chosen, but in the case of the hydrovinylation of cy-clooeta-1,3-diene in the presence of $P(menthyl)_2Pr^i$ the catalyst is still quite active at a 1:3.8 ratio [3, 39].

3.3.3.3 The Product

The reactions which have been reported are listed in Table 1 along with representative catalysts. In the presence of the appropriate ligand and under suitable conditions, many of the reactions proceed with a surprising chemoselectivity, regioselectivity, and enantioselectivity. The main side reactions are the isomerization of the primary hydrovinylation product or its further reaction with a second molecule of ethylene and the oligomerization or polymerization of the individual alkenes. These side reactions frequently become of significance only after the consumption of one of the reacting alkenes or at elevated temperatures. The hydrovinylation products are presented briefly below and this is followed by a more detailed discussion of the enantioselective control.

Styrene is converted into 3-phenyl-1-butene with remarkably high selectivity in the presence of nickel and palladium catalysts modified by P-donor ligands. After consumption of the styrene, the same catalysts isomerize the primary product mainly to 2-phenyl-2-butene. In contrast, the product of the reaction catalyzed by ligand-free palladium catalysts, e. g., [(PhCH:CH₂)PdCl₂]₂ at elevated temperatures is mainly 1-phenyl-1-butene [9, 13–15, 18]. Alkene-substituted styrene derivatives, e. g., stilbene, are much less reactive but ring-substituted derivatives can be readily hydrovinylated: the yield varies considerably with the position of the substituent. Divinylbenzene reacts with almost exclusive dihydrovinylation and, for example, *p*-divinylbenzene is converted into **15**. Recent interest has centered on the hydrovinylation of *p*-isobutylstyrene and *p*-chlorostyrene since the products are potential precursors to the α-arylpropanoic acid derivatives ibuprofen and suprofen, while the ready conversion of 2-vinylnaphthalene to 3-naphthyl-1-butene suggests that naproxen should also be accessible [6, 7, 8 h, 38]. These and related compounds are important nonsteroidal antiflammatory agents [42].

Table 1. The hydrovinylation of activated alkenes and cyclic 1,3-dienes.

Alkene	Primary product	Typical catalyst precursor [a]	Ref.
(vinylferrocene)	(1-methylallylferrocene)	$[(\eta^3\text{-}C_3H_5)NiCl]_2\text{–}Et_3Al_2Cl_3\text{–}PR_3$	[40]
(2-vinylfuran)	(furan product)	$RhCl_3 \cdot 3H_2O$	[11]
		$Ni(PBu^t_3)_2Cl\text{–}Et_2AlCl$	[11]
(styrene)	(3-phenyl-1-butene)	$RhCl_3 \cdot 3H_2O$	[9, 10]
		$Ni(S{:}C(NEt_2)C(NEt_2){:}S)_2\text{–}Et_2AlCl\text{–}P(OPh)_3$	[22]
		$Ni(acac)_2\text{–}Et_3Al/BF_3 \cdot OEt_2\text{–}P(OPh)_3$	[24]
		$ArNi(PR_3)_2Br\text{–}BF_3 \cdot OEt_2$	[29–32]
		$[ArNi(PR_3)_2(MeCN)]^+BF_4^-$	[28]
		$(\eta^3\text{-}C_3H_5)Ni(PR_3)O_2CCF_3\text{–}BF_3 \cdot OEt_2$	[8h]
		$[(\eta^3\text{-}C_3H_5)NiCl]_2\text{–}Et_3Al_2Cl_3\text{–}PR_3$	[4–7, 8c, 8i, 37, 38]
		$(cod)_2Ni\text{–}Et_3Al_2Cl_3\text{–}PR_3$	[38]
		$Pd(PhCN)_2Cl_2\text{–}PBu_3\text{–}AgBF_4$	[16]
		$PhPd(PPh_3)_2X\text{–}H_2O$	[17]
		$[(\eta^3\text{-}C_3H_5)PdCl]_2\text{–}BF_3\text{–}PPh_3$	[18]
		$[(\eta^3\text{-}C_3H_5)Pd(Ph_2PC_2H_4CO_2Et)]^+SbF_6^-$	[19, 20]
		$[(\eta^3\text{-}2\text{-}MeC_3H_4)Pd(cod)]^+BF_4^-\text{–}PPh_2O\text{-}menthyl$	[20]

Table 1. (Continued)

Alkene	Primary product	Typical catalyst precursor [a]	Ref.
R-substituted styrene; R = 2-Me, 3-Me, 4-Me, 3-Et, 4-Et, 2-Cl, 3-Cl, 4-Cl, 4-OMe, 4-CH$_2$CHMe$_2$	R-substituted 1-phenyl-but-3-ene (branched)	NiX$_2$–AlEt$_3$/BF$_3$ · OEt$_2$–P(OPh)$_3$	[21, 23]
		ArNi(PPh$_3$)$_2$Br–BF$_3$ · OEt$_2$	[29]
		[ArNi(PR$_3$)$_2$(MeCN)]$^+$BF$_4^-$	[28]
		[(η^3-C$_3$H$_5$)NiCl]$_2$–Et$_2$AlCl–PR$_3$	[6, 7, 38]
R-substituted styrene (meta)	R-substituted branched product	[ArNi(PR$_3$)$_2$(MeCN)]$^+$BF$_4^-$	[28]
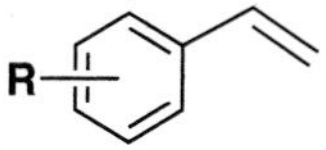 divinylbenzene; R = 2-CH:CH$_2$, 3-CH:CH$_2$, 4-CH:CH$_2$	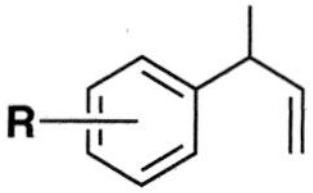 R = CH$_2$:CHC(Me)H	[(η^3-C$_3$H$_5$)NiCl]$_2$–Et$_2$AlCl–PR$_3$	[8 h]
2-vinylnaphthalene	branched naphthalene product	[ArNi(PR$_3$)$_2$(MeCN)]$^+$BF$_4^-$	[28]
6-MeO-2-vinylnaphthalene	branched MeO-naphthalene product	[(η^3-C$_3$H$_5$)NiCl]$_2$–Et$_2$AlCl–PR$_3$	[6]
α-methylstyrene	branched product	Ni(acac)$_2$–AlEt$_3$/BF$_3$ · OEt$_2$–P(OPh)$_3$	[21, 23, 24]
		ArNi(PPh$_3$)$_2$Br–BF$_3$ · OEt$_2$	[29]

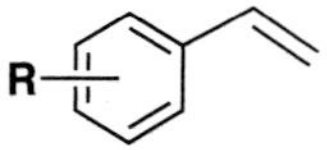

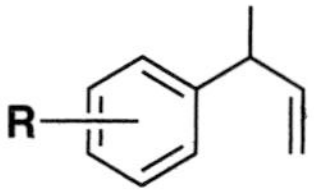

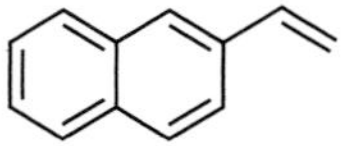

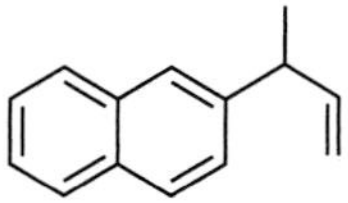

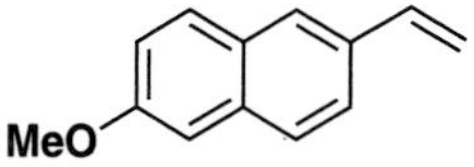

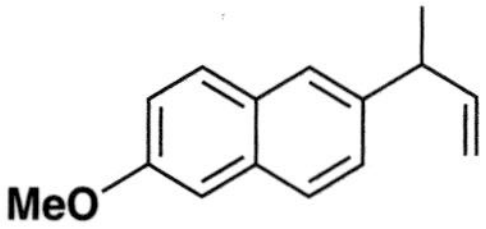

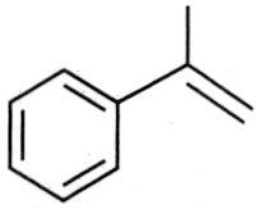

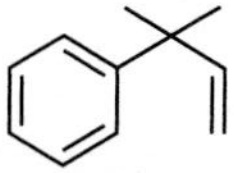

Table 1. (Continued)

Alkene	Primary product	Typical catalyst precursor[a]	Ref.
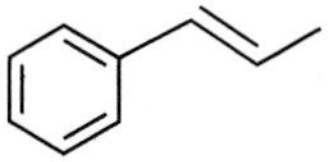 (structure)	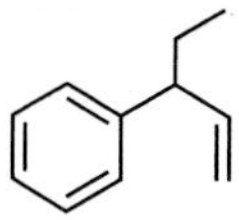 (structure)	$ArNi(PPh_3)_2Br–BF_3 \cdot OEt_2$	[29]
PhCH:CHBr	PhC(Me)HCH:CHBr	$ArNi(PPh_3)_2Br–BF_3 \cdot OEt_2$	[29]
PhCH:CHPh	PhC(Me)HCH:CHPh	$[ArNi(PR_3)_2(MeCN)]^+BF_4^-$	[28]
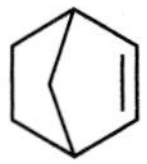 (structure)	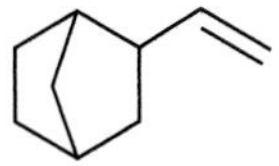 (structure)	$Ni(acac)_2–AgBF_4–PR_3$	[38]
(structure)	(structure)	$[(\eta^3\text{-}C_3H_5)NiCl]_2–Et_3Al_2Cl_3–PR_3$	[3–7, 8 c, 8 e, 38]
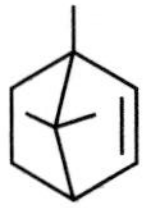 (structure)	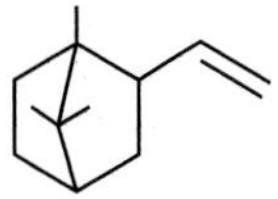 (structure)	$[(\eta^3\text{-}C_3H_5)NiCl]_2–Et_2AlCl–PR_3$	[3, 8 c]
(structure)	(structure)	$Co(acac)_2–Et_3Al_2Cl_3–PR_3$ $Ni(acac)_2–Et_3Al_2Cl_3–PR_3$	[12] [12]
(structure)	(structure)	$[(\eta^3\text{-}C_3H_5)NiCl]_2–Et_3Al_2Cl_3–PR_3$ $[(\eta^3\text{-}C_3H_5)NiCl]_2–Et_2AlCl–PR_3$	[3, 8 c] [6, 7, 8 g]
(structure)	(structure)	$[(\eta^3\text{-}C_3H_5)NiCl]_2–Et_2AlCl–PR_3$	[8g]

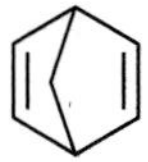

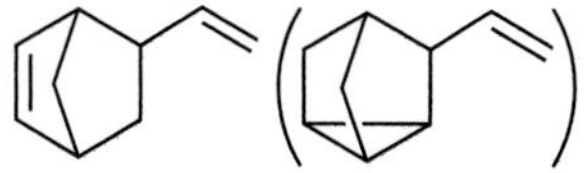

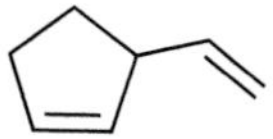

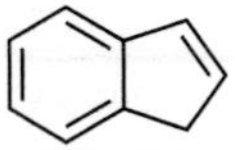

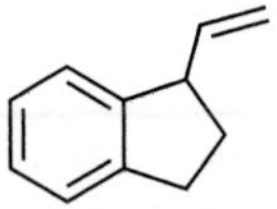

Table 1. (Continued)

Alkene	Primary product	Typical catalyst precursor [a]	Ref.
 (cyclohexa-1,3-diene structure)	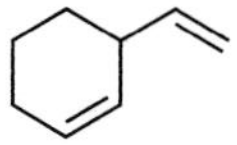 (vinylcyclohexene structure)	Ni(acac)$_2$–Et$_2$AlBr/AlEt$_3$–PR$_3$	[25, 26]
		Ni(PR$_3$)$_2$Cl$_2$–Et$_2$AlCl	[27, 41]
		(cod)$_2$Ni–Et$_2$AlCl–PR$_3$	[27, 33–35]
		(cod)$_2$Ni–Et$_2$AlBr/AlEt$_3$–PR$_3$	[25, 26]
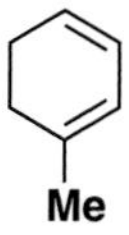 (methyl-cyclohexadiene structure, Me)	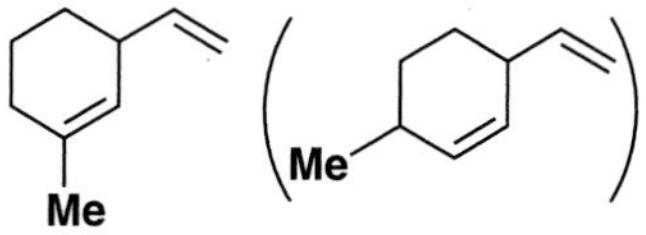 (two vinyl-methylcyclohexene structures, Me)	Ni-cat. (not specified)	[25]
 (cycloheptadiene structure)	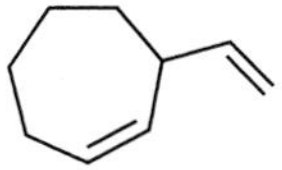 (vinylcycloheptene structure)	Ni(PBu$_3$)$_2$Cl$_2$–Et$_2$AlCl/AlEt$_3$	[25–27]
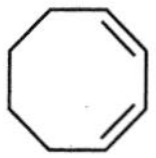 (cyclooctadiene structure)	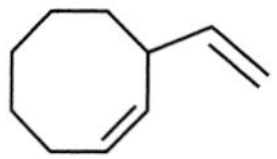 (vinylcyclooctene structure)	Ni(PBu$_3$)$_2$Cl$_2$–Et$_2$AlBr/AlEt$_3$	[25–27, 41]
		[(η^3-C$_3$H$_5$)NiCl]$_2$–Et$_3$Al$_2$Cl$_3$–PR$_3$	[3, 8 a, 8 b, 39]

[a] acac = acetylacetonate; cod = cycloocta-1,5-diene.

The nickel-catalyzed hydrovinylation of bicycloheptene has been used as a standard reaction to test the efficacy of a new ligand. The reaction occurs with complete diastereoselectivity to give *exo*-2-vinylbicycloheptane (**16**) and none of the *endo*-isomer is formed. The same species, however, catalyze the isomerization of the primary product to *cis*- and *trans*-2-ethylidenebicycloheptane (**17**) and the codimerization with further ethylene to the butenyl derivatives **18** and **19**. The product distribution is dependent upon the nature of the ligand [3, 8 c, 40].

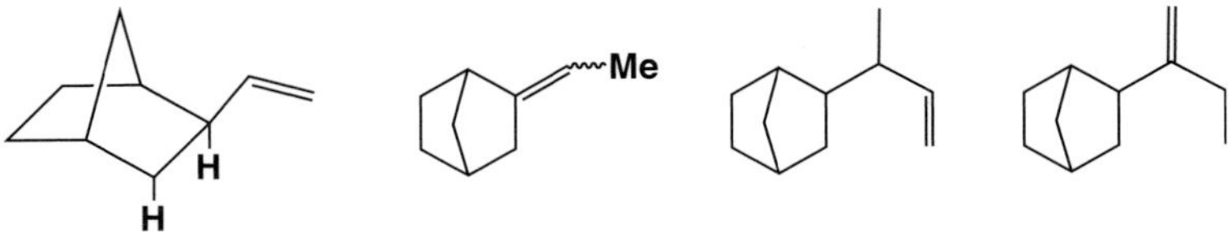

| 16 | 17 | 18 | 19 |

The reaction has been extended to bicycloheptadiene and to bornene. In the former case, monohydrovinylation is the main reaction and is accompanied by isomerization and the formation of C_{11}-codimers whereas in the latter case only the isomerization product, 3-ethylidenebornane, could be isolated. Of interest in this reaction is the observation of an enantioselective hydrovinylation: the (+)-enantiomer of bornene in the racemic starting material reacts preferentially and the unreacted substrate becomes enriched in the (–)-enantiomer [3, 8 c].

The nickel-catalyzed hydrovinylations of cyclo-1,3-pentadiene, -hexadiene, -heptadiene and -octadiene have been reported. Cyclopentadiene has only been successfully reacted using $[(\eta^3\text{-}C_3H_5)NiCl]_2$–$Et_3Al_2Cl_3$–$PR_3$ or related catalysts [6, 7, 8 g, 38] and ligand-free systems [8 g] or the combination $Ni(PBu_3)_2$-Cl_2–$Et_2AlBr/AlEt_3$ [25, 41] are inactive. The product of the reaction, 3-vinylcyclopent-1-ene, is readily converted into chaulmoogric acid **20** (eq. (3)), which is of interest as a bacteriostatic drug [43].

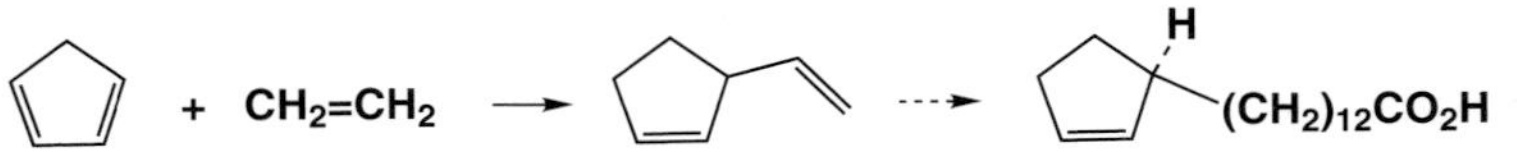

$$\tag{3}$$

20

The rate of reaction of the other three dienes studied decreases with increasing ring size [27]; in the case of cycloocta-1,3-diene, hydrovinylation is accompanied by isomerization or reaction with a second ethylene molecule, and the yield of 3-vinyl-cyclooct-1-ene never exceeds 50 %.

3.3.3.3.1 Enantioselective Control

The main interest in the hydrovinylation reaction lies in the generation of a new asymmetric center (eq. (1)) and considerable effort has been invested in obtaining high enantioselectivity by modifying the metal atom with optically active ligands. Selected results have been brought together in Table 2, in which only those

Table 2. Selected enantioselective hydrovinylation reactions.

Product	ee (%)	T (°C)	Ligand	Catalyst[a]	Ref.
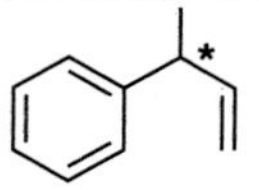	95.2	−70	(R,R)-1	A	[5–7, 38]
	91	−70	(S,S)-1	A	[40]
	6	−70	(R,S)-1	A	[40]
	60	−60	(R)-2	B	[8 i, 37]
	22	−60	P(menthyl)$_2$Pr^i	A	[8 c]
	58	rt [b]	PPh$_2$O-menthyl	C	[20]
	32	rt	PPh$_2$OC(Me)HCO$_2$Et$_2$	D	[20]
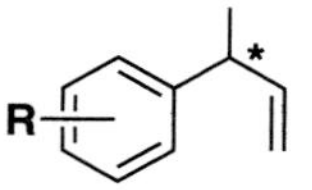 (R, see Table 1)	80–95	−50 to −70	(R,R)-1	A	[7, 8 h, 38]
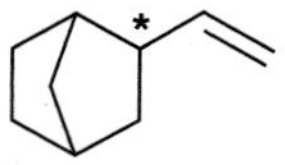	53	−65	(R,R)-1	A	[38, 40]
	40	−65	(R,S)-1	A	[38, 40]
	3	−70	(R)-3	A	[40]
	65	−70	P(menthyl)$_2$Pr^i	A	[8 c, 8 e][c]
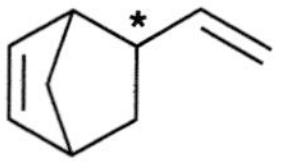	77.5	−65	P(menthyl)$_2$Pr^i	A	[3, 8 c]
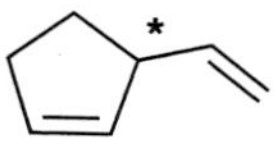	93	−70	(R,R)-1	A	[8 g, 38]
	90	0	(R,R)-1	A	[40]
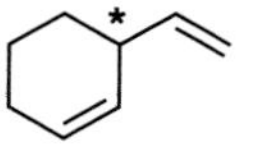	93	0	13	E	[33]
	85	40	13	E	[33]
	47[d]	−70	11	E	[33, 34]
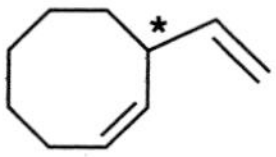	53	−75	P(menthyl)$_2$Me	A	[3, 8 a, 39]

[a] A, [(η^3-C$_3$H$_5$)NiCl$_2$–Et$_3$Al$_2$Cl$_3$–ligand;
    B, [(η^3-C$_3$H$_5$)NiCl]$_2$–MAO–ligand;
    C, [(η^3-2-MeC$_3$H$_4$)Pd(cod)]$^+$BF$_4^-$–ligand;
    D, [(η^3-2-MeC$_3$H$_4$)Pd(PPh$_2$OC(Me)HCO$_2$Et)]$^+$SbF$_6^-$;
    E, (cod)$_2$Ni–Et$_2$AlCl–ligand.
[b] rt, room temperature.
[c] See [8 e], p. 8.
[d] The original value of 73.5 % [34] has been revised [33].

reactions having high chemoselectivity have been included since isomerization of the primary product to achiral compounds can falsify the results due to a kinetic racemate separation associated with the difference in the rate of isomerization of the enantiomers. A particularly convincing example of this effect has been observed during the isomerization of 3-phenyl-1-butene by an $[(\eta^3\text{-}2\text{-MeC}_3H_4)\text{-}Pd(cod)]^+$ BF_4^-–P(O-menthyl)$_3$ catalyst: a 3:1 mixture of 3-phenyl-1-butene (*ee* 20 %) and 2-phenyl-2-butene is converted in 72 h to a 1:2 mixture having an *ee* of 38 % [20].

The hydrovinylation of cyclopentadiene in the presence of an $[(\eta^3\text{-C}_3H_5)\text{-}NiCl]_2$–Lewis acid–PR$_3$ catalyst shows complete chemo- and regioselectivity and, with a suitable choice of catalyst components, an optical yield of 94 % can be obtained. The effect of varying the ligand upon the optical yield shows no obvious correlation between structure and optical yield [3, 6–8 a, g, 39]. Particularly surprising is the relative ineffectiveness of the Horner-type phosphines having both a chiral P-atom and a chiral substituent, e. g. P(menthyl)(Bu^i)Et, but it is conceivable that the influences of the two centers of induction (chiral P-atom/ menthyl group) are opposed to each other. However, very low optical induction has also been observed during hydrovinylation of cycloocta-1,3-diene in the presence of "true" Horner phosphines, such as PBu^i(Ph)Me [39].

Acceptable optical yields are obtained in the presence of the 1-substituted azaphospholene (*R*,*R*)-**1**. The effect of varying the complex anion in the presence of this particular ligand upon the enantioselectivity has been studied by reacting $[(\eta^3\text{-C}_3H_5)\text{NiCl}]_2$ with either a silver salt (AgX; X = BF$_4$, SO$_3$CF$_3$, ClO$_4$, PF$_6$, SbF$_6$) or Et$_n$AlCl$_{3-n}$ and Et$_3$Al$_2$Cl$_3$ in CH$_2$Cl$_2$ at –70 °C. The results have been compared with the molar conductivity of the catalyst solution in CH$_2$Cl$_2$ at –40 °C and are shown in Figure 2. The enantioselectivity is high for those systems in which the complex anion can be expected to interact with the metal atom and low for those systems in which it is likely that the ions are separated. These results suggest that effective enantiomeric control is associated with the occupation of a coordination site at the nickel atom by the complex anion. The mechanistic implications are discussed in the following section.

These results, however, should be contrasted with those observed earlier for the hydrovinylation of styrene using a similar catalyst activated by P(menthyl)$_2$Pr^i [4]: the optical yield was found to decrease in the order

$$\text{SbF}_6^-(ee\ 37\ \%) > \text{PF}_6^- \sim \text{Et}_3\text{Al}_2\text{Cl}_3 \sim \text{BF}_4^- > \text{CF}_3\text{SO}_3^- \sim \text{ClO}_4^-\ (ee\ 12\ \%)$$

Moreover, the absolute configuration of the product in the presence of chlorate is opposite to that observed in the other cases. However, this could be the result of an enantioselective isomerization of the primary product to achiral 2-phenyl-2-butene under the reaction conditions (–10 °C).

The high activity of the nickel catalysts frequently enables the hydrovinylation reaction to be carried out at low temperatures, thereby allowing full implementation of the small differences in the free activation enthalpy for the formation of the diastereomeric intermediates. The increase in the diastereomeric excess with decreasing reaction temperature for the hydrovinylation of *p*-divinylbenzene to

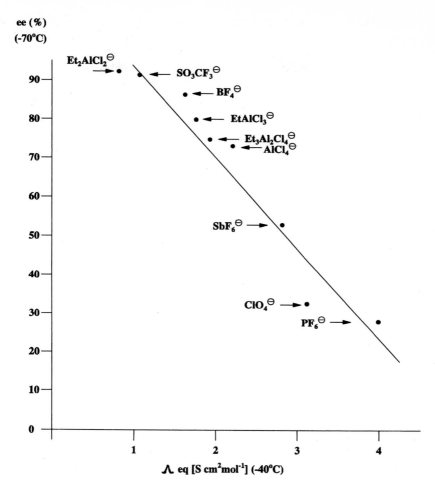

Figure 2. The enantioselectivity of the hydrovinylation of cyclopentadiene to
(–)-(*R*)-3-vinyl cyclopent-1-ene as a function of the molar conductivity of
[(*η*³-C₃H₅)NiCl]₂–(*R*,*R*)-**1**–AgX or Et*ₙ*AlCl₃₋*ₙ* in CH₂Cl₂ [6, 7, 8 g].

1,4-bisbutenylbenzene in the presence of (*R*,*R*)-**1** reaches a maximum of 80 % at
ca. –50 °C [8 h]. A similar effect is observed during the hydrovinylation of bicycloheptene in the presence of (*R*,*R*)-**1** [40] and of cyclohexa-1,3-diene in the presence of Ph₂PN(Me)CH(Me)Ph [34]. In other cases, e. g., the hydrovinylation of
cyclopentadiene [8 g] or of styrene [40] in the presence of (*R*,*R*)-**1**, or of cyclo-
hexa-1,3-diene in the presence of **13** (threophos) [33], the optical yield at 0 °C
is so high that cooling the reaction has no significant effect, while the results
for reactions involving bicycloheptadiene [8 c] or cycloocta-1,3-diene [8 a, b, 39]
in the presence of P(menthyl)₂R are unreliable due to significant isomerization
of the primary product and the formation of codimers at all temperatures. The

high enantioselectivity obtained at very low temperatures in these last cases may result in part from a restriction of rotation of the donor ligand about the P–Ni bond which has been demonstrated by variable-temperature NMR spectroscopy and which causes the ligand to adopt a conformer in which the isopropyl group associated with the menthyl substituent takes up a position above the coordination plane (see Section 3.3.3.4) [48, 49].

3.3.3.4   The Mechanism

It is frequently assumed that the mechanism of the hydrovinylation reaction is identical for catalysts containing the same metal, irrespective of the nature of the metal precursor. However, it is questionable whether this assumption can be extended to different metals and it should not, for example, be assumed that the nickel-catalyzed reactions have mechanisms identical to those of the palladium-catalyzed reactions.

Although important details remain unclear, it is generally accepted that the key intermediate in the nickel-catalyzed linear oligomerization of alkenes in general, and of hydrovinylation in particular, is a nickel hydride species which is bonded to a donor ligand, an electronegative group X (generally a halide), and alkene molecules. The Lewis acid promoter is assumed to interact with the halide atom and the observation that these reactions in general proceed satisfactorily in polar solvents (CH_2Cl_2, PhCl) suggests that polar species are involved. It has still to be decided whether interaction with the Lewis acid results in polarization of the Ni–X bond or whether complete electron transfer occurs to give a close ion pair. Although the formation of an ionic species would create an additional coordination site, there is in the case of nickel no supporting evidence whereas it has been shown by X-ray crystallography that the product (**21**) of the reaction between $(\eta^3$-$C_3H_5)Ni(PCy_3)Cl$ (Cy = cyclohexyl) and $MeAlCl_2$ contains a chlorine atom which bridges the two metal atoms [3]. Furthermore, as mentioned in the previous section, the optical induction in the hydrovinylation of cyclopentadiene in the presence of the ionic species generated from $(\eta^3$-$C_3H_5)Ni[(R,R)$-**1**$]Cl$–$AgPF_6$ is much lower than that obtained in the presence of $(\eta^3$-$C_3H_5)Ni[(R,R)$-**1**]-Cl–Et_2AlCl, suggesting not only that the complex anion remain attached to the nickel atom, but that it also plays an important role in determining the geometry of the intermediates generated during the catalysis.

$$\left\langle\!\!\left\langle-\!\!\!\!\begin{array}{c} \overset{PCy_3}{\diagup} \\ Ni \\ \diagdown \\ Cl\!\!-\!\!AlMeCl_2 \end{array}\right.\right.$$

21

The hydrovinylation reaction is suggested to proceed by an extension of the conventional Cossee-type mechanism: addition of a Ni–H species to the alkene, insertion of a second alkene molecule into the resulting Ni–alkyl bond followed by β-H transfer with elimination of the product and regeneration of the hydride.

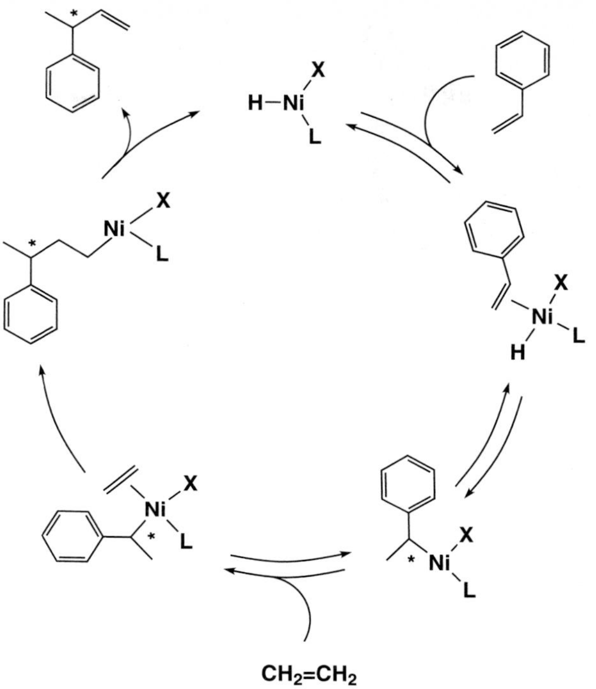

Figure 3. A schematic representation of the mechanism of the nickel-catalyzed hydrovinylation of styrene [8 h].

This is shown *schematically* in Figure 3 for the hydrovinylation of styrene; the individual steps will be discussed further for the reaction catalyzed by $[(\eta^3\text{-}C_3H_5)NiCl]_2\text{-}(R,R)\text{-}\mathbf{1}\text{-EtAlCl}_2$.

A detailed mechanism must account for the high stereoselectivity ($> 90\,\%$ *ee*) combined with high regioselectivity and high chemoselectivity ($> 90\,\%$ 3-phenyl-1-butene). The formation of 3-phenyl-1-butene in the catalytic reaction indicates that initially a styrene molecule, and not the sterically less demanding ethylene molecule, complexes to the metal. This presumably has an electronic origin and has been confirmed at least for zerovalent nickel complexes of the type (alkene)-$Ni[P(OC_6H_4Me\text{-}2)_3]_2$ using equilibrium constant data [44, 45]. The arrangement of the styrene molecule in the intermediate dictates the stereochemistry of the hydrovinylation product and insight has been obtained by using structural information to construct model compounds. The crystal structure of $(R,R)\text{-}\mathbf{1}$ (and of $(R,S)\text{-}\mathbf{1}$) has been determined by X-ray diffraction and has been used to construct the model of the $HNi[(R,R)\text{-}\mathbf{1}]Cl\text{-}AlEtCl_2$ species shown in Figure 4.

The arrangement of the groups around the central metal atom is governed by the size of the complex anion (X) and the relatively rigid geometry of the bulky 1-substituted azaphospholene molecule, which limits rotation of the N-bonded

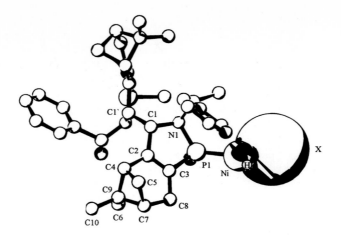

Figure 4. A model of the HNi[(*R*,*R*)-**1**]X species (X = EtAlCl₃) [5] (reproduced by permission from G. Wilke, *Angew. Chem.* **1988**, *100*, 189–211).

CH(Ph)Me group about the N–C axis to ca. 40°, hence forcing this group to occupy a position above the Ni atom. Rotation of the two halves of the aza-phospholene ligand about C1–C1′ is also constrained (to ca. 50°) and as a result the methylene groups at C5 and C8 intrude into the coordination sphere of the Ni atom, forcing the complex anion (X) to occupy an opposing site. The styrene molecule can be expected to approach the Ni atom by a pathway which will mini-mize the interaction between the phenyl substituent and both the complex anion (X) and the N-bonded substituents of the ligand. A possible square-planar arrange-ment is shown below (**22**) but a trigonal-pyramidal geometry cannot be excluded.

22   **23**   **24**

The regiochemistry of the hydrovinylation product (3-phenyl-1-butene) requires the exclusive addition of the Ni atom to the phenyl-substituted olefinic C atom (Ni→C2) and of the hydrogen atom to the terminal C atom. The pathway for this addition, which is presumably accompanied by an anticlockwise rotation of the styrene molecule about the Ni–alkene axis in **22**, is not clear but has pre-cedence in the preferred Ni→C2 addition which is observed in the initial step of the dimerization of propene using related catalysts [46]. That the further reaction involves an ethylene molecule and not a second styrene molecule will certainly be the result of the further steric restriction placed upon the system with the formation of the Ni–CH(Ph)Me fragment (**23**). Elimination of the 3-phenyl-1-butene molecule presumably proceeds by β-H transfer and it is conceivable that the suppression of the insertion of further ethylene molecules is either the result of an immediate β-H transfer with elimination of the product, or of the stabilization of the Ni–CH$_2$CH$_2$CH(Ph)Me fragment by the formation of a relatively strong agostic β-H interaction with the metal atom (**24**) which prevents the complexation of a further ethylene molecule. It should be mentioned in this context that a density functional calculation for a hypothetical EtNi(acac) species suggests that here a β-agostic bond has a strength of 10 kcal mol^{-1} [47].

The difference in enantioselectivity and catalytic activity for the hydrovinylation of styrene in the presence of (*R*,*R*)-**1** (*ee* 95 % at –70 °C, TON 1800 cycles/Ni atom h) or of (*R*,*S*)-**1** (*ee* 8 %, TON 50 cycles/Ni atom h) [5, 40] will be associated with differences in the spatial environment around the Ni atom in the active species. This is shown in Figure 5.

Whereas with (*R*,*R*)-**1** (Figure 5 a) the N-bonded substitutents and the CMe$_2$ bridge of the pinene fragment lie on opposite sides of the PNC$_3$ ring, with (*R*,*S*)-**1** (Figure 5 b) they lie on the same side. As a result, one would expect that in the presence of (*R*,*S*)-**1** not only will the approach of the styrene molecule to the metal atom be energetically more difficult than in the presence of (*R*,*R*)-**1** but that also the steric differentiation of the two sides of the Ni plane will be less pronounced and hence enantioselectivity will be lost.

The crucial role of the bulky substituent at C1 in (*R*,*R*)-**1** in maintaining the rigidity of the intermediate species is supported by a molecular modeling investigation and it has also been shown experimentally that whereas the introduction of a phenyl group at C1 is sufficient to produce a highly enantioselective catalyst, the introduction of an H atom or a Me group results in deactivation of the catalyst [8 i, 37].

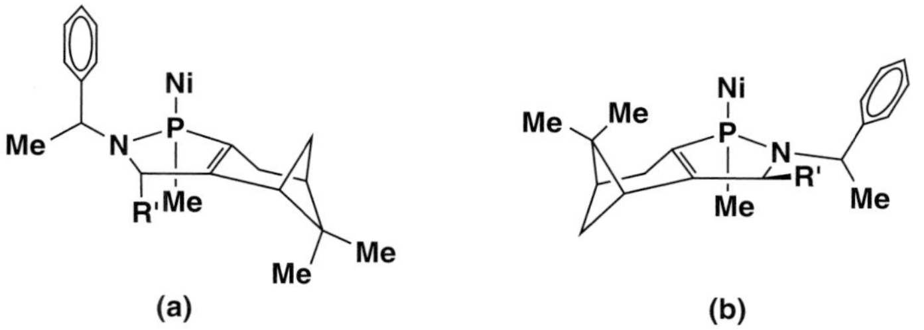

(a)   (b)

Figure 5. A comparison of the spatial environment of the Ni atom upon complexation to (*R*,*R*)-**1** (a) and (*R*,*S*)-**1** (b). The global minimum obtained from molecular modeling is shown in each case. The second half of the azaphospholene molecule is designated as R'; the chirality at the P atom is (*R*) in (a) and (*S*) in (b) [5]. (Adapted from K. Angermund, A. Eckerle, F. Lutz, *Z. Naturforsch. Teil B* **1988**, *50*, 488–502).

Arguments similar to those presented above for the hydrovinylation of styrene will dictate the stereochemical course of the reactions involving the other alkenes investigated. Thus, interference of the methylene bridge of a complexed bicycloheptene molecule (**25**) with the substituents on the donor atom and with the complex anion will direct the course of the hydrovinylation reaction to the *exo*-isomer of (+)-(1*S*,2*S*,4*R*)-2-vinylbicycloheptane (eq. (4)) [3], while the arrangement of the five-membered ring with respect to the coordination plane will result in the conversion of cyclopentadiene into (–)-(*R*)-3- vinylcyclopentane.

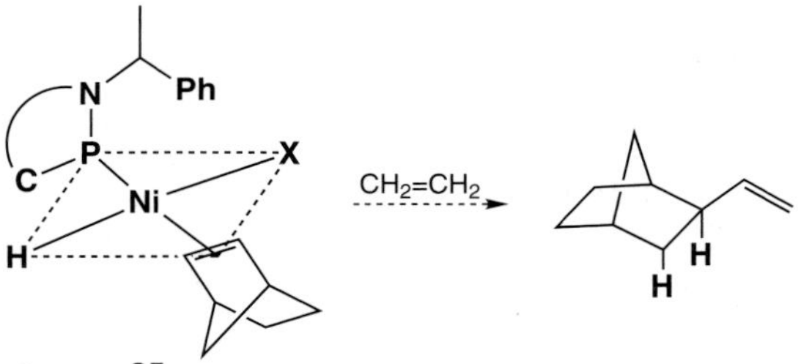

$$CH_2=CH_2 \quad\longrightarrow \qquad\qquad\qquad\qquad\qquad\qquad (4)$$

25

The blocking of a coordination position at the Ni atom by a substituent attached to the donor ligand, which probably plays a decisive role in inducing enantioselectivity in the hydrovinylation reactions involving the azaphospholene **1**, may well be a general phenomenon and has also been observed in complexes containing menthyl-substituted phosphine ligands: a crystal structure determination of (η^3-C_3H_5)Ni[P(menthyl)(Me)Bu^t]Cl (**26**), shows that the isopropyl group occupies a position below the coordination plane and that the unique H atom lies between the secondary C atom and the metal atom [48]. Furthermore, variable-temperature NMR spectroscopic studies have shown that rotation about the Ni–P bond in **26** is restricted [48]. Similar steric effects have been observed for (η^3-1,3-Me$_2$C$_3$H$_3$)-Ni[P(menthyl)$_2$Me]Me [49] and a number of related rhodium, nickel, and palladium complexes [50–52].

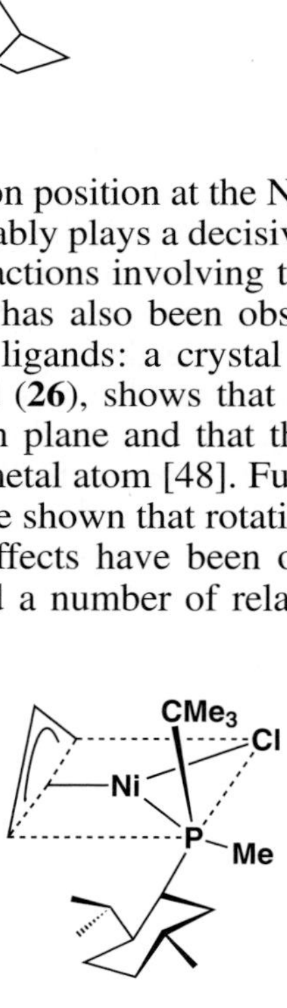

26

It is generally accepted that the nickel-catalyzed hydrovinylation of cyclic 1,3-dienes proceeds in an analogous manner to that discussed for styrene with an initial 1,2-addition of the Ni–H species. However, it should be stressed that an initial 1,4-addition has not been excluded. The observation of two isomeric products from the reaction involving hexadeuterocyclopentadiene suggests that the intermediate cycloalkenyl–Ni system rearranges through an (η^3-cyclopentenyl)Ni species [8 g]. The involvement of η^3-allyl species has also been discussed for the reactions involving the other cyclic dienes [8 a, b, 25, 27] and it is even conceivable that the initial intermediate formed in the hydrovinylation of styrene (**23**) stabilizes itself as an (η^3-1-MeCHC$_6$H$_5$)Ni species. A similar mechanism has been discussed in detail for the codimerization of ethylene with buta-1,3-diene to give hexa-1,4-diene which is catalyzed by a variety of transition metals [2], and for the Pd-catalyzed hydrovinylation of styrene [20].

The origin of the initial Ni–H species in the catalysis is a source of speculation. It has been suggested that the (η^3-allyl)Ni precursors react with insertion of an ethylene molecule followed by β-H transfer (e. g., eq. (5)), while in the case of the zerovalent nickel species the ethylaluminum component could react directly either with alkyl transfer or with an intermediate Ni(CH$_2$Cl)Cl species formed by the oxidative addition of dichloromethane, e. g., eq. (6) [3, 5, 6]. Related organopalladium compounds, e. g. ClCH$_2$Pd(Cy$_2$PC$_2$H$_4$PCy$_2$)Cl, have been characterized by X-ray diffraction [54–56].

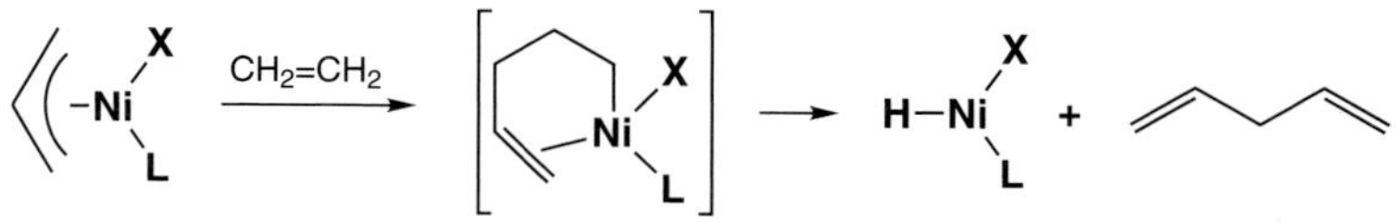

$$(5)$$

$$\text{(cod)}_2\text{Ni} + \text{CH}_2\text{Cl}_2 \xrightarrow[\text{-2 cod}]{\text{2 PPh}_3} [\text{ClCH}_2\text{Ni(PPh}_3)_2\text{Cl}] \xrightarrow{-[\text{CH}_2]} \text{Ni(PPh}_3)_2\text{Cl}_2$$

$$(6)$$

$$\xrightarrow[\text{(X = EtAlCl}_2)]{\text{Et}_2\text{AlCl}} \text{EtNi(PPh}_3)_2\text{X} \xrightarrow{-\text{C}_2\text{H}_4} \text{HNi(PPh}_3)_2\text{X}$$

Although a number of stable nickel hydride compounds have been isolated, e. g., HNi(PCy$_3$)$_2$Cl [8 g, 57, 58], only two examples are known which react further with an alkene and neither of these is catalytically active [5, 8 f, 59]. It is significant, however, that in eq. (7) the expected Ni→C2 addition is observed.

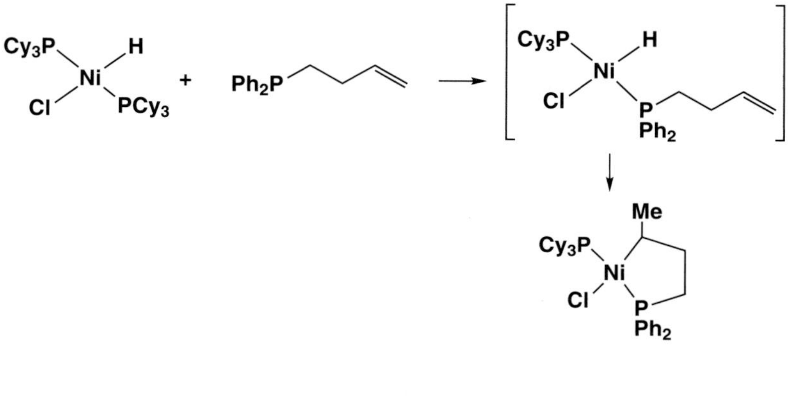

$$(7)$$

An example of the further reaction of a Ni–alkyl species with an alkene is shown in eq. (8) [60].

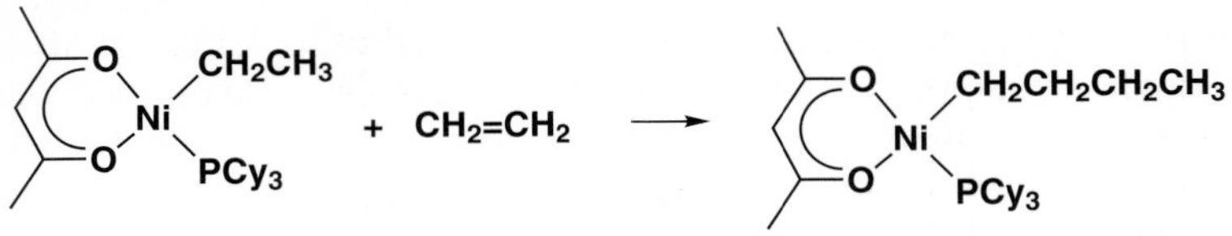

(8)

In the case of palladium, a number of neutral and ionic hydrido complexes, e. g., HPd(PCy$_3$)$_2$Cl and [HPd(PBu$^t_3$)$_2$(MeCN)]$^+$BPh$_4^-$ [61, 62], have been isolated and the latter have been shown to react with cyclic 1,3-dienes to give ionic (η^3-allyl)Pd compounds [63].

It has been suggested that (η^3-1-MeCHC$_6$H$_5$)metal species play a role in the hydrovinylation of styrene [20, 21, 24, 28]. Although both η^1- and η^3-benzyl complexes of nickel have been isolated, e. g., PhCH$_2$Ni(PMe$_3$)$_2$Cl, (η^3-PhCH$_2$)-Ni(PMe$_3$)Cl, and [(η^3-PhCH$_2$)Ni(PPh$_3$)$_2$]$^+$PF$_6^-$ [64, 65], the only derivative prepared from styrene has not been fully characterized [66, 67]. In contrast, palladium- and platinum-containing compounds have been prepared by the protonation of complexed styrene (eq. (9)) and the crystal structure, with an *anti* configuration of the Me group, has been confirmed by X-ray diffraction for the platinum compound derived from *p*-bromostyrene [68, 69].

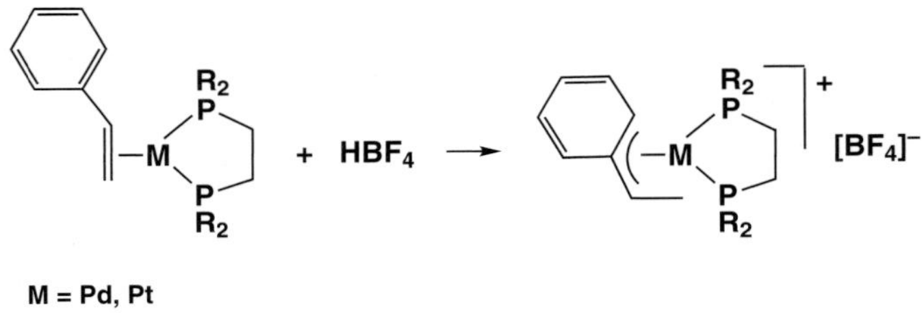

(9)

M = Pd, Pt

The relevance of these observations to catalysis is, however, questionable since NMR spectroscopic studies indicate that in particular the (η^3-benzyl)Ni complexes undergo a facile suprafacial exchange which, if projected to the catalytic reaction, would result in the loss of enantioselectivity.

3.3.3.5   Outlook

Progress in the enantioselective hydrovinylation of alkenes has been slow: although the reaction has been investigated for over 25 years, it is still confined to a handful of alkenes and, whereas spectacular selectivity has been obtained in some cases, the optimization of each new system has been largely empirical and dependent upon the synthesis of new ligand types. However, it can be anticipated that, with the aid of molecular modeling, it will be possible to predict the space-filling requirements of a ligand for the hydrovinylation of a particular alkene. This will necessitate a more detailed understanding of the mechanism and here the help of both theoreticians and experimental organometallic chemists will be needed – the former to define more precisely the course of the reaction and the latter to design stoichiometric reactions which model the individual steps in the mechanism (it is, for example, surprising how little is known about the effect of chiral ligands upon the chemistry of nickel–alkyl species), and to develop rational syntheses of suitable ligands.

The high activity of some of the nickel catalysts, which allows the reactions to be carried out at low temperature, will presumably preclude their use on a technical scale and these systems will have to be modified to give acceptable results at higher temperatures. In this respect, it should be noted that the cationic systems $[ArNi(PR_3)_2(MeCN)]^+$ [28] and $[(\eta^3\text{-allyl})Pd(PR_3)_2]^+$ [19] give satisfactory results at ambient temperatures, whereby only the latter has as yet been modified for enantioselective synthesis.

3.3.3.6 Postscript

Significant progress has been made in the last few years in optimizing the Ni- and Pd-catalysts and this has been reviewed in part [70–74]. Most attention has been given to the systems $[(\eta^3\text{-}C_3H_5)NiBr]_2$-Ligand-AgX or NaBAr$_4$ [75–79] and $[(\eta^3\text{-allyl})Pd(Ligand)]^+$ BF$_4^-$ [80–85] and less attention to catalysts derived from $[Ni(MeCN)_6][BF_4]_2$ [86, 87] and $[mesitylNi(PR_3)_2(NCMe)]^+$ BF$_4^-$ [88]. The use of $Pd(OAc)_2$-CF_3SO_2H-$Bu_2PC_3H_6PBu_2$ has been patented [89]. A heterogenized version of a Pd catalyst has been prepared by reaction of $[(\eta^3\text{-}2\text{-MeC}_3H_4)Pd(cod)]^+$ BF$_4^-$ with a phosphine-substituted carbosilane dendrimer and used to catalyze the hydrovinylation of styrene whereby isomerization of the product can be minimized by carrying out the reaction in a membrane reactor [81].

Attempts continue to optimize the optical yield of the product of the hydrovinylation of styrene [75–77, 80, 82, 83, 85, 88] and vinylnaphthalene [75, 76, 82] and their derivatives. In the case of the Ni catalysts the best results (*ee* 80–90 %) are obtained with the system $[(\eta^3\text{-}C_3H_5)NiBr]_2$-Ligand-NaB(C$_6H_3$(CF$_3$-3,5)$_2$)$_4$ in the presence of the azaphospholene **1** [77] or the hemilabile ligand **27** [75, 76]. There is an indication that the disadvantage of carrying the reaction out at low temperatures can be avoided by using liquid or supercritical CO_2 as the reaction medium [77] (cf. Section 3.1.13). Similar enantioselectivities under mild conditions have been obtained using the $[(\eta^3\text{-allyl})Pd(Ligand)]^+$ X$^-$ system in the presence of P-chiral ligands, e. g., **28** [80, 82, 85] or the Cr-complex **29** [83], whereby isomerization of the product can be suppressed by terminating the reaction before completion.

27   28   29

Interestingly, the monodentate phosphine-modified Ni and Pd catalysts respond differently to variation of the counterion: the best results for the Ni catalysts are obtained in the presence of weakly coordinating counterions such as OTf^- whereas the analogous Pd catalysts require the presence of a noncoordinating counterion such as BAr_4^- or SbF_6^-. The latter are also the preferred anions for Ni-catalysts modified by **1** or the hemilabile ligand **27** [75–77, 80].

The title reaction has been extended to the codimerization of propene and vinylarene derivatives using a Ni-catalyst [78]. The same catalyst, as well as the analogous Pd-system, is also active for the intramolecular hydrovinylation of α,ω-diolefins to give cyclic products (eq. (10)) [79].

$$(10)$$

It has also been reported that a classical metathesis catalyst, $RCH{=}Ru(PCy_3)_2$ (CO)Cl, catalyzes the reaction between ethylene and alkynes to give mainly hydrovinylation products [90].

References

[1] Studiengesellschaft Kohle mbH, NL Appl. 6.409.179 (1965); DE Appl. Aug. 10, 1963; *Chem. Abstr.* **1965**, *63*, 5770h.
[2] A. C. L. Su, *Adv. Organomet. Chem.* **1979**, *17*, 269.
[3] B. Bogdanović, *Angew. Chem.* **1973**, *85*, 1013.
[4] B. Bogdanović, *Adv. Organomet. Chem.* **1979**, *17*, 105.
[5] G. Wilke, *Angew. Chem.* **1988**, *100*, 189.
[6] G. Wilke in *Organometallics in Organic Synthesis 2* (Eds.: H. Werner, G. Erker), Springer, Berlin, **1989**, pp. 1–20.
[7] G. Wilke, K. Angermund, G. Fink, C. Krüger, T. Leven, A. Mollbach, J. Monkiewicz, S. Rink, H. Schwager, K. H. Walter, in *New Aspects of Organic Chemistry II*, Kondansha, Tokyo, **1992**, pp. 1–18.
[8] Ph. D. Theses, Ruhr-Universität Bochum: (a) B. Meister, **1971**; (b) B. Henc, **1971**; (c) A. Lösler, **1973**; (d) H. Brandes, **1979**; (e) H. Kuhn, **1983**; (f) T. Leven, **1988**; (g) S. Rink, **1989**; (h) P. Eckerle, **1992**; (i) A. Eckerle, **1994**.
[9] H. Umezaki, Y. Fujiwara, K. Sawara, S. Teranishi, *Bull. Chem. Soc. Jpn.* **1973**, *46*, 2230.
[10] T. Alderson, E. L. Jenner, R. V. Lindsey, *J. Am. Chem. Soc.* **1965**, *87*, 5638.
[11] U. M. Dzhemilev, L. Y. Gubaidullin, G. A. Tolstikov, *Bull. Acad. Sci. USSR* **1976**, 2009.
[12] S. M. Pillai, G. L. Tembe, M. Ravindranathan, *J. Mol. Catal.* **1993**, *84*, 77.
[13] K. Kawamoto, A. Tatani, T. Imanaka, S. Teranishi, *Bull. Chem. Soc. Jpn.* **1971**, *44*, 1239.
[14] K. Kawamoto, T. Imanaka, S. Teranishi, *Bull. Chem. Soc. Jpn.* **1970**, *43*, 2512.
[15] M. G. Barlow, M. J. Bryant, R. N. Haszeldine, A. G. Mackie, *J. Organomet. Chem.* **1970**, *21*, 215.
[16] Mitsubishi Chem. Ind. (S. Hattori, K. Tatsuoka, T. Shimizu), JP 72 25.133 (1972); *Chem. Abstr.* **1973**, *78*, 3922.
[17] H. Nozima, N. Kawata, Y, Nakamura, K. Maruya, T. Mizoroki, A. Ozaki, *Chem. Lett.* **1973**, 1163.

[18] T. Ito, K. Takahashi, Y. Takami, *Nippon Kagaku Kaishi* **1974**, 1097; *Chem. Abstr.* **1974**, *81*, 77567.

[19] G. J. P. Britovsek, W. Keim, S. Mecking, D. Sainz, T. Wagner, *J. Chem. Soc., Chem. Commun.* **1993**, 1632.

[20] G. J. P. Britovsek, Dissertation, Techn. Hochschule Aachen, **1993**.

[21] G. A. Mamedaliev, A. G. Azizov, *Polym. J. (Tokyo)* **1985**, *17*, 1075.

[22] A. G. Azizov, D. B. Akhmedov, S. M. Aliyev, *Neftekhimiya* **1984**, *24*, 353; *Chem. Abstr.* **1984**, *101*, 110309.

[23] A. G. Azizov, G. A. Mamedaliev, S. M. Aliev, V. S. Aliev, *Azerb. Khim. Zh.* **1978**, 3–8; *Chem. Abstr.* **1979**, *90*, 6002.

[24] A. G. Azizov, G. A. Mamedaliev, S. M. Aliev, V. S. Aliev, *Azerb. Khim. Zh.* **1979**, 3; *Chem. Abstr.* **1980**, *93*, 203573.

[25] B. Adler, J. Beger, C. Duschek, C. Gericke, W. Pritzkow, H. Schmidt, *J. Prakt. Chem.* **1974**, *316*, 449.

[26] J. Beger, C. Duschek, C. Gericke, *J. Prakt. Chem.* **1974**, *316*, 952.

[27] G. Peiffer, X. Cochet, F. Petit, *Bull. Soc. Chim. Fr. II*, **1979**, 415.

[28] R. Ceder, G. Muller, J. I. Ordinas, *J. Mol. Catal.* **1994**, *92*, 127.

[29] N. Kawata, K. Maruya, T. Mizoroki, A. Ozaki, *Bull. Chem. Soc. Jpn.* **1974**, *47*, 413.

[30] N. Kawata, K, Maruya, T. Mizoroki, A. Ozaki, *Bull. Chem. Soc. Jpn.* **1971**, *44*, 3217.

[31] Mitsubishi Yuka Fine Chem. Co. (S. Kitatsume, S. Otaba), JP 8691.138 (1986); *Chem. Abstr.* **1986**, *105*, 227505.

[32] Tokyo Inst. Technol. (A. Ozaki, T. Mizoroki), DE-OS 2.211.745 (1973); *Chem. Abstr.* **1973**, *78*, 110835.

[33] G. Buono, C. Siv, G. Peiffer, C. Triantaphylides, P. Denis, A. Mortreux, F. Petit, *J. Org. Chem.* **1985**, *50*, 1781.

[34] G. Buono, G. Peiffer, A. Mortreux, F. Petit, *J. Chem. Soc., Chem. Commun.* **1980**, 937.

[35] X. Cochet, A. Mortreux, F. Petit, *C. R. Hebd. Seances Acad. Sci., Ser. C* **1978**,*288*, 105.

[36] Soc. Chim. Charbonnages (M. Petit, A. Mortreux, F. Petit, G. Buono, G. Peiffer), FR 2.550.201 (1985); *Chem. Abstr.* **1986**, *104*, 149172.

[37] K. Angermund, A. Eckerle, F. Lutz, *Z. Naturforsch. Teil B* **1995**, *50*, 488.

[38] Studiengesellschaft Kohle mbH (G. Wilke, J. Monkiewicz, H. Kuhn), DE-OS 3.618.169 (1987); *Chem. Abstr.* **1988**, *109*, 6735.

[39] B. Bogdanovic', B. Henc, B. Meister, H. Pauling, G. Wilke, *Angew. Chem.* **1972**, *84*, 1070. *Angew. Chem., Int. Ed. Engl.* **1972**, *11*, 1023.

[40] J. Monkiewicz, G. Wilke, unpublished results, **1987**.

[41] R. G. Miller, T. J. Kealy, A. L. Barney, *J. Am. Chem. Soc.* **1967**, *89*, 3756.

[42] H. R. Sonawane, N. S. Bellur, J. R. Ahuja, D. G. Kulkarni, *Tetrahedron: Asymmetry* **1992**, *3*, 163.

[43] M. Hooper, *Chem. Soc. Rev.* **1987**, *16*, 437.

[44] C. A. Tolman, W. C. Seidel, *J. Am. Chem. Soc.* **1974**, *96*, 2774.

[45] C. A. Tolman, *J. Am. Chem. Soc.* **1974**, *96*, 2780.

[46] See, for example, P. W. Jolly in *Comprehensive Organometallic Chemistry* (Eds.: G. Wilkinson), F. G. A. Stone, E. W. Abel), Pergamon Press, Oxford, **1982**, Vol. 8, pp. 618–623.

[47] L. Fan, A. Krzywicki, A. Somogyvari, T. Ziegler, *Inorg. Chem.* **1994**, *33*, 5287.

[48] H. Brandes, R. Goddard, P. W. Jolly, C. Krüger, R. Mynott, G. Wilke, *Z. Naturforsch. Teil B* **1984**, *39*, 1139.

[49] B. L. Barnett, C. Krüger, *J. Organomet. Chem.* **1974**, *77*, 407.

[50] D. Valentine, J. F. Blount, K. Toth, *J. Org. Chem.* **1980**, *45*, 3691.

[51] K. Kan, Y. Kai, N. Yasuoka, N. Kasai, *Bull. Chem. Soc. Jpn.* **1977**, *50*, 1051.

[52] K. Kan, K. Miki, Y. Kai, N. Yasuoka, N. Kasai, *Bull. Chem. Soc. Jpn.* **1978**, *51*, 733.

[53] M. Barkowsky, Ph. D. Thesis, Ruhr-Universität Bochum, **1991**.
[54] A. Döhring, R. Goddard, G. Hopp, P. W. Jolly, N. Kokel, C. Krüger, *Inorg. Chim. Acta* **1994**, *222*, 179.
[55] G. Ferguson, B. L. Ruhl, *Acta Crystallogr., Sect. C* **1984**, *40*, 2020.
[56] W. A. Herrmann, W. R. Thiel, C. Broßmer, K. Öfele, T. Priermeier, W. Scherer, *J. Organomet. Chem.* **1993**, *461*, 51.
[57] M. L. H. Green, T. Saito, P. J. Tanfield, *J. Chem. Soc. A* **1971**, 152.
[58] K. Jonas, G. Wilke, *Angew. Chem.* **1969**, *81*, 534.
[59] U. Müller, W. Keim, C. Krüger, P. Betz, *Angew. Chem.* **1989**, *101*, 1066; W. Keim, *Angew. Chem.* **1990**, *102*, 251.
[60] G. T. Crisp, S. Holle, P. W. Jolly, *Z. Naturforsch. Teil B* **1982**, *37*, 1667.
[61] P. M. Maitlis, P. Espinet, M. J. H. Russell, in *Comprehensive Organometallic Chemistry* (Eds.: G. Wilkinson, F. G. A. Stone, E. W. Abel), Pergamon Press, Oxford, **1982**, Vol. 6, pp. 340–342.
[62] M. Sommovigo, M. Pasquali, P. Leoni, P. Sabatino, D. Braga, *J. Organomet. Chem.* **1991**, *418*, 119.
[63] D. J. Mabott, P. M. Maitlis, *J. Chem. Soc., Dalton. Trans.* **1976**, 2156.
[64] E. Carmona, J. M. Marin, M. Paneque, M. L. Poveda, *Organometallics* **1987**, *6*, 1757.
[65] E. Carmona, M. Paneque, M. L. Poveda, *Polyhedron* **1989**, *8*, 285.
[66] J. R. Ascenso, M. A. A. F. de C. T. Carrondo, A. R. Dias, P. T. Gomes, M. F. M. Piedade, C. C. Romao, A. Revillon, I. Tkatchenko, *Polyhedron* **1989**, *8*, 2449.
[67] J. R. Ascenso, A. R. Dias, P. T. Gomes, C. C. Romao, Q. T. Pham, D. Neibecker, I. Tkatchenko, *Macromolecules* **1989**, *22*, 998.
[68] L. E. Crascall, S. A. Litster, A. D. Redhouse, J. L. Spencer, *J. Organomet. Chem.* **1990**, *394*, C35.
[69] L. E. Crascall, J. L. Spencer, *J. Chem. Soc., Dalton Trans.* **1992**, 3445.
[70] S. Hashiguchi, R. Noyori, *Kagaku, Zokan (Kyoto)* **1995**, *124*, 203.
[71] F. Kakiuchi, *Kagaku (Kyoto)* **1998**, *53*, 71.
[72] N. Nomura, J. Jin, H. Park, T. V. RajanBabu, M. Valluri, M. A. Avery, *Chemtracts* **1999**, *12*, 52.
[73] T. V. Rajanbabu, in *Comprehensive Asymmetric Catalysis* (Eds.: E. N. Jacobsen, A. Pfaltz, H. Yamamoto), Springer, Berlin, **1999**, pp. 417–427.
[74] T. V. RajanBabu, N. Nomura, J. Jin, B. Radetich, H. Park, M. Nandi, *Chem. Eur. J.* **1999**, *5*, 1963.
[75] N. Nomura, J. Jin, H. Park, T. V. RajanBabu, *J. Am. Chem. Soc.* **1998**, *120*, 459.
[76] M. Nandi, J. Jin, T. V. RajanBabu, *J. Am. Chem. Soc.* **1999**, *121*, 9899.
[77] A. Wegner, W. Leitner, *J. Chem. Soc., Chem. Commun.* **1999**, 1583.
[78] J. Jin, T. V. RajanBabu, *Tetrahedron* **2000**, *56*, 2145.
[79] B. Radetich, T. V. RajanBabu, *J. Am. Chem. Soc.* **1998**, *120*, 8007.
[80] R. Bayersdörfer, B. Ganter, U. Englert, W. Keim, D. Vogt, *J. Organomet. Chem.* **1998**, *552*, 187.
[81] N. J. Hovestad, E. B. Eggeling, H. J. Heidbüchel, J. T. B. H. Jastrzebski, U. Kragl, W. Keim, D. Vogt, G. van Koten, *Angew. Chem.* **1999**, *111*, 1763; *Angew. Chem., Int. Ed.* **1999**, *38*, 1655.
[82] J. Albert, J. M. Cadena, J. Granell, G. Muller, J. I. Ordinas, D. Panyella, C. Puerta, C. Sanudo, P. Valerga, *Organometallics* **1999**, *18*, 3511.
[83] U. Englert, R. Haerter, D. Vasen, A. Salzer, E. B. Eggeling, D. Vogt, *Organometallics* **1999**, *18*, 4390.
[84] G. J. P. Britovsek, K. J. Cavell, W. Keim, *J. Mol. Catal. A:* **1996**, *110*, 77.
[85] Hoechst A.-G. (W. Keim, D. Vogt, R. Bayersdörfer), DE 19.512.881 (1996); *Chem. Abstr.* **1996**, *125*, 248773.

[86] A. L. Monteiro, M. Seferin, J. Dupont, R. F. de Souza, *Tetrahedron Lett.* **1996**, *37*, 1157.
[87] V. Fassina, C. Ramminger, M. Seferin, A. L Monteiro, *Tetrahedron* **2000**, *56*, 7403.
[88] G. Muller, J. I. Ordinas, *J. Mol. Catal. A:* **1997**, *125*, 97.
[89] Shell Oil Co. (E. Drent), US 5.227.561 (1993); *Chem. Abstr.* **1994**, *120*, 31520.
[90] C. S. Yi, D. W. Lee, Y. Chen, *Organometallics* **1999**, *18*, 2043.

3.3.4   Carbon Dioxide as a C_1 Building Block

Eckhard Dinjus, Roland Fornika, Stephan Pitter,
Thomas Zevaco

3.3.4.1   Introduction

The use of carbon dioxide (CO_2) as a raw material in chemical syntheses is a research area of extraordinary scientific, economic, and ecological interest [1–3]. The removal of CO_2 from emissions of industrial processes in order to reduce the generally and controversially discussed greenhouse effect encourages chemists to initiate research in this field [2]. The possibility of recycling large amounts of CO_2 would be rather more attractive than storage if economical and ecologically beneficial processes are developed for the conversion of CO_2 into useful products. For synthetic chemists two different approaches are possible to achieve this goal: firstly, the conversion of carbon dioxide into bulk chemicals, allowing the fixation of large amounts of CO_2; secondly, the synthesis of fine chemicals from CO_2 and other readily available substrates. Carbon monoxide (CO) and phosgene ($COCl_2$) are currently used as C_1-building block in many industrial processes, but for reasons of working safety and ecological doubt CO_2 is an ideal raw material in many respects: it is nontoxic, easy to store, to transport, and to handle, and – another important aspect – cheap!

    The most important chemical process running on Earth is the fixation of carbon dioxide by green plants using solar energy. Photosynthesis and other enzymic examples of carboxylation with essential metals represent the natural carbon dioxide activation processes which have been optimized over all the years of development [1 a, f, g]. In photosynthesis, carbon dioxide is reduced by water into carbohydrates using sunlight as energy source. CO_2 can also be reduced to carbon monoxide or hydrogenated to methanol or methane with heterogeneous or enzymic and thus *homogeneous* catalysts. In spite of the large amount of CO_2 available, only a few processes using carbon dioxide as a C_1-building block have been developed in the synthetic chemical industry up to now. The most important processes are the synthesis of urea by reaction with ammonia, the synthesis of salicylic acid (Kolbe–Schmidt reaction) as a process for forming a new C–C bond and as an example of the use of the whole or intact carbon dioxide molecule for synthesis (cf. Figure 1). For oxidative coupling reactions many *stoichiometric* processes are known and detachment of the reaction components from the metal

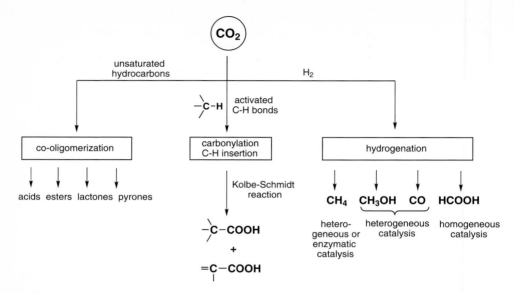

Figure 1. Examples of industrially useful reactions of carbon dioxide with energy-rich co-substrates (without additional use of other energy sources).

center leads to products of great interest. Further work is necessary in this research area to transfer these reactions into a *catalytic* cycle.

The binding of carbon dioxide to a transition metal center, which can be brought about in various ways, generally involves activation of the molecule and several spectroscopic methods are suitable allowing the characterization of CO_2 complexes [30]. In order to obtain a better understanding of carbon dioxide activation several CO_2 complexes have been investigated and described but the formation of a transition metal–CO_2 complex is not a necessary prerequisite for catalytic processes converting CO_2 into usable chemical products [1 b].

Owing to the generally high activation barriers for reactions involving the highly oxidized and thermodynamically stable CO_2 molecule, catalysts are required in most of these reactions. Apart from hydrogenation of CO_2, C–C coupling reactions are hitherto the domain of homogeneous catalytic reactions, e. g., catalyst development for synthesis of lactones and pyrones. Examples of both above-mentioned approaches to CO_2 activation will be given in this section.

Homogeneous organometallic catalysts possess an adjustable molecular structure and offer high selectivity for the formation of a wide range of small to large products. Industrial applications of catalytic processes so far have used heterogeneous catalysts by reason of getting higher reaction rates and quite easy separation from the reaction product. The creation of a highly effective homogeneous catalytic system therefore requires increased reactivity, elimination of slow mass transfer and diffusion, reactants in high concentrations, and a weak solvatation sphere around the catalyst.

The catalytic efficiency is also often determined by the nature of the coordinating ligands. The electronic and steric nature of the ligand has a remarkable influence on the activity of the catalytic systems and many attempts have been made to obtain a comprehensive system of ligand classification that allows correlation of catalytic activity and ligand structure. In this section it will be shown how homogeneous catalysts for CO_2 transforming reactions can be developed and optimized on the basis of these concepts by combination of experimental and theoretical work.

3.3.4.2 Catalytic C–C Bond-Forming Reactions

3.3.4.2.1 Palladium-Catalyzed Synthesis from 1,3-Dienes and CO_2

There is an ongoing interest in catalytic C–C bond-forming reactions of CO_2 [3] and much work has been invested in palladium-catalyzed synthesis of δ-lactone **2** from butadiene **1** and CO_2 [3 e, 3 f, 4]. Table 1 presents the catalyst development for this catalytic coupling reaction, and the optimum conditions as known up to now are summarized in eq. (1).

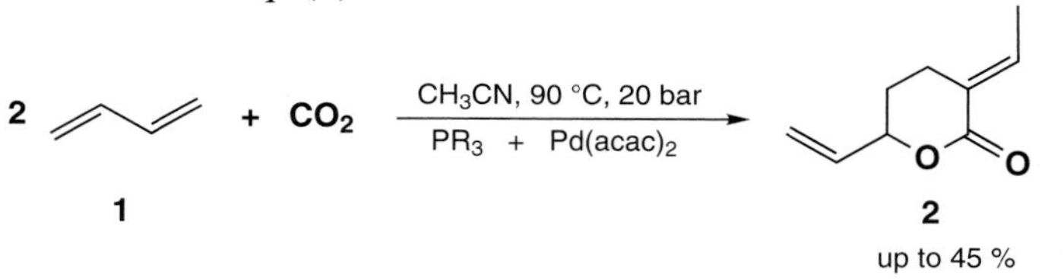

$$ \text{(1)} $$

up to 45 %

After the pioneering work of Inoue et al. [7] and Musco [5], the most detailed study of the cyclotrimerization of butadiene and CO_2 has been carried out by Behr using catalysts formed *in situ* from Pd(acac)$_2$ (acac = acetylacetonate) and three equivalents of a suitable phosphine ligand [4].

These studies also revealed the formation of several other coupling products from which isomeric C$_9$ γ-lactones and isomeric octadienyl esters of nonatrienecarbonic acid have been isolated. A number of more and less effective variations of phosphine–palladium-based catalysts were reported [3 e, f, 4, 8]. Efforts to establish an enantioselective catalytic synthesis of the δ-lactone by use of chiral coligands, remained unsuccessful however [8 e, f].

Basic trialkyl phosphines are best suited as ligands for the palladium-catalyzed cyclotrimerization of **1** and CO_2, and a strong influence of the ligand structure on the performance of the catalyst is observed. As noted earlier [4 c], the Tolman concept [10 a] of electronic (Σ_x^i) and steric (Θ) parameters is obviously not sufficient to explain the observed ligand effects. The steric parameter E_R, recently developed for phosphine ligands on the basis of molecular mechanics [11], also failed to show any correlation with the experimental results. The understanding of these

Table 1. Selected catalytic systems for butadiene–CO$_2$ reactions.

Catalyst	Adducts	Solvent	Temp. [°C] (time [h])	Lactones	Esters	Butadiene dimers	Ref.
Ni(cod)$_2$	P(*i*-Pr)$_3$	CH$_3$CN	90 (15)	1.5	0.1		[4 c]
Pd(acac)$_2$	P(*i*-Pr)$_3$	CH$_3$CN	90 (15)	40.3	1.3		[4 b]
Pd(acac)$_2$	P(*n*-Bu)$_3$	CH$_3$CN	90 (15)	3.1	14.5		[4 b]
Pd(acac)$_2$	P(*i*-Pr)$_3$ 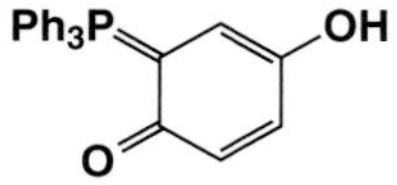 + (Ph$_3$P / OH / O quinone structure)	CH$_3$CN	90 (15)	58.8	2.8		[4 c]
[Pd(PPh$_3$)$_2$(*p*-benzoquinone)]	NEt$_3$, H$_2$O	CH$_3$CN	60 (18)	62.3	0.9	22	[6]
[Pd(PPh$_3$)$_2$(*p*-benzoquinone)]	PPh$_3$, *N*-ethylpiperidine, hydroquinone, *p*-benzoquinone	CH$_3$CN	60 (3)	51	1.7	2.8	[6]
[Pd(PPh$_3$)$_2$(*p*-benzoquinone)]	PPh$_3$, *N*-ethylpiperidine, hydroquinone, *p*-benzoquinone	CH$_3$CN	60 (18)	81	3.5	2.6	[6]
[(η^3-2-Me-C$_3$H$_4$)Pd(OAc)]	P(*i*-Pr)$_3$	C$_6$H$_6$	70 (20)	27	19	25	[5 b]

effects is, however, a necessary prerequisite for the development of new and more effective catalysts. Recent results from an investigation that combines classical ligand concepts and a simple molecular modeling approach show a strong dependence between solid-state structure parameters of transition metal–phosphine complexes and their catalytic activity [9].

The accepted mechanistic suggestion [3 a] as shown in Scheme 1 is related to butadiene oligomerization and telomerization (with nucleophiles): analogously, two molecules of butadiene undergo a C–C coupling at a low-valent palladium complex (Structure **3**) with formation of a bis(allyl)palladium intermediate (**5**). The necessary coordination site is made available by dissociation of one ligand (e. g., a phosphine ligand). The following steps are CO$_2$ insertion resulting intermediate **6** in and reductive elimination of the product molecule with simultaneous isomerization (**6** is also reported to be the intermediate for coupling products other than **2** [3 f]). The effects of the various additives (see Table 1) on the catalytic cycle have not yet been understood in full detail.

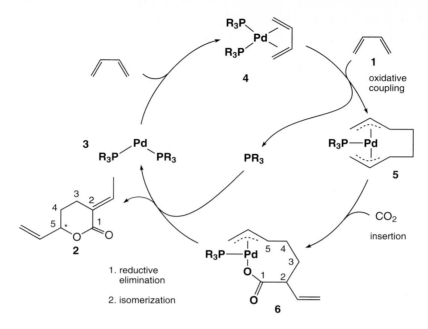

Scheme 1. Key steps of the catalytic cycle for the formation of **2**.

The observation that nitrile solvents are mostly necessary to achieve a high catalytic efficiency ([4c]; see also Table 1) led to the development of palladium catalysts with hemilabile ligands of the general formula $R_2P(CH_2)_nCN$ [13a]. Pitter et al. showed that these P,N ligands enable a conversion in a number of alternative solvents such as tetrahydrofurane (THF) or benzene [13b]. The nitrile group of the hemilabile ligand obviously compensates the polar function of the usually applied solvent, acetonitrile. Also, in the absence of any solvents (homogeneous catalysis in liquid butadiene/CO_2) a butadiene conversion of up to 95 % is achieved; this is advantageous for future process development, since no supplementary solvent is needed and the separation of the solvent from the product is unnecessary. Reasonable yields of **2** are only obtained with a spacer length of more than five CH_2 units, in accordance with the hemilabile character of the P,N ligand including a chelating coordination mode at the palladium atom.

Progress in process development for the synthesis of **2** recently was made by Behr and co-workers. Extraction of a palladium-phosphine catalyst by use of 1,2,4-butanetriol as extractant offers an effective separation from **2** and also an easy catalyst recycling [14]. Pitter et al. have shown that immobilization of homogeneous palladium catalysts on a polystyrene support is an alternative to the homogeneously catalyzed synthesis which enables easy catalyst recovery [15].

A bis(dicyclohexylphosphino)butane (DCPB)-based palladium catalyst was found to catalyze the analogous reaction between isoprene and CO_2 [11]. A mixture of lactones **7** and **8** is obtained but the yield of co-oligomerization products is significantly lower (8 %) than for the reaction of 1,3-butadiene.

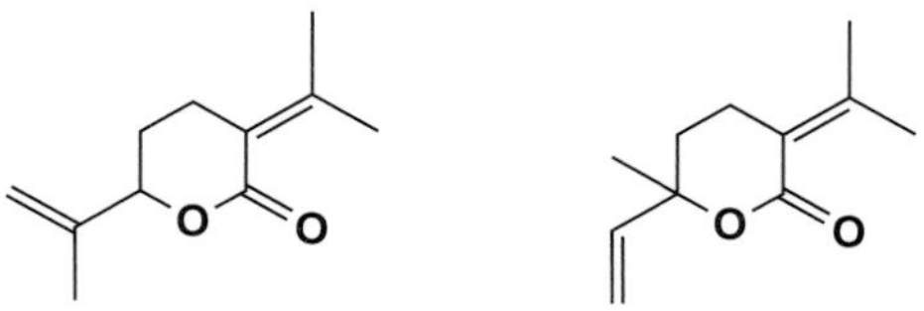

7   8

3.3.4.2.2   Nickel Catalyzed Cotrimerization of Alkynes and CO_2 to 2-Pyrones

Investigations in Homogeneous Solution Under Classical Conditions

The formation of 2-pyrones (**9**) from CO_2 and alkynes was first described by Inoue and co-workers using $Ni(cod)_2$ and chelating phosphines as catalysts [16]. Yields were very low, however, even under drastic reaction conditions. It was shown that the catalytic system $Ni(cod)_2/PR_3$ in acetonitrile/THF gave higher turnover numbers and a very high selectivity under mild conditions (cf. eq. (2)) [17]. The catalytic conversion of alkynes with CO_2 represents the sole example until now of a homogeneous catalytic reaction which yields C–C bond formation with CO_2 and selective formation of cyclooligomers using a cheap 3d-metal complex catalyst. The variation of alkyne substituents allows the synthesis of a wide range of 2-pyrones (mono- and disubstituted alkynes and alkynes with functional groups such as –OR and –OOR [18]).

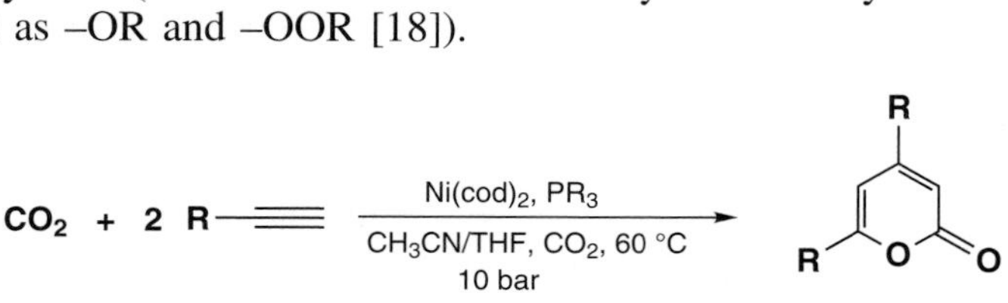

$$CO_2 + 2 \ R-\!\!\!\equiv\!\!\!- \xrightarrow[\substack{CH_3CN/THF, CO_2, 60\,°C \\ 10\ bar}]{Ni(cod)_2, PR_3}$$

(2)

9

By systematic variation of the phosphine ligands it was found that the most efficient catalyst systems are formed from basic phosphines with small cone angles. The optimum ligand-to-metal ratio ranges between one and two. A phosphine excess decreases the catalytic activity, probably due to the formation of inactive coordinatively saturated nickel complexes. The reaction is inhibited at CO_2 pressures above 30 bar in conventional solvents. The decrease in catalytic activity may be caused by the formation of inactive $Ni(CO_3)$ and $Ni(CO)_4$ at high CO_2 pressure. Analogously, Tsuda and co-workers synthesized novel polymeric materials (eq. (3)) by reaction of long-chain a,ω-dialkynes also by nickel catalysis [18 e].

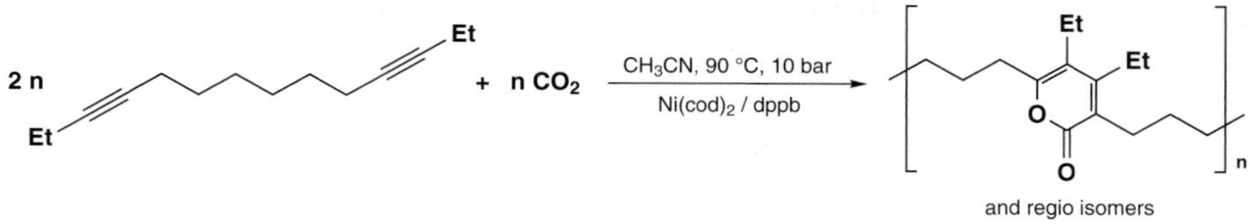

Principle mechanisms of pyrone formation are summarized in Scheme 2. The probable pathway via Structures **10–11–12** is based on related stoichiometric reactions with model complexes [19] and X-ray structural investigations on precatalysts, and is consistent with the experimental details. In **11**, the sp^2 center next to nickel is suitable for the insertion of further alkynes, yielding the intermediate **12**. Reductive elimination of the product **9** and addition of a further alkyne molecule closes the cycle. Analogous complexes to the intermediate **11** were shown to be versatile stoichiometric reagents for transformations to unsaturated acids and esters, by the groups of Hoberg, Dinjus, and Walther [20].

Dunach and co-workers synthesized unsaturated carboxylates instead of pyrones, using electrochemically generated Ni^0 centers from alkynes and CO_2, and they proposed the same initial coupling product, **11**. The formation of unsat-

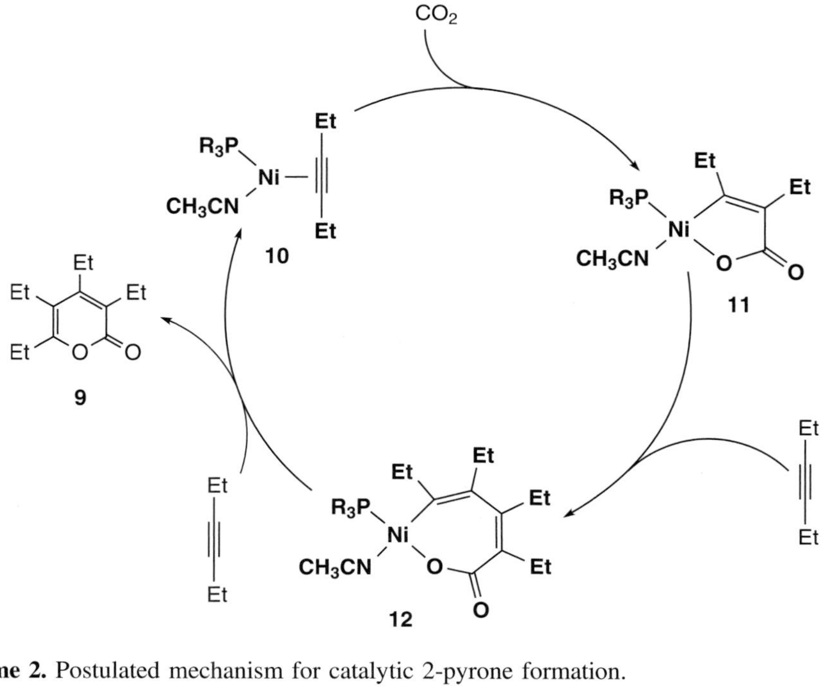

Scheme 2. Postulated mechanism for catalytic 2-pyrone formation.

urated acids is a catalytic process relating to Ni^0 but the necessary presence of Mg^{2+} ions is realized by use of a sacrificial Mg anode [21].

In 1993 Reetz et al. [22 a] and later Dinjus [22 b] reported on a nickel-catalyzed 2-pyrone synthesis in supercritical (sc) CO_2 by means of ([Ni(cod)$_2$]/$Ph_2P(CH_2)_4PPh_2$=dppb) as catalyst (see also [14, 22]). The utilization of supercritical fluids, in particular CO_2, recently thoroughly reviewed in [23], shows – apart from the well-known technologies – many advantages in chemical reactions with respect to its special properties such as variable density, high fluidity, and miscibility with other gases. The use of (2 PMe_3/Ni(cod)$_2$) as catalyst enables faster 2-pyrone formation in sc CO_2 than with dppb [24].

3.3.4.3   Transition Metal Catalyzed Formation of Formic Acid and its Derivatives from CO_2 and H_2

Transition metal catalyzed C–C bond-forming reactions involving CO_2 as a C_1-building block offer an interesting approach to highly functionalized organic molecules. The catalytic addition of hydrogen to CO_2 also provides an important starting point for the utilization of CO_2, as several technically important basic chemicals can be produced in this way (Scheme 3). Equivalents necessary for the reduction of CO_2 are also available from direct electron transfer processes. Both cases yield formic acid as product with the oxidation number +2.

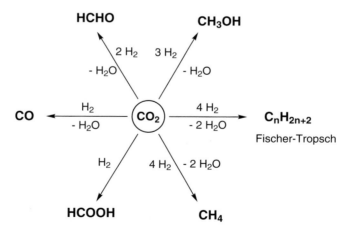

Scheme 3. Theoretical possibilities for the reduction of CO_2.

3.3.4.3.1 Direct Synthesis of Formic Acid and Formates

Thermodynamic Situation in the Hydrogenation of CO_2

The formation of formic acid from carbon dioxide and hydrogen is an exothermic but strongly endergonic process under standard conditions. The equilibrium in eq. (4) lies therefore far to the left ($\Delta H^{\ominus} = -31.6$ kJ mol^{-1}; $\Delta G^{\ominus} = +32.9$ kJ mol^{-1}).

$$CO_2\,(g) \;+\; H_2\,(g) \quad \overset{\text{catalyst}}{\rightleftharpoons} \quad HCOOH\,(l) \qquad\qquad (4)$$

This unfavorable situation is ruled by the large difference in entropy between the two gaseous reactants and a liquid product that forms very strong intermolecular hydrogen bonds. A suitable set of reaction conditions for the formation of formic acid from CO_2 and H_2 has to decrease this entropic gap. High pressure and relatively low temperatures will obviously help to shift the equilibrium to the right. Even more important is the choice of the right solvent, as solvation will not only lower the entropy of the reactants by enclosing them in a solvent cage, but may also break up the strong hydrogen bonds between HCO_2H molecules. The small negative value of the Gibbs free energy in aqueous solutions strongly supports these considerations. Base addition will work in the same direction, especially if amines are used, as they are known to form stable adducts with carbon dioxide. Another possibility of shifting the equilibrium (cf. eq. (4)) to the right is by trapping formic acid in the form of derivatives such as esters or *N,N*-dimethylformamide (DMF). The first report on the direct formation of formic acid from carbon dioxide and hydrogen was published as early as 1970 by Haynes et al. [25]. Wilkinson's catalyst $(Ph_3P)_3RhCl$ and other Group 9 and 10 transition metal complexes were used as catalysts [25 b–d]. A positive effect of small amounts of water was also described. Industrial research groups became interested in this reaction [26 a–e].

Investigations Under Classical Conditions

The first homogenously catalyzed example was demonstrated by Inoue et al. in 1976 [26 f]. They used rhodium(I) phosphine complexes, including Wilkinson's catalyst for the catalytic hydrogenation of CO_2 in benzene solution in the presence of tertiary amines. Inoue's catalyst showed a better performance when small amounts of water were added but the TON did not reach more than 150, even under drastic reaction conditions. Other investigations showed the possibility of obtaining higher yields when an isopropanol/amine mixture containing up to 20 % water was used [26 d]. Aqueous solutions often have higher rates and yields than the systems in organic solvents. The accelerating effect of small amounts of water in organic solvents allows several mechanistic explanations [26, 27]. It is possible that a donating interaction between water and the CO_2 carbon atom increases the nucleophilicity of the CO_2 oxygen atoms and that the capacity of the CO_2 to bind to a metal center is intensified in this way. Calculations by *ab initio* SCF methods confirm that a CO_2–water interaction, as described, is more stable than either of the two species [25]. Carbon monoxide is shown not to act as an intermediate in CO_2 hydrogenation. The addition of CO as reactor co-gas using $RuCl_3/PPh_3/NEt_3$ as the catalytic active system shows a drastic decrease in activity forming a catalytically inactive $Ru(CO)_3(PPh_3)_2$ species [28 d].

Rhodium formate complexes **14** have been inferred as possible key intermediates during the catalytic cycle of CO_2 hydrogenation in DMSO/NEt$_3$ mixtures [29 a, b]. Therefore, the complexes [{R$_2$P–(X)–PR$_2$}Rh(hfacac)] (Structure **13**) have been introduced as stable model compounds for **14** [30].

13   *14*

Different complexes of Structure **13** were synthesized in order to further improve the catalytic activity by variation of the ligand structure.

Very fast formation of HCO_2H is observed when a solution of [{Ph$_2$P(CH$_2$)$_4$PPh$_2$}Rh(hfacac)] (2.5×10^{-3} mol dm^{-3}) in DMSO/NEt$_3$ (5:1) is stirred under H_2/CO_2 (1:1, 4 MPa) in a stainless steel autoclave at 25 °C [31].

Complexes **13** are ideally suited for a systematic study of structural changes in rhodium–phosphine chelates upon small changes in the ligand structure, as there is no steric interaction between the phosphine ligand and the hfacac moiety [30–32].

The influence of the ligand on the coordination sphere of rhodium complexes **13** in the solid state is prevalent in solution also, as seen from the linear correlation between the P–Rh–P angles and the ^{103}Rh chemical shifts [31, 32]. The chemical shift of the ^{103}Rh nucleus has been determined from 2D-(^{31}P,^{103}Rh)–{^{1}H}-NMR experiments. For the series of ligands R$_2$P(CH$_2$)$_n$PR$_2$ of complexes **13** a linear increase of the relative catalytic activity in CO_2 hydrogenation with increasing δ-values is observed. The fact that larger ligands coordinated to the rhodium center accelerate the catalytic activity is reflected by the results of CAMD calculations. The elimination of the product (formic acid) seems to be the rate-determining step. Up to 2200 mol of HCO_2H per mol of rhodium with turnover frequencies as high as 374 h^{-1} can be achieved with the *in situ* catalyst [Rh(cod)H]$_4$/dppb [29 b].

The accessible molecular surface (AMS) model is introduced as a unique approach for the description of steric ligand effects in homogeneous rhodium-catalyzed hydrogenation of CO_2 to formic acid [33].

Hydrogenation of CO_2 in Aqueous Solution

As CO_2 removal from process waste gases is predominantly carried out in water, the hydrogenation of CO_2 in aqueous solution is a very attractive starting point for the utilization of the raw material CO_2. Only a few attempts have been made in recent decades to carry out catalytic hydrogenation of CO_2 in water as solvent [35–37]. Transition metal complexes incorporating phosphine ligands which have been proved as catalysts in organic solvents are not suitable for use in aqueous solution for reasons of nonsolubility under these conditions. Only when complexes of rhodium, containing the water-soluble phosphine $P(C_6H_4\text{-}m\text{-}SO_3Na)_3$ (TPPTS, cf. Section 3.1.1.1) [38] were used homogeneous catalytic systems could be obtained, which show higher activities and better yields as catalysts in organic solvents [39]. For the hydrogenation of CO_2 in aqueous solution, catalysts formed *in situ* from suitable precursors and TPPTS are used, but the most effective system until now is found with the water-soluble analog of Wilkinson's catalyst $[ClRh(TPPTS)_3]$. Equation (5) presents the reaction conditions leading to a TON of 3440 and a TOF of 1365 h^{-1} [39]. It is noteworthy that the amine concentration is never exeeded by formic acid concentration in aqueous systems, and formic acid formation is absolutely suppressed without addition of any amine [39 a].

$$CO_2 \;+\; H_2 \quad \xrightarrow[\substack{40\ \text{bar, 12 h, r. t.} \\ H_2O,\ Me_2NH}]{[ClRh(TPPTS)_3]} \quad HCOOH \tag{5}$$

$$TON = 3440$$

Homogeneous catalytic hydrogenation of HCO_3^- to HCO_2^- in aqueous solution has been reported for the first time [36 a, b].

Mechanistic Investigations of the Hydrogenation of CO_2

The key step in the catalytic formation of formic acid from carbon dioxide and dihydrogen is the formation of a new formate C–H bond. The formate unit on the metal center could be theoretically realized firstly by insertion of CO_2 into a metal–hydride bond and secondly by hydride transfer to coordinated CO_2. In all cases where the catalytically active intermediates in the hydrogenation of CO_2 to formic acid has been proved by spectroscopic methods, the formation of a formate complex was found. Many stoichiometric reactions give hints to the generally accepted mechanism of CO_2 insertion based on experimental and theoretical work [40–43]. Extensive mechanistic studies on the formation of formic acid have been carried out with the rhodium complex $[Rh(Me_2PPh)_3\text{-}(nbd)]BF_4$ (**15**) in THF under increased pressure in absence of amines, using IR and NMR spectroscopical methods, [27]. These investigations proved the formation of a cationic dihydro complex (**16**) leading to compounds **17** and **18** with a η^1- and η^2-bound formate ligand by insertion of CO_2 into the Rh–H bond (cf. Scheme 4). This mechanism is not transferable to the considerably more effective rhodium catalyzed hydrogenation of CO_2 to formic acid in DMSO/NEt$_3$ mixtures. For the most active catalytic systems containing a Rh/P$_2$ ratio of 1:1 (P$_2$ = chelating bisphosphane), the mechanism presented in Scheme 5 has been postulated. The catalytic cycle starts with an electronically unsaturated 14-electron species, the neutral hydrido complex of Structure **20**, which has been already described in the literature [45, 46].

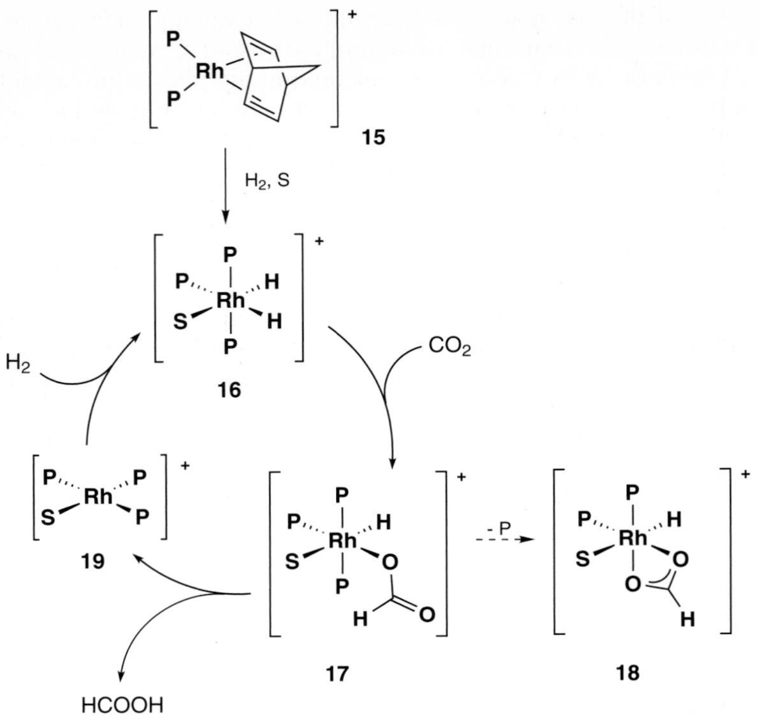

Scheme 4. Catalytic cycle for the hydrogenation of CO_2 with catalyst [(nbd)Rh(Me$_2$PPh)$_2$] [BF$_4$] in THF according to [27]. S = solvent.

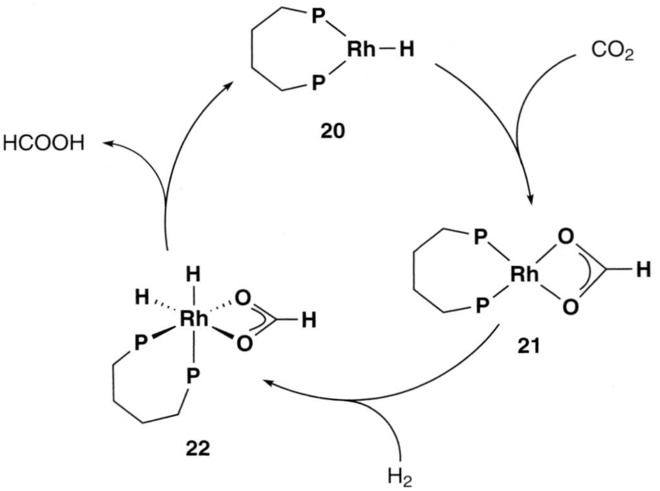

Scheme 5. Postulated mechanism for the hydrogenation of CO_2 to formic acid with the most active catalytic system dppb/Rh with a ligand/metal ratio of 1:1. P = PPh$_2$.

These mechanistic investigations are supported by theoretical studies on the model compound $[(PH_3)_3RhH]$ and they confirm the formation of a formate unit coordinating in a η^1-binding mode in the presence of three phosphine ligands [47]. Recent *ab initio* calculations pointed out that the η^2-formate unit is the most stable coordination mode in complexes, but the species incorporating the η^1-bonded formate seems to be the more reactive intermediate [48].

In spite of the elimination of formic acid in a couple of steps changing the oxidation number of the rhodium metal center from +1 to +3 and *vice versa*, the reaction could take place by an alternative mechanistic pathway via σ-metathesis between the coordinated formate unit and the nonclassical bound hydrogen molecule [48, 49]. Initial rate measurements of a complex of the type **13** show that kinetic data are consistent with a mechanism involving a rate-limiting product formation by liberation of formic acid from an intermediate that is formed via two reversible reactions of the actual catalytically active species, first with CO_2 and then with H_2. The calculations provide a theoretical analysis of the full catalytic cycle of CO_2 hydrogenation. From these results s-bond metathesis seems to be an alternative low-energy pathway to a classical oxidative addition/reductive elimination sequence for the reaction of the formate intermediate with dihydrogen [48 a].

Hydrogenation of CO_2 Under Supercritical Conditions

Carbon dioxide in its supercritical state is a reaction medium of great interest. Noyori and co-workers [50] recently discovered that ruthenium(II)–phosphine complexes of structure $[(X)_2Ru(PMe_3)_4]$ (**23** (X = H) and **24** (X = Cl)) can act as highly active catalysts for an effective transition metal catalyzed hydrogenation of CO_2 to formic acid in a supercritical mixture of CO_2, H_2, and NEt_3 without use of any further solvent. According to eq. (4) a TON of 7200 per mol Ru in $scCO_2$/ Et_3N at 50 °C is reached. In the supercritical state the reaction rate is about 18 times higher than under comparable conditions in THF. This observation is explained by particular properties of the supercritical phase relating to miscibility and mass transport of the reactants. The catalysts were also selected because of their good solubility, which is similar to that in hexane. In order to guarantee a homogeneous supercritical phase during the reaction process, the reaction conditions (pressure and temperature) have to be more drastic than those described up to now. It is noteworthy that traces of water are indicated as necessary for reaching a high reaction rate, too [51].

23   24

After reaching the equilibrium concentration, the HCO_2H/NEt_3 ratio is about 1.6:1. The formic acid reacts with dimethylamine, if present in the reaction mixture, to give DMF; without addition of any amine, no formation of formic acid takes place. The high performance of the formic acid production in the case of ruthenium complex catalysis is bound to the supercritical CO_2 phase. If the reaction is carried out in liquid CO_2 with comparable CO_2 concentrations, but at 15 °C instead of 50 °C under otherwise identical reaction conditions, the TON decreases from 7200 to 20 and the TOF from 1400 to 1.3 h^{-1}. This means that the remarkable catalytic activity and efficiency are based on the characteristic properties of $scCO_2$, such as the extremely high miscibility with hydrogen [51–53] and a good mass-transfer capability.

Under these conditions, $scCO_2$ becomes an excellent medium for its own hydrogenation [7 a, 27, 29 c, 31, 39 a, 50 b]. If it is possible to develop a continuous-flow system to solve the problems of extracting and recycling amine and catalyst after the release of formic acid, the industrial realization of this high-pressure process can be expected.

3.3.4.3.2   Synthesis of Formic Acid Derivatives

Under suitable conditions the hydrogenation of CO_2 can lead to amides or esters in the presence of amines or alcohols, with formation of free formic acid in between. These derivatives of formic acid are stable under standard conditions in the presence of the catalyst.

Synthesis of Alkyl Formates

In 1970 the transition metal catalyzed formation of alkyl formates from CO_2, H_2, and alcohols was first described. Phosphine complexes of Group 8 to Group 10 transition metals and carbonyl metallates of Groups 6 and 8 show catalytic activity (TON 6–60) and in most cases a positive effect by addition of amines or other basic additives [26 a, 54–58]. A more effective catalytic system has been found when carrying out the reaction in the supercritical phase (TON 3500) [54 a]. Similarly to the synthesis of formic acid, the synthesis of methyl formate in $scCO_2$ is successful in the presence of methanol and ruthenium(II) catalyst systems [54 b].

The reaction mixture forms a single supercritical phase. The time dependence of the product formation shows that formic acid is formed first with a subsequent esterification which must take place thermally. Amine is an inhibitor of esterification but its presence is required for reasonable yields in the hydrogenation step.

The use of supercritical conditions for the homogeneous hydrogenation of CO_2 with following thermal esterification leads to high yields of methyl formate under mild conditions. Therefore, it might be possible to develop an industrial procedure for the synthesis of methyl formate with CO_2 as C_1-building block when catalysts and reaction conditions are optimized.

Catalytic Production of Dimethylformamide (DMF) from scCO$_2$

DMF as a useful polar solvent is produced industrially on a large technical scale (250 000 tons/year) by carbonylation of dimethylamine in the presence of methanol [59]. Using Raney nickel as catalyst, the synthesis of DMF from dimethylamine, CO$_2$, and hydrogen was first discovered by Farlow and Adkins [60]. The formation of DMF from dimethylamine, H$_2$, and CO$_2$ is thermodynamically favorable under standard conditions; thermodynamic data are given for aqueous reactants and liquid products in eq. (6) [61]. The enthalpy of DMF production ($\Delta H^0 = -56.5$ kJ mol^{-1}, $\Delta G^0 = -0.75$ kJ mol^{-1}, $\Delta S^0 = -119$ kJ mol^{-1}K^{-1}) is more favorable than that for methyl formate ($\Delta H^0 = -15.3$ kJ mol^{-1}).

$$CO_2 + H_3C\overset{\overset{H}{|}}{N}CH_3 + H_2 \longrightarrow H\overset{\overset{O}{||}}{C}\underset{\underset{CH_3}{|}}{N}CH_3 + H_2O \qquad (6)$$

Homogeneous catalysis of this reaction was first reported by Haynes et al. in 1970 [62]; with palladium catalyst in benzene as a solvent and at 5.6 MPa (H$_2$/CO$_2$) and 100 °C they could realize a TON of 1200 within 17 h. More efficient was a ruthenium complex with dppe as a chelating phosphine ligand described by Kiso and Saeki [63] with a TON of 3400 in hexane and similar conditions. Jessop et al. [64] found that, in the presence of a catalytic amount of RuCl$_2$- [P(CH$_3$)$_4$] as a catalyst precursor, scCO$_2$ reacts with H$_2$ and dimethylamine to give DMF with a TON up to 370 000 within 37 h. As a source of dimethylamine they used liquid dimethylammonium dimethylcarbonate, but dimethylamine gave identical results. The conversion of dimethylamine reached 94 %, and the selectivity for DMF was 99 %. The TON of 370 000 is superior to the largest TON of 3400 for DMF formation from CO$_2$ in a conventional liquid solvent. The authors discuss the production of DMF from scCO$_2$ proceeding in two steps on the basis of the composition of the product as a function of reaction time.

Complexes prepared from RuCl$_3$ and bidentate phosphine ligands (dppm, dppe, dppp, dmpe) have been shown to be the most active catalyst precursors in DMF synthesis from CO$_2$ in the presence of dihydrogen and dimethylamine known up to now. The highest TOF of 360 000 h^{-1} can be reached with RuCl$_2$(dppp)$_2$ as precursor. To avoid difficulties in separating the homogeneous catalysts from the products a solvent-free reaction design is extended to this hydrogenation reaction. The advantages of both homogeneous and heterogeneous catalysts were combined by anchoring catalytically active metal complexes via organic groups within an oxide network. For this, silyl ether complex analogs of group VIII metal complexes have been incorporated into a silica matrix by the sol–gel method, resulting in stable hybrid-gel catalysts, which can be easily separated from the reaction mixture by filtration (TOF up to 1860 h^{-1}) [54 b].

The fast Ru-catalyzed hydrogenation of CO$_2$ to formic acid is followed by the slower thermal condensation of formic acid and dimethylamine. Dimethylamine acts as a base to stabilize the formic acid in the first step and serves as a reactant in the second step.

The driving force for this process is probably the existence of a two-phase system with a supercritical phase and a liquid phase. The overall combination of the steps in a one-pot procedure is also responsible for the high rate of DMF production (cf. eqs. (7) and (8)). With this improved catalytic efficiency and the lower toxicity of CO_2 compared with CO, the reaction of CO_2 with hydrogen and dimethylamine could become competitive with the carbonylation of dimethylamine as an industrial method for DMF production.

$$CO_2 \ + \ H_2 \ \xrightarrow[\text{base}]{\text{Ru catalyst}} \ HCOOH \tag{7}$$

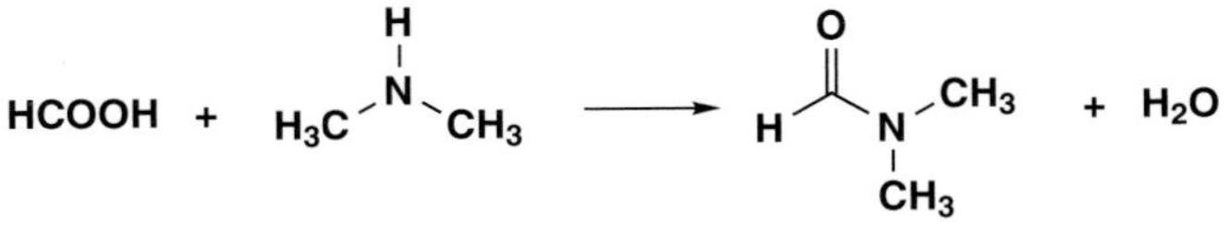

$$HCOOH \ + \ H_3C\underset{}{\overset{\underset{\displaystyle N}{\overset{\displaystyle H}{|}}}{}}CH_3 \ \longrightarrow \ H\underset{\underset{\displaystyle CH_3}{\overset{\displaystyle |}{N}}}{\overset{\displaystyle O}{\overset{\displaystyle \|}{C}}}\!\!\diagdown\!CH_3 \ + \ H_2O \tag{8}$$

Catalytic Syntheses of Formoxysilanes from CO_2

Independently, in 1981 two groups reported the hydrosilylation of carbon dioxide into formoxysilanes of the type $R_2R'SiOCHO$ (R,R' = alkyl) catalyzed by transition metal complexes, preferably based on ruthenium (eq. (9)) [65].

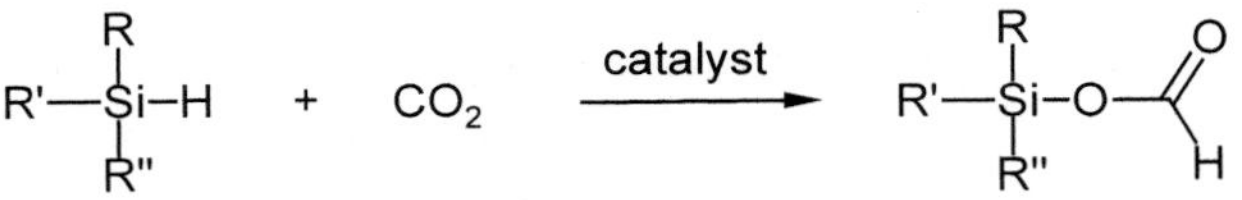

$$R'\!\!-\!\!\underset{R''}{\overset{R}{Si}}\!\!-\!\!H \ + \ CO_2 \ \xrightarrow{\text{catalyst}} \ R'\!\!-\!\!\underset{R''}{\overset{R}{Si}}\!\!-\!\!O\!\!-\!\!\overset{O}{\underset{H}{C}} \tag{9}$$

Ligand-modified Ir, Ru, or Pd catalysts achieve TONs up to 465 and TOFs up to 232, and yields up to 90 % [65 a, b, 66 a, 67, 69].

Jessop reported on the utilisation of sc CO_2 as substrate and solvent for the hydrosilylation reaction, but conversion and selectivity are comparatively low [68]. Most recently, Pitter and co-workers reported the use of *trans*-[Ru^{II}Cl-(MeCN)$_5$][Ru^{III}Cl$_4$(MeCN)$_2$] derived from partial reduction of Ru^{III} in acetonitrile solution as a highly efficient catalyst for the synthesis of several formoxysilane derivatives under moderate conditions [69]. Analogously, Et_2SiH_2, Ph_2SiH_2, and p-$C_6H_4(Me_2SiH)_2$ yield $Et_2Si(OOCH)_2$, $Ph_2Si(OOCH)_2$, and p-$C_6H_4(Me_2Si-OOCH)_2$, respectively. Such multifunctional formoxysilanes are discussed as potential cross-linking agents in RTV silicones. Interestingly, the catalysts have been found to be almost completely recyclable, thus giving rise to much higher TONs [69].

3.3.4.4 Catalyzed Formation of Organic Carbonates

Organic carbonates can be roughly classified into three groups (Structures **25–27**).

$$
\begin{array}{ccc}
R_1-O-\overset{\overset{\displaystyle O}{\|}}{C}-O-R_2 &
\underset{R_1 \quad R_2}{\overset{\overset{\displaystyle O}{\|}}{\underset{O\diagdown C\diagup O}{}}} &
\left[O-\overset{\overset{\displaystyle O}{\|}}{C}-O-[R]\right]_n \\
\textbf{25} & \textbf{26} & \textbf{27} \\
\text{Linear Carbonates} & \text{Cyclic Carbonates} & \text{Polycarbonates}
\end{array}
$$

Linear dialkyl carbonates and in particular dimethyl carbonate (DMC) are used in many industrial applications in the industry [70, 71]. The usual way of synthesizing carbonates involves reactive C_1 agents such as phosgene or CO [72]. Although these methods are from an economical viewpoint more than profitable, the development of an environmentally friendly industrial process involving CO_2 as a C_1-building block attracts an ever-increasing interest.

Besides a heterogeneous catalytic system [73] involving hydroxo tin(IV) compounds, a homogeneous system for the synthesis of DMC from CO_2 was recently reported by Sakakura et al. [74]. This synthesis is also based on the use of a dehydrating agent (an orthoester [74 a] or, less expensively, an acetal [74 c]) and methanol as substrates. A dibutyltin(IV) alkoxide, $Bu_2Sn(OMe)_2$, acts as catalyst and tetrabutylphosphonium iodide as a co-catalyst. The authors proposed the formation of an active tin–methylcarbonato species, $Bu_2Sn(OMe)(CO_3Me)$, as the key intermediate of the reaction [74 b]. It should be mentioned that the formation of DMC from methanol and CO_2 in the presence of tin derivatives had already been reported in earlier work by Kitzlink [75]. Dibutyltin dialkoxides have been known for a long time to react with CO_2, hence forming tin–carbonato derivatives [76, 74 b, 77]. In comparison, tin aryloxides display no noticeable reactivity toward CO_2.

The second main class of organic carbonates, the cyclic carbonates, also known as dioxolanones, have found many applications as versatile intermediates in organic synthesis [70]. They also represent promising building blocks for the production of polyurethane- and polycarbonate-based polymers. The catalytic synthesis of cyclic carbonates is the topic of a regularly increasing number of publications [78] and has led to some noteworthy industrial applications [70]. A general mechanism is summarized in Scheme 6. The catalyst possessing both basic and acidic sites favours the approach of the reactants, epoxide and carbon dioxide, to result in an intermediate (Structure **28**). Some remarkable epoxide–Lewis acid adducts recently were characterized structurally by Darensbourg et al. in the case of cadmium(II) pyrazolylcarboxylato derivatives [78 f]. The coordinated epoxide would then undergo a nucleophilic attack of the "activated" CO_2 molecule. The desired cyclic carbonate is formed via an "intramolecular" ring closure from the alkoxo-alkylcarbonato compound **29**, the active metal center being available afterwards to perform the next cycle.

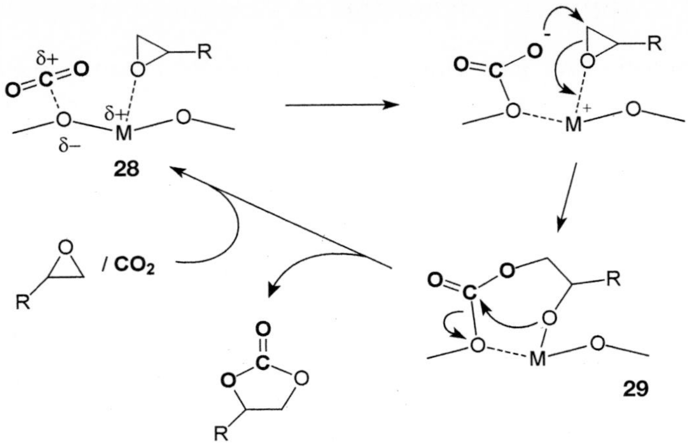

Scheme 6. Key steps of the catalytic synthesis of cyclic carbonates involving CO_2.

The last class of organic carbonates, the polycarbonates, finds ever-increasing use in our modern consumer society [80]. The usual industrial method of synthesizing polycarbonates involves phosgene and bisphenol A derivatives. Two principal classes of catalysts involving CO_2 as a C_1-building block have been reported in the literature. One comprises zinc(II) carboxylates which are obtained from dicarboxylic acids of the type $HO_2C(CH_2)_nCO_2H$, with n varying from 1 to 10 [79, 81]; their common synthesis is based on the reaction of zinc oxide with dicarboxylic acids in aprotic media. The structure of most of these dicarboxylates is still under debate and, despite comprehensive studies on the reactivity of the $Zn(II)/CO_2$/epoxide system, the overall mechanism remains a subject of discussion. First attempts to use $scCO_2$ as reagent and solvent have been newly reported by Beckmann's group [82]; the use of a fluorinated half-ester from maleic acid and tridecafluorooctanol as ligand allows a better solubility of the catalyst in the $scCO_2$ and results in a higher TON and a far better selectivity. Other developments for soluble zinc dicarboxylates with carboxylates bearing unsaturated spacers like ferrocenes or constrained alkenes have been reported [83].

The second class of catalysts are zinc(II) mono- or dialkoxides obtained from polyhydric phenols and dialkylzinc with partly polymeric structures. This system, extensively studied by Kuran [84], is an optimization of the water/diethylzinc and polyphenol/diethylzinc systems developed by Inoue [85]. The use of soluble zinc phenoxides and their analogous cadmium complexes as catalyst for the copolymerization of CO_2 and epoxide was studied extensively by the Darensbourg group [86]. This work focused on the use of mononuclear phenoxide derivatives with bulky substituents, e. g., phenyl- and *tert*-butyl groups, on the aromatic ring to a homogeneous catalytic system and thus enhance the activity of the Zn^{II} phenoxides. The catalysts developed are stabilized through ancillary neutral ligands and include, to give an example, $(2,6-(C_6H_5)_2C_6H_3O)Zn(THF)_2$, $(2,4,6-(C_4H_9)_3C_6H_2O)Zn(THF)_2$.

The high reactivity of these two classes of catalysts, carboxylate and alkoxide derivatives, has been confirmed by recent work of Coates and co-workers [87]. They reported the synthesis of two new types of Zn^{II} diimido complexes (**30** and **31**) as shown in Scheme 7 and successfully utilized both types of complexes in the copolymerization of CO_2 with epoxides. Their high activities and selectivies in regard to the carbon dioxide insertion (up to 96 % carbonate linkages) are unprecedented.

Both catalytic systems, alkoxides and carboxylates, are often described as "efficient" catalysts for the copolymerization of CO_2 and epoxides but some drawbacks which hamper a widespread industrial utilization need to be pointed out. The phenoxides, though displaying good selectivities, have up to now only been tested with model substrates, e. g., propylene- and cyclohexene oxides, and the carboxylates, though active, present low-to-fair selectivities. Cyclization

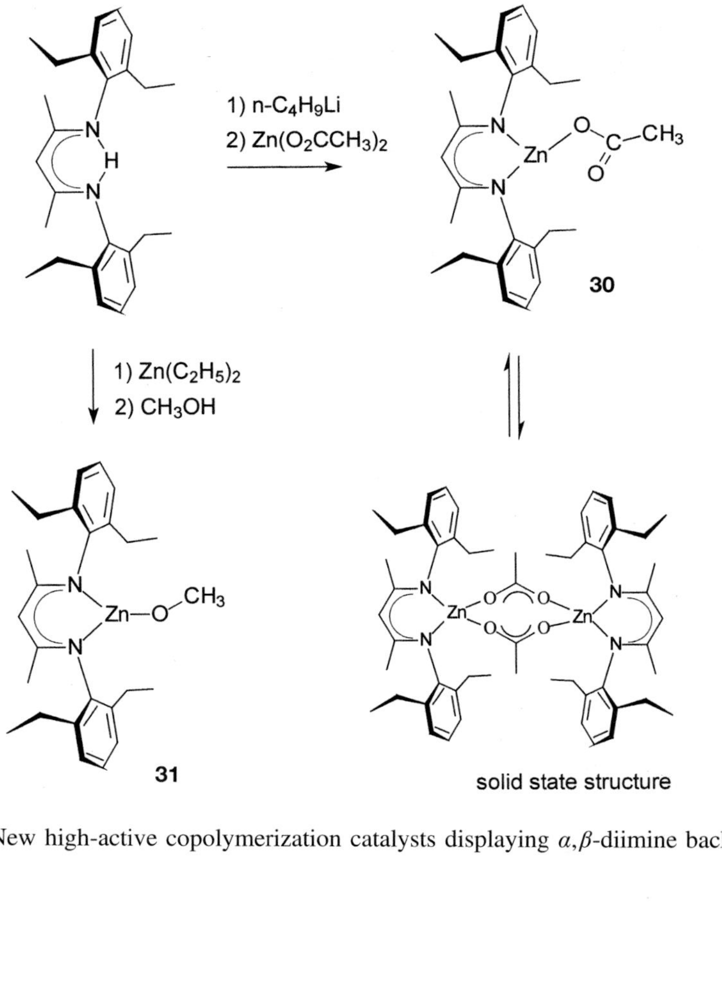

Scheme 7. New high-active copolymerization catalysts displaying α,β-diimine backbone.

is an important side-reaction due to the thermodynamic stability of the five-membered dioxolan-2-one; this phenomenon is more likely to occur at higher temperatures. Their presence in the reaction mixture, although they are attractive building blocks for other organic syntheses, complicates the separation and purification procedure of the desired polycarbonates.

3.3.4.5   Summary and Outlook

Recent research has shown that CO_2 can be utilized as a C_1-building block for the synthesis of both bulk and fine chemicals. There are quite a few examples of very efficient processes using homogeneous catalysts and some of their mechanisms are fairly well understood. Mostly the catalysts consist of transition metal–phosphine complexes and in some special cases it was shown that classical ligand concepts, together with the use of modern computing methods, may lead to a better understanding of ligand effects and finally to the development of more active catalysts. Compared with conventional processes utilizing other C_1 sources, it has to be pointed out that alternative routes with CO_2 require a marketable basis, as for example in the production of polycarbonates.

The use of ecologically harmless $scCO_2$ as solvent and substrate in chemical reactions is a particularly intriguing prospect. Increased governmental and environmental restrictions on solvent emission make this supercritical fluid more and more attractive as a reaction medium because it can be easily separated from the product and recycled more efficiently than conventional liquid solvents. The special properties (miscibility, transport properties, etc.) of sc CO_2 require a development of suitably adjusted catalysts. A simple transformation of catalyst properties from conventional solvents to $scCO_2$ will mostly fail, and will not lead to higher catalytic efficiency. Supported catalysts could perhaps play a particular role in this field as the possibility of product extraction by depressurization of the supercritical phase and subsequent compression of the CO_2 (solvent/substrate) should permit the development of a profitable continuous process.

The optimization of existing processes, the investigation of current reaction pathways, and the search for novel catalytic reactions involving CO_2 as a chemical feedstock still remain an important and motivating research area.

References

[1] (a) *Organic and Bio-Organic Chemistry of Carbon Dioxide* (Eds.: S. Inoue, N. Yamazaki), John Wiley, New York, **1982**; (b) A. Behr, *Carbon Dioxide Activation by Metal Complexes*, VCH, Weinheim, **1988**; (c) *Catalytic Activation of Carbon Dioxide* (Ed.: W. M. Ayers), ASC Symposium Series 363, American Chemical Society, Washington DC, **1988**; (d) *Enzymatic and Model Carboxylation and Reduction Reactions for Carbon Dioxide Utilization* (Eds.: M. Aresta, J. V. Schloss), NATO ASI Series C, 314, Kluwer Academic Press, Dordrecht, **1990**; (e) *Electrochemical and Electrocatalytic Reactions of Carbon Dioxide* (Eds.: B. P. Sullivan, K. Krist, H. E. Guard), Elsevier, Amster-

dam, **1993**; (f) M. M. Halmann, *Chemical Fixation of Carbon Dioxide*, CRC Press, Boca Raton, **1993**; (g) *Carbon Dioxide Chemistry: Environmental Issues* (Eds.: J. Paul, C.-M. Pradier), Royal Society of Chemistry, London, **1994**; (h) W. Leitner, E. Dinjus, F. Gaßner, CO_2 Chemistry, in *Aqueous Phase Organometallic Catalysis – Concepts and Applications* (Eds.: B. Cornils, W. A. Herrmann), Wiley/VCH, Weinheim, **1998**, Chapter 6.15, p. 486.

[2] (a) *Greenhouse Gas Emissions from Power Stations*, IEA Greenhouse Gas R&D Programme, Cheltenham, **1993**; (b) C.-D. Schönwiese, B. Diekmann, *Der Treibhauseffekt. Der Mensch verändert das Klima*, Rowohlt, **1990**; (c) W. Seifritz, *Der Treibhauseffekt*, Carl Hanser Verlag, Munich, **1991**; (d) *The Handbook of Environmental Chemistry, Vol. 1, Part A* (Ed.: O. Hurtzinger), Springer Verlag, Berlin, **1980**; (e) R. Kümmel, S. Papp, *Umweltchemie*, VEB Deutscher Verlag für Grundstoffindustrie, Leipzig, **1990**; (f) E. T. Sundquist, *Science* **1993**, *259*, 934; (g) P. S. Zurer, *Chem. Eng. News* **1991**, *69(13)*, 7; (h) J. J. Sarmiento, *Chem. Eng. News* **1993**, *71(22)*, 30.

[3] (a) A. Behr, *Angew. Chem.* **1988**, *100*, 681; *Angew. Chem. Int. Ed. Engl.* **1988**, *27*, 661; (b) I. S. Kolomnikov, T. V. Lysak, *Russ. Chem. Rev. (Engl. Transl.)* **1990**, *59*, 344; (c) D. Walther, *Nachr. Chem. Tech. Lab.* **1992**, *40*, 1214; (d) M. Aresta, E. Quaranta, I. Tommasi, *New J. Chem.* **1994**, *18*, 133; (e) A. Behr, *Asp. Hom. Catal.* **1988**, *6*, 59; (f) P. Braunstein, D. Matt, D. Nobel, *Chem. Rev.* **1988**, *88*, 747; (g) P. G. Jessop, T. Ikaria, R. Noyori, *Chem. Rev.* **1995**, *95*, 259; (h) W. Leitner, *Coord. Chem. Rev.*, **1996**; (i) D. Walther, E. Dinjus, J. Sieler, *Z. Chem.* **1983**, *23*, 237; (k) D. J. Darensbourg, R. A. Kudaroski, *Adv. Organomet. Chem.* **1983**, *22*, 129; (l) D. Walther, *Coord. Chem. Rev.* **1987**, *79*, 135; (m) P. G. Jessop, T. Ikaria, R. Noyori, *Chem. Rev.* **1995**, *95*, 259; (n) W. Leitner, *Angew. Chem.* **1995**, *107*, 2391; (o) W. Leitner, *Coord. Chem. Rev.* **1996**, *153*, 257; (p) V. Haack, E. Dinjus, S. Pitter, *Angew. Makromol. Chem.* **1998**, *257*, 19.

[4] (a) A. Behr, K.-D- Juszak, W. Keim, *Synthesis* **1983**, 574; (b) A. Behr, K.-D. Juszak, *J. Organomet. Chem.* **1983**, *255*, 263; (c) A. Behr, R. He, K.-D. Juszak, C. Krüger, Y.-H. Tsay, *Chem. Ber.* **1986**, *119*, 991.

[5] (a) A. Musco, C. Prego, V. Tartiari, *Inorg. Chim. Acta* **1978**, *28*, L147; (b) A. Musco, *J. Chem. Soc., Perkin Trans. 1* **1980**, 693.

[6] I. C. I. (J. A. Daniels), EP 0.050.445 (1982); *Chem. Abstr.* **1982**, *97*, 127500w.

[7] (a) Y. Sasaki, Y. Inoue, H. Hashimoto, *J. Chem. Soc., Chem. Commun.* **1976**, 605; (b) Y. Inoue, Y. Itoh, H. Kazama, H. Hashimoto, *Bull. Chem. Soc. Jpn.* **1980**, *53*, 3329.

[8] (a) P. Braunstein, D. Matt, D. Nobel, *J. Am. Chem. Soc.* **1988**, *110*, 3207; (b) Université de Strasbourg Louis Pasteur (P. Braunstein, D. Matt, D. Nobel), FR 2.617.163 (1988); (c) Shell (E. Drent) EP 0.234.668 (1987); (d) Montedison S. p. A. (A. Musco, R. Santi, G. P. Chiusoli), DE 2.838.610 (1979); (e) A. R. Elsagir, PhD Thesis, University of Jena (1997); (f) M. Nauck, PhD Thesis, University of Heidelberg (1998).

[9] Hüls AG (W. Keim, A. Behr, B. Hegenrath, K.-D- Juszak), DE 3.317.013 (1984); *Chem. Abstr.* **1985**, *102*, 78723g.

[10] C. A. Tolman, *Chem. Rev.* **1977**, *77*, 313.

[11] T. L. Brown, *Inorg. Chem.* **1992**, *31*, 1286.

[12] E. Dinjus, W. Leitner, *Appl. Organomet. Chem.* **1995**, *9*, 43.

[13] (a) E. Dinjus, S. Pitter, H. Görls, B. Jung, *Z. Naturforsch. Teil B* **1996**, *51*, 934; (b) E. Dinjus, S. Pitter, *J. Mol. Cat.* **1997**, *125*, 39.

[14] A. Behr, M. Heite, *Chem. Ing. Tech.* **2000**, *72*, 58.

[15] (a) N. Holzhey, S. Pitter, E. Dinjus, *J. Organomet. Chem.* **1997**, *541*, 243; (b) Forschungszentrum Karlsruhe (S. Pitter, N. Holzhey, E. Dinjus), DE 197.25.735 (1998); (c) S. Pitter, N. Holzhey, *J. Mol. Cat.* **1999**, *146*, 25.

[16] Y. Inoue, Y. Itoh, H. Hashimoto, *Chem. Lett.* **1977**, 855.

[17] D. Walther, E. Dinjus, H. Schönberg, J. Sieler, *J. Organomet. Chem.* **1987**, *334*, 377.

[18] (a) T. Tsuda, S. Morikawa, R. Sumiya, T. Saegusa, *J. Org. Chem.* **1988**, *55*, 3140; (b) T. Tsuda, S. Morikawa, T. Saegusa, *J. Chem. Soc., Chem. Commun.* **1989**, 9; (c) T. Tsuda, S. Morikawa, K. Kunisada, N. Nagahama, *Synth. Commun.* **1989**, *19*, 1575; (d) T. Tsuda, S. Morikawa, N. Haseguwa, *J. Org. Chem.* **1990**, *55*, 2978; (e) T. Tsuda, *Polym. Mater. Sci. Eng.* **1999**, *80*, 449.

[19] (a) G. Burkhart, H. Hoberg, *Angew. Chem.* **1982**, *92*, 75; *Angew. Chem. Int. Ed. Engl.* **1982**, *21*, 76; (b) H. Hoberg, A. Schäfer, G. Burkhart, C. Krüger, M.-J. Ramso, *J. Organomet. Chem.* **1984** , *266*, 203; (c) D. Walther, G. Bräunlich, R. Kaupe, J. Sieler, *J. Organomet. Chem.* **1992**, *436*, 109.

[20] (a) D. Walther, E. Dinjus, H. Schönberg, J. Sieler, *J. Organomet. Chem.* **1987**, *334*, 377; (b) O. Lindqvist, L. Anderson, *Z. Anorg. Chem.* **1988**, *560*, 119; (c) E. Dinjus, J. Kaiser, J. Sieler, D. Walther, *Z. Anorg. Allg. Chem.* **1981**, *483*, 63; (d) J. Kaiser, J. Sieler, U. Braun, L. Golic, E. Dinjus, D. Walther, *J. Organomet. Chem.* **1982**, *224*, 81; (e) D. Walther, E. Dinjus, V. Herzog, *Z. Chem.* **1982**, *22*, 303; (f) D. Walther, E. Dinjus, J. Sieler, J. Kaiser, O. Lundquist, L. Andersen, *J. Organomet. Chem.* **1982**, *240*, 289; (g) D. Walther, E. Dinjus, V. Herzog, *Z. Chem.* **1983**, *23*, 188; (h) D. Walther, E. Dinjus, V. Herzog, *Z. Chem.* **1984**, *24*, 260; (i) D. Walther, E. Dinjus, *Z. Chem.* **1984**, *24*, 296; (k) D. Walther, E. Dinjus, *Z. Chem.* **1982**, *22*, 228; (l) E. Dinjus, D. Walther, H. Schütz, W. Schade, *Z. Chem.* **1983**, *23*, 303; (m) D. Walther, E. Dinjus, J. Sieler, N. N. Thanh, W. Schade, I. Leban, *Z. Naturforsch. Teil B* **1983**, *38*, 835; (n) D. Walther, E. Dinjus, *Z. Chem.* **1984**, *24*, 63; (o) D. Walther, E. Dinjus, H. Görls, J. Sieler, O. Lindquist, L. Andersen, *J. Organomet. Chem.* **1985**, *286*, 103; (p) D. Walther, E. Dinjus, J. Sieler, L. Andersen, O. Lindquist, *J. Organomet. Chem.* **1984**, *276*, 99; (q) E. Dinjus, D. Walther, H. Schütz, *Z. Chem.* **1983**, *23*, 408; (r) H. Hoberg, D. Schaefer, *J. Organomet. Chem.* **1982**, *236*, C28; (s) H. Hoberg, Y. Peres, A. Milchereit, *J. Organomet. Chem.* **1986**, *307*, C38; (t) H. Hoberg, Y. Peres, A. Milchereit, *J. Organomet. Chem.* **1986**, *307*, C41; (u) H. Hoberg, D. Schaefer, *J. Organomet. Chem.* **1983**, *238*, 383; (v) G. Burkhart, H. Hoberg, *Angew. Chem.* **1982**, *94*, 75; (w) H. Hoberg, D. Schaefer, G. Burkhart, *J. Organomet. Chem.* **1982**, *228*, C21; (x) H. Hoberg, D. Schaefer, G. Burkhart, C. Krüger, M. J. Ramao, *J. Organomet. Chem.* **1984**, *266*, 203; (y) H. Hoberg, D. Schaefer, *J. Organomet. Chem.* **1983**, *251*, C51.

[21] S. Derieu, J.-C. Clinet, E. Dunach, E., J. Perichon, *J. Org. Chem.* **1992**, *58*, 2578 and references cited therein.

[22] (a) M. T. Reetz, W. Könen, T. Strack, *Chimia* **1993**, *97*, 493; (b) E. Dinjus, Reactions under Extreme and Nonclassical Conditions, COST, Lahnstein, March **1995**; (c) E. Dinjus, R. Fornika, M. Scholz in *Chemistry under Extreme or Non-classical Conditions* (Eds.: R. v. Eldik, C. D. Hubbard), Spektrum Akademischer Verlag, Heidelberg, **1997**.

[23] *Chemical Syntheses Using Supercritical Fluids* (Eds.: P. G. Jessop, W. Leitner), Wiley-VCH, Weinheim, **1999**.

[24] E. Dinjus, C. Geyer, F. Plenz, unpublished results.

[25] (a) P. Haynes, L. H. Slaugh, J. F. Kohnle, *Tetrahedron Lett.* **1970**, 365; (b) K. Kudo, H. Phala, N. Sugita, Y. Takezaki, *Chem. Lett.* **1977**, 1495; (c) S. Schreiner, J. Y. Yu, L. Vaska, *J. Chem. Soc., Chem. Commun.* **1988**, 602.

[26] (a) Mitsubishi Co. (Y. Hashimoto, Y. Inoue), JP 138.614 (1976); *Chem. Abstr.* **1977**, *87*, 67853v; (b) Tjin Ltd. (T. Yamaji), JP 166.146 (1981); *Chem. Abstr.* **1982**, *96*, 122211x; (c) Tjin Ltd. (T. Yamaji), JP 140.948 (1981); *Chem. Abstr.* **1982**, *96*, 68352d; (d) BP Ltd. (D. J. Drury, J. E. Hamlin), EP 95.321 (1983); *Chem. Abstr.* **1984**, *100*, 174262k; (e) BP Ltd. (A. G. Kent), EP 151.510 (1985); *Chem. Abstr.* **1986**, *104*, 109029h; (f) Y. Inoue,

H. Izumida, Y. Sasaki, H. Hashimoto, *Chem. Lett.* **1976**, 863; (g) C. P. Lau, Y. Z. Chen, *J. Mol. Catal.* **1995**, *101*, 33.

[27] J.-C. Tsai, K. M. Nicholas, *J. Am. Chem. Soc.* **1992**, *114*, 5117.

[28] M. T. Ngyen, T.-K. Ha, *J. Am. Chem. Soc.* **1984**, *106*, 599.

[29] (a) T. Burgemeister, F. Kastner, W. Leitner, *Angew. Chem.* **1993**, *105*, 781; *Angew. Chem. Int. Ed. Engl.* **1993**, *32*, 739; (b) W. Leitner, E. Dinjus, F. Gaßner, *J. Organomet. Chem.* **1994**, *475*, 257; (c) E. Graf, W. Leitner, *J. Chem. Soc., Chem. Commun.* **1992**, 623.

[30] (a) P. J. Fennis, P. H. M. Budzelaar, J. H. G. Frijns, A. G. Orpen, *J. Organomet. Chem.* **1990**, *393*, 287; (b) W. Leitner, E. Dinjus, R. Fornika, H. Görls, *J. Organomet. Chem.* **1996**, *511*, 145; (c) R. Fornika, PhD Thesis, Universität Jena (1994).

[31] R. Fornika, H. Görls, R. Seemann, W. Leitner, *J. Chem. Soc., Chem. Commun.* **1995**, 1479.

[32] (a) R. Benn, H. Brenneke, R.-D. Reinhardt, *Z. Naturforsch. Teil B* **1985**, *40*, 1763; (b) R. Benn, H. Brenneke, A. Rufinska, *J. Organomet.Chem.* **1987**, *320*, 115.

[33] (a) K. Angermund, W. Baumann, E. Dinjus, R. Fornika, H. Görls, M. Kessler, C. Krüger, W. Leitner, M. Lutz, *Chem. Eur. J.* **1997**, *3*, 755; (b) W. Leitner, M. Buehl, R. Fornika, Ch. Six, W. Baumann, E. Dinjus, M. Kessler, C. Krueger, A. Rufinska, *Organometallics* **1999**, *18*(7), 1196.

[34] W. Leitner, E. Dinjus, F. Gaßner, *J. Organomet. Chem.* **1994**, *475*, 257.

[35] K. Kudo, N. Sugita, Y. Takeszaki, *Nippon Kagaku Kaishi* **1977**, 302.

[36] C. J. Stadler, S. Chao, D. P. Summers, M. S. Wrighton, *J. Am. Chem. Soc.* **1983**, *105*, 6318.

[37] (a) M. M. Taqui Khan, S. B. Halligudi, S. Shukla, *J. Mol. Catal.* **1989**, *53*, 305; (b) M. M. Taqui Khan, S. B. Halligudi, S. Shukla, *J. Mol. Catal.* **1989**, *57*, 47.

[38] (a) Ruhrchemie AG (R. Gärtner, B. Cornils, H. Springer, P. Lappe), DE 3.235.030 (1982); *Chem. Abstr.* **1984**, *101*, 55331t; (b) Ruhrchemie AG (L. Bexten, B. Cornils, D. Kupies), DE 3.431.643 (1984); *Chem. Abstr.* **1986**, *105*, 117009n; (c) W. A. Herrmann, C. W. Kohlpaintner, *Angew. Chem.* **1993**, *105*, 1588; *Angew. Chem. Int. Ed. Engl.* **1993**, *32*, 1524.

[39] (a) F. Gaßner, W. Leitner, *J. Chem. Soc., Chem. Commun.* **1993**, 1465; (b) F. Gaßner, PhD Thesis, Universität Jena (1994).

[40] (a) J. R. Pugh, M. R. Bruce, B. P. Sullivan, T. J. Mayer, *Inorg. Chem.* **1991**, *30*, 86; (b) K. K. Pandey, K. H. Garg, S. K. Tiwari, *Polyhedron* **1992**, 947; (c) J. C. Berthet, M. Ephritikhine, *New J. Chem.* **1992**, *16*, 767; (d) D. Nietlispach, H. W. Bosch, H. Berke, *Chem. Ber.* **1994**, *127*, 2403.

[41] (a) B. P. Sullivan, T. J. Meyer, *Organometallics* **1986**, *5*, 1500; (b) D. J. Darensbourg, M. J. Darensbourg, L. Y. Groh, P. Wiegreffe, *J. Am. Chem. Soc.* **1987**, *109*, 7539; (c) D. J. Darensbourg, H. P. Wiegreffe, P. W. Wiegreffe, *J. Am. Chem. Soc.* **1990**, *112*, 9252.

[42] (a) A. Dedieu, C. Bo, F. Ingold, in Ref. [3a], p. 22; (b) N. Koga, K. Morokuma, *Chem. Rev.* **1991**, *91*, 283.

[43] (a) S. Sakaki, K. Ohkubo, *Inorg. Chem.* **1988**, *27*, 2020; (b) C. Bo, A. Dedieu, *Inorg. Chem.* **1989**, *28*, 304.

[44] (a) S. Sakaki, K. Ohkubo, *Organometallics* **1989**, *8*, 2973; (b) S. Sakaki, K. Ohkubo, *Inorg. Chem.* **1989**, *28*, 2583.

[45] (a) V. W. Day, M. F. Fredrich, G. S. Reddy, A. J. Sivak, W. R. Pretzer, E. L. Muetterties, *J. Am. Chem. Soc.* **1977**, *99*, 8091; (b) A. J. Sivak, E. L. Muetterties, *J. Am. Chem. Soc.* **1979**, *101*, 4878.

[46] (a) M. D. Fryzuk, *Can. J. Chem.* **1983**, *61*, 1347; (b) M. D. Fryzuk, T. Jones, F. W. B. Einstein, *Organometallics* **1984**, *3*, 185; (c) M. D. Fryzuk, W. E. Piers, S. J. Rettig, F. W. B. Einstein, T. Jones, T. A. Albright, *J. Am. Chem. Soc.* **1989**, *111*, 5709; (d)

M. D. Fryzuk, W. E. Piers, *Organometallics* **1990**, *9*, 986; (e) M. D. Fryzuk, W. E. Piers, F. W. B. Einstein, T. Jones, *Can. J. Chem.* **1989**, *67*, 883.
[47] S. Sakaki, Y. Musahi, *J. Chem. Soc., Dalton Trans.* **1994**, 3047.
[48] F. Hutschka, A. Dedieu, W. Leitner, *Angew. Chem.* **1995**, *107*, 1905; *Angew. Chem. Int. Ed. Engl.* **1995**, *34*, 1742.
[49] P. G. Jessop, R. H. Morris, *Coord. Chem. Rev.* **1992**, *121*, 155.
[50] (a) P. G. Jessop, Y. Hisao, T. Ikariya, R. Noyori, *J. Am. Chem. Soc.* **1996**, *118*, 344; (b) P. G. Jessop, T. Ikariya, R. Noyori, *Science* **1995**, *269*, 1065; (c) P. G. Jessop, T. Ikariya, R. Noyori, *Nature (London)* **1994**, *368*, 231; (d) T. Ikariya, P. G. Jessop, R. Noyori, *JP Appl.* 274.721 (1993).
[51] C. Y. Tsang, N. B. Streett, *Chem. Eng. Sci.* **1989**, *36*, 993.
[52] S. M. Howdle, M. Poliakoff, *J. Chem. Soc., Chem. Commun.* **1989**, 1099.
[53] S. M. Howdle, M. A. Healy, M. Poliakoff, *J. Am. Chem. Soc.* **1990**, *112*, 4804.
[54] (a) P. G. Jessop, Y. Hsiao, T. Ikariya, R. Noyori, *J. Chem. Soc., Chem. Commun.* **1995**, 707; (b) O. Kröcher, R. A. Köppel, A. Baiker, *Chimia* **1997**, 48.
[55] (a) I. S. Kolomnikov, T. S. Lobeeva, M. E. Vol'pin, *Izv. Akad. Nauk. Ser. Khim.* **1970**, 2650; (b) T. S. Lobeeva, M. E. Vol'pin, *Izv. Akad. Nauk. Ser. Khim.* **1972**, 2329.
[56] Y. Inoue, Y. Sasaki, H. Hashimoto, *J. Chem. Soc., Chem. Commun.* **1975**, 718.
[57] (a) D. Darensbourg, C. Ovalles, M. Pala, *J. Am. Chem. Soc.* **1983**, *105*, 5937; (b) D. Darensbourg, C. Ovalles, *J. Am. Chem. Soc.* **1984**, *106*, 3750; (c) D. Darensbourg, C. Ovalles, *J. Am. Chem. Soc.* **1987**, *109*, 330.
[58] G. O. Evans, C. J. Newell, *Inorg. Chim. Acta* **1978**, *31*, L387.
[59] H. Bipp, U. K. Kicezka, *Ullmann's Encycl. Ind. Chem. 5th ed.*, **1989**, Vol. A12, pp. 1–12.
[60] M. W. Farlow, H. Adkins, *J. Am. Chem. Soc.* **1935**, *57*, 2272.
[61] S. Schreiner, J. Y. Yu, L. Vaska, *J. Chem. Soc., Chem. Commun.* **1988**, 602.
[62] P. Haynes, H. Slaugh, J. F. Kohnle, *Tetrahedron Lett.* **1970**, 365.
[63] Y. Kiso, K. Saeki, *Kokai Tokkyo Koho*, JP 36.617 (1977).
[64] P. G. Jessop, Y. Hsiao, T. Ikariya, R. Noyori, *J. Am. Chem. Soc.* **1994**, *116*, 8851.
[65] (a) H. Koinuma, F. Kawakami, H. Kato, H. Hirai, *J. Chem. Soc., Chem. Comm.* **1981**, 213; (b) G. Süss-Fink, J. Reiner, *J. Organomet. Chem.* **1981**, *221*, C36.
[66] (a) A. Jansen, H. Görls, S. Pitter, *Organometallics* **2000**, *19*, 135; (b) Forschungszentrum Karlsruhe (S. Pitter, A. Jansen, E. Dinjus), DE 199.11.616 (2000).
[67] T. C. Eisenschmid, R. Eisenberg, *Organometallics* **1989**, *8*, 1822.
[68] P. G. Jessop, *Top. Catal.* **1998**, *3*, 95.
[69] S. Pitter, A. Jansen, unpublished results.
[70] A.-A. G. Shaikh, S. Sivaram, *Chem. Rev.* **1996**, *96*, 681.
[71] M. A. Pacheco, C. L. Marshall, *Energy & Fuels* **1997**, *11*, 2.
[72] (a) G. Illuminati, U. Romano, R. Tesei, DE 2.528.412 (1979); (b) F. Merger, F. Towee, L. Schroff, EP 0.000.162 (1979); (c) A. Bomben, M. Selva, P. Tundo, *Recl. Trav. Chim. Pays-Bas* **1996**, *115*, 256.
[73] A. Wagner, W. Löffler, B. Haas, WO 94/22805, **1994**.
[74] (a) T. Sakakura, Y. Saito, M. Okano, J.-C. Choi, T. Sako, *J. Org. Chem.* **1998**, *63*, 7095; (b) T. Sakakura, Y. Saito, M. Okano, J.-C. Choi, T. Sako, *J. Am. Chem. Soc.* **1999**, *121*, 3793; (c) T. Sakakura, Y. Saito, M. Okano, J.-C. Choi, T. Sako, *J. Org. Chem.* **1999**, *64*, 4506.
[75] J. Kizling, *Collect. Czech. Chem. Comm.* **1993**, *58*, 1399; (b) J. Kizling, I. Pastucha, *Collect. Czech. Chem. Commun.* **1994**, *59*, 2116; (c) J. Kizling, I. Pastucha, *Collect. Czech. Chem. Commun.* **1995**, *60*, 687.
[76] (a) In A. G. Davies, *Organotin Chemistry*, Weinheim, VCH, **1997**; (b) A. J. Bloodworth, A. G. Davies, S. C. Vasishtha, *J. Chem. Soc. (C)* **1967**, 1309; (c) A. G. Davies, P. G.

Harrison, *J. Chem. Soc. (C)* **1967**, 1313; (d) A. G. Davies, D. C. Kleinschmidt, P. R. Palan, S. C. Vasishtha, *J. Chem. Soc. (C)* **1971**, 3972.

[77] J. Kümmerlen, A. Sebald, H. Reuter, *J. Organomet. Chem.* **1992**, *427*, 309.

[78] For some recent examples of CO_2 insertion into epoxides, see: (a) K. Kasuga, N. Kabata, *Inorg. Chim. Acta* **1997**, *257*, 277; (b) T. Yano, H. Matsui, T. Koike, H. Ihiguro, H. Fujihara, M. Yoshihara, T. Maeshima, *J. Chem. Soc., Chem. Commun.* **1997**, 1129; (c) K. Yamaguchi, K. Ebitani, T. Yoshida, H. Yoshida, K. Kaneda, *J. Am. Chem. Soc.* **1999**, *121*, 4526; (d) K. Kasuga, S. Nagao, T. Fukumoto, M. Handa, *Polyhedron* **1996**, *15*, 69; (e) W. J. Kruper, D. V. Dellar, J. *Org. Chem.* **1995**, *60*, 725; (f) D. Darensbourg, M. W. Holtcamp, B. Khandelwal, K. K. Klausmeyer, J. H. Reibenspies, *J. Am. Chem. Soc.* **1995**, *117*, 538.

[79] (a) S. A. Motika, T. L. Pickering, A. Rokicki, B. K. Stein, US 5.026.676 (1991); (b) H.-N. Sun, US 4.783.445 (1988); (c) H.-N. Sun, US 4.789.727 (1988); (d) A. Rokicki, US 4.943.677 (1990); (e) W. E. Carroll, S. A. Motika, US 4.960.862 (1990); (f) H. Kawachi, S. Minami, J. N. Armor, A. Rokicki, B. K. Stein, US 4.981.948 (1991); (g) S. Inoue, M. Kanbe, T. Takada, N. Miyazaki, M. Yokokawa, US 3.953.383.

[80] (a) H. Schnell, Chemistry and Physics of Polycarbonates, in *Encyclopedia of Polymer Science and Technology, Vol. 10,* John Wiley, New York, **1964**; (b) H. Schnell, *Angew. Chem.* **1966**, *73*, 629; (c) W. Kuran, in *Polymeric Material Encyclopedia, Vol. 9,* CRC Press, Boca Raton, **1996**.

[81] (a) S. Inoue, *Makromol. Chem., Rapid Commun.* **1980**, *1*, 775; (b) K. Soga, K. Uenishi, S. Ikeda, *J. Polym. Sci.: Polym. Chem. Ed.* **1979**, *17*, 415.

[82] (a) E. J. Beckmann, T. Hoefling, D. Stofesky, M. Reid, R. Enick, *J. Supercrit. Fluids* **1992**, *5*, 237; (b) M. Super, E. Berluche, C. Costello, E. J. Beckman, *Macromolecules* **1997**, *30*, 368.

[83] (a) T. A. Zevaco, H. Görls, E. Dinjus, *Polyhedron* **1998**, *17*, 613; (b) T. A. Zevaco, H. Görls, E. Dinjus, *Inorg. Chem. Commun.* **1998**, *1*, 170; (c) T. A. Zevaco, H. Görls, E. Dinjus, *Polyhedron* **1998**, *17*, 2199.

[84] (a) W. Kuran, S. Psynkiewicz, *Makromol. Chem.* **1979**, *180*, 1253; (b) W. Kuran, A. Rokicki, D. Romanowska, *J. Polym. Sci., Polym. Chem. Ed.* **1979**, *17*, 2003; (c) W. Kuran, T. Listos, *Macromol. Chem. Phys.* **1994**, *195*, 1011; (d) P. Gorecki, W. Kuran, J. *Polym. Chem.: Polym. Lett Ed.* **1985**, *23*, 299.

[85] (a) S. Inoue, H. Koinuma, T. Tsuruta, *Makromol. Chem.* **1969**, *130*, 210; (b) S. Inoue, H. Koinuma, T. Tsuruta, *Makromol. Chem.* **1971**, *143*, 97.

[86] J. Darensbourg, M. W. Holtcamp, *Coord. Chem. Rev.* **1996**, *153*, 155; (b) D. J. Darensbourg, N. W. Stafford, T. Katsuaro, *J. Mol. Catal. A* **1995**, *104*, L1–L4.

[87] (a) M. Cheng, E. B. Lobkovsky, G. W. Coates, *J. Am. Chem. Soc.* **1998**, *120*, 11018; (b) M. Cheng, N. A. Darling, E. B. Lobkovsky, G. W. Coates, *J. Chem. Soc., Chem. Commun.* **2000**, 2007.

[88] (a) T. Aida , S. Inoue, *Acc. Chem. Res.* **1996**, *29*, 39; (b) S. Hiroshi, K. Chikara, T. Aida, S. Inoue, *Macromolecules* **1994**, *27*, 2013; (c) Y. Watanabe, T. Yasuda, T. Aida, S. Inoue, *Macromolecules* **1992**, *25*, 1396; (d) T. Aida , S. Inoue, *J. Am. Chem. Soc.* **1983**, *105*, 1304.

3.3.5   Reductive Carbonylation of Nitro Compounds

Markus Dugal, Daniel Koch, Guido Naberfeld, Christian Six

3.3.5.1   Introductory Remarks

Reductive carbonylation of nitro compounds, especially nitroaromatic compounds according to eq. (1), has been the subject of thorough industrial research starting in 1962 and continuing until the beginning of the 1990s due to the demand for a new, phosgene-free method for the production of isocyanates [1] and the discussions on the chlorine cycle in industry.

$$Ar-NO_2 + 3\,CO \longrightarrow Ar-NCO + 2\,CO_2 \tag{1}$$

The "dream reaction" leading to industrially relevant isocyanates would be a low-cost one-step synthesis starting from the corresponding nitro precursors [2, 3]. Arising problems favored an alternative two-step reaction via urethanes, which seemed to represent a feasible technical method to reduce the costs of isocyanate production by about 25–30 % [4]. All the announcements referring thereto have been shown to be invalid, simply because the abundant observations claimed in numerous patents and other publications led to an inadequate and optimistic evaluation, although the chemistry was very poorly understood at that time. After 1969 research on homogeneous reductive carbonylation of nitro compounds using compounds of ruthenium, rhodium, and palladium as catalysts increased in academic laboratories. The most recent period is characterized by in-depth studies in academia on the one hand on the mechanism of activation and catalysis and the nature of the catalytic species, and a declining interest in industry on the other hand, at least when speaking of the manufacture of large-scale diisocyanates which typically find use in the polyurethane industry.

The purpose of this section is to summarize the results of this continuous development, focusing on the most interesting compounds: isocyanates and urethanes. Analogous reactions of this type leading to different products will just be mentioned in passing [1, 5–7].

3.3.5.2   Synthesis of Isocyanates

3.3.5.2.1   Manufacture with Phosgene

At least 90 % of the worldwide production of isocyanates is accounted for by two aromatic isocyanates, toluene diisocyanate (TDI), a distilled compound, and polymethylene polyphenylene polyisocyanate (PMDI), an undistilled isocyanate mixture with a low vapor pressure (Structures **1–3**). Together these two products currently amount to a total annual world production of 3.3 Mt. They are mainly used for the production of a broad spectrum of polyurethanes, e. g., foams, elastomers, and coatings [8, 9].

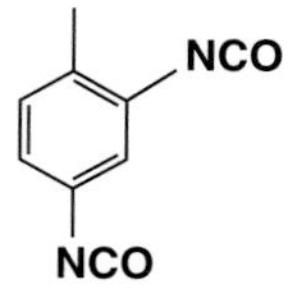

1

2,4 - TDI

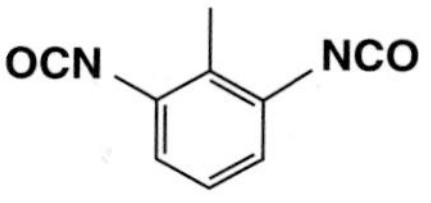

2

2,6 - TDI

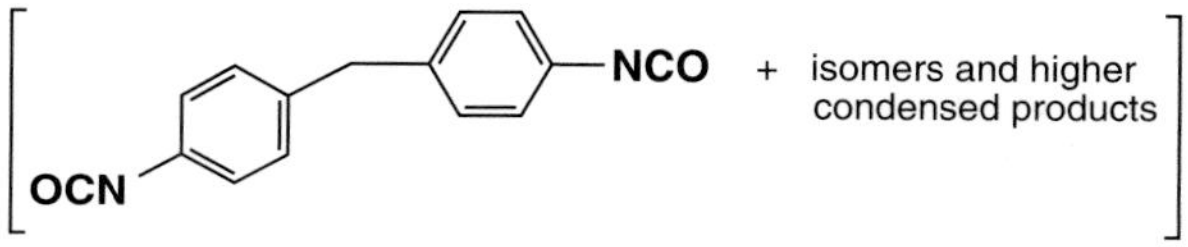

+ isomers and higher condensed products

3

PMDI

Aromatic isocyanates are produced commercially by phosgenation of the corresponding amine base [10]. Phosgene excess, HCl, and solvent are recycled. Phosgene, which is produced catalytically on charcoal from carbon monoxide and chlorine, is a highly active, poisonous, and corrosive gas. Therefore, numerous attempts have been made to develop phosgene-free processes, i. e., methods for the production of isocyanates without handling chlorine, one of them being the homogeneous catalytic carbonylation of nitroaromatic compounds.

3.3.5.2.2 Attempts with Carbon Monoxide

Numerous patents [3, 11–13] and other publications [3, 14–16] describe the direct carbonylation of nitroaromatic compounds to isocyanates or alternatively a modified carbonylation to urethanes in the presence of alcohol, followed by a thermal transformation to isocyanates [4, 17–19] (eq. (2)).

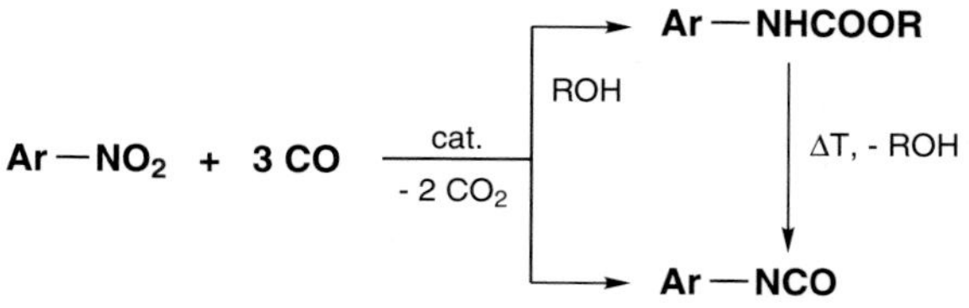

$$
\text{Ar}-\text{NO}_2 \;+\; 3\,\text{CO} \;\xrightarrow[\;-\,2\,\text{CO}_2\;]{\text{cat.}}\;
\begin{array}{c}
\xrightarrow{\text{ROH}} \text{Ar}-\text{NHCOOR} \\[2mm]
\Big\downarrow \Delta T,\, -\,\text{ROH} \\[2mm]
\longrightarrow \text{Ar}-\text{NCO}
\end{array}
\tag{2}
$$

Ar = aromatic group
ROH = aliphatic alcohol

Direct Carbonylation to Isocyanates

In the first reported direct *N*-carbonylation of nitroaromatics to isocyanates, simple Pd- or Rh-based systems were used to catalyze the reaction of aromatic mononitro compounds with carbon monoxide [11, 12]. Later, it became possible to work without the drastic reaction conditions that had been required initially, by using Lewis acid co-catalysts [13]. Various catalysts and catalyst mixtures, normally based on Ru, Rh, or Pd complexes with co-catalysts, were described in numerous patents and publications [1, 3, 14–16]. The careful choice of the composition of the triad consisting of metal salt, co-catalysts and ligand (preferably aromatic amines) led to efficient catalyst systems [14 a–e] for the direct reductive carbonylation process. A quite active Pd–phenanthroline–H$^+$ system with non-coordinating carboxylic acids such as 2,4,6-trimethlybenzoic acid as proton source is worth mentioning [14 d].

However, although promising results have been achieved with mononitro compounds, dinitro compounds can be converted only with low selectivities and using high catalyst concentrations. Furthermore, in spite of extensive investigations of the reaction mechanism (see Section 3.3.5.3.1), questions that still remain unanswered are, whether the active catalytic species is a heterogeneous one or a soluble species generated *in situ,* and what the function of the co-catalyst is.

The difficulties of utilizing the direct reductive carbonylation of nitroaromatic compounds for the production of industrially relevant isocyanates are documented by three publications discussing different palladium-based catalysts. A metallacyclic complex from the reaction of nitrobenzene with carbon monoxide in the presence of Pd–*o*-phenanthroline decomposes to phenyl isocyanate only in moderate yield, which may be an indication of an intrinsic limitation related to the mechanistic pathway of the catalytic reaction [20]. Heteropoly-compounds with high redox potentials effectively modify the thoroughly investigated catalyst PdCl$_2$, resulting in good selectivities but poor conversion of nitrobenzene to phenyl isocyanate [14 e]. Another study has focused on the reductive carbonylation of 2,4-dinitrotoluene to 2,4-TDI [14 b]. Although the conversions and selectivities reported are prohibitive for commercial use, for the first time a deeper understanding of parts of the reaction pathways has been obtained.

Summarizing, from the investigations in this field it can be concluded that from an industrial viewpoint the direct carbonylation of nitroaromatics to isocyanates represents no economically feasible alternative, for the conventional phosgenation process, for the following reasons:

(1) High catalyst concentrations are necessary due to generally low turnover numbers, while insufficient stability and unsolved problems in catalyst recycling, especially in continuous processes, remain.
(2) Despite promising results for the model reaction of mononitroaromatic compounds to monoisocyanates, the selectivities for the industrially important reaction of dinitrotoluene to TDI are unacceptable (for PMDI see [21]).

Reductive Carbonylation to Urethanes

Simultaneously with the disclosure of the direct carbonylation to isocyanates, in 1962 ICI patents claimed the formation of urethanes from nitroarenes by reductive carbonylation in the presence of alcohols. This approach can be pictured as a direct carbonylation step followed by a trapping reaction of the isocyanate produced with an alcohol with the formation of urethanes. Subsequent work used complexes of ruthenium [22], rhodium [23] and palladium [24], which also showed good selectivity and high yield for dinitroaromatic substrates in some cases [25]. Despite the fact that these interesting reactions were discovered more than 40 years ago a clear picture of the mechanisms with group VIII catalysts is still not in sight. It turns out that the nature of the catalyst and the alcohol itself seems to have a strong impact on the elementary steps of this interesting transformation. For instance, it was discovered that the presence of alcohol often had a strong influence on the activity of a system. Further, it was found that temperature effects in the presence or absence of alcohol was often not compatible with the hypothesis of a common carbonylation mechanism. All these results suggested that the alcohol interacts with the catalyst and does not simply trap the isocyanate, which implies that the catalytic cycle is different from the catalytic cycle of the direct carbonylation [28–30]. Obviously, more work is needed to rationalize the current data.

Taken altogether, the two-stage process based on the reductive carbonylation to urethanes operates under milder conditions, with lower catalyst concentrations and a good selectivity to the intermediate urethane, but causes a new problem that for a long time was underestimated, i. e., the thermal cracking to isocyanates [26], especially to TDI. Although it was apparently practicable as an industrial process, the carbonylation reaction conditions and the high temperatures required for splitting the urethane obviously limited its applicability as a general synthetic method. Therefore, the announcement of the construction of a pilot plant for TDI [4] turned out to be falsely optimistic.

More recently a new approach employing BCl_3 as reactant (Scheme 1) has been proposed as a substitute for the industrial thermal cracking process [27]. In most cases, quantitative conversion to the product isocyanates was achieved under mild conditions but with the disadvantage of a high BCl_3 consumption.

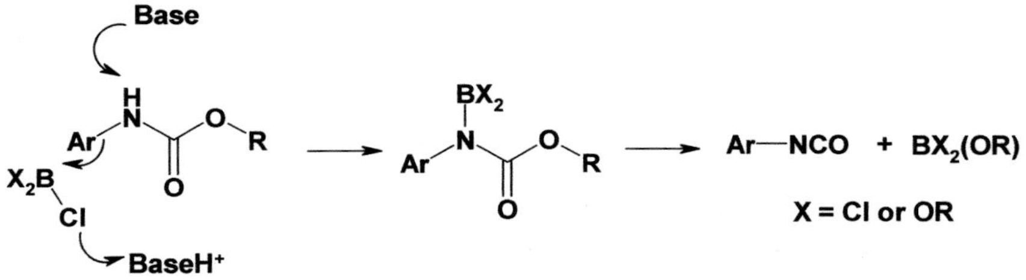

Scheme 1

3.3.5.3   Thermodynamics, Kinetics, and Mechanism

3.3.5.3.1   Direct Carbonylation of Nitroaromatics

In this section, available kinetic and mechanistic data for the reaction of nitro-aromatic compounds with carbon monoxide are summarized. This reaction is thermodynamically favorable, being characterized by high equilibrium constants [32] and negative reaction enthalpies (eqs. (3) and (4)) [1, 14 b]. Without catalysis high activation barriers (high temperatures, high pressure) have to be overcome, leading to azo derivatives [33]. Due to the high exothermicity of the reaction an optimized catalyst system and optimized reaction conditions are required to ensure selective isocyanate formation.

$$+ \ 3\,CO \longrightarrow \qquad + \ 2\,CO_2 \tag{3}$$

$\Delta H = -129$ kcal/mol

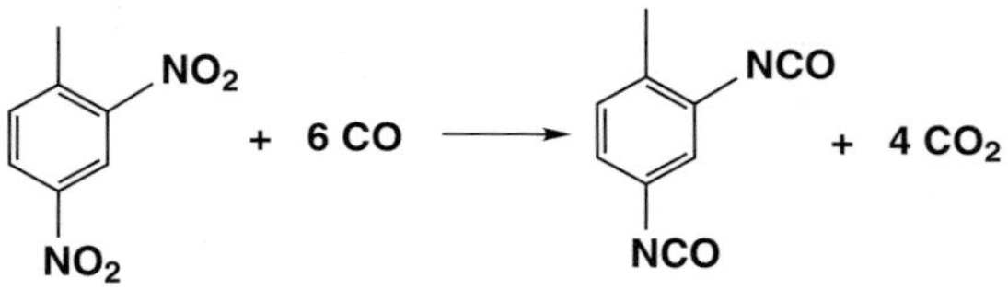

$$+ \ 6\,CO \longrightarrow \qquad + \ 4\,CO_2 \tag{4}$$

$\Delta H = -228$ kcal/mol

For the direct carbonylation with group VIII transition metal catalysts two main types of mechanisms have been proposed so far, involving the formation of a metal–imido (e. g., Structure **4**) or a metallacyclic intermediate (e. g., Structure **5**) [3].

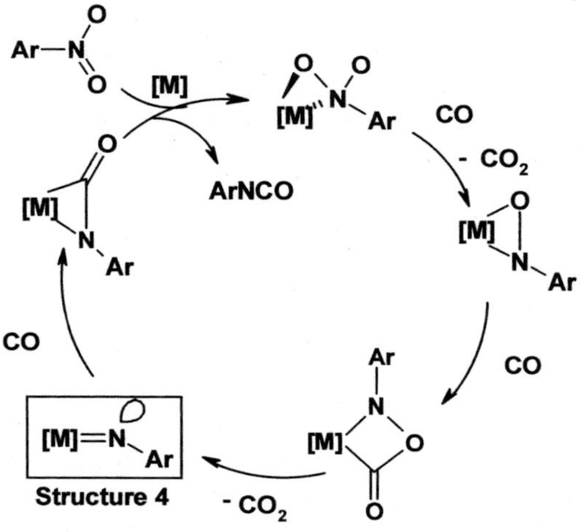

Scheme 2. Mechanism for the direct carbonylation involving a metal–imido intermediate [3].

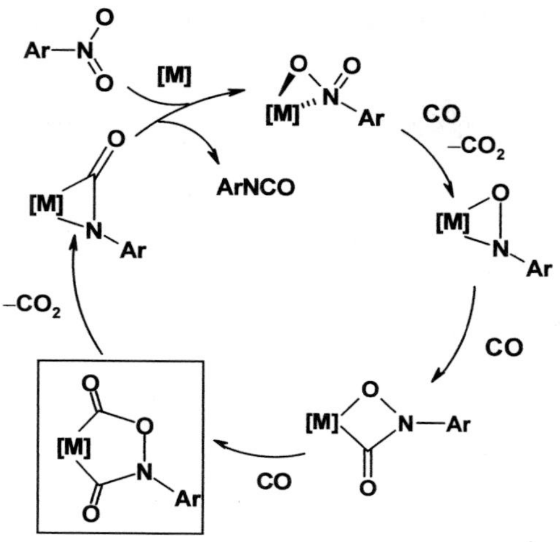

Structure 5

Scheme 3. Mechanism for the direct carbonylation involving a metallacyclic intermediate [3].

The Metal–Imido Mechanisms

In an early publication [16] the carbonylation of nitroaromatics was described as a stepwise deoxygenation of the nitro group, generating an excited singlet nitrene (probably stabilized by coordination on a metal center). Based on this description, the formation of a metal–imido intermediate was usually assumed in most of the proposed mechanisms until the mid-1980s [5, 34–38].

The intermediacy of an imido species does in fact rationalize the formation of most of the minor typical by-products isolated after carbonylation reactions: after the deoxygenation of the nitro group the resulting excited singlet nitrene is spontaneously intercepted by carbon monoxide to form an isocyanate. In the case of lack of carbon monoxide, intersystem crossing to the ground-state triplet nitrene occurs, which is responsible for unwanted side reactions. Therefore, optimized reaction conditions (high carbon monoxide pressure and temperature) are obligatory to ensure reasonable selectivity.

The Metallacyclic Mechanism

Despite several experimental facts [3] rendering a transient metal–imido species a likely source for many products of the carbonylation reaction, its role as an actual intermediate in the catalytic transformation of simple nitroaromatic substrates has never been proven. Accordingly, a type-**5** mechanism (Scheme 3, involving no such intermediate) could also be operative for the formation of isocyanate. In this case, an imido complex could also be generated by a parallel minor pathway

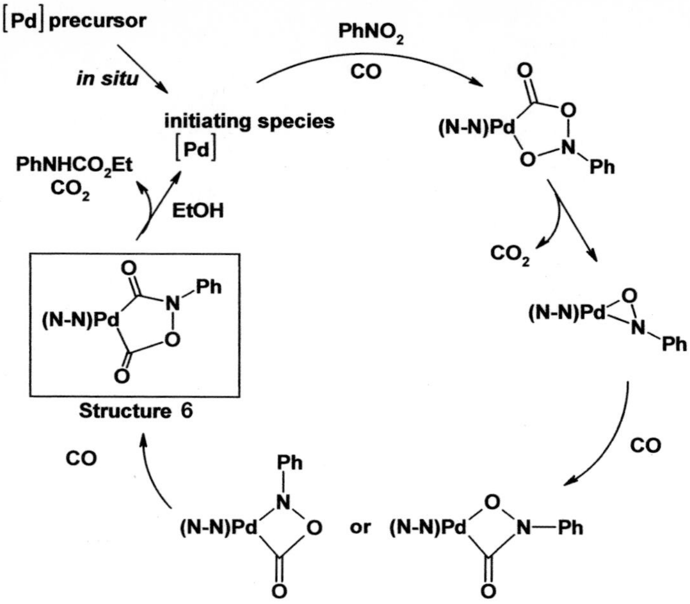

Scheme 4. Reaction cycle to the metallacyclic complex and products [20].

and account for the by-products isolated. Such a mechanism, although proposed very early [39, 40], has gained more consideration just recently from investigations conducted on the ([Pd]/phen/H⁺) system [20, 41]. A surprisingly stable 1:1 intermediate metallacyclic complex (Structure **6**, N–N = *o*–phenanthroline) could be isolated from the reaction of nitrobenzene, carbon monoxide, and Pd–*o*-phenanthroline and structurally characterized [42, 43].

General evidence for that kind of mechanism comes from *ab initio* theoretical calculations performed on a related platinum complex [3, 44] and from the reactivity of four- and five-membered heterometallacyles [45] structurally close to some of the intermediates in postulated mechanisms (e.g., Scheme 3). Moreover, related metallacycles have often been isolated from the reaction medium after nitroaromatic carbonylation, indicating that such species can easily be generated under typical carbonylation conditions [46, 47].

3.3.5.3.2 Indirect Carbonylation of Nitroaromatics

Conceptually, the indirect carbonylation of nitroaromatics can be pictured as a direct carbonylation reaction, followed by a scavenger reaction of the highly reactive intermediate isocyanate by the alcohol in a subsequent step before by-product formation comes into play. The latter is known to occur spontaneously at ambient temperature [48, 49] and is catalyzed efficiently by many compounds having either Lewis acidity or basicity [50–53]. Since this follow-up reaction is very much favored on thermodynamic grounds, the complete indirect carbonylation process is even more exothermic than the direct one [3].

For a long time, the indirect carbonylation reaction was believed to proceed via that modified direct carbonylation mechanism. In the early 1970s, such a belief was also supported by the demonstration that the described scavenger reaction, known to be feasible with free isocyanates, could be applied as well to isocyanates complexed on various metal centers [54, 55]

Around the mid-1980s, however, more and more experimental facts accumulated that indicated distinct mechanistic differences between the direct and indirect carbonylation reactions. For instance, it was discovered that the presence of alcohol often had a strong influence on the activity of a given system when compared with the corresponding direct process [56–58]. Moreover, the reported influence of temperature using the same catalysts, whether in the presence of alcohol or not, was often not likely to be compatible with the hypothesis of a common carbonylation mechanism for both processes [59]. Finally, it was reported in many instances that the nature of the alcohol itself was decisive regarding the yield in carbamate [56]. In some cases, depending on the catalyst used, alcohols having active hydrogen acted as a molecular hydrogen source in the medium and led to a noticeable increase in the formation of aniline or other hydrogenated products compared with alcohols commonly used in these processes [60, 61]. All of these facts indicate that the alcohol interacts with the catalyst during the carbonylation process and does not simply trap the intermediate isocyanate. Therefore mechanisms in which the alcohol took part in the formation of the actual active species were considered.

In Scheme 5 (a) and (b) for instance, the alcohol intervenes very early in the catalytic cycle and it is essential for the efficient carbonylation/deoxygenation of the substrate [56, 62–64]. Among the mechanisms proposed, only Scheme 5 (c) [3] remains somewhat related to the simple scheme mentioned earlier invoking isocyanate as the primary reaction product, subsequently trapped by alcohol.

In the mechanisms according to Scheme 5 (a)–(c) the initial steps (nitro activation and first deoxygenation) are believed to be similar to those delineated for direct carbonylation (cf. Schemes 2 and 3). None of these, however, includes a step where a metal–imido intermediate is generated and subsequently carbonylated to give the isocyanate. Since the early studies on imido complexes, the carbonylation of such an intermediate in relation to competitive protonation to give an amido species was thought questionable when a proton donor like alcohol was present [65–68]. In this respect, the mechanism (Scheme 5 (a)) initially advanced for $Ru_3(CO)_{12}$/TBACl [3, 56, 69] and other cluster-based systems was the first serious proposal for indirect carbonylation, despite presenting very little experimental support. Now, Scheme 5 (b) is clearly established for [(dppe)Ru(CO)$_3$] [3, 31, 70–72] and appears to be the mechanism operative with $Ru_3(CO)_{12}$ in the presence of dppe [3], and possibly also with other cluster-based systems for which Scheme 5 (a) had formerly been proposed. This mechanism finds indubitably the strongest experimental support among the proposed mechanisms. Remarkably, Scheme 5 (c), which has been discussed for the ([Pd]/phen/H$^+$) [3]

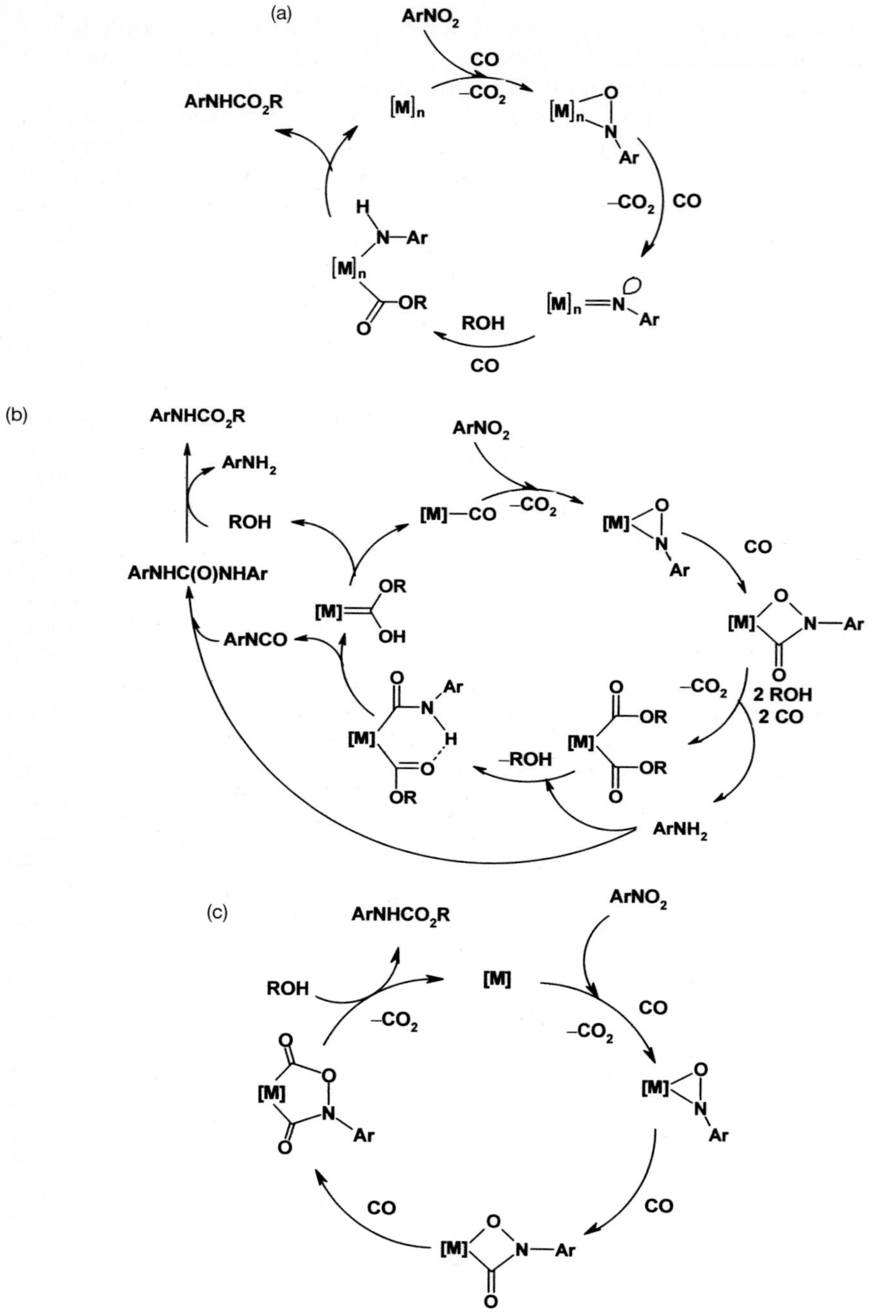

(a)

(b)

(c)

Scheme 5. Mechanisms for the indirect carbonylation involving an interaction of the alcohol with the catalyst (a and b) or an isocyanate intermediate (c) [3].

system, is the only mechanism that allows a catalyst to retain its activity for iso-cyanate production without the presence of alcohol. Indeed, no free isocyanate can possibly be generated by mechanism 5 (a) or (b) under direct conditions. A Brønsted acid promoter was present in most catalytic systems for which mechanism 5 (c) has been put forward. In that respect, the absence of carbamoyl or alkoxycarbonyl intermediates in Scheme 5 (c) is consistent with the presence of protons, which are known to disfavor the formation of such complexes [73–76].

Now, if one wants to tie together all the mechanistic data available for indirect carbonylation reactions on group VIII catalysts, no unifying picture currently emerges and, depending on the nature of the catalytic system used, the mechanism according to either Scheme 5 (b) or (c) appears very likely to be operative.

3.3.5.4   Outlook

Although in principle it is a practicable industrial process, catalytic reductive carbonylation of nitroaromatic compounds has not become a general synthetic method on a technical scale so far: this type of reaction remains a laboratory tool for special products, although excellent selectivities are already observed. The situation will change, if the comprehensive studies of the catalytic cycle, especially from a kinetic and mechanistic viewpoint, should lead to the design of a continuous catalytic process with significant improvement in catalyst load, lifetime, and turnover frequency in combination with a practice-oriented concept in catalyst recovery or regeneration. Results of relevant investigations are summarized here, focusing on industrially relevant aspects. Summing up, a "Golden Age" cannot be predicted yet for a large-scale industrial application of homogeneous catalytic carbonylation with noble metallacyclic complexes of nitroaromatic compounds to the corresponding isocyanates. The classic phosgenation route remains the only economically attractive route for industrial production of commodity isocyanates.

References

[1] S. Cenini, M. Pizzotti, C. Crotti, Corrado, *Aspects Homogen. Catal.* **1988**, *6*, 97, and references cited therein.
[2] *Plastics Handbook – Polyurethanes, Vol. 7, 3rd ed.* (Eds.: G. Oertel, L. Abele), Carl Hanser, Munich, **1993**.
[3] For a comprehensive review see: F. Paul, *Coord. Chem. Rev.* **2000**, *203*, 269.
[4] Anon., *Chemical Week* **1997**, March 9, 43.
[5] A. F. M. Iqbal, *Chem. Technol.* **1974**, *4*(9), 566.
[6] H. M. Colquhoun, D. J. Thompson, M. V. Twigg (Eds.), *Carbonylation – Direct Synthesis of Carbonyl Compounds,* Plenum, New York, **1991**, p. 164.
[7] H. Ulrich, *Chemistry and Technology of Isocyanates,* Wiley, Chichester, **1996**, pp. 333–334, 375–379.
[8] J. K. Backus et al., *Encycl. Polym. Sci. Eng.,* **1988**, *13*, 243.

[9] H. Ulrich, Isocyanates, Organic, in *Ullmann's Encycl. Ind. Chem.*, *6th ed.*, Wiley-VCH, Weinheim, **2001** (electronic version).

[10] H. J. Twichett, *Chem. Soc. Rev.* **1974**, *3*(2), 209.

[11] American Cyanamid (W. B. Hardy, R. P. Bennet), DE 1.237.103 (1963).

[12] American Cyanamid (W. B. Hardy, R. P. Bennet), US 3.461.149 (1965).

[13] Olin Mathieson Corp. (G. F. Ottmann, E. H. Kober, D. F. Gavin), US 3.523.962; (E. H. Kober, W. J. Schnabel, T. C. Kraus, G. F. Ottmann), US 3.523.965; (W. J. Schnabel, E. H. Kober, T. C. Kraus), US 3.714.216 (1967), (E. H. Kober, W. J. Schnabel), DE 2.018.299 (1970); (G. F. Ottmann, W. J. Schnabel, E. Smith), 3.728.370 (1970); (P. D. Hammond, J. A. Scott), US 3.812.169; (P. D. Hammond, W. C. Clarke, W. I. Denton), US 3.832.372 (1972).

[14] (a) V. I. Manov-Yuvenskii, B. A. Redoshkin, B. K. Nefedov, G. P. Beyaeva, *Bull. Acad. Sci. USSR Div. Chem. Sci.* **1980**, *29*, 117; (b) R. Ugo, R. Psaro, M. Pizotti, P. Nardi, C. Dossi, A. Andretta, G. Caparella, *J. Organomet. Chem.* **1991**, *417*, 211; (c) Y. Izumi; Y. Satoh, K. Urabe, *Chem. Lett.* **1990**, 795; (d) S. Cenini, F. Ragaini, M. Pizotti, F. Porta, G. Mestroni, E. Alessio, *J. Mol. Catal.* **1991**, *64*, 179; (e) Y. Izumi, Y. Satoh, H. Kondoh, K. Urabe, *J. Mol. Catal.* **1992**, *72*, 37.

[15] W. B. Hardy, R. P. Bennett, *Tetrahedron Lett.* **1967**, 961.

[16] F. J. Weigert, *J. Org. Chem.* **1973**, *38*, 1316.

[17] ICI (B. A. Mountfield), GB 993.704 (1962); (A. Ibbotson), GB 1.080.094; (G. A. Gamlen, A. Ibbotson), GB 1.092.157 (1965).

[18] Mitsui Toatsu (F. Zunistein sen. et al.), DE 2.555.557 (1974).

[19] Shell (E. Drent, P. W. van Leeuwen), EP 0.086.281 (1981).

[20] P. Leconte, F. Metz, A. Mortreux, J. A. Osborn, F. Paul, F. Petit, A. Pillot, *J. Chem. Soc., Chem. Commun.* **1990**, 1616.

[21] The only economic process for PMDI is the reaction via aniline-formaldehyde and subsequent phosgenation.

[22] S. Cenini, C. Crotti, M. Pizzotti, F. Porta, *J. Org. Chem.* **1988**, *53*, 1243.

[23] C. V. Rode, S. P. Gupta, R. V. Chaudhari, C. Pirozhkov, A. L. Lapidus, *J. Mol. Catal.* **1994**, *91* 195.

[24] A. Bontempi, E. Alessio, G. Chanos, G. Mestroni, *J. Mol. Catal.* **1987**, *42*, 67.

[25] Montedison (E. Alessio, G. Mestroni), EP 0.169.650 (1985).

[26] M. Z. A. Badr, M. M. Aly, S. A. Mahgoub, A. A. Attallah, *Rev. Roum. Chim.* **1992**, *37*, 489.

[27] D. C. D. Butler, H. Alper, *Chem. Commun.* **1998**, 2575.

[28] S. Bhaduri, H. Khwaja, N. Sapre, K. Sharma, A. Basu, P. G. Jones, G. Carpenter, *J. Chem. Soc., Dalton Trans.* **1990**, 1313.

[29] S. Bhaduri, H. Khwaja, K. Sharma, P. G. Jones, *J. Chem. Soc., Chem. Commun.* **1989**, 515.

[30] G. Mestroni, G. Zassinovich, E. Alessio, M. Tornatore, *J. Mol. Catal.* **1989**, *49*, 175.

[31] J. D. Gargulak, W. L. Gladfelter, *J. Am. Chem. Soc.* **1994**, *116*, 3792; J. D. Gargulak, A. J. Berry, M. D. Noirot, W. L. Gladfelter, *J. Am. Chem. Soc.* **1992**, *114*, 8933.

[32] K. Schwetlick, K. Unverferth, H. Tietz, *SYSpur Rep.* **1981**, 3.

[33] G. D. Buckley, N. H. Ray, *J. Chem. Soc.* **1949**, 1154; E. Glaser, R. van Beneden, *Chem.-Ing.-Tech.* **1957**, *29*, 512.

[34] T. Kajimoto, J. Tsuji, *Bull. Chem. Soc. Jpn.* **1969**, *42*, 827.

[35] B. K. Nefedov, V. I. Manov-Yuvenskii, S. S. Novikov, *Doklady Chem. (Proc. Acad. Sci. USSR)* **1977**, *234*, 347.

[36] L. V. Gorbunova, I. L. Knyazeva, E. A. Davydova, G. A. Abakumov, *Bull. Acad. Sci. USSR Div. Chem. Sci.* **1980**, *29*, 761.

[37] F. Lefebvre, P. Gelin, B. Elleuch, C. Naccache, Y. Ben Taarit, *Bull. Chim. Soc. Fr.* **1984**, 361.

[38] V. I. Manov-Yuvenskii, K. B. Petrovskii, A. L. Lapidus, *Bull. Acad. Sci. USSR Div. Chem. Sci.* **1986**, *34*, 1561.
[39] K. Unferverth, K. Schwetlick, *React. Kinet. Catal. Lett.* **1977**, *6*, 231; K. Unferverth, R. Höntsch, K. Schwetlick, *J. Prakt. Chem.* **1979**, *321*, 928.
[40] K. Unferverth, R. Höntsch, K. Schwetlick, *J. Prakt. Chem.* **1979**, *321*, 86.
[41] F. Paul, J. Fischer, P. Ochsenbein, J. A. Osborn, *Organometallics* **1998**, *11*, 2199.
[42] A. S. o Santi, B. Milani, G. Mestroni, L. Randaccio, *J. Organomet. Chem.* **1997**, *545–546*, 89.
[43] N. Masciocchi, F. Ragaini, S. Cenini, A. Sironi, *Organometallics* **1998**, *17*, 1052.
[44] P. Fantucci, M. Pizzotti, F. Porta, *Inorg. Chem.* **1991**, *30*, 2277.
[45] F. Paul, J. Fischer, P. Ochsenbein, J. A. Osborn, *Angew. Chem., Int. Ed. Engl.* **1993**, *32*, 1638.
[46] F. Ragaini, S. Cenini, *Organometallics* **1994**, *13*, 1178; F. Ragaini, S. Cenini, F. Demartin, *J. Chem. Soc., Chem. Commun.* **1992**, 1467.
[47] L. Dahlenburg, C. Prengel, *Inorg. Chim. Acta* **1986**, *122*, 55.
[48] O. Agherghinei, C. Prisacariu, A. A. Caraculacu, *Rev. Roum. Chim.* **1991**, *36*, 9.
[49] A. A. Caraculacu, I. Agerghinei, M. Gaspar, C. Prisacariu, *J. Chem. Soc., Perkin Trans.* **1990**, 1343.
[50] D. P. N. Satchell, R. S. Satchell, *Chem. Soc. Rev.* **1975**, *4*, 231.
[51] G. Hazzard, S. A. Lammiman, N. L. Poon, D. P. N. Satchell, R. S. Satchell, *J. Chem. Soc., Perkin Trans. II* **1985**, 1029.
[52] J. J. Tondeur, G. Vandendunghen, M. Watelet, *Chim. Nouv.* **1992**, *10*, 1148.
[53] K. Schwetlick, R. Noak, F. Stebner, *J. Chem. Soc., Perkin Trans. II* **1994**, 599.
[54] K. von Werner, W. Beck, *Chem. Ber.* **1971**, *104*, 2907.
[55] K. von Werner, W. Beck, *Chem. Ber.* **1972**, *105*, 3947.
[56] S. Cenini, C. Crotti, M. Pizzotti, F. Porta, *J. Org. Chem.* **1988**, *53*, 1243.
[57] S. Bhaduri, H. Khwaja, K. Sharma, P. G. Jones, *J. Chem. Soc., Chem. Commun.* **1989**, 515.
[58] H. A. Alper, K. E. Hashem, *J. Am. Chem. Soc.* **1981**, *103*, 6514.
[59] S. Bhaduri, H. Khwaja, N. Sapre, K. Sharma, A. Basu, P. G. Jones, G. Carpenter, *J. Chem. Soc., Dalton Trans.* **1990**, 1313.
[60] C.-H. Liu, C.-H. Cheng, *J. Organomet. Chem.* **1991**, *420*, 119.
[61] G. Mestroni, G. Zassinovich, E. Alessio, M. Tornatore, *J. Mol. Catal.* **1989**, *49*, 175.
[62] A. Bassoli, B. Rindone, S. Cenini, *J. Mol. Catal.* **1991**, *66*, 163.
[63] A. Bassoli, B. Rindone, S. Tollari, S. Cenini, C. Crotti, *J. Mol. Catal.* **1990**, *60*, 155.
[64] E. Bolzacchini, R. Lucini, S. Meinardi, M. Orlandi, B. Rindone, *J. Mol. Catal. A Chem.* **1996**, *110*, 227.
[65] S. Cenini, M. Pizzotti, F. Porta, G. La Monica, *J. Organomet. Chem.* **1975**, *88*, 237.
[66] W. Beck, M. Bauder, G. La Monica, S. Cenini, R. Ugo, *J. Chem. Soc., Part A* **1971**, 113.
[67] D. E. Wigley, Prog. *Inorg. Chem.* **1994**, *42*, 239.
[68] A. L. Lapidus, S. D. Pirozhkov, A. R. Tumanova, A. V. Dolidze, A. M. Yukhimenko, *Bull. Acad. Sci. USSR Div. Chem. Sci.* **1992**, *41*, 1672.
[69] S. Cenini, M. Pizzotti, C. Crotti, F. Porta, G. La Monica, *J. Chem. Soc., Chem. Commun.* **1984**, 1286.
[70] A. J. Kunin, M. D. Noirot, W. L. Gladfelter, *J. Am. Chem. Soc.* **1989**, *111*, 2739.
[71] S. J. Skoog, W. L. Gladfelter, *J. Am. Chem. Soc.* **1997**, *119*, 11049.
[72] J. D. Gargulak, W. L. Gladfelter, *J. Am. Chem. Soc.* **1994**, *116*, 3792.
[73] G. Cavinato, L. Toniolo, *J. Organomet. Chem.* **1993**, *444*, C65.
[74] J. E. Byrd, J. Halpern, *J. Am. Chem. Soc.* **1971**, *93*, 1634.
[75] R. J. Angelici, *Acc. Chem. Res.* **1972**, *5*, 335.
[76] C. R. Green, R. J. Angelici, *Inorg. Chem.* **1972**, *11*, 2095.

3.3.6   New Approaches in C–H Activation of Alkanes

Ayusman Sen

3.3.6.1   Introduction

Alkanes are by far the most abundant but the least reactive members of the hydro-carbon family; the known reserves of methane alone approach those of petroleum [1]. Unfortunately, a significant portion of the methane produced is not utilized because of the difficulty associated with the transportation of a flammable, low-boiling gas. Its possible use as an automobile fuel is also limited by the intrinsic disadvantages of gaseous fuels, i.e., low energy content per unit volume and the hazards associated with handling and distribution. Consequently, the selective cat-alytic activation and functionalization of C–H bonds of methane in particular, and alkanes in general, to form useful organics constitute a "Holy Grail" in chemistry. In this context, Table 1 presents thermodynamic data indicating which alkane functionalizations are feasible and, therefore, worth pursuing.

The lack of reactivity of alkanes stems from their unusually high bond energies (C–H bond energy of methane: 104 kcal/mol) and most reactions involving the homolysis of a C–H bond occur at fairly high temperatures or under photolytic conditions. Moreover, the selectivity in these reactions is usually low because of the subsequent reactions of the intermediate products, which tend to be more reactive than the alkane itself. Using methane as an example, its homolytic C–H bond energy is 10 kcal/mol higher than that in methanol. Therefore, unless methanol can be removed as soon as it is formed, any oxidation procedure that involves hydrogen-atom abstraction from the substrate C–H bond would normally cause rapid over-oxidation of methanol. For example, the radical-initiated chlori-nation of methane invariably leads to multiple chlorinations [2] (chlorination, however, is more specific in the presence of superacids [3]). In order to achieve

Table 1. $\Delta G^{\ominus}$ at 298 K for selected alkane functionalizations.

Reaction			$\Delta G^{\ominus}$ [kcal/mol] [a]
$2\ CH_4$ (g)	$\rightarrow$	C_2H_6 (g) + H_2 (g)	+ 16.4
$2\ CH_4$ (g) + $^1/_2\ O_2$ (g)	$\rightarrow$	C_2H_6 (g) + H_2O (l)	− 40.3
C_2H_6 (g)	$\rightarrow$	$CH_2{=}CH_2$ (g) + H_2 (g)	+ 24.1
C_2H_6 (g) + $^1/_2\ O_2$ (g)	$\rightarrow$	$CH_2{=}CH_2$ (g) + H_2O (l)	− 32.6
CH_4 (g) + CO (g)	$\rightarrow$	CH_3CHO (l)	+ 14.3
CH_4 (g) + CO_2 (g)	$\rightarrow$	CH_3CO_2H (l)	+ 13.2
CH_4 (g) + $^1/_2\ O_2$ (g)	$\rightarrow$	CH_3OH (l)	− 27.6
CH_4 (g) + CO (g) + $^1/_2\ O_2$ (g)	$\rightarrow$	CH_3CO_2H (l)	− 48.3

[a] 1 kcal/mol = 4.184 kJ/mol.

the selective functionalization of alkanes, it is therefore necessary in most instances to promote a pathway that does not involve C–H bond homolysis as one of the steps. The problem is compounded by the fact that practically economical processes usually require the direct use of dioxygen as the oxidant. Because of its triplet electronic configuration, reactions between dioxygen and alkanes most often involve unselective radical pathways (cf. Section 2.8.1) [4].

Apart from the selectivity with respect to the degree of oxidation, a second selectivity issue arises for C_3 and higher alkanes: the selectivity with respect to the particular C–H bond that is functionalized. Again, since the homolytic bond energies decrease in the order: primary C–H > secondary C–H > tertiary C–H bonds, radical pathways involving C–H bond homolysis almost always show a marked preference for the functionalization of tertiary C–H bonds. This is in contrast to many commodity chemicals that are terminally functionalized [5].

In principle, the above selectivity problems can be avoided in suitably designed homogeneous metal-ion-catalyzed oxidation procedures. Transition metals, particularly those whose most stable oxidation states differ by $2e^-$, often promote nonradical pathways even in the presence of dioxygen [6]. Moreover, since metal–carbon bond strengths parallel those of C–H bonds and because of steric factors, the preferential functionalization of primary C–H bonds becomes possible [7]. As a bonus, metal-ion catalyzed reactions usually operate at low temperatures (~ 100 °C or below) [8].

Below, we describe homogeneous catalytic systems for the catalytic activation and functionalization of C–H bonds of alkanes. The account only highlights some of the recent advances in the area, focusing especially on oxidative functionalizations involving dioxygen as the oxidant: these are of particular importance since the vast majority of commercially important organic chemicals (alcohols, aldehydes, ketones, acids) are derived from alkanes through one or more oxidative functionalization steps [5]. Some reviews have appeared [9]. For convenience the reactions are classified into three pathways: radical, oxidative addition, and electrophilic, although the mechanism is not known in every case and several pathways may be operating simultaneously.

3.3.6.2 Radical Pathways

This involves the metal as a $1e^-$ oxidant, as shown in eqs. (1) and (2). From a thermodynamic standpoint, the $1e^-$ oxidation of alkanes is generally less favorable than the corresponding $2e^-$ oxidation [10] and, therefore, requires the use of either very strong oxidants or relatively high temperatures. Sometimes, as shown in eq. (2), an auxiliary ligand on the metal may participate in the C–H bond-breaking step. Equation (2) appears to represent nature's preferred route to alkane C–H activation. For example, it is generally accepted that in the enzyme cytochrome *P*-450, the species responsible for alkane C–H cleavage is a porphyrinato–$Fe^V{=}O$ complex [11]. The C–H activating species in methane monooxygenase has been less well characterized but a high-valent Fe=O species similar to that in cytochrome *P*-450 has been postulated [11]. The high specificity observed in enzymic systems is presumably a result of steric restraints. More commonly, however, the organic free radicals generated will participate in a multitude of reaction pathways leading to a large number of products [4]. Thus, most commercial metal-catalyzed processes belonging to this group, such as the Co^{III}-catalyzed oxidation of cyclohexane, are generally carried out at fairly low conversion levels ($< 10\%$) to enhance selectivity [4 a].

$$M^{N+} + R\text{-}H \; \rightleftharpoons \; M^{(N-1)+} + R^{\bullet} + H^{+} \tag{1}$$

$$M^{N+}{=}O + R\text{-}H \; \rightleftharpoons \; M^{(N-1)+}\text{-}OH + R^{\bullet} \tag{2}$$

In an effort to mimic the chemistry of cytochrome *P*-450, a large amount of work has been performed on alkane oxidations mediated by transition metal–porphyrin complexes [12]. Particularly noteworthy are the shape-selective oxidations of terminal methyl groups using bulky porphyrin ligands [13]. Additionally, Hill and others have published work on the polyoxometallate-catalyzed alkane functionalizations [14]. Here again, a high-valent metal–oxo species is thought to be responsible for the C–H activation step. Unfortunately, with some exceptions, dioxygen cannot be used as the oxidant; instead, hydrogen peroxide and related organic and inorganic peroxo species are usually used. This further underscores the problem of simultaneous activation of the alkane C–H bond and dioxygen in a practically useful catalytic system. One notable exception is a system described by Lyons and Ellis which directly utilizes dioxygen to oxidize isobutane and propane [15]. Polyhalogenated metalloporphyrin complexes are used as catalysts and only the weak tertiary and secondary C–H bonds are attacked. Although a high-valent metal–oxo species was initially proposed as the C–H activating agent, recent work tends to support a radical pathway initiated by metal-catalyzed decomposition of alkyl hydroperoxides [16].

Several interesting variations on the above radical chemistry have been described recently. One such system is copper salt catalyzed alkane oxidation by dioxygen in the presence of an aldehyde [17]. The proposed mechanism involves the initial autoxidation of the aldehyde to the corresponding peracid, which is the real oxidant for the Cu^{II}-mediated oxidation of the alkane (eqs. (3)–(5)). The ratio of alkane oxidized to aldehyde converted is relatively low because much of the peracid formed reacts with the aldehyde to form two molecules of carboxylic acid.

$$R'CHO + O_2 \longrightarrow R'CO_3H \tag{3}$$

$$Cu^{II} + R'CO_3H \longrightarrow Cu^{III}\text{-}O\bullet + R'CO_2H \tag{4}$$

$$Cu^{III}\text{-}O\bullet + R\text{-}H \longrightarrow Cu^{II} + R\text{-}OH \tag{5}$$

Related to the above is the "Gif" system discovered by Barton [18]. In essence, it involves $Fe^{II} + O_2$ + reducing agent or $Fe^{III} + H_2O_2$. The mechanism is unsettled although a high-valent Fe=O species has been implicated in the C–H cleavage step. The reactivity profile appears to be inconsistent with the generation of free radicals, e. g., secondary C–H bonds are attacked in preference to tertiary C–H bonds. Instead, Barton has postulated the [2 + 2] addition of a C–H bond across the Fe=O bond as the key step in this system. If so, this may be regarded as an example of ligand-assisted electrophilic C–H activation (cf. eq. (13 b), see below). A (perhaps) related system involving high-valent Ru=O species has been reported by Drago [19]. This system converts methane to methanol and formaldehyde using H_2O_2 as the oxidant.

The sulfoxidation of alkanes to alkane sulfonic acids using a combination of sulfur dioxide and dioxygen and catalyzed by vanadium compounds has been reported [20]. The mechanism involves intermediacy of alkyl radicals which are trapped by sulfur dioxide and then further oxidized to the product.

The final variation on metal-mediated radical chemistry of alkanes involves mercury-sensitized photochemical dimerization of alkanes [21]. The high selectivity in the reaction arises from the fact that the sequence of steps (eqs. (6)–(9)) occurs in the gas phase and the dimerization product is invariably less volatile than the starting alkane. Using this procedure, Crabtree has even achieved the cross-dimerization of alkanes with functional organics.

$$\text{Hg} + \text{h}\nu \longrightarrow \text{Hg}^* \tag{6}$$

$$\text{Hg}^* + \text{R-H} \longrightarrow \text{R}^{\bullet} + \text{H}^{\bullet} + \text{Hg} \tag{7}$$

$$\text{H}^{\bullet} + \text{R-H} \longrightarrow \text{R}^{\bullet} + \text{H}_2 \tag{8}$$

$$2\,\text{R}^{\bullet} \longrightarrow \text{R-R} \tag{9}$$

3.3.6.3   Oxidative Addition Pathways

The second C–H cleavage pathway involves the oxidative addition of the C–H bond to a low-valent metal center (eq. (10)), and was initially reported by Bergman, Graham, and Jones [22]. Unlike the systems described in the previous section, there is a strong preference for attack at the primary C–H bond.

$$\text{M}^{N+} + \text{R-H} \rightleftharpoons \text{M}^{(N+2)+} \overset{\text{R}}{\underset{\text{H}}{<}} \tag{10}$$

A two-center version of the oxidative addition reaction described above has also been observed by Wayland with porphyrinato Rh–Rh-bonded dimers [23]. By using a sterically encumbered ligand, such as tetramesitylporphyrin (TMP), the Rh–Rh bond energy is considerably reduced, permitting the formation of a (TMP)Rh–R and a (TMP)Rh–H species (Scheme 1).

$$(\text{por})\text{Rh}^{\text{II}}\text{-Rh}^{\text{II}}(\text{por}) \rightleftharpoons 2\,(\text{por})\text{Rh}^{\text{II}\bullet} \xrightarrow{\text{R-H}}$$

$$[(\text{por})\text{Rh-R-H-Rh}(\text{por})] \longrightarrow (\text{por})\text{Rh}^{\text{III}}\text{-R} + (\text{por})\text{Rh}^{\text{III}}\text{-H}$$

Scheme 1

The presence of reactive low-valent metal species prevents the simultaneous presence of most oxidizing agents that are capable of functionalizing the bound hydrocarbyl group in the oxidative addition product. Thus, it is difficult to construct a "one-pot" catalytic oxidation procedure, although nonoxidative catalytic functionalizations based on eq. (10) have been demonstrated. For example, first Crabtree and then Tanaka and Goldman have reported the efficient transfer dehydrogenation of alkanes to olefins under photochemical, as well as thermal, conditions (eq. (11)) [24]. Typically, a second olefin, such as *t*-butylethylene or norbornene, was the hydrogen acceptor. A particularly notable recent achievement has been the selective dehydrogenation of long-chain alkanes to *a*-olefins [24 a]. The related photochemical carbonylation of alkanes to aldehydes and the analogous isocyanide insertions have also been reported [24 c, d, 25]. Photons are required since the carbonylation of alkanes to aldehydes is thermodynamically disfavored (see Table 1).

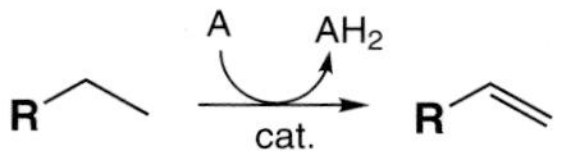

$$R\diagdown\diagup \xrightarrow[\text{cat.}]{\overset{A \quad AH_2}{\diagdown\diagup}} R\diagdown\diagdown \qquad (11)$$

cat. = [Ir(PR$_3$)$_2$(solv)$_2$H$_2$]$^+$ or Rh(PR$_3$)$_3$Cl

Another reaction of some synthetic utility is the insertion of olefins into aromatic C–H bonds [9 d]. This reaction is catalyzed by ruthenium compounds and requires a coordinating group (typically, ketone) on the aromatic ring. The group binds to the metal and the *ortho* C–H bonds are activated due to the resulting chelate effect.

Although oxidizing agents are not tolerated by most systems that activate C–H bonds through an oxidative addition pathway, they are compatible with boranes. In a series of elegant papers, Hartwig has demonstrated the selective formation of

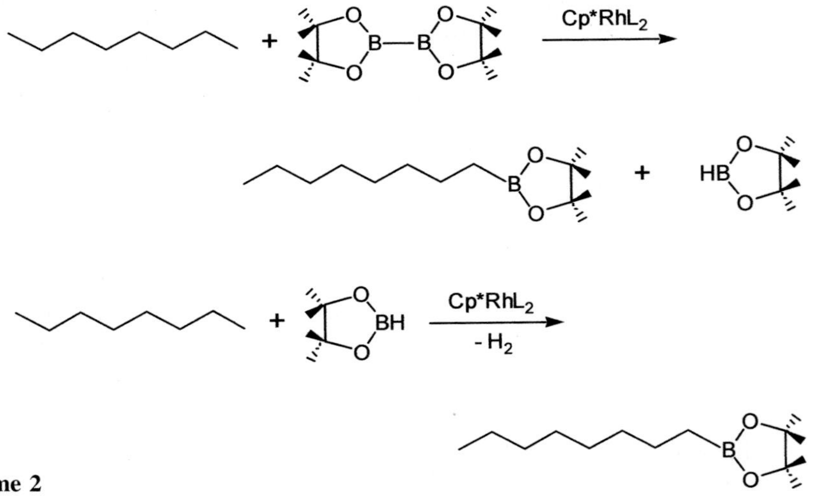

Scheme 2

terminal alkyl boranes starting with an alkane, a diboron compound, and a catalyst (Scheme 2) [26]. The mechanism is believed to involve the oxidative addition of a B–B (or a B–H) bond, as well as a terminal C–H bond of the alkane, and is followed by the reductive elimination of alkyl borane.

3.3.6.4 Electrophilic Pathways

The activation of C–H bonds by an electrophilic pathway is shown schematically in eq. (12) and has been observed with a number of late transition metal ions [9]. A driving force for the reaction shown in eq. (12) is the stabilization of the leaving group, H^+, by solvation in polar solvents. The related four-center electrophilic activation by transition, lanthanide, and actinide metal centers has also been reported, (eqs. (13a) and (13b)) [9 b, c, g, 27]. In these instances, a ligand on the metal assists the reaction by acting as the base.

$$M^{N+} + R\text{-}H \rightleftharpoons M^{N+}\text{-}R^- + H^+ \tag{12}$$

$$L_nM\text{-}X + H\text{-}R \rightleftharpoons L_nM^{\delta+} \underset{X^{\delta-}}{\overset{R^{\delta-}}{\diamond}} H^{\delta+} \rightleftharpoons L_nM\text{-}R + H\text{-}X \tag{13a}$$

$$L_nM{=}X + H\text{-}R \rightleftharpoons L_nM^{\delta+} \underset{X^{\delta-}}{\overset{R^{\delta-}}{\diamond}} H^{\delta+} \rightleftharpoons L_nM{-}X \overset{\overset{R\ H}{|\ \ |}}{} \tag{13b}$$

The most significant advantage of the C–H activation pathway shown in eq. (12) is that the late transition metal electrophiles are compatible with oxidants, including dioxygen. Therefore, in principle, it should be possible to design a catalytic oxidation procedure that is based on an initial electrophilic C–H cleavage step, as shown in Scheme 3 and first demonstrated by Shilov and his colleagues using the Pt^{II} ion as the C–H activating species (see below) [28].

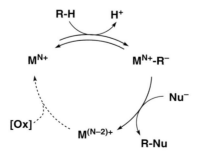

Scheme 3. Initial electrophilic C–H cleavage step. Ox = $2e^-$ oxidant; Nu^- = nucleophile.

For two reasons much of the work in this area has been carried out in strong acids. First, the conjugate bases of strong acids are poorly coordinating, thereby enhancing the electrophilicity of the metal ion. Second, the esterification of the alcohol, the primary product of alkane oxidation, protects it from overoxidation. One impressive achievement in this area is the Hg^{II}-catalyzed oxidation of methane to methyl sulfate in pure sulfuric acid, described by Catalytica, Inc., workers [29]. Both high selectivity and high conversion have been achieved. The sulfuric acid serves both as the solvent and the reoxidant for the metal. Although an electrophilic mechanism similar to Scheme 3 has been claimed, further studies indicate that a radical pathway, occurring at least in parallel, cannot be ruled out [30]. More recently, a 2,2′-bipyrimidyl complex of Pt^{II} has been employed for the same reaction [31]. Again, an electrophilic mechanism has been suggested for this reaction. Theoretical examination of this and related C–H activation chemistry by Pt^{II} suggests that the mechanism of C–H activation is either σ-bond metathesis (cf. eq. (7a)) or oxidative addition, depending on the anionic ligand present [32]. Overall, as reported, the system does not appear to be commercially viable since dioxygen cannot be directly employed as the oxidant. Moreover, for ethane and higher alkanes, significant amounts of decomposition products are formed through the sulfuric acid-induced dehydrations [30a]. Other noteworthy results in the area of electrophilic C–H activation in strong acids are the Pd^{II} and Pt^{II} catalyzed insertion of acetylenes into aromatic C–H bonds [33], and the Pd^{II}/Cu^{II} catalyzed carbonylation of alkanes, including methane, in trifluoroacetic acid [34]. In this case, the oxidant was the peroxydisulfate ion. These and related reactions [35] build upon an earlier report of electrophilic activation and functionalization of alkanes by the Pd^{II} ion in trifluoroacetic acid [36].

Electrophilic C–H activations can also be effected in water. At first glance, water would appear to be particularly unpromising as a solvent for such reactions. Because of their extremely poor coordinating ability alkanes should not be able to compete with water for coordination sites. Moreover, the intermediate metal–alkyl species would be prone to hydrolytic decomposition. In one respect, however, water is an almost ideal medium for C–H functionalization: the O–H bond energy exceeds the corresponding C–H bond energy of even methane. Indeed, the selective oxidation of methane to methanol is carried out by methane monooxygenase in aqueous medium.

Shilov and his co-workers were the first to demonstrate metal-mediated alkane functionalization in water [28]. They showed that simple Pt^{II} complexes, such as $PtCl_4^{2-}$, will activate and oxidize the C–H bonds of alkanes in the presence of an oxidizing agent, most notably Pt^{IV} salts. Although Shilov suggested a $Pt^{II/0}$ cycle in accordance with Scheme 3, subsequent work with model systems suggest that a $Pt^{II/IV}$ cycle is more likely (Scheme 4) [9b]. Additionally, the exact nature of the C–H activation step remains uncertain [32]. Sen [30a, 37], and also Bercaw and Labinger [38], have followed up on aspects of this work and have shown that a wide variety of substrates including methane can be functionalized with unusual selectivity. Thus, although the homolytic C–H bond energy of methane is 10 kcal/mol higher than that in methanol, a C–H bond of methanol would not be expected to be significantly more susceptible to *electrophilic* cleavage than that of methane. Indeed, Sen has observed that in water at 100 °C, the rate constant for the oxidation of methane to methanol by the $PtCl_4^{2-}/PtCl_6^{2-}$ combination (the Pt^{IV} species acts merely as a reoxidant for the $Pt^0 \rightarrow Pt^{II}$ step; see Scheme 4) is only one-seventh of that for methanol oxidation by the same system [30 a]. The observed similarity in rates is even more striking, given the much higher binding ability of methanol to the Pt^{II} center. Moving to substrates with C–H bonds somewhat weaker than those in methane results in actual *reversal* of commonly observed selectivity. Thus, the relative rate of C–H bond activation by the Pt^{II} ion decreases in the order $H–CH_2CH_3 > H–CH_2CH_2OH > H-CH(OH)CH_3$, i. e., an order that is exactly the *opposite* of that expected on the basis of homolytic C–H bond energies [30 a]. On a practical level, this shows that the direct conversion of ethane to 1,2-ethanediol (ethylene glycol) is possible.

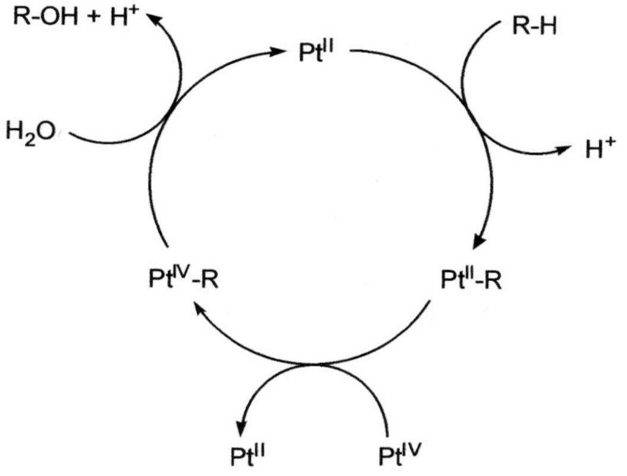

Scheme 4

The preferential oxidation of the methyl group of ethanol by the $PtCl_4^{2-}/PtCl_6^{2-}$ combination in water at 90 °C was first reported by Bercaw and Labinger [38 a] and subsequently confirmed by Sen [30 a], who observed the exclusive oxidation of the methyl group in ethanol resulting in the formation of 1,2-ethanediol as the sole product. A chelate effect which results in a less strained transition state for the oxidation of the methyl group of ethanol may be responsible for the observed selectivity (Scheme 5) [37]. Indeed, for *n*-propanol, the methyl group is the preferred site of attack and 1,3-propanediol is formed. Thus, the remote oxidation of highly flexible linear and branched alkyl chains with unprecedented regioselectivity becomes possible. The order of reactivity is α-C–H$\ll\beta$-C–H$<\gamma$-C–H$<\delta$-C–H for alcohols and α-C–H$\ll\beta$-C–H$<\gamma$-C–H$\geq\delta$-C–H for the acids [37]. These reactions are also very specific with respect to the degree of oxidation: only hydroxylation is observed, and further oxidation to the corresponding aldehyde or carboxylic acid functionality does not occur. This is a result of the strained transition state that is involved in the activation of a C–H bond α to the hydroxyl group (see Scheme 5).

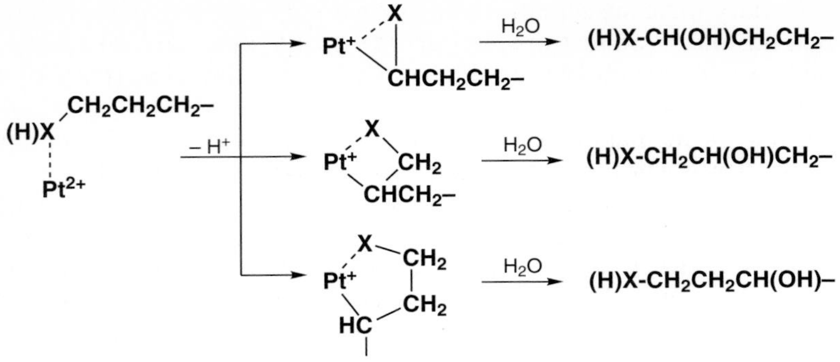

Scheme 5. Transition state during the activation of a C–H bond (X = ligating atom; note that in acids there is an extra atom between X and the C–H bond being attacked).

The activation and functionalization of C–H bonds by the Pt^{II} ion is particularly attractive because of the unusual regioselectivity, high oxidation level specificity, and mildness of reaction conditions. Moreover, Sen has recently reported that, in the presence of copper chloride at 120–160 °C, Shilov chemistry can be made catalytic with dioxygen as the ultimate oxidant [39]. A number of aliphatic acids were tested, and turnover numbers of up to 15/hour with respect to platinum were observed. H/D exchange studies also confirm the marked preference for the activation of primary C–H bonds in the presence of weaker secondary C–H bonds. This study constituted the first example of the direct use of dioxygen in the catalytic oxidation of unactivated primary C–H bonds under mild conditions that does not involve the use of a co-reductant (e. g., sacrificial metals, $2H^+$ + $2e^-$, dihydrogen, or carbon monoxide; see below).

Recently, Sen has reported two catalytic systems, one heterogeneous and the other homogeneous, which simultaneously activate dioxygen and alkane C–H bonds, resulting in direct oxidations of alkanes. In the first system, metallic palladium was found to catalyze the oxidation of methane and ethane by dioxygen in aqueous medium at 70–110 °C in the presence of carbon monoxide [40]. In aqueous medium, formic acid was the observed oxidation product from methane while acetic acid, together with some formic acid, was formed from ethane [40 a]. *No* alkane oxidation was observed in the absence of added carbon monoxide. The essential role of carbon monoxide in achieving "difficult" alkane oxidation was shown by a competition experiment between ethane and ethanol, both in the presence and absence of carbon monoxide. In the absence of added carbon monoxide, only ethanol was oxidized. When carbon monoxide was added, almost half of the products were derived from ethane. Thus, the more inert ethane was oxidized *only* in the presence of added carbon monoxide.

Studies indicate that the overall transformation encompasses three catalytic steps in tandem (Scheme 6) [9 a, 40]. The first is the water-gas shift reaction involving the oxidation of carbon monoxide to carbon dioxide with the simultaneous formation of dihydrogen. It is possible to by-pass this step by replacing carbon monoxide with dihydrogen. The second catalytic step involves the combination of dihydrogen with dioxygen to yield hydrogen peroxide (or its equivalent). The final step involves the metal-catalyzed oxidation of the substrate by hydrogen peroxide (or its equivalent).

Whereas acetic acid was formed in good yield from ethane, the analogous formation of formic acid from methane proceeded only in low yield because of the general instability of the latter acid under the reaction conditions. Since formic acid is a much less desirable product from methane than is methanol, the possibility of halting the oxidation of methane at the methanol stage was examined.

Simply changing the solvent in the Pd-based catalytic system from water to a mixture of water and a perfluorocarboxylic acid (some water is necessary for the reaction; see Scheme 6) had no significant effect on product composition: formic acid was still the principal product from methane. However, the addition of Cu^I or Cu^{II} chloride to the reaction mixture had a dramatic effect. Methanol and its ester now became the preferred products, with virtually no acetic and little formic acid being formed [40 b]. The activation parameters for the overall reaction determined under the condition when the rate was first order in both methane and carbon monoxide were: $A = 2 \times 10^4$ s^{-1}; $E_a = 15.3$ kcal mol^{-1}. Since methyl trifluoroacetate is both volatile and easily hydrolyzed back to the acid and methanol, it should be possible to design a system where the acid is recycled and methanol is the end product. Lee and co-workers have recently reported on the further characterization of the catalyst in this bimetallic Pd/Cu system [41].

Free alkyl radicals appear not to be intermediates. Thus, primary C–H bonds are at least as reactive (usually much more) than secondary, tertiary, or benzylic C–H bonds, or C–H bonds a to an alcohol functionality. For example, alkane oxidation proceeds much faster than the oxidation of the corresponding alcohol. Even the relatively unselective HO· radical shows a significantly higher preference for attack on secondary than on primary C–H bonds.

The reactivity pattern suggests the presence of a strongly electrophilic oxidant. This is supported by the following observations [43]. For a series of *para*-substituted phenols, the rate of reaction decreased with increasing electronegativity of the *para* substituent, with an approximately linear correlation between the electron affinity of the substituent and the ratio of the log of the rate of oxidation of sub-

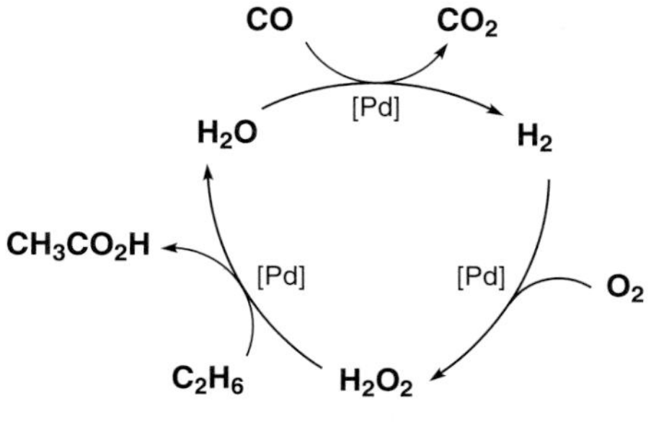

Scheme 6. Catalytic steps in tandem.

$$L_xRh\text{-}R \quad\begin{cases} \xrightarrow[k_{Nu}]{Nu^-} \quad R\text{-}Nu \\[2em] \xrightarrow[k_{CO}]{CO} \quad L_xRh\text{-}COR \xrightarrow{Nu^-} RCO\text{-}Nu \end{cases}$$

Scheme 7   (Nu = OH, $C_3F_7CO_2$)

stituted phenol to the parent phenol. This is consistent with an initial electrophilic attack at the ring. Additionally, the ease of oxidation decreased in the order $(CH_3)_2S > (CH_3)_2SO > (CH_3)_2SO_2$, which further supports the conclusion that the system acts as an electrophilic oxidant.

In the metallic Pd-based system, the role of the metal is two-fold. First, it generates hydrogen peroxide (or its equivalent) in accordance with Scheme 6 [9 a, 40, 42, 43]. Second, it causes nonspecific over-oxidation of the organic substrate using the hydrogen peroxide thus generated. This latter reaction is suppressed when $CuCl_2$ is added. In the bimetallic $Pd/CuCl_2$-based system, experiments suggest that the principal role of metallic palladium is to generate hydrogen peroxide *in situ* and the species responsible for the remote hydroxylation of the substrate by hydrogen peroxide is Cu chloride [42].

In the second (slower) system, $RhCl_3$, in the presence of several equivalents of Cl^- and I^- ions, was found to catalyze the direct functionalization of methane in the presence of carbon monoxide and dioxygen at 80–85 °C [44]. The reaction proceeded in water to give acetic acid as the principal product [44 a]. However, a much higher rate was observed in a 6:1 (v/v) mixture of perfluorobutyric acid and water, the products being methanol and acetic acid [44 b]. It is possible to form *either* methanol *or* acetic acid selectively by a simple change in the solvent system. The ratio of alcohol derivative to the corresponding higher acid may be assumed to be a function of the relative rates of nucleophilic attack versus carbon monoxide insertion into a common Rh–alkyl bond (i. e., k_{Nu}/k_{CO}; see Scheme 7). While, to a first-order approximation, k_{CO} is likely to be independent of the solvent, k_{Nu} would depend on the nature of the nucleophile derived from the solvent. Presumably, the perfluorobutyrate ion is a better nucleophile than water since more of the alcohol derivative was formed in perfluorobutyric acid–water mixture than in pure water. This also explains why acetic acid was once again the major product when the perfluorobutyrate ion was tied up as the ester. Consistent with the mechanistic scenario shown in Scheme 7 was also the observation that the ratio of acetic acid to methanol derivative formed from methane increased with increasing pressure of CO although the overall reaction was sharply inhibited at high CO pressures.

In addition to Sen's work on the Rh-catalyzed oxidative carbonylation of methane, Grigoryan has also reported a similar reaction in acetic acid [45]. Predictably, the reaction rate is between that observed in pure water and in the perfluorocarboxylic acid–water mixture. Finally, Otsuka has reported the oxidative carbonylation of methane to acetic acid by rhodium-doped iron phosphate [46]. The Pd/Cu and the Rh-based systems show similar selectivity patterns that are, for the most part, without precedent. For example, in both cases, methane is *significantly more reactive* (at least five times) than methanol [9 a, 40, 44]. For the Rh-based system, even methyl iodide was found to be less reactive than methane [44 b]!

A more interesting reactivity pattern exhibited by these two systems is their preference for C–C cleavage over C–H cleavage for higher alkanes [40 b, 44 b]. Indeed, we are unaware of any other catalytic system that effects the oxidative cleavage of alkane C–C bonds under such mild conditions. For example, the Rh-based system converts ethane to a mixture of methanol, ethanol, and acetic acid, with a ratio of products formed through C–H relative to C–C cleavage of approx. 0.6 on a per-bond basis [44 b]. As with methanol, control experiments indicated ethane is more reactive than ethanol. Additionally, neither ethanol nor acetic acid is the precursor to methanol. Finally, part of the acetic acid is even formed by initial C–C cleavage of ethane followed by carbonylation of the resultant C_1 fragment. For C_4 and higher alkanes, C–C cleavage products were *virtually all* that were observed; especially noteworthy was the formation of ethanol from *n*-butane, which indicates that vicinal diols are not the precursors to the C–C cleavage products. The above reactivity profile exhibited by the two systems, together with other observations, appears to be inconsistent with the intermediacy of *free* alkyl radicals in the oxidation process.

A curious aspect of the Pd- and Rh-based systems is that, apart from their ability to activate both dioxygen and the alkane, both require a co-reductant (carbon monoxide) [40, 44]. Thus, there is a striking resemblance to monooxygenases [11]. In nature, while the dioxygenases utilize the dioxygen molecule more efficiently, it is the monooxygenases that carry out "difficult" oxidations, such as alkane oxidations. In the latter, one of the two oxygen atoms of dioxygen is reduced to water in a highly thermodynamically favorable reaction and the free energy gained thereby is employed to generate a high-energy oxygen species, such as a metal–oxo complex, from the second oxygen atom (eq. (14)). The "Gif" system of Barton [18] is also designed on this premise. In at least the Pd-based Sen system [40], the co-reductant, carbon monoxide, is employed to generate dihydrogen (eq. (15)), the latter being formally equivalent to $2H^+ + 2e^-$ that is employed in the biological systems (cf. eqs. (14) and (16)).

$$O_2 \; + \; 2\,H^+ \; + \; 2\,e^- \; \longrightarrow \; H_2O \; + \; [O] \qquad (14)$$

$$CO \; + \; H_2O \qquad\qquad \longrightarrow \; CO_2 \; + \; H_2 \qquad (15)$$

$$O_2 \; + \; H_2 \qquad\qquad \longrightarrow \; H_2O \; + \; [O] \qquad (16)$$

How general is this requirement for a co-reductant (e. g., CO or H_2) in achieving "difficult" catalytic hydrocarbon oxidations by dioxygen? Sen's work has provided two examples of catalytic systems that operate in this manner (i. e., as monoxygenase analogs) [40, 44]. There have been other recent publications on catalytic systems for the oxidation of hydrocarbons, including olefins and aromatics, that also call for either CO or H_2 as the coreductant [48]. While, from a practical standpoint, it is more desirable for both oxygen atoms of O_2 to be used for substrate oxidation, with the exception of the Shilov system, there appears to be no currently known catalytic system that operates as an artificial "dioxygenase" under mild conditions toward "difficult" substrates, such as those possessing unactivated primary C–H bonds.

3.3.6.5   Conclusions

While few of the catalytic systems discussed above meet the criteria for successful commercial processes, it is clear that impressive progress has been made in recent years in the field of alkane activation and functionalization. Who, for example, would have believed that it is possible to functionalize methane in water at or below 100 °C with reasonable turnover rates [40, 44]! A rich new area of organometallic chemistry and catalysis that is both fundamentally interesting and useful beckons. Future progress will depend on a better understanding of the organometallic chemistry of metal complexes with nontraditional, "hard", ligands (as opposed to soft, easily oxidizable, ligands that are traditionally used for metal complexes that catalyze "reductive" chemistry, such as hydrogenation and carbonylation [8]). Radically different reactivity patterns may be anticipated. For example, Pt^{IV} complexes of the type, $PtCl_5R^{2-}$, which are the proposed intermediates in alkane hydroxylation by the $PtCl_4^{2-}/PtCl_6^{2-}$ combination, react with water to form alcohols [37 a, 38 a, 49]. The formation of an alcohol by hydrolysis implies a Pt–C bond polarity that is the *opposite* of that normally observed for metal–alkyls (metal–alkyls generally yield alkane and metal hydroxide upon hydrolysis).

References

[1] (a) M. G. Axelrod, A. M. Gaffney, R. Pitchai, J. A. Sofranko, in *Natural Gas Conversion II* (Eds.: H. E. Curry-Hyde, R. F. Howe), Elsevier, Amsterdam, **1994**, p. 93; (b) C. D. Masters, D. H. Root, E. D. Attanasi, *Science* **1991**, *253*, 146; (c) C. Starr, M. F. Searl, S. Alpert, *Science* **1992**, *256*, 981.

[2] (a) J. March, *Advanced Organic Chemistry*, Wiley, New York, **1985**, p. 620 and references therein; (b) M. L. Poutsma, in *Free Radicals* (Ed.: J. K. Kochi), Wiley, New York, **1973**, Vol. II, p. 159.

[3] G. Olah, *Acc. Chem. Res.* **1987**, *20*, 422.

[4] Reviews: (a) G. W. Parshall, S. D. Ittel, *Homogeneous Catalysis*, Wiley, New York, **1992**, p. 237; (b) J. A. Howard, in ref. [2 b], p. 3.

[5] D. E. Collins, F. A. Richey, in *Riegel's Handbook of Industrial Chemistry* (Ed.: J. A. Kent), Van Nostrand Reinhold, New York, **1992**, p. 800.

[6] Reviews: (a) R. S. Drago, *Coord. Chem. Rev.* **1992**, *117*, 185; (b) L. I. Simándi, *Catalytic Activation of Dioxygen by Metal Complexes*, Kluwer Academic, Dordrecht, **1992**, p. 74.

[7] (a) H. E. Bryndza, L. K. Fong, R. A. Paciello, W. Tam, J. E. Bercaw, *J. Am. Chem. Soc.* **1987**, *109*, 1444; (b) R. G. Bergman, *Science* **1984**, *223*, 902.

[8] Review: G. W. Parshall, S. D. Ittel, *Homogeneous Catalysis*, Wiley, New York, **1992**.

[9] General reviews on the problem of C–H activation and functionalization in solution: (a) A. Sen, *Acc. Chem. Res.* **1998**, *31*, 550; (b) S. S. Stahl, J. A. Labinger, J. E. Bercaw, *Angew. Chem. Int. Ed.* **1998**, *37*, 2181; (c) W. D. Jones, *Top. Organomet. Chem.* **1999**, *3*, 9; (d) F. Kakiuchi, S. Murai, *Top. Organomet. Chem.* **1999**, *3*, 47; (e) A. Sen, *Top. Organomet. Chem.* **1999**, *3*, 81; (f) R. H. Crabtree, *Chem. Rev.* **1995**, *95*, 987; (g) B. A. Arndtsen, R. G. Bergman, T. A. Mobley, T. H. Peterson, *Acc. Chem. Res.* **1995**, *28*, 154; (h) J. A. Labinger, *Fuel Process. Technol.* **1995**, *42*, 325; (i) *Selective Hydrocarbon Oxidation and Functionalization* (Eds.: J. A. Davies, P. L. Watson, A. Greenberg, J. F. Liebman), VCH, New York, **1990**; (j) *Activation and Functionalization of Alkanes* (Ed.: C. L. Hill), Wiley, New York, **1989**; (k) A. E. Shilov, *Activation of Saturated Hydrocarbons by Transition Metal Complexes*, D. Reidel, Dordrecht, **1984**.
[10] See ref. [9 k], p. 125.
[11] Reviews: (a) S. E. Groh, M. J. Nelson, in ref. [9 i], p. 305; (b) *Oxygenases and Model Systems* (Ed.: T. Funabiki), Kluwer, Dordrecht, **1997**, Ch. 5–8; (c) J. S. Valentine, in *Bioinorganic Chemistry* (Eds.: I. Bertini, H. B. Gray, S. J. Lippard, J. S. Valentine), University Science Books, Mill Valley, CA, **1994**, p. 253; (d) D. Mansuy, P. Battioni, in *Bioinorganic Catalysis* (Ed.: J. Reedijk), Marcel Dekker, New York, **1993**, p. 395; (e) L. Que, in *Bioinorganic Catalysis* (Ed.: J. Reedijk), Marcel Dekker, New York, **1993**, p. 347; (f) *Cytochrome P-450* (Eds.: T. Omura, Y. Ishimura, Y. Fujii-Kuriyama), VCH, New York, 1993, p. 17; (g) K. E. Liu, S. J. Lippard, *Adv. Inorg. Chem.* **1995**, *42*, 263; (b) A. L. Feig, S. J. Lippard, *Chem. Rev.* **1994**, *94*, 759.
[12] Reviews: (a) *Metalloporphyrins in Catalytic Oxidations* (Ed.: R. A. Sheldon), Marcel Dekker, New York, **1994**; (b) *Metalloporphyrins Catalyzed Oxidations* (Eds.: F. Montanari, L. Casella), Kluwer, Dordrecht, **1994**; (c) B. Meunier, in *Catalytic Oxidations with Hydrogen Peroxide as Oxidant* (Ed.: G. Strukul), Kluwer, Dordrecht, **1992**, p. 153.
[13] K. S. Suslick, in ref. [9 j], p. 219.
[14] Reviews: (a) G. Strukul, in ref. [12 c], p. 177; (b) C. L. Hill, in ref. [12 c], p. 253; (c) C. L. Hill, A. M. Khenkin, M. S. Weeks, Y. Hou, *ACS Symp. Ser.* **1993**, *523*, 67; (d) R. A. Sheldon, *Topics Curr. Chem.* **1993**, *164*, 21. Also see: (e) D. Mansuy, J.-F. Bartoli, P. Battioni, D. K. Lyon, R. G. Finke, *J. Am. Chem. Soc.* **1991**, *113*, 7222; (d) R. Neumann, M. Dahan, *Nature* **1997**, *388*, 353.
[15] J. E. Lyons, P. E. Ellis, in ref. [12 a], p. 297.
[16] M. W. Grinstaff, M. G. Hill, J. A. Labinger, H. B. Gray, *Science* **1994**, *264*, 1311.
[17] N. Komiya, T. Naota, Y. Oda, S.-I. Murahashi, *J. Mol. Catal. A: Chem.* **1997**, *117*, 21.
[18] D. H. R. Barton, D. Doller, *Acc. Chem. Res.* **1992**, *25*, 504.
[19] A. S. Goldstein, R. S. Drago, *J. Chem. Soc., Chem. Commun.* **1991**, 21.
[20] Y. Ishii, K. Matsunaka, S. Sakaguchi, *J. Am. Chem. Soc.* **2000**, *122*, 7390.
[21] R. H. Crabtree, S. H. Brown, C. A. Muedas, P. Krajnik, R. R. Ferguson, *Adv. Chem. Ser.* **1992**, *230*, 197.
[22] (a) Refs. [9 c, g]; (b) W. D. Jones, F. J. Feher, *Acc. Chem. Res.* **1989**, *22*, 91.
[23] (a) X.-X. Zhang, B. B. Wayland, *J. Am. Chem. Soc.* **1994**, *116*, 7897; (b) B. B. Wayland, S. Ba, A. E. Sherry, *J. Am. Chem. Soc.* **1991**, *113*, 5305.
[24] (a) F. Liu, E. B. Pak, B. Singh, C. M. Jensen, A. S. Goldman, *J. Am. Chem. Soc.* **1999**, *121*, 4086; (b) J. A. Maguire, A. Petrillo, A. S. Goldman, *J. Am. Chem. Soc.* **1992**, *114*, 9492; (c) J. A. Maguire, W. T. Boese, M. E. Goldman, A. S. Goldman, *Coord. Chem. Rev.* **1990**, *97*, 179; (d) M. Tanaka, T. Sakakura, *Adv. Chem. Ser.* **1992**, *230*, 181; (e) M. J. Burk, R. H. Crabtree, *J. Am. Chem. Soc.* **1987**, *109*, 8025.
[25] (a) A. J. Kunin, R. Eisenberg, *Organometallics* **1988**, *7*, 2124; (b) W. D. Jones, in ref. [9 i], p. 113.
[26] (a) H. Chen, S. Schlecht, T. C. Semple, J. F. Hartwig, *Science* **2000**, *287*, 1995; (b) K. M. Waltz, J. F. Hartwig, *J. Am. Chem. Soc.* **2000**, *122*, 11358.

[27] Reviews: (a) I. P. Rothwell, in ref. [9 i], p. 43; (b) P. L. Watson, in ref. [9 i], p. 79; (c) R. F. Jordan, *Adv. Organomet. Chem.* **1991**, *32*, 325; (c) L. M. Slaughter, P. T. Wolczanski, T. R. Klinckman, T. R. Cundari, *J. Am. Chem. Soc.* **2000**, *122*, 7953.

[28] Ref. [9 k], p. 142.

[29] R. A. Periana, D. J. Taube, E. R. Evitt, D. G. Löffler, P. R. Wentrcek, G. Voss, T. Masuda, *Science* **1993**, *259*, 340.

[30] (a) A. Sen, M. A. Benvenuto, M. Lin, A. C. Hutson, N. Basickes, *J. Am. Chem. Soc.* **1994**, *116*, 998. Also see: (b) I. P. Stolarov, M. N. Vargaftik, D. I. Shishkin, I. I. Moiseev, *J. Chem. Soc., Chem. Commun.* **1991**, 938.

[31] R. A. Periana, D. J. Taube, S. Gamble, H. Taube, T. Satoh, H. Fujii, *Science* **1998**, *280*, 560.

[32] (a) T. M. Gilbert, I. Hristov, T. Ziegler, *Organometallics* **2001**, *20*, 1183; (b) H. Heiberg, L. Johansson, O. Gropen, O. B. Ryan, O. Swang, M. Tilset, *J. Am. Chem. Soc.* **2000**, *122*, 10831; (c) P. E. M. Siegbahn, R. H. Crabtree, *J. Am. Chem. Soc.* **1996**, *118*, 4442.

[33] C. Jia, D. Piao, J. Oyamada, W. Lu, T. Kitamura, Y. Fujiwara, *Science* **2000**, *287*, 1992.

[34] (a) Y. Fujiwara, K. Takaki, Y. Taniguchi, *Synlett* **1996**, 591; (b) K. Nakata, Y. Yamaoka, T. Miyata, Y. Taniguchi, K. Takaki, Y. Fujiwara, *J. Organomet. Chem.* **1994**, *473*, 329; (c) K. Nakata, T. Miyata, T. Jintoku, A. Kitani, Y. Taniguchi, K. Takaki, Y. Fujiwara, *Bull. Chem. Soc. Jpn.* **1993**, *66*, 3755.

[35] (a) K. Nomura, S. Uemura, *J. Chem. Soc., Chem. Commun.* **1994**, 129; (b) M. N. Vargaftik, I. P. Stolarov, I. I. Moiseev, *J. Chem. Soc., Chem. Commun.* **1990**, 1049.

[36] Review: A. Sen, *Platinum Metals Rev.* **1991**, *35*, 126.

[37] (a) A. C. Hutson, M. Lin, N. Basickes, A. Sen, *J. Organomet. Chem.* **1995**, *504*, 69; (b) N. Basickes, A. Sen, *Polyhedron* **1995**, *14*, 197; (c) L.-C. Kao, A. Sen, *J. Chem. Soc., Chem. Commun.* **1991**, 1242; (d) A. Sen, M. Lin, L.-C. Kao, A. C. Hutson, *J. Am. Chem. Soc.* **1992**, *114*, 6385.

[38] (a) G. A. Luinstra, L. Wang, S. S. Stahl, J. A. Labinger, J. E. Bercaw, *J. Organomet. Chem.* **1995**, *504*, 75; (b) G. A. Luinstra, L. Wang, S. S. Stahl, J. A. Labinger, J. E. Bercaw, *Organometallics* **1994**, *13*, 755; (c) J. A. Labinger, A. M. Herring, D. K. Lyon, G. A. Luinstra, J. E. Bercaw, I. T. Horváth, K. Eller, *Organometallics* **1993**, *12*, 895; (d) G. A. Luinstra, J. A. Labinger, J. E. Bercaw, *J. Am. Chem. Soc.* **1993**, *115*, 3004.

[39] M. Lin, C. Shen, E. A. Garcia-Zayas, A. Sen, *J. Am. Chem. Soc.* **2001**, *123*, 1000.

[40] (a) M. Lin, A. Sen, *J. Am. Chem. Soc.* **1992**, *114*, 7307; (b) M. Lin, T. E. Hogan, A. Sen, *J. Am. Chem. Soc.* **1997**, *119*, 6048.

[41] E. D. Park, S. H. Choi, J. S. Lee, *J. Catal.* **2000**, *194*, 33.

[42] C. Shen, E. A. Garcia-Zayas, A. Sen, *J. Am. Chem. Soc.* **2000**, *122*, 4029.

[43] A. Pifer, T. Hogan, B. Snedeker, R. Simpson, M. Lin, C. Shen, A. Sen, *J. Am. Chem. Soc.* **1999**, *121*, 7485.

[44] (a) M. Lin, A. Sen, *Nature* **1994**, *368*, 613; (b) M. Lin, T. E. Hogan, A. Sen, *J. Am. Chem. Soc.* **1996**, *118*, 4574.

[45] (a) E. G. Chepaikin, G. N. Boyko, A. P. Bezruchenko, A. A. Leshcheva, E. H. Grigoryan, *J. Mol. Cat. A: Chem.* **1998**, *129*, 15; (b) E. G. Chepaikin, A. P. Bezruchenko, A. A. Leshcheva, G. N. Boiko, I. V. Kuzmenkov, E. H. Grigoryan, A. E. Shilov, *J. Mol. Cat. A: Chem.* **2001**, *169*, 89.

[46] Y. Wang, M. Katagiri, K. Otsuka, *J. Chem. Soc., Chem. Commun.* **1997**, 1187.

[47] I. Yamanaka, M. Soma, K. Otsuka, *Chemistry Lett.* **1996**, 565.

[48] Representative examples: (a) I. Tabushi, *Coord. Chem. Rev.* **1988**, 86, 1; (b) M. Otake, *Chemtech* **1995**, 36; (c) T. Miyake, M. Hamada, Y. Sasaki, M. Oguri, *Appl. Catal. A: General* **1995**, *131*, 33; (d) T. Teranishi, N. Toshima, *J. Chem. Soc., Dalton Trans.* **1995**, 979; (e) Y. Wang, K. Otsuka, *J. Catal.* **1995**, *155*, 256.

[49] L. A. Kusch, V. V. Lavrushko, Yu. S. Misharin, A. P. Moravsky, A. E. Shilov, *Nouv. J. Chim.* **1983**, *7*, 729.

3.3.7 Pauson–Khand Reaction

Wolfgang A. Herrmann

3.3.7.1 Introduction

When P. L. Pauson et al. reported in 1971 the surprising formation of cyclopent-2-en-1-ones from precursor compounds as simple as carbon monoxide, olefins, and alkynes [1], there was hardly any hope of catalytic procedures to come. Rather, an almost stoichiometric amount of octacarbonyldicobalt $Co_2(CO)_8$ was necessary to achieve this obviously metal-mediated C–C coupling reaction. Three carbon–carbon bonds form during this remarkable process (eq. (1)). Typical reaction conditions were heating in suitable solvents such as isooctane or toluene. In most cases, the easily available alkyne complexes **1** were reacted at 60–120 °C, usually in an atmosphere of carbon monoxide, with the corresponding alkene (eq. (2)). So far, dicobaltoctacarbonyl has been the mediator reagent of choice. However, good results were also reported with other organometallic compounds. For example, yields up to 80 % resulted from the first intramolecular Pauson–Khand reactions (PKRs) that were mediated by $Ru_3(CO)_{12}$ [34].

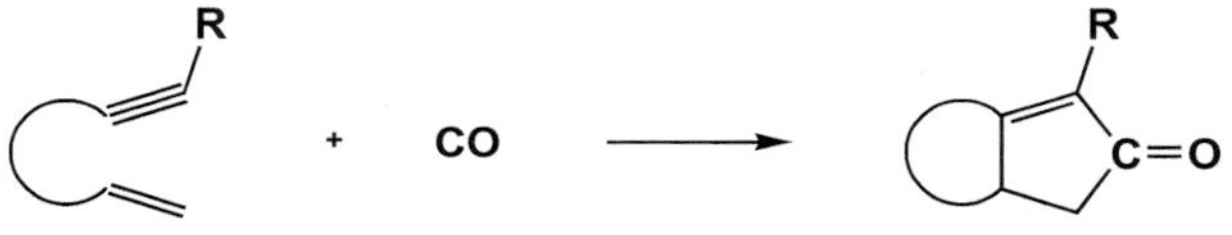

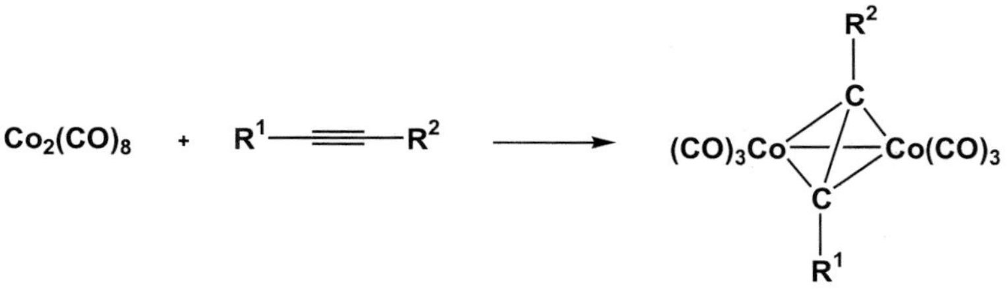

The PKR represented by eq. (1) and – in more general terms – eq. (3) has a number of advantages for the synthetic organic chemist. There are many natural products that contain cyclopentenone units. Three C–C-bonds are sequentially formed around one (or several?) cobalt centers. There is a high degree of regioselectivity (*vide infra*). Particularly, the larger of the two alkyne substituents (R^L) prefers the α-position of the carbonyl group. Beyond that, asymmetric varieties in inter- and intramolecular PKRs are possible. In the meantime, catalytic versions with decent activities but excellent selectivities are available [2–4].

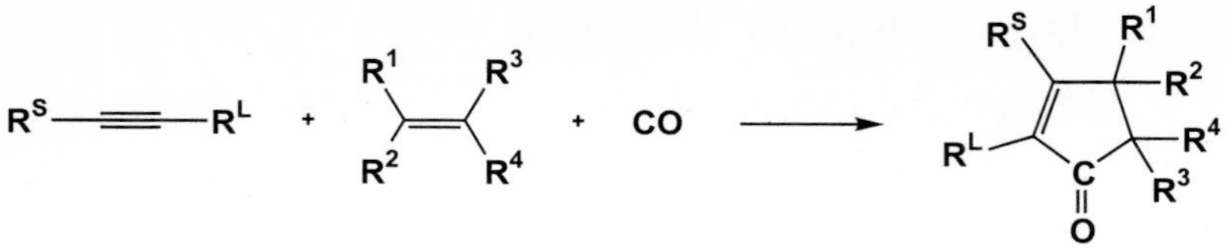

$$R^S \!\!-\!\!\!\equiv\!\!\!-\!R^L \;+\; \begin{array}{c} R^1 \quad R^3 \\ \diagup \\ R^2 \quad R^4 \end{array} \;+\; CO \;\longrightarrow\; \begin{array}{c} R^S \quad R^1 \\ \diagup \\ R^L \quad R^2 \\ R^4 \\ R^3 \\ O \end{array} \qquad (3)$$

S = small
                } substituent
L = large

The key disadvantage of the PKRs is the multiplicity of reactions that carbonyl-cobalt complexes can undergo with olefins and alkynes. The majority of thermal degradation products, e. g. cluster compounds, are not or only slightly reactive in the desired C–C coupling reactions. Therefore, the yields of many PKRs were disappointingly low. An improvement was the use of amine *N*-oxides $R_3N{\rightarrow}O$ that effect easy CO elimination to form vacant coordination sites at cobalt [5]. This is a version of the so-called Hieber base reaction. A significant acceleration of the PKR was thus achieved (in hours instead of days). Note that the CO elimination from $Co_2(CO)_8$ is thought to be the rate-determining step of the PKR sequence. Presumably the best protocol for the *stoichiometric* PKR was recently presented by Sugihara et al. [6], who added organic amines to the reaction mixtures. After several minutes, excellent yields were obtained in PKRs following eqs. (4) and (5).

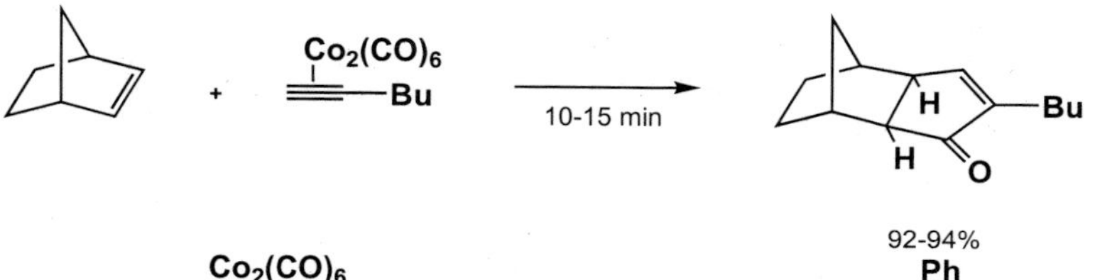

$$\qquad (4)$$

92-94%

$$\qquad (5)$$

89-90%

Phosphine sulfides $R_3P{=}O$ also accelerate PKRs in the presence of $Co_2(CO)_8$; only atmospheric pressure of carbon monoxide is necessary in this case [35].

3.3.7.2   The Catalytic Option

Only catalytic versions fulfill the economic demands of industry. The handling of huge amounts of cobalt carbonyls in the stoichiometric versions is cumbersome, expensive, and environmentally unpleasant. A first breakthrough was made by Rautenstrauch at Firmenich S. A., Geneva [7, 8]. He showed that the dihydrojas-

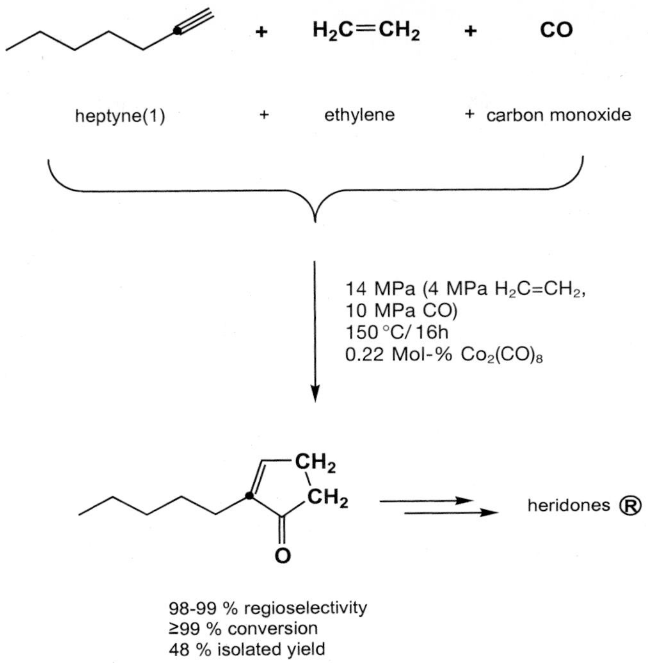
+ H2C=CH2 + CO
heptyne(1) + ethylene + carbon monoxide
14 MPa (4 MPa H2C=CH2,
10 MPa CO)
150 °C/ 16h
0.22 Mol-% Co2(CO)8
CH2
CH2
O
heridones ®
98-99 % regioselectivity
≥99 % conversion
48 % isolated yield

Scheme 1

monate precursor was formed under quasi-catalytic conditions in acceptable yields when pressure conditions (carbon monoxide, ethylene) were applied (Scheme 1) [4 d, 7, 8].

Subsequent work of the Korean group of Jeong and Chung [9] led to further improvements via catalyst modification – e. g. addition of phosphites – or the application of supercritical (sc) media [10]. Supercritical carbon dioxide at 20 MPa and 90 °C at a CO partial pressure of 3 MPa gave an 82 % product yield after 24 hours according to eq. (6) (cf. Section 3.1.13).

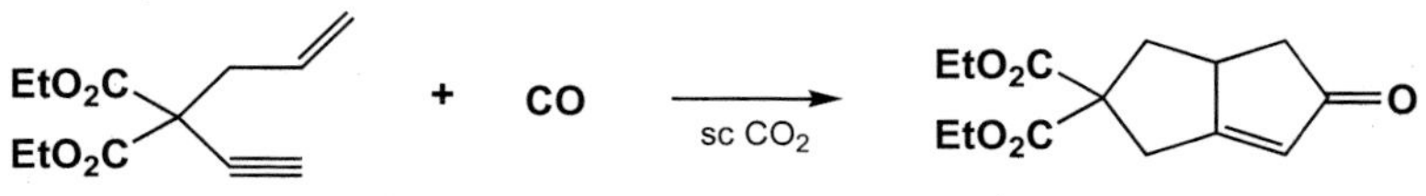
EtO2C
EtO2C
+ CO
sc CO2
EtO2C
EtO2C
O

(6)

Good yields also resulted from the photoactivation of $Co_2(CO)_8$ at only low CO pressure [11]. However, from experience with the photochemistry of this particular class of metal carbonyl compounds there is not the slightest chance of an upscaled application of this procedure.

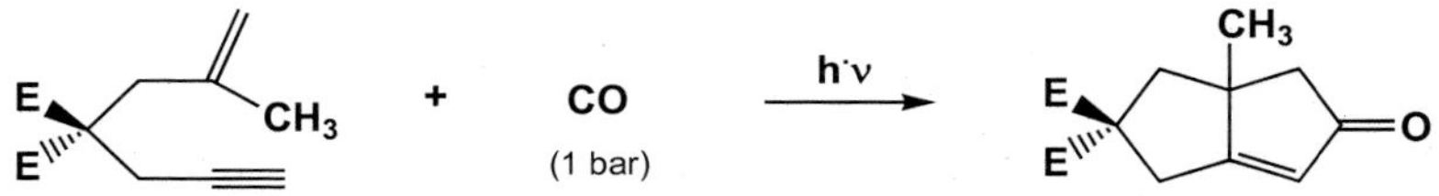
E
E
CH3
+ CO
(1 bar)
h·ν
CH3
E
E
O

(7)

Two Japanese research groups reported also on an efficient Ru-based catalytic PKR: 2 mol% of the commercial $Ru_3(CO)_{12}$ effected intramolecular PKRs in good yields at 140–160 °C under a CO pressure of 1–1.5 MPa [12, 13].

3.3.7.3   Related Reactions

Metallacyclopentenes typically undergo CO insertion to form the corresponding cyclopentenones. This principle was exploited by Negishi et al. [14] to generate zirconocene-type intermediates **2** from enynes and convert them into the PK products **3** (Scheme 2).

Iminoclopentenones **4** result from enynes and isocyanides upon treatment with stoichiometric amounts of [Ni(cod)₂] in the presence of tris(*n*-butyl)phosphane [15]. On this basis, Buchwald et al. developed a route that first traps the titana-cyclopentenes with the isocyanide, followed by hydrolysis; cf. Scheme 2 [16]. The sequence became catalytic when $(C_5H_5)_2Ti[P(CH_3)_3]_2$ was used as a catalyst, with trialkylsilyl cyanides acting as sources for the isocyanides [17].

A major step was made when titanocenes were employed for the cyclization: a number of functional groups are tolerated, the C–C coupling works even with disubstituted olefins, only low CO pressure is necessary, and the yields are beyond 85 %. Stereoselectivity is induced by *ansa*-titanocenes such as [(*S,S*)-(ebthi)-Ti(CO)₂] as chiral catalyst. *Ees* of up to 96 % are thus achieved (see eq. (8)). However, the "catalyst" is still required in amounts of 5–20 mol% [18].

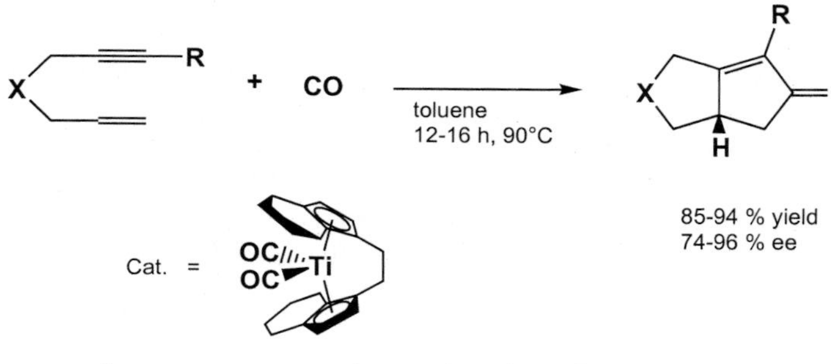

$$ \text{(8)} $$

85-94 % yield
74-96 % ee

Methylene cyclopentenones can be made with *allenes* – instead of alkenes – as precursor compounds. Thus, iron-, molybdenum-, cobalt- and titanium-mediated PK-type C–C coupling reactions lead preferentially to the *β*-methylene cyclopentenones [25, 26]. A side reaction is the polymerization of allenes in the presence of Co₂(CO)₈ if the allenes are not at least disubstituted. An example using Co₂(CO)₈/ *N*-methylmorpholine *N*-oxide (6 equiv.) in THF/CH₂Cl₂ at –80/+20 °C is shown in eq. (9) (cf. [25]).

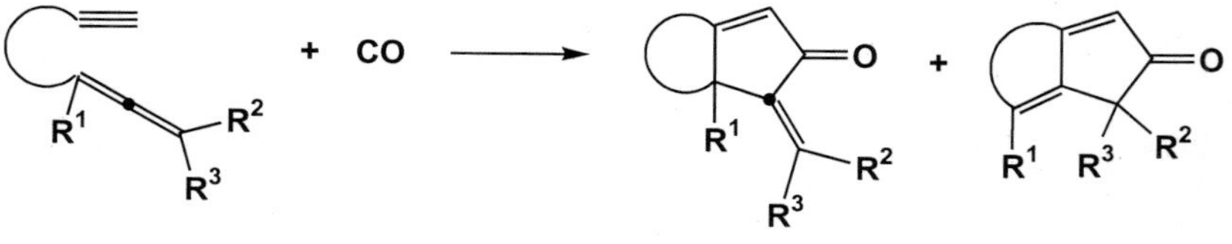

$$ \text{(9)} $$

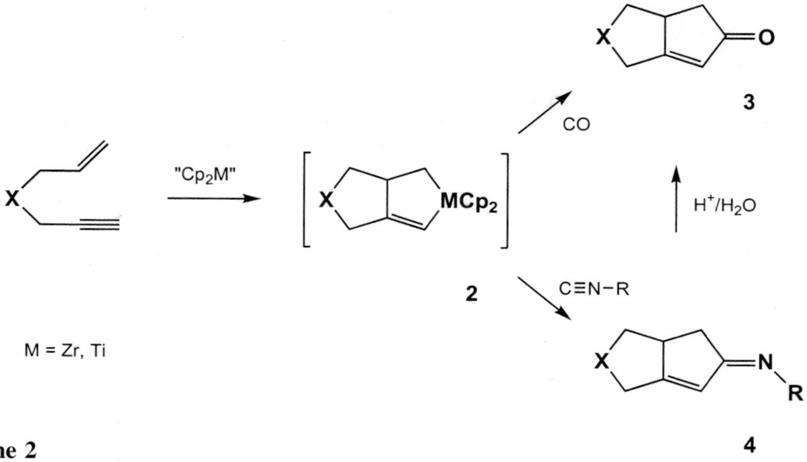

Scheme 2

3.3.7.4   **Stereoselective PKRs and *Hetero*-Reactions**

Enantioselectivities up to 44 % were reached in intermolecular PKRs when chiral aminoxides $R^*_3N{\rightarrow}O$ were used [19]. Although the mechanism is not known, it seems likely that the chiral *N*-oxide discriminates between the prochiral carbonyl cobalt units, either oxidizing one carbon monoxide selectively to produce a vacant site for the alkene insertion, or stabilizing a vacant site on one of the cobalts preferentially. This approach was modified by application of chiral precursor substrates [20]. Albeit the synthesis of the latter is cumbersome, the concept was successfully applied in several total syntheses, for example of hirsutene [21], brefeldine A [22], β-cuparenone [23], and (+)-15-norpentalenene [24] (eq. (10)). Stoichiometric amounts of the mediator compound $Co_2(CO)_8$ are still necessary in this useful version of the Pauson–Khand reaction.

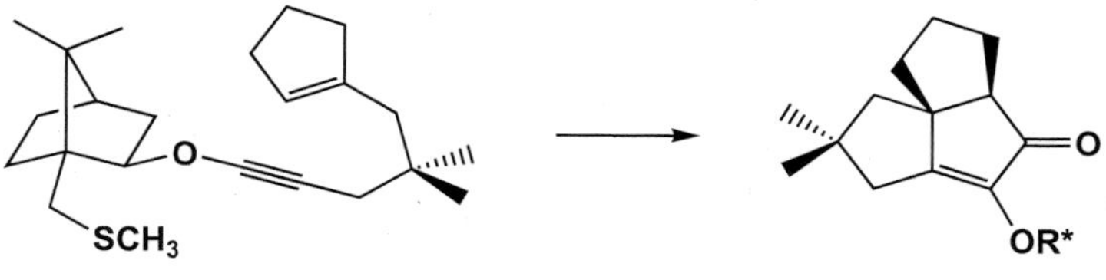

$$(10)$$

d.s. 89/11
yield 63%

γ-Butyrolactones are formed upon cyclization of enones with carbon monoxide mediated by a titanocene species (eq. (11)) [27]. The yield is nearly quantitative when toluene is employed at 105 °C for 15–18 hours under a slight pressure (1.2 bar) of carbon monoxide. Formally, this PKR-type coupling process is a [2+2+1] addition reaction.

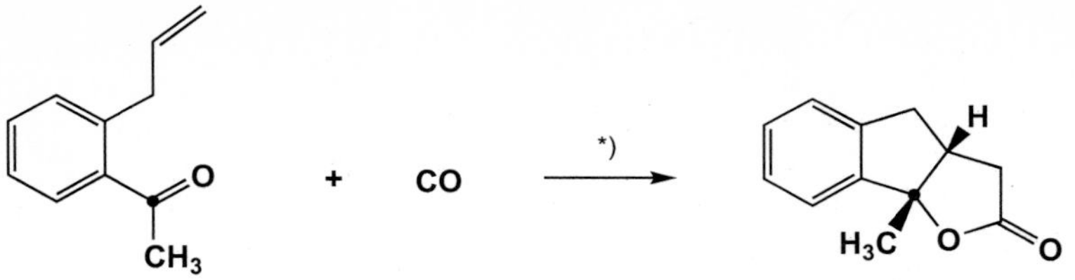

$$\text{(11)}$$

*) Mediator reagent: $(C_5H_5)_2Ti[P(CH_3)_3]_2$

3.3.7.5   Degenerate (Intermittent) and Domino PK Reactions

According to observations made by Krafft et al. [28], monocyclic products are formed in the presence of air–oxygen; $Co_2(CO)_8$ mediates the cyclization process, which is stopped oxidatively at the alkyne during the first C–C bond formation. Equation (12) gives a typical example of this "degenerate" PKRs (sometimes called "intermittent"). Several consecutive PKR can be used to synthesize complicated hydrocarbons elegantly that are otherwise not available or only with difficulty. In such cases, it is not of primary interest how much of the mediator, e. g. $Co_2(CO)_8$, is necessary. An excellent example is the construction of the ferrestrane framework according to Scheme 3, as reported by Thommen and Keese in 1997 [29].

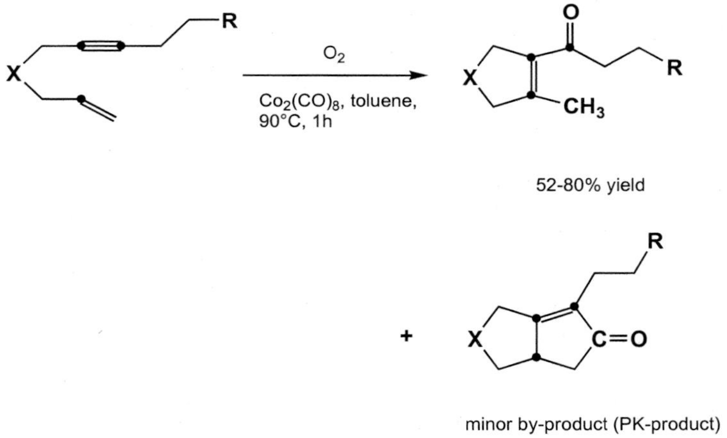

O_2

$Co_2(CO)_8$, toluene, 90°C, 1h

52-80% yield

$$\text{(12)}$$

+

minor by-product (PK-product)

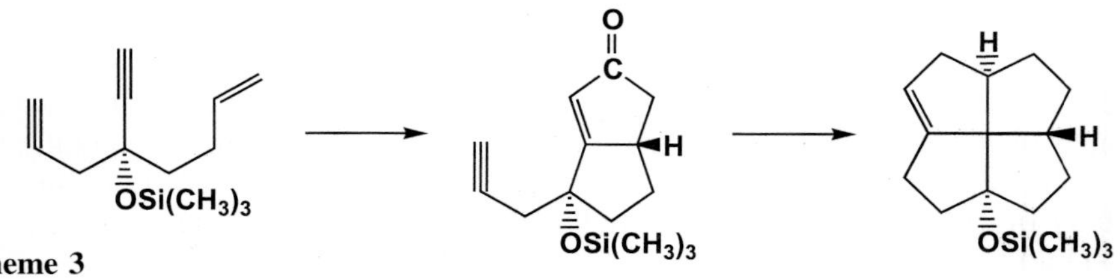

Scheme 3

3.3.7.6 Substitution Effects, Selectivity, and Mechanism

Best efficiencies under the standard "quasi-catalytic" PK conditions are gained with electron-poor alkynes, strained olefins (e. g., norbornene), and open-chain olefins with low substitution (e. g., ethylene). Ethylene has the advantage that it can be applied in large excess by pressurizing the reaction system. Equations (13) and (14) are standard examples.

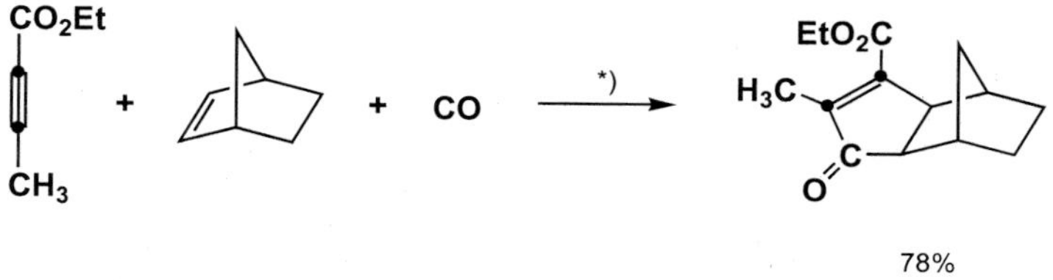

$$(13)$$

78%

*) N-methyl-morpholine-N-oxide (NMO)
CH₂Cl₂/ 0-25°C / Co₂(CO)₈

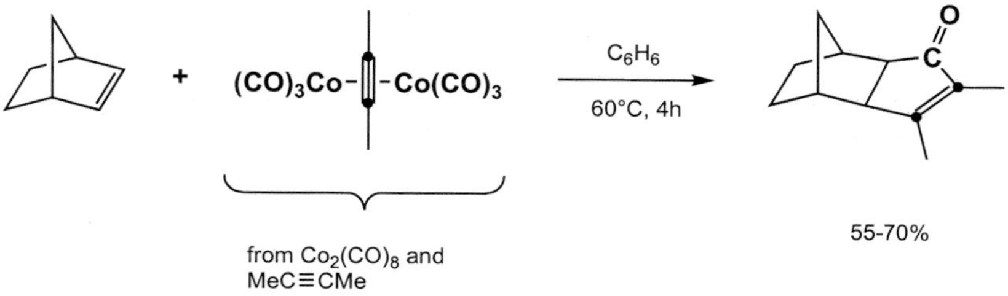

$$(14)$$

55-70%

from Co₂(CO)₈ and
MeC≡CMe

Steric effects play a major role in the selectivity of the PK cyclization, too. This can be seen especially with highly unsymmetrically substituted alkynes: the large substituent L preferentially shows up adjacent to the carbonyl functionality (Figure 1).

There is a plausible mechanistic proposal for the standard PKR on the basis of Co₂(CO)₈ as mediator and/or catalyst. Clearly the well-known μ-alkyne dicobalt

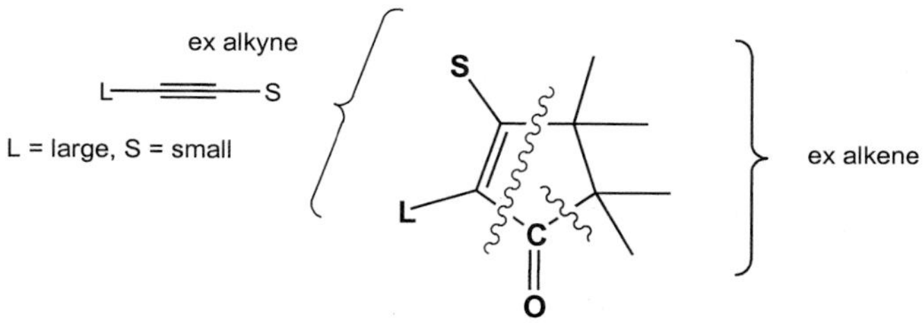

Figure 1. Effect of ligand structure.

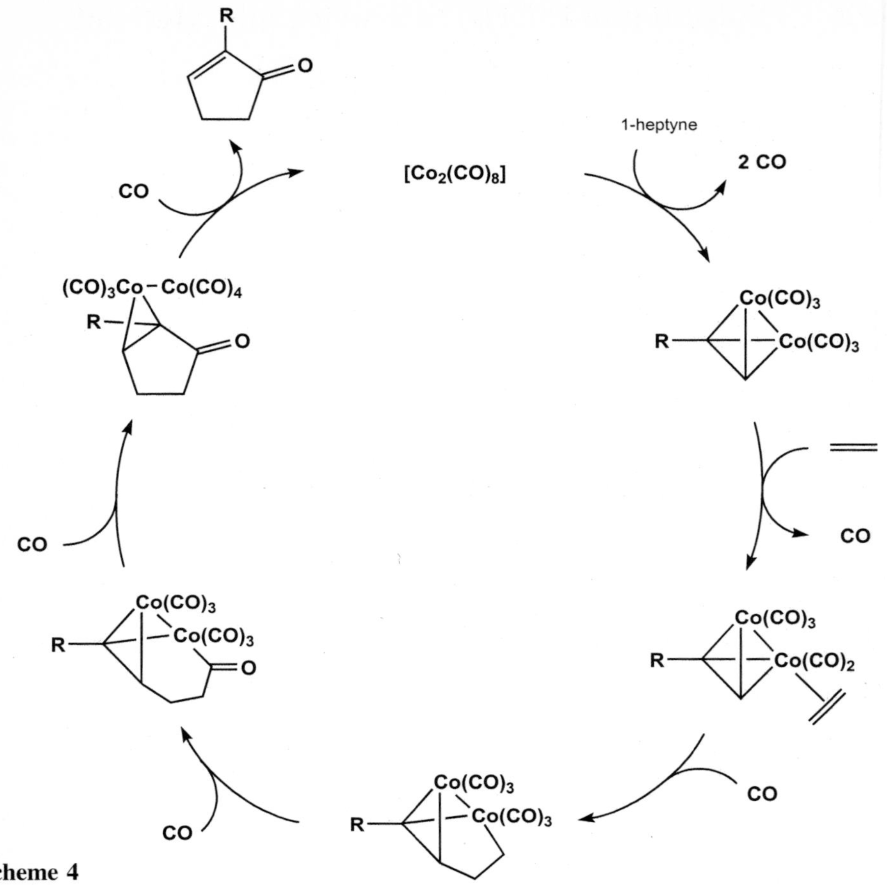

R

O

1-heptyne

[Co$_2$(CO)$_8$]

2 CO

CO

(CO)$_3$Co–Co(CO)$_4$

R

O

Co(CO)$_3$

R — Co(CO)$_3$

CO

CO

Co(CO)$_3$

R — Co(CO)$_3$

O

Co(CO)$_3$

R — Co(CO)$_2$

CO

CO

Co(CO)$_3$

R — Co(CO)$_3$

CO

Scheme 4

complexes (R–C≡C–R)Co$_2$(CO)$_6$ are initially formed via CO dissociation. This is thought to be the rate-determining step. An alkene addition with subsequent C–C coupling (alkyne/alkene coupling) then follows. The first C–C coupling step takes place between the less bulky end of the alkyne (S) and the alkene, and it is this step that explains the regioselectivity with respect to the alkyne. The carbonylation step comes prior to the π-decomplexation of the cyclopentenone, which final step regenerates the carbonylcobalt species. Scheme 4 summarizes the generally accepted textbook mechanism [2–4, 30].

The main side reactions are cyclotrimerization (of the alkyne), co-cyclotrimerization (alkyne/alkyne/alkene), and formation of cyclopentadienones **5** and **6**. In addition, spirofuranones such as **7** and **8** can form, with the latter being the Diels–Alder product of **7** with ethylene. It is generally observed that high ethylene and CO pressures disfavor the formation of side products. Further details on the side-product formation were given in [4 d].

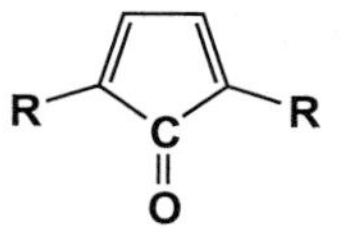

5 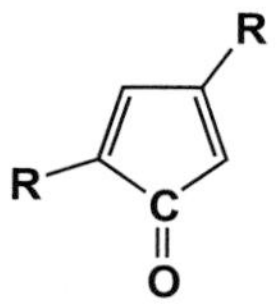   6 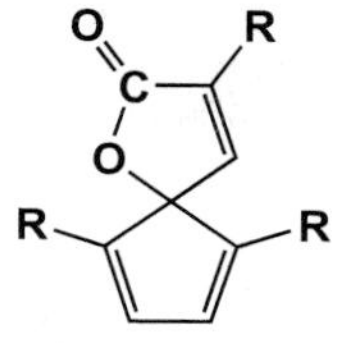   7 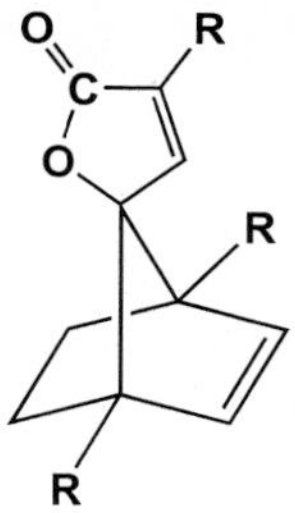   8

It is noteworthy that the cobalt cluster $Co_4(CO)_{12}$ – for many years thought to be inactive in PKRs – does also work in more or less *stoichiometric* versions under modest conditions, e. g., 70 °C, 0.1 MPa CO [31]. However, the reported turnovers (TON) are below 15. Additives such as cyclohexylamine seem to facilitate disproportionation of the zerovalent cobalt. It is thus doubtful whether Co^0 is retained throughout the catalytic cycle, albeit this still is the textbook opinion on the mechanism.

3.3.7.7   Commercial Perspectives

Cyclopentenones are building blocks of fragrances. For this reason, intensive research has been performed at Firmenich S. A., Geneva. Thus, the methyl-dihydrojasmonates – the so-called hediones® – are important perfumery chemicals [32]. Their production is well above 1500 tons per year. The conventional synthesis is based on an aldol reaction between cyclopentanone and *n*-valeraldehyde. The resulting aldol is then converted into 2-pentylcyclopent-2-en-1-one **10** (R = *n*-pentyl). Michael addition of dimethyl malonate and saponification/decarboxylation gives the hediones® in an approx. 95:5 *trans/cis* mixture. The PK approach under the best conditions gave 48 % **10** (toluene, 0.22 mol% $Co_2(CO)_8$ based on 1-heptyne). The overall turnover number (TON) was ca. 220, the overall turnover frequency (TOF) ca. 150/h. The regioselectivity was excellent, ranging as high as 98–99 % [4 d, 7, 8]. The Pauson–Khand approach has not yet led to an industrial synthesis of methyldihydrojasmonate. An enantioselective ruthenium-catalyzed hydrogenation of the vinylog β-oxoester was recently reported by Rautenstrauch anc co-workers [33].

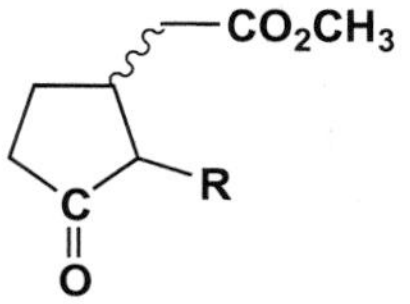

9 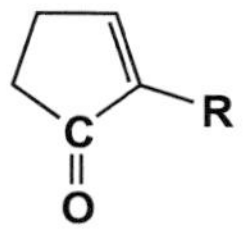   10

3.3.7.8   Outlook

Compared with the status reported in ref. [4 d], significant progress has been made in the quasi-catalytic PK approach, particularly in the synthesis of natural products. In addition, the regio- and stereoselectivity has been improved by various additives such as specific solvents, drop-in ligands (e. g., chiral amine oxides), and combinations of cobalt salts with reducing agents (e. g., $Co(acac)_2/Na[BH_4]$ or $CoBr_2/Zn$) (cf. [34, 35]). The greatest challenge, however, remains a truly *catalytic version* of this unique triple C–C coupling process. It is clear that the chemistry occurring around the mediator metal, so far most efficiently cobalt, must become more favorable. In particular, the degradation of the active carbonyl-cobalt species with its detrimental effect upon the catalytic performance must be avoided. Much scientific imagination should be applied to the design of a (binuclear) cobalt complex exhibiting an anchoring ligand to stabilize the system toward the formation of clusters and the bulk metal. Bimetallic catalysts taking care of the different types of C–C coupling chemistry combined in PKRs are another possibility for a solution for the as-yet poor catalytic performance. Supercritical reaction conditions [36] should be given further attention, too.

References

[1] (a) I. U. Khand, G. R. Knox, P. L. Pauson, W. E. Watts, *J. Chem. Soc., Chem. Commun.* **1971**, 36; (b) I. U. Khand, G. R. Knox, P. L. Pauson, W. E. Watts, *J. Chem. Soc., Perkin I* **1973**, 975; (c) I. U. Khand, G. R. Knox, P. L. Pauson, W. E. Watts, M. I. Foreman, *J. Chem. Soc., Perkin I* **1973**, 977; (d) P. L. Pauson, I. U. Khand, *Ann. NY Acad. Sci.* **1977**, *295*, 2.

[2] (a) N. E. Schore, *Chem. Rev.* **1988**, *88* 1081; (b) N. E. Schore, in *Comprehensive Organic Synthesis* (Eds.: B. M. Trost, I. Fleming, L. A. Paquette), *Vol. 5,* Pergamon, Oxford, **1991**; (c) N. E. Schore, *Org. React.* **1991**, *40*, 1.

[3] (a) L. S. Hegedus, *Organische Synthese mit Übergangsmetallen*, VCH, Weinheim, **1995**; (b) R. Noyori, *Asymmetric Catalysis in Organic Synthesis*, Wiley, New York, **1994**; (c) P. L. Pauson, *Tetrahedron* **1985**, *41*, 5855; (d) P. L. Pauson, in *Organometallics in Organic Synthesis* (Eds.: A. de Meijere, H. tom Dieck), Springer, Berlin, **1987**, p. 233; (e) N. E. Schore, *Org. React.* **1991**, *40*, 1.

[4] Recent review articles: (a) A. J. Fletches, S. D. R. Chustie, *J. Chem. Soc., Perkin Trans. 1*, **2000**, 1657; (b) O. Geis, H.-G. Schmalz, *Angew. Chem.* **1998**, *110*, 955; *Angew. Chem. Int. Ed.* **1998**, *37*, 911; (c) K. M. Brummond, J. L. Kent, *Tetrahedron* **2000** 56, 3263; (d) V. Rautenstrauch, in: *Applied Homogeneous Catalysis with Organometallic Compounds* (Eds.: B. Cornils, W. A. Herrmann), 1st ed., Wiley-VCH, Weinheim, **1996**.

[5] (a) S. Shambayati, W. E. Crowe, S. L. Schreiber, *Tetrahedron Lett.* **1990**, *31*, 5289; (b) N. Jeong, Y. K. Chung, B. Y. Lee, S. H. Lee, S.-E. Yoo, *Synlett* **1991**, 204; (c) A. R. Gordon, C. Johnstone, W. J. Kerr, *Synlett* **1996**, 1083.

[6] (a) T. Sugihara, M. Yamada, H. Ban, M. Yamaguchi, C. Kaneko, *Angew. Chem.* **1997**, *109*, 2884; *Angew. Chem. Int. Ed. Engl.* **1997**, *36*, 2801; (b) T. Rajesh, M. Periasamy, *Tetrahedron Lett.* **1998**, *39*, 117.

[7] V. Rautenstrauch, P. Mégard, J. Conesa, W. Küster, *Angew. Chem.* **1990**, *102*, 1441; *Angew. Chem. Int. Ed. Engl.* **1990**, *29*, 1413.

[8] Firmenich S.A. (V. Rautenstrauch, W. Keim), CH 681.224 (1990).

[9] (a) N. Jeong, S. H. Hwang, Y. Lee, Y. K. Chung, *J. Am. Chem. Soc.* **1994**, *116*, 3159; (b) B. Y. Lee, Y. K. Chung, N. Jeong, Y. Lee, S. H. Hwang, *J. Am. Chem. Soc.* **1994**, *116*, 8793; (c) N. Y. Lee, Y. K. Chung, *Tetrahedron* Lett. **1996**, *37*, 3145; (d) N. Jeong, S. H. Hwang, Y. W. Lee, Y. S. Lim, *J. Am. Chem. Soc.* **1997**, *119*, 10549.

[10] N. Jeong, S. H. Hwang, *Angew. Chem.* **2000**, *112*, 650; *Angew. Chem. Int. Ed.* **2000**, *39*, 636.

[11] B. L. Pagenkopf, T. Livinghouse, *J. Am. Chem. Soc.* **1996**, 118, 2285.

[12] T. Kondo, N. Suzuki, T. Okada, T.-A. Mitsudo, *J. Am. Chem. Soc.* **1997**, *119*, 6187.

[13] T. Morimoto, N. Chatani, Y. Fukumoto, S. Murai, *J. Org. Chem.* **1997**, *62*, 3762.

[14] (a) E.-I. Negishi, S. J. Holmes, J. M. Tour, J. A. Miller, *J. Am. Chem. Soc.* **1985**, *107*, 2568; (b) E.-I. Negishi, F. E. Cederbaum, T. Takahashi, *Tetrahedron Lett.* **1986**, *27*, 2829; (c) E.-I. Negishi, S. J. Holmes, J. M. Tour, J. A. Miller, F. E. Cederbaum, D. R. Swanson, T. Takahashi, *J. Am. Chem. Soc.* **1989**, *111*, 3336.

[15] (a) K. Tamao, K. Kobayashi, Y. Ito, *J. Am. Chem. Soc.* **1988**, *110*, 1286; (b) K. Tamao, K. Kobayashi, Y. Ito, *Synlett* **1992**, 539.

[16] R. B. Grossmann, S. L. Buchwald, *J. Org. Chem.* **1992**, *57*, 5803.

[17] (a) S. C. Berk, R. B. Grossmann, S. L. Buchwald, *J. Am. Chem. Soc.* **1993**, *115*, 4912; (b) S. C. Berk, R. B. Grossmann, S. L. Buchwald, *J. Am. Chem. Soc.* **1994**, *116*, 8593; (c) F. A. Hicks, S. C. Berk, S. L. Buchwald, *J. Org. Chem.* **1996**, *61*, 2713.

[18] F. A. Hicks, S. L. Buchwald, *J. Am. Chem. Soc.* **1996**, *118*, 11688.

[19] W. J. Kerr, G. G. Kirk, D. Middlemiss, *Synlett* **1995**, 1085.

[20] (a) J. Castro, A. Moyano, M. A. Pericàs, A. Riera, A. E. Greene, *Tetrahedron: Asymmetry* **1994**, *5*, 307; (b) V. Bernardes, X. Verdaguer, N. Kardos, A. Riera, A. Moyano, M. A. Pericàs, A. E. Greene, *Tetrahedron Lett.* **1994**, *35*, 575; (c) X. Verdaguer, A. Moyano, M. A. Pericàs, A. Riera, V. Bernardes, A. E. Greene, A. Alvarez-Larena, J. F. Piniella, *J. Am. Chem. Soc.* **1994**, *116*, 2153; (d) S. Fonquerna, A. Moyano, M. A. Pericàs, A. Riera, *Tetrahedron* **1995**, *51*, 4239; (e) S. Fonquerna, A. Moyano, M. A. Pericàs, A. Riera, *J. Am. Chem. Soc.* **1997**, *119*, 10225; (f) E. Montenegro, M. Poch, A. Moyano, M. A. Pericàs, A. Riera, *Tetrahedron Lett.* **1998**, *39*, 335.

[21] J. Castro, H. Sörensen, A. Riera, C. Morin, A. Moyano, M. A. Pericàs, A. E. Greene, *J. Am. Chem. Soc.* **1990**, *112*, 9388.

[22] V. Bernardes, N. Kann, A. Riera, A. Moyano, M. A. Pericàs, A. E. Greene, *J. Org. Chem.* **1995**, *60*, 6670.

[23] J. Castro, A. Moyano, M. A. Pericàs, A. Riera, A. E. Greene, A. Alvarez-Larena, J. F. Piniella, *J. Org. Chem.* **1996**, *61*, 9016.

[24] J. Tormo, A. Moyano, M. A. Pericàs, A. Riera, *J. Org. Chem.* **1997**, *62*, 4851.

[25] (a) M. Ahmar, F. Antras, B. Cazes, *Tetrahedron Lett.* **1995**, *36*, 4417; (b) M. Ahmar, O. Chabanis, J. Gauthier, B. Cazes, *Tetrahedron Lett.* **1997**, *38*, 5277; (c) M. Ahmar, C. Locatelli, D. Colombier, B. Cazes, *Tetrahedron Lett.* **1997**, *38*, 5281.

[26] (a) R. Aumann, H.-J. Weidenhaupt, *Chem. Ber.* **1987**, *120*, 23; (b) J. L. Kent, H. Wan, K. M. Brummond, *Tetrahedron Lett.* **1995**, *36*, 2407.

[27] (a) N. M. Kablaoui, F. A. Hicks, S. L. Buchwald, *J. Am. Chem. Soc.* **1996**, *118*, 5818; (b) N. M. Kablaoui, F. A. Hicks, S. L. Buchwald, *J. Am. Chem. Soc.* **1997**, *119*, 4424.

[28] M. E. Krafft, A. M. Wilson, O. A. Dasse, B. Shao, Y. Y. Chung, Z. Fu, L. V. R. Bongaga, M. K. Mollmann, *J. Am. Chem. Soc.* **1996**, *118*, 6080.

[29] M. Thommen, R. Keese, *Synlett* **1997**, 231.

[30] I. L Scott, *J. Org. Chem.* **1992**, *20*, 5277.

[31] M. E. Krafft, L. V. R. Boñaga, *Angew. Chem.* **2000**, *112*, 3822; *Angew. Chem. Int. Ed.* **2000**, *39*, 3676.

[32] Reviews: (a) E. P. Demole, in *Fragrance Chemistry: The Science of the Sense of Smell* (Ed.: E. T. Theimer), Academic Press, New York, **1982**, p. 374; (b) G. Ohloff, *Riechstoffe und Geruchssinn: Die molekulare Welt der Düfte*, Springer, Berlin, **1990**, pp. 149–152.

[33] D. A. Dobbs, K. P. M. Vanhessche, E. Brazi, V. Rautenstrauch, J.-Y. Lenoir, J.-P. Genêt, S. H. Bergens, *Angew. Chem.* **2000**, *112*, 2080; *Angew. Chem. Int. Ed.* **2000**, *39*, 1992.

[34] T. Kondo, N. Suzuki, T. Okada, T. Mitsudo, *J. Am. Chem. Soc.* **1997**, *119*, 6187.

[35] M. Hayashi, Y. Hashimoto, Y. Yamamoto, I. Usuki, K. Saigo, *Angew. Chem.* **2000**, *112*, 645; *Angew. Chem. Int. Ed.* **2000**, *39*, 631;

[36] Review: P. G. Jessop, T. Ikariya, R. Noyori, *Chem. Rev.* **1999**, *99*, 475.

3.3.8   Cyclooligomerization of Alkynes

Helmut Bönnemann, Werner Brijoux

3.3.8.1   Introduction

The transition metal-catalyzed cyclotrimerization of acetylene (eq. (1)) was discovered by Berthelot [1] back in the mid-19th century using heterogeneous systems.

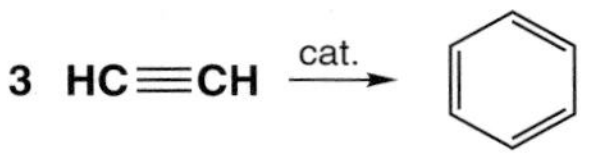

$$3 \; HC\equiv CH \xrightarrow{\text{cat.}} \bigcirc \qquad (1)$$

The merits of *homogeneous* catalysts in this field were demonstrated most convincingly by Reppe [2]. In 1973, Yamazaki and Wakatsuki [3] first reported the homogeneous catalytic cycloaddition of alkynes and nitriles. Since then Vollhardt [4] has developed a number of elegant synthetic organic applications. Bönnemann and co-workers [5] focused their work on the development of highly reactive organocobalt catalyst for the homogeneous synthesis of a-substituted pyridines according to eq. (2).

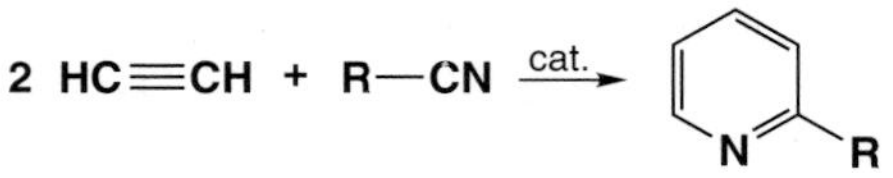

$$2 \; HC\equiv CH + R{-}CN \xrightarrow{\text{cat.}} \bigcirc_{N \quad R} \qquad (2)$$

In 1988 Schore [6] published a comprehensive review of cycloaddition reactions of alkynes mediated by transition metal complexes and their application in organic synthesis.

This section focuses on the transition metal-catalyzed formation of ring systems using alkynes in the homogeneous phase. Because of the great number of possible products, this survey is restricted to five- and six-membered heterocycles, and six- and eight-membered carbocycles.

3.3.8.2   Survey of the Catalysts

Alkyne cyclotrimerization occurs at various homogeneous and heterogeneous transition metal and Ziegler-type catalysts [7]. Substituted benzenes have been prepared in the presence of iron, cobalt, and nickel carbonyls [8] as well as trialkyl- and triarylchromium compounds [9]. Bis(acrylonitrile)nickel [10] and bis(benzonitrile)palladium chloride [11] catalyze the cyclotrimerization of tolane to hexaphenylbenzene. $NiCl_2$ reduced by $NaBH_4$ has been utilized for the trimerization of 3-hexyne to hexaethylbenzene [12]. Ta_2Cl_6(tetrahydrothiophene)$_3$ and Nb_2Cl_6(tetrahydrothiophene)$_3$ as well as η^5-Cp-, η^5-Ind-, and η^5-Flu-rhodium complexes (Cp = cyclopentadienyl; Ind = indenyl; Flu = fluorenyl) were found to cyclotrimerize terminal alkynes [13]. Hexaisopropyl- and hexa-*t*-butylbenzene can only be synthesized in the presence of highly active cobalt catalysts [14]. Appropriate catalysts for alkyne cyclotrimerizations were surveyed by Jhingan and Maier in 1987 [15].

In contrast to carbocyclic alkyne cyclotrimerizations, the catalytic pyridine synthesis from alkynes and nitriles relies exclusively on cobalt catalysts with a few exceptions where rhodium [16] and iron complexes [17] could be applied. The cobalt-catalyzed pyridine synthesis can even be carried out in a one-potreaction generating the catalyst from $CoCl_2 \cdot 6\ H_2O/NaBH_4$ + nitrile/alkyne *in situ* [18].

This system may be recommended for the quick exploration of new synthetic applications in research laboratories which do not specialize in organometallic techniques because the cobalt salts can be used in the hydrated form under air, and no sophisticated ligands are necessary. Cobalt(I) halide complexes of the type $[XCoL_3]$ having a moderate activity in the synthesis of 2-alkylpyridines are also easily accessible (eq. (3)) [19].

$$CoX_2 + L + Red. \longrightarrow XCoL_3 \qquad\qquad (3)$$

X = Cl, Br, I; Red. = $NaBH_4$, Zn
L = $P(C_6H_5)_3$, $P(OEt)_3$, $P(OC_3H_7)_3$

Two types of pre-prepared organocobalt complexes proved to be most effective catalysts for the co-cyclization of alkynes and nitriles: the allylcobalt type, where the organic group is η^3-bonded to the metal (Structure **1**), and also the η^5-Cp- and η^5-Ind-cobalt half-sandwich compounds (**2**, **3**). During the catalytic cycle in the case of the η^3-allylcobalt catalyst a 12-electron system is regenerated, whereas in the case of the η^5-Cp- and η^5-Ind-cobalt complexes the catalytic reaction involves a 14-electron moiety. In fact, the cobalt-catalyzed pyridine synthesis was one of the first examples where η^5-Cp groups were used as controlling ligands in homogeneous catalysis [5 f, g]. The modification of the basic η^5-Cp ligand systems by additional substituents, R, transferring electron-donating or -withdrawing effects to the η^5-Cp group, results in strong changes in catalyst activity and selectivity. In addition, η^6-borininato ligands may be used as 6π-electron ligands for cobalt (**4**).

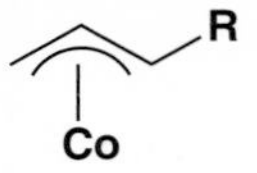

Yamazaki's complex (Structure **5**) contains two alkyne molecules linked together to form a five-membered metallacycle. Arene-solvated cobalt atoms, obtained by reacting cobalt vapor and arenes, have been used by Italian workers to promote the conversion of a,ω-dialkynes and nitriles giving alkynyl-substituted pyridines [20]. η^6-Tolueneiron(0) complexes have also been utilized for the co-cyclotrimerization of acetylene and alkyl cyanides or benzonitrile giving a-substituted pyridine derivatives. However, the catalytic transformation to the industrially important 2-vinylpyridine fails in this case: acrylonitrile cannot be co-cyclotrimerized with acetylene at the iron catalyst [17].

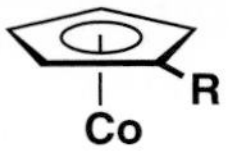

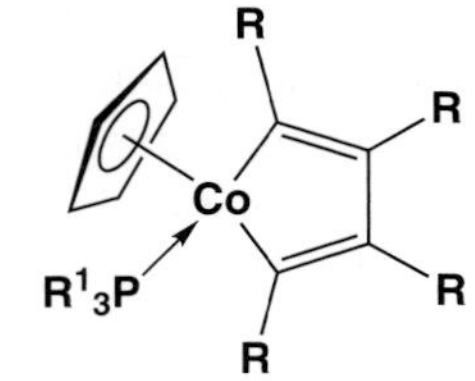

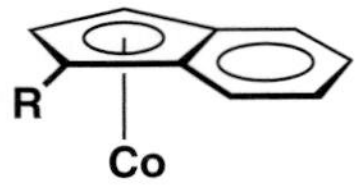

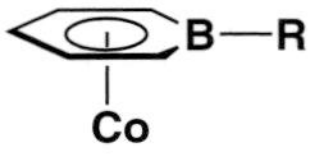

In 1989 Oehme et al. reported a photoassisted synthesis of a-substituted pyridines under mild conditions using η^5-Cp-cobalt complexes as the catalyst [21].

Bönnemann and co-workers [22] and others [23] have tried acetylacetonato- and η^5-Cp-rhodium as well as resin-attached η^5-Cp-rhodium complexes as catalysts in the pyridine synthesis [23]. However, rhodium catalysts are generally less effective than the analogous cobalt systems.

3.3.8.3   Five- and Six-Membered Heterocycles

The classic pyridine syntheses have been extensively reviewed by Abramovitch [24]. Many of them rely on the condensation of aldehydes or ketones with ammonia in the vapor phase. However, these processes suffer from unsatisfactory selectivity. Soluble organocobalt catalysts allow a selective one-step access to pyridine and a wide range of a-substituted derivatives from acetylene and the corresponding cyano compounds (eq. (2)).

The *homogeneous* catalytic [2+2+2]-cycloaddition of alkynes and nitriles was first discovered by Yamazaki and Wakatsuki [3] using the phosphine-stabilized cobalt(III) complex (Structure **5**). At the same time, Bönnemann and co-workers [5] observed the co-cyclization (eq. (2)) at cobalt catalysts prepared *in situ*, as well as using phosphine-free organocobalt(I) diolefin complexes.

The substituent on the alkyne and the cyano group can be widely varied. The basic catalytic reaction (eq. (2)) was developed into a general synthetic method for the selective preparation of pyridines. Only small amounts of benzene derivatives are formed as the by-product.

3.3.8.3.1 Pyridine

The parent compound has been prepared under mild conditions using the homogeneous η^6-1-phenylborininatocobalt cod (1,5-cyclooctadiene) catalyst (eq. (2), R = H) [26]. However, the turnover number was very limited (about 100). The strong incentive for further developments lies in the fact that both HCN and acetylene are cheap bulk chemicals in industry. The introduction of boron into the carbocyclic ligand attached to the cobalt enhances the catalytic activity considerably, probably via the suppression of the protolytic 1,4-addition of HCN to the olefinic cobaltacycle; the resulting cyano-substituted 1,3- dienes cannot be displaced from the cobalt center by acetylene and the catalytic cycle is stopped (eq. (4)).

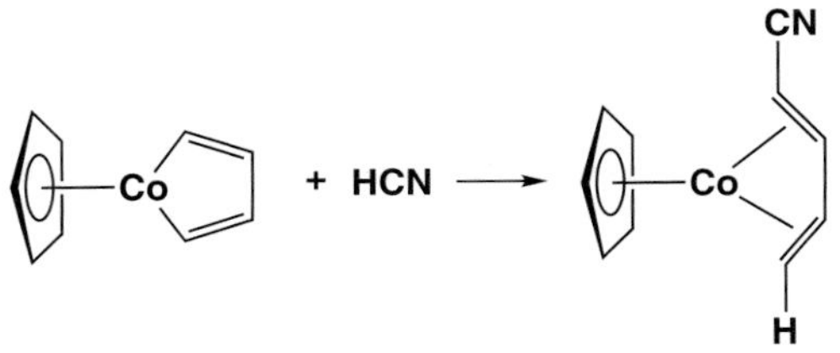

$$\text{(structure)} + \text{HCN} \longrightarrow \text{(structure)} \qquad (4)$$

3.3.8.3.2 Alkyl-, Alkenyl-, and Arylpyridines

A two-step process for the production of α-picoline has been commercialized by DSM in the Netherlands. Acrylonitrile is first reacted with a large excess of acetone [27] (eq. (5)). In the liquid phase a monocyanoethylation product is formed initially, catalyzed by a primary amine and a weak acid. The ring closure in the vapor phase giving α-picoline is catalyzed by a palladium contact.

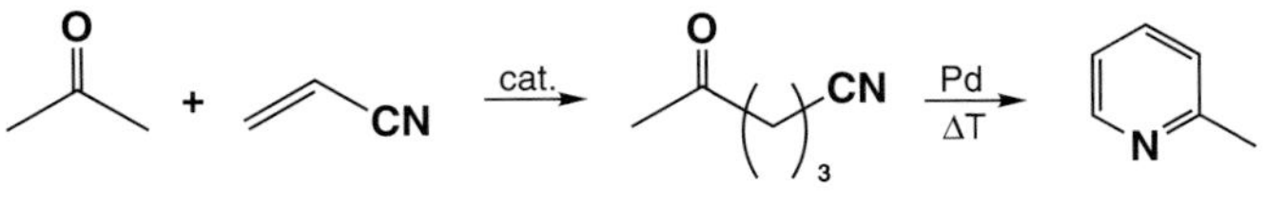

$$\text{(structure)} \xrightarrow[]{\text{cat.}} \text{(structure)} \xrightarrow[\Delta T]{Pd} \text{(structure)} \qquad (5)$$

Nippon Steel has developed an interesting liquid-phase process for α-picoline from ethylene and ammonia [28]. The catalyst is reminiscent of the well-known Wacker process, viz. the Pd^{2+}/Cu^{2+} redox system (eq. (6)).

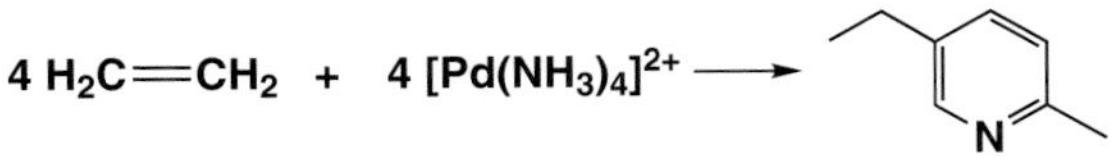

$$4\, H_2C{=}CH_2 \;+\; 4\,[Pd(NH_3)_4]^{2+} \longrightarrow \text{(structure)} \qquad (6)$$

The preferred catalysts for the one-step co-cyclization of acetylene and acetonitrile (or alkyl cyanides in general) to give α-picoline (or 2-alkylpyridines) are η^5-Cp-cobalt cod or η^5-trimethylsilyl-Cp-cobalt cod (eq. (2)). The α-picoline synthesis is best performed in pure nitrile without any additional solvent [5 d].

A significant outlet for α-picoline is the production of 2-chloro-6-(trichloro-methyl)pyridine, which is used as a nitrification inhibitor in agricultural chemistry and in the manufacture of the defoliant 4-amino-2,5,6-trichloropicolinic acid. However, the major commercial outlet for α-picoline is still its use as a starting material for the two-step production of 2-vinylpyridine. The total yield of 2-vinyl-pyridine formed via eq. (7) can be as high as 90 %.

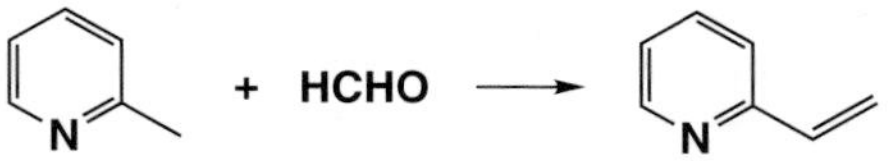

$$\text{(structure)} \quad + \quad \text{HCHO} \quad \longrightarrow \quad \text{(structure)} \tag{7}$$

2-Vinylpyridine may also be obtained in almost quantitative yields from 2-alkylaminopyridine derivatives (directly available through cobalt catalysis) using a supported (e. g., Al_2O_3) alkali metal hydroxide [29].

α-Ethylpyridine, α-undecylpyridine, and other α-alkylpyridines can be prepared in an analogous way from acetylene and the alkyl cyanides. The preferred catalyst is the η^5-trimethylsilyl-Cp-cobalt system. 2-Undecylpyridine is formed similarly (94 % yield) and can be easily separated from the reaction mixture [30]. The hydrochlorides and methiodides of a number of 2-alkylpyridines have an effect on the aqueous surface tension and show antibacterial properties. The salts of 2-pentadecylpyridine show the best results [31].

Starting from optically active nitriles, Botteghi and co-workers [32] have applied the cobalt-catalyzed reaction for the prepartion of optically active 2-substituted pyridines (eq. (8)). The chiral center is maintained during the alkyne–nitrile co-cyclization reaction. This reaction has recently been extended to the synthesis of bipyridyl compounds having optically active substituents [33] and provides an access to chiral ligands of potential interest in transition metal-catalyzed asymmetric synthesis.

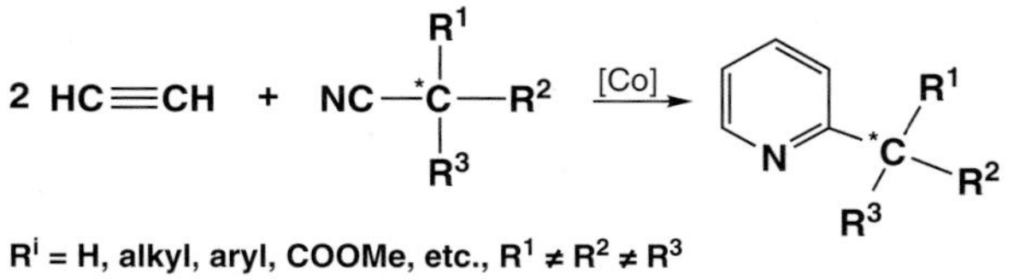

$$2 \ HC\equiv CH \ + \ NC-\overset{R^1}{\underset{R^3}{C}}-R^2 \ \xrightarrow{[Co]} \ \text{(structure)} \tag{8}$$

R^i = H, alkyl, aryl, COOMe, etc., $R^1 \neq R^2 \neq R^3$

The reaction of monosubstituted alkynes with nitriles (see eq. (9)) gives a mixture of isomeric trialkylpyridines (collidines). Collidines have been prepared using η^5-Cp-cobalt cod at 130 °C with high turnover numbers [5 d, h].

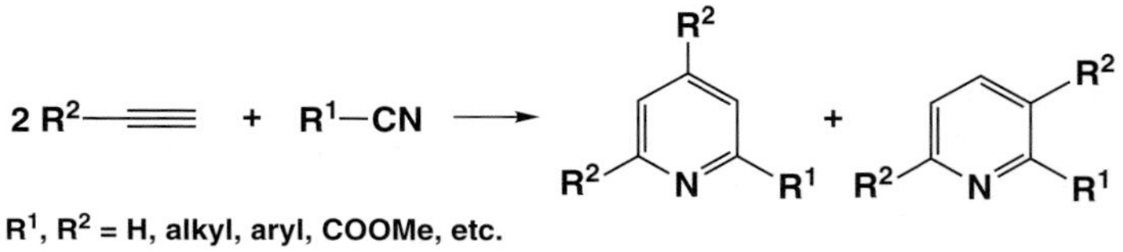

$$2 \ R^2\!\!\!-\!\!\!\equiv \ + \ R^1-CN \ \longrightarrow \ \text{(structure)} \ + \ \text{(structure)} \tag{9}$$

R^1, R^2 = H, alkyl, aryl, COOMe, etc.

The catalytic reaction may also be carried out using two different alkynes. For example, the co-cyclization of acetylene and propyne with acetonitrile yields a mixture of dimethylpyridines (lutidines) in addition to a-picoline and the isomeric collidines. The co-cyclization, however, turned out to be nonselective. For experimental details see [5 g]. The cobalt-catalyzed co-cyclization of benzonitrile and acetylene at η^5-Cp-cobalt cod gives 2-phenylpyridine in high yield.

The most interesting application from an industrial point of view is the cobalt-catalyzed one-step synthesis of 2-vinylpyridine from acetylene and acrylonitrile (eq. (10)). In this way the fine chemical can be manufactured using equal amounts by weight of the comparatively inexpensive components, acetylene and acrylonitrile. The 2-vinylpyridine synthesis must be carried out in pure acrylonitrile below 130–140 °C, otherwise acrylonitrile and the product 2-vinylpyridine undergo thermal polymerization [34]. Therefore only very active catalysts can be applied in the reaction of eq. (10). The best results were obtained using η^6-1-phenyl-borininatocobalt cod as the catalyst (productivity: 2.78 kg 2-vinylpyridine per g cobalt [5 e].

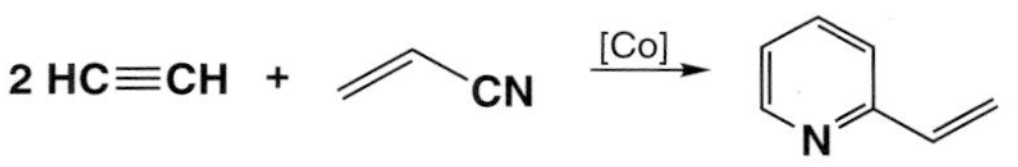

$$2\ HC\equiv CH\ +\ \text{(acrylonitrile)}\ \xrightarrow{[Co]}\ \text{(2-vinylpyridine)} \qquad (10)$$

The outlet for 2-vinylpyridine is the manufacture of copolymers for the use in tire cord binders [35]. 2-Vinylpyridine is also an additive in dyeing processes for acrylic fibers: 1–5 % of copolymerized 2-vinylpyridine provides the reactive sites for the dye.

3.3.8.3.3   2-Amino- and 2-Alkylthiopyridines

A wide variety of substituents at the cyano group is tolerated by the cobalt catalyst. For example, monomeric cyanamide reacts with acetylene in the presence of η^6-borininato cobalt half-sandwich complexes ([Co-B]) to give 2-aminopyridine [5 e] (eq. (11)).

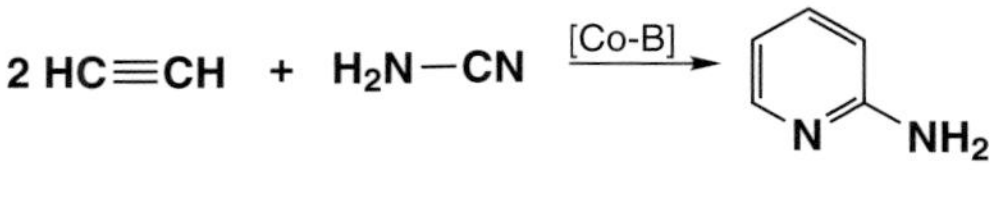

$$2\ HC\equiv CH\ +\ H_2N-CN\ \xrightarrow{[Co-B]}\ \text{(2-aminopyridine)} \qquad (11)$$

2-Aminopyridine is prepared conventionally by the substitution of the pyridine ring via the so-called Chichibabin reaction using sodium amide in dimethylaniline [36]. 2-Aminopyridine is used in the manufacture of several chemotherapeutics and of dyes for acrylic fibers, and as an additive for lubricants [37]. Alkyl thiocyanates react [38] to give 2-alkylthiopyridines (eq. (12)) which are otherwise accessible only by multistep synthetic pathways [39]. The catalytic reaction (eq. (12)) seems to offer on easy entry into the pyrithione systems.

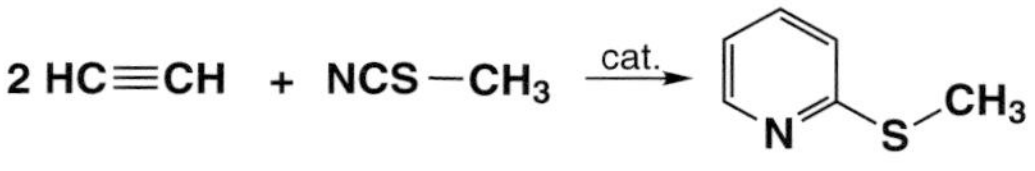

$$2\ HC\equiv CH\ +\ NCS-CH_3\ \xrightarrow{cat.}\ \text{(2-methylthiopyridine)} \qquad (12)$$

In classical access, 2-chloropyridine-*N*-oxide reacts with sodium hydrogensulfide to give pyrithione which, in the form of its zinc salt, is added to hair cosmetics as a general antifungal agent [40].

3.3.8.3.4   Bipyridines

The industrial route for 2,2'-bipyridyl consists in the dehydrodimerization of pyridine on Raney nickel using a process developed by Imperial Chemical Industries [41]. 2,2'-Bipyridyl reacts with ethylene bromide to give 1,1'-ethylene-2,2'-bipyridylium dibromide (diquat, which is widely used as a herbicide).

The cobalt-catalyzed synthesis enables 2,2'-bipyridyl to be prepared directly from 2-cyanopyridine and acetylene in a 72 % yield with a 2-cyanopyridine conversion of 21 % (eq. (13)) [5 d].

$$
\text{(structure)} \quad + \quad 2\ HC\equiv CH \quad \xrightarrow{[Co]} \quad \text{(structure)} \tag{13}
$$

Starting from readily available cyanopyridines, reaction with alkynes leads to substituted bipyridyls. Polynuclear pyridine derivatives can also be synthesized [42]. Substituted alkynes give two positional isomers.

3.3.8.3.5   Miscellaneous

An interesting variation is the reaction of a,ω-diynes on η^5-Cp-cobalt diene complexes. 1,7-Octadiyne initially undergoes an intramolecular process to give, in the presence of an excess of nitrile, derivatives of tetrahydroisoquinoline (eq. (14)).

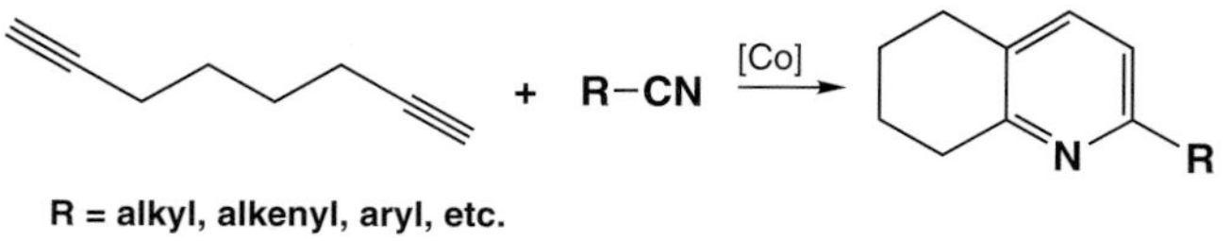

$$
\text{(structure)} \quad + \quad R-CN \quad \xrightarrow{[Co]} \quad \text{(structure)} \tag{14}
$$

R = alkyl, alkenyl, aryl, etc.

The annelated pyridine is also obtained with η^5-Cp-cobalt dicarbonyl as catalyst [43]. Using this variant of the cobalt-catalyzed cycloaddition, Schleich and co-workers [44] opened up a new route to pyridoxine (vitamin B$_6$) as its hydrochloride (eq. (15)).

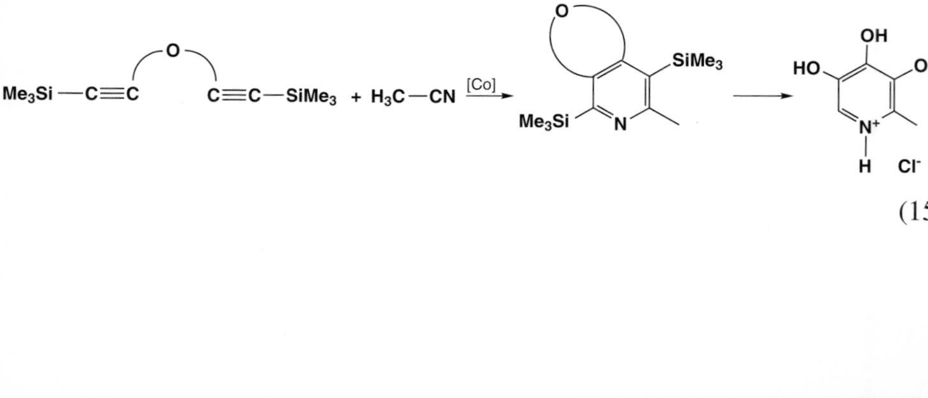

$$
Me_3Si-C\equiv C \quad C\equiv C-SiMe_3 \ + \ H_3C-CN \quad \xrightarrow{[Co]} \quad \text{(structure)} \quad \longrightarrow \quad \text{(structure)} \tag{15}
$$

Applying the versatility of the cobalt-catalyzed pyridine formation (eq. (2)), Vollhardt [45] has varied the basic reaction extensively. Using rather sophisticated alkyne and nitrile precursors with η^5-Cp-cobalt dicarbonyl as the catalyst, the preparation of a number of polyheterocyclic systems having physiological interest was brought about. Using eq. (14) a synthetic route to the isoquino[2,1-*b*][2,6]naphthyridine nucleus (eq. (16)) was developed [46].

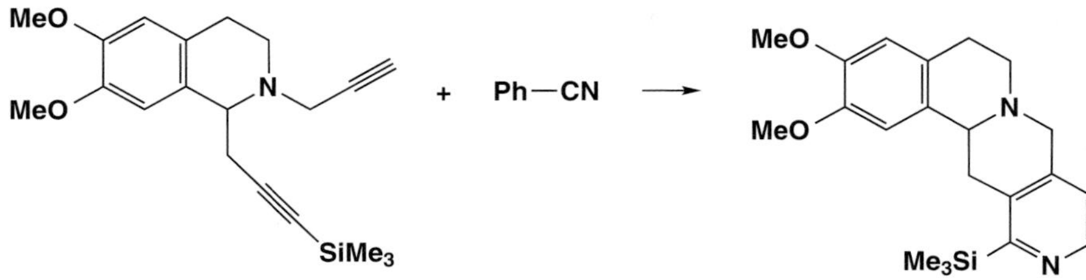

MeO
MeO
N
SiMe3
+
Ph—CN
MeO
MeO
N
Me3Si
N
Ph

(16)

6-Heptyne nitrile was incorporated into the indole system giving a pyridine derivative related to the ergot alkaloids [47].

3.3.8.3.6 Dihydroindoles

The cobalt-catalyzed co-cyclization of alkynes with heterofunctional substrates is not limited to nitriles. η^5-Cp-cobalt half-sandwich complexes are capable of co-oligomerizing alkynes with a number of C=C, C=N, C=O, or C=S bonds in a Diels–Alder-type reaction. Chen has observed that these cycloadditions are best performed in the presence of a small amount of ketones or esters [48]. This modified cycloaddition may be used for the formation of dihydroindole systems at the η^5-Cp-cobalt catalyst (eq. (17)).

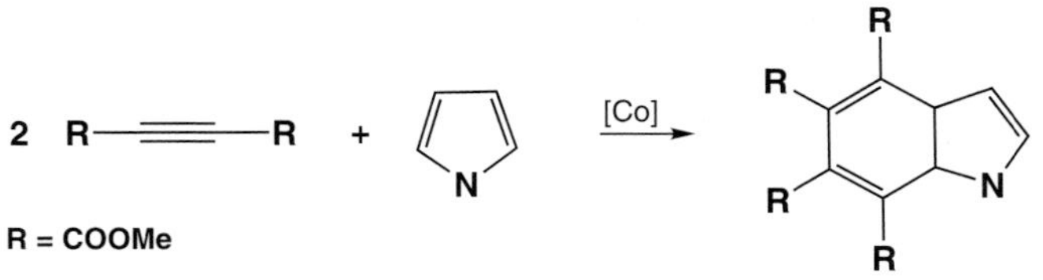

2 R—R
+
N
[Co]
R = COOMe
R
R
R
R
N

(17)

Similar cycloaddition reactions of C=C bonds have been described starting from substituted pyrrole and imidazole derivatives (eq. (18)) [49].

N
X
CH2
n
+
R1—R2
R1
R2
N
X
CH2
n
R1, R2 = SiMe3, CH2CH3, COOMe, OMe; X = O, H2; n = 2, 3

(18)

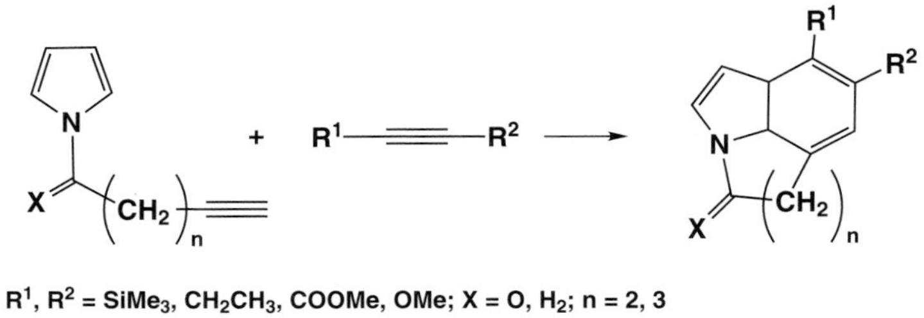

The resulting heterocycles in the complex may be further reduced or desilylated (either in the complex or after demetallation). Further synthetic potential exists in the use of the primary products, obtained by cobalt-mediated cycloadditions, as synthons in organic chemistry. For example, indole derivatives have been co-cyclized at the η^5-Cp-cobalt catalyst to give 4a,9a-dihydro-9H-carbazoles or, after oxidation, precursors for strychnine [50]. Remarkably, the cycloaddition of acrolein in the presence of a small amount of methyl acetate occurs at the carbonyl, rather than at the C=C double bond, to give vinylpyran selectively (eq. (19)) [48].

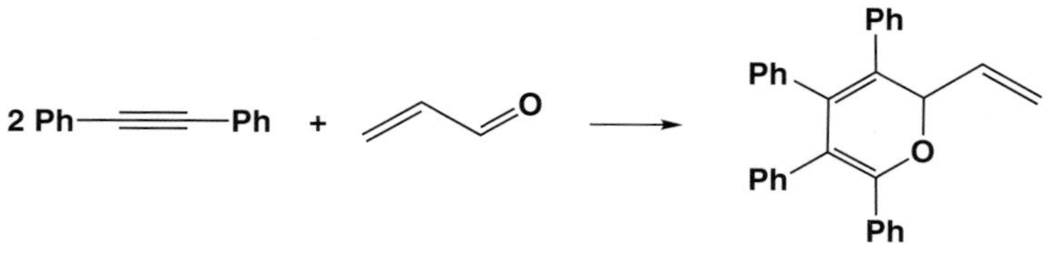

(19)

3.3.8.3.7 Related Reactions

Yamazaki [51] has reviewed a number of stoichiometric cycloaddition reactions at the cobaltacyclopentadiene ring system that lead to a plethora of heterocycles. For example, five- and six-membered heterocycles containing nitrogen, sulfur, selenium, and phosphorus have been made accessible by this route (eq. (20)) [52]. The co-cyclization of substituted alkynes and isocyanates in the presence of a rhodium metallocycle gives 2-pyridones (eq. (21)) [53].

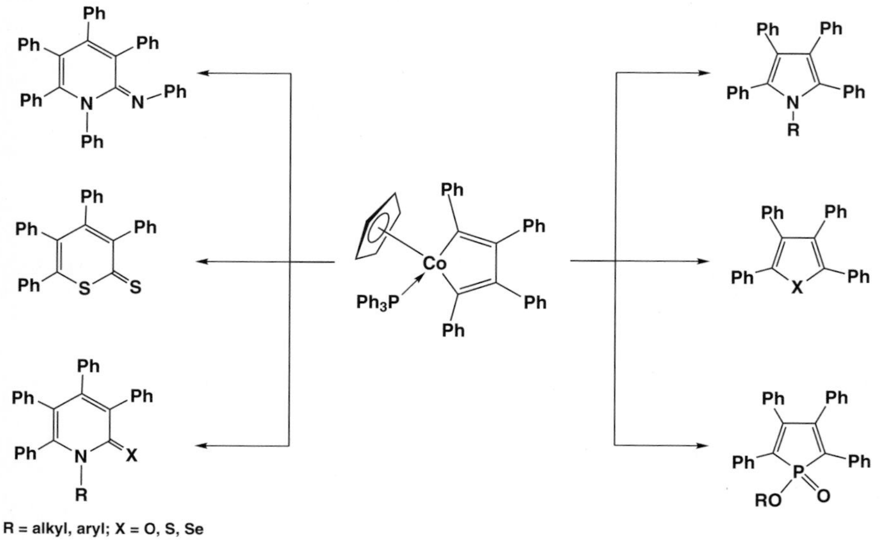

(20)

R = alkyl, aryl; X = O, S, Se

$$2\ R^1\!\!-\!\!\equiv\!\!-COOMe\ +\ R^2\!\!-\!\!NCO\ \xrightarrow{[Rh]}\ \text{(pyridone product)}\ \text{(+ arene)}\quad(21)$$

R^1, R^2 = alkyl, aryl, etc.

$$2\ R\!\!-\!\!\equiv\!\!-R\ +\ {}^1\!/_8\ S_8\ \xrightarrow{[Co]}\ \text{(thiophene product)}\quad(22)$$

R = H, alkyl, aryl, COOMe, etc.

The stability of the η^5-Cp-cobalt core even allows elemental sulfur to be incorporated in the alkyne co-cyclization. In the presence of excess alkynes the η^5-Cp-cobalt half-sandwich complexes catalyze the formation of thiophenes (eq. (22)) [54]. Kajitani [55] has expanded this work to rhodium complexes and included selenium as an additional heteroatom.

3.3.8.4 Six- and Eight-Membered Carbocycles

The cyclotrimerization of acetylene to benzene (eq. (1)) is highly exothermic. The free energy of this process was estimated to be 595 kJ per mol of product [56]. Monosubstituted acetylenes give 1,2,4- or 1,3,5-trisubstituted benzene derivatives (eq. (23)). The regioselectivity of the cyclization may be controlled by the electronic properties and the steric demand of the catalyst, as well as by the reaction conditions. Because of the inherent sensitivity of organometallic catalysts to heteroatoms, this reaction is mainly limited to alkyl-, alkenyl- or aryl-substituted acetylenes. Educts containing polar heteroatoms are only processed by very few homogeneous catalysts such as the organocobalt systems discussed above.

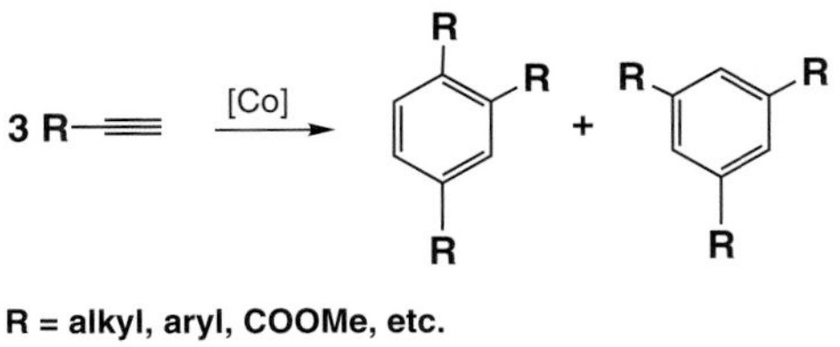

$$3\ R\!\!-\!\!\equiv\ \xrightarrow{[Co]}\ \text{(1,2,4-product)}\ +\ \text{(1,3,5-product)}\quad(23)$$

R = alkyl, aryl, COOMe, etc.

3.3.8.4.1 Substituted Benzenes and Cyclohexadienes

In practice the alkyne cyclotrimerization to benzene and its derivatives may be performed using both homogeneous and heterogeneous catalysts. Most catalysts reported in the survey of the catalysts (cf. Section 3.3.8.2) may be applied, giving good yields, to the cyclotrimerization of unsymmetrically substituted terminal and also internal alkynes. As mentioned above, in the case of terminal alkynes

1,2,4- and 1,3,5-trisubstituted benzenes are formed (eq. (23)). For example, a chromium(VI) catalyst trimerizes propyne to give pseudocumene and mesitylene in a 4:1 ratio [57]. The cyclotrimerization of 1-hexyne, 1-octyne, methyl propiolate, and phenylacetylene at organorhodium half-sandwich complexes was investigated by Ingrosso and co-workers [13 b]. In the case of the alkyl-substituted acetylenes the regioselectivity of the trimerization was found to be independent of the rhodium catalyst applied. The cyclization of methyl propiolate at the η^5-Flu-rhodium catalyst gave a higher proportion of the symmetrically substituted benzene derivatives than were found at the η^5-Ind-rhodium complex. The η^5-Ind-rhodium-bis(ethylene) complex was found to be unusually selective in the cyclotrimerization of 3,3-dimethyl-1-butyne, giving a 76 % yield of 1,2,4-tri-*t*-butylbenzene [58]. The 1,3,5-isomer is available from 3,3-dimethyl-1-butyne in the presence of $PdCl_2$ [59]. This type of catalytic alkyne reaction has been reviewed by Maitlis [60]. The mechanism for the trimerization of disubstituted alkynes at Mo or W vinyl complexes was studied by Davidson et al. [61]. They found two different reaction pathways which were dependent on the substituents at the vinyl groups and the acetylene.

The regiochemical product distribution of the co-cyclization of two or three different alkynes occurs statistically. In some cases carefully controlled reaction conditions allow isolation of a main product from mixed cyclotrimerizations. For example, 1,2,3,4-tetraphenyl-5,6-diethylbenzene can be obtained from cobalt-catalyzed reaction of tolane and 3-hexyne in good yield [62]. The first example of an intermolecular, regiospecific cross-benzannulation reaction catalyzed by $Pd(PPh_3)_4$ was reported by Yamamoto [63]. The reaction of 2-alkyl-but-1-ene-3-yne with disubstituted diynes leads exclusively in high yields to 1,4-dialkyl-2-ethynylbenzene. No other isomers are formed. The selective synthesis of radiolabeled toluene and *p*-xylene via co-cyclotrimerization was obtained using a heterogeneous chromium catalyst described in [57, 64].

Instead of a second or third alkyne, an alkene C=C double bond may be incorporated into the cyclotrimerization reaction. Iron [65], rhodium [66], nickel [67], palladium [68], or cobalt [69] catalysts have been used to form cyclohexadienes. However, the preparative use of this catalytic co-cyclization is disturbed by consecutive side reactions of the resulting dienes such as cycloaddition or dehydrogenation. Itoh, Ibers and co-workers [70] have reported the straight palladium-catalyzed co-cyclization reaction of $C_2(CO_2Me)_2$ and norbornene (eq. (24)).

$$2\ R\!\!-\!\!\equiv\!\!-\!\!R\ +\ \text{(norbornene)}\ \xrightarrow{[Pd]}\ \text{(product)} \tag{24}$$

R = COOMe

Jonas and Tadic [71] have investigated the homogeneous cobalt-catalyzed co-cyclotrimerization of acetylene and olefins. The reaction with η^5-Ind-cobalt bis(ethylene) as the catalyst was carried out with ethylene, α-olefins and 2-butene as well as cyclohexene and cyclooctene (eq. (25)).

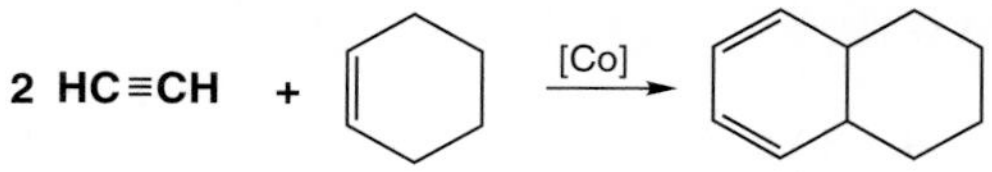

$$2 \ HC \equiv CH \ + \ \ [\text{ring}] \ \xrightarrow{[\text{Co}]} \ [\text{product}] \qquad (25)$$

The reaction according to eq. (25) leads exclusively to *cis*-hexahydronaphthalene (*cis*-hexaline), a product which is otherwise accessible only by multistep synthetic pathways [72]. Macomber [73] reported the [2+2+2] cycloaddition reaction of diphenylacetylene or $C_2(CO_2Me)_2$ and *endo*-dicyclopentadiene or norbornylene, respectively, in the presence of η^5-Cp-cobalt dicarbonyl or η^5-methyl-Cp-cobalt dicarbonyl in refluxing toluene.

Intramolecular cyclohexadiene syntheses have been developed by Vollhardt [74]. Enediynes with a terminal double bond react in isooctane at 100 °C in the presence of η^5-Cp-cobalt dicarbonyl giving a three-ring system [75] according to eq. (26).

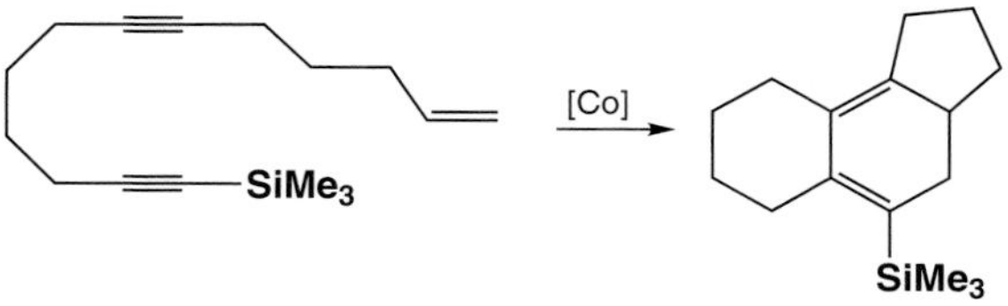

$$[\text{enediyne with SiMe}_3] \ \xrightarrow{[\text{Co}]} \ [\text{tricyclic product with SiMe}_3] \qquad (26)$$

Malacria and co-workers reported an improved diastereoselectivity of the Co-catalyzed cycloaddition of substituted linear enediynes by introducing substituents such as esters, sulfoxide, or phosphine oxide at the terminal position of either the double or the triple bond [76].

With an appropriate precursor C-ring, dienyl steroids have been made accessible in a remarkably highly stereoselective process [77]. Intramolecular cycloaddition reactions of enediynes containing a terminal alkyne group have also been observed by Vollhardt [78] (eq. (27)).

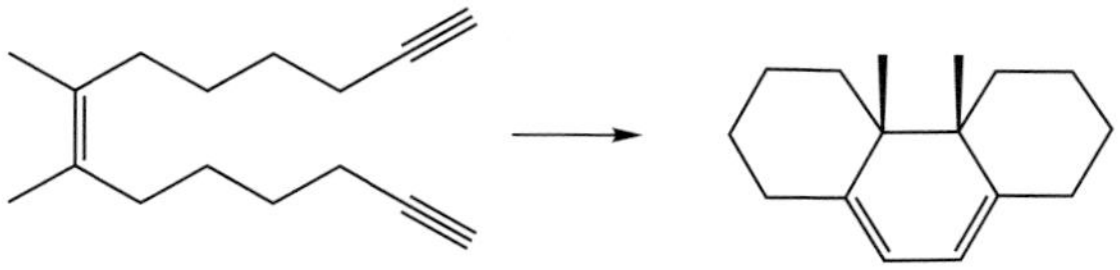

$$[\text{enediyne}] \ \longrightarrow \ [\text{tricyclic product}] \qquad (27)$$

Supercritical water exhibits better solvent properties for apolar organic compounds than water itself and was applied by Jerome and Parsons [79] as well as Dinjus and co-workers [80] as the solvent for the Co-mediated cyclotrimerization of monosubstituted acetylenes to benzene derivatives. Eaton et al. published the cyclotrimerization of acetylenes bearing functional groups in a water/methanol (80:20) mixture using an R-Cp cobalt cod complex as the catalyst. The water solubility of the Co complex was achieved by the special substituent $R=CO(CH_2)_2CH_2OH$ on the Cp ligand [81].

3.3.8.4.2   Phenylenes

o-Diethynylbenzene, available from *o*-diiodobenzene, can easily co-cyclize with internal alkynes to 2,3-disubstituted diphenylenes [82] with η^5-Cp-cobalt dicarbonyl as the catalyst (eq. (28)).

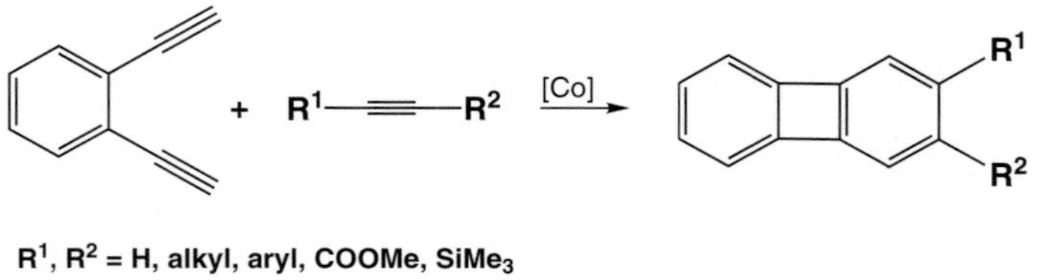

$$\text{(28)}$$

R^1, R^2 = H, alkyl, aryl, COOMe, SiMe$_3$

In the case of $R^1 = R^2 = SiMe_3$ the successive synthesis of polyphenylenes has been reported. Subsequent iodination of the trimethylsilyl group generates a new *o*-diiodoarene as the educt for the subsequent *o*-diethynylarene, which can react with further bis(trimethylsilyl)acetylene forming terphenylene, and so on. Multiphenylenes synthesized in this way have been claimed to represent a new type of organic semiconductor [83].

3.3.8.4.3   Cyclooctatetraenes

As early as 1948, Reppe et al. reported the discovery of the "cyclic polymerization of acetylene" to cyclooctatetraene (eq. (29)) using nickel catalysts [84]. This discovery represented a true landmark in transition metal catalysis.

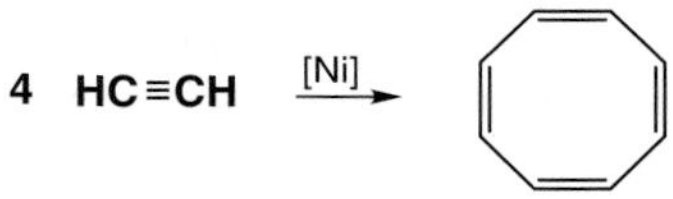

$$4 \ \ HC{\equiv}CH \ \ \xrightarrow{[Ni]} \ \ \text{(29)}$$

Recently, researchers of this reaction propose a bis(cyclooctatetraene) dinickel complex as the active catalyst for the cyclotetramerization of acetylene [85].

Monosubstituted alkynes may be included in this cyclization giving 1,2,4,7-, 1,2,4,6-, and 1,3,5,7-tetrasubstituted cyclooctatetraene derivatives [86]. A special case is the cyclotetramerization of 1-phenylpropyne giving the octasubstituted C_8 product besides the hexasubstituted benzene derivative [87] (eq. (30)).

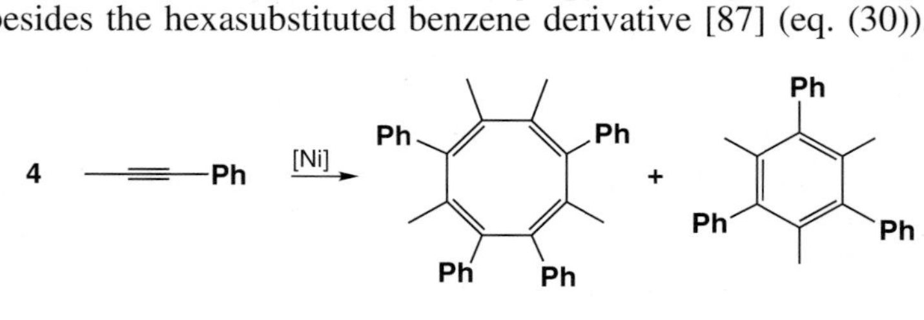

$$\text{(30)}$$

References

[1] M. Berthelot, *Liebigs Ann. Chem.* **1866**, *141*, 173.

[2] (a) W. Reppe, *Chemie und Technik der Acetylen-Druck-Reaktionen,* 2nd ed., VCH, Weinheim, **1952**; (b) J. W. Copenhaver, M. H. Bigelow, *Acetylene and Carbon Monoxide Chemistry,* Reinhold, New York, **1949**; (c) C. W. Bird, *Transition Metal Intermediates in Organic Synthesis,* Academic Press, New York, **1967**.

[3] (a) H. Yamazaki, Y. Wakatsuki, *Tetrahedron Lett.* **1973**, 3383; (b) H. Yamazaki, Y. Wakatsuki, *J. Organomet. Chem.* **1977**, *139*, 157; (c) Y. Wakatsuki, H. Yamazaki, *J. Organomet. Chem.* **1977**, *139*, 169; (d) Y. Wakatsuki, H. Yamazaki, *J. Chem. Soc., Dalton Trans.* **1978**, 1278.

[4] (a) K. P. C. Vollhardt, *Acc. Chem. Res.* **1977**, *10*, 1; (b) A. Naiman, K. P. C. Vollhardt, *Angew. Chem.* **1977**, *89*, 758; *Angew. Chem., Int. Ed. Engl.* **1977**, *16*, 708; (c) J. R. Fritch, K. P. C. Vollhardt, *Angew. Chem.* **1980**, *92*, 570; *Angew. Chem., Int. Ed. Engl.* **1980**, *19*, 559; (d) G. Ville, K. P. C. Vollhardt, M. J. Winter, *J. Am. Chem. Soc.* **1981**, *103*, 5267; (e) J. P. Tane, K. P. C. Vollhardt, *Angew. Chem.* **1982**, *94*, 642; *Angew. Chem. Int. Ed. Engl.* **1982**, *21*, 617; (f) J. R. Fritch, K. P. C. Vollhardt, *Organometallics* **1982**, *1*, 590; (g) J. S. Drage, K. P. C. Vollhardt, *Organometallics* **1982**, *1*, 1545; (h) D. J. Brien, A. Naiman, K. P. C. Vollhardt, *J. Am. Chem. Soc.* **1982**, *104*, 133.

[5] (a) H. Bönnemann, R. Brinkmann, H. Schenkluhn, *Synthesis* **1974**, 575; (b) Studiengesellschaft Kohle mbH (H. Bönnemann, H. Schenkluhn), US 4.006.149 (1975); (c) H. Bönnemann, *Angew. Chem.* **1978**, *90*, 517; *Angew. Chem., Int. Ed. Engl.* **1978**, *17*, 505; (d) H. Bönnemann, W. Brijoux, *Asp. Homogeneous Catal.* **1984**, *5*, 75; (e) H. Bönnemann, W. Brijoux, R. Brinkmann, W. Meurers, R. Mynott, W. von Philipsborn, T. Egolf, *J. Organomet. Chem.* **1984**, *272*, 231; (f) H. Bönnemann, *Angew. Chem.* **1985**, *97*, 264; *Angew. Chem., Int. Ed. Engl.* **1985**, *24*, 248; (g) H. Bönnemann, W. Brijoux, *Adv. Heterocycl. Chem.* **1990**, *48*, 177; (h) J. S. Viljoen, J. A. K. du Plessis, *J. Mol. Catal.* **1993**, *79*, 75; (i) J. A. K. du Plessis, J. S. Viljoen, *J. Mol. Catal. A: Chem.* **1995**, *99*, 71.

[6] N. E. Schore, *Chem. Rev.* **1988**, *88*, 1081.

[7] (a) F. W. Hoover, O. W. Webster, C. T. Handy, *J. Org. Chem.* **1961**, *26*, 2234; (b) B. Franzus, P. J. Canterino, R. A. Wickliffe, *J. Am. Chem. Soc.* **1959**, *81*, 1514; (c) Studiengesellschaft Kohle mbH (K. Ziegler), DE 1.233.374 (1967), GB 831.328 (1960); (d) J. J. Eisch, W. C. Kaska, *J. Am. Chem. Soc.* **1966**, *88*, 2213; (e) E. F. Lutz, *J. Am. Chem. Soc.* **1961**, *83*, 2551; (f) A. F. Donda, G. Moretti, *J. Org. Chem.* **1966**, *31*, 985; (g) T. Masuda, Y.-X. Deng, T. Higashimura, *Bull. Chem. Soc. Jpn.* **1983**, *56*, 2798; (h) V. A. Sergeev, Y. A. Chernomordik, V. S. Kolesov, V. V. Gavrilenko, V. V. Korshak, *Zh. Org. Khim.* **1975**, *11*, 777; (i) W. Schönfelder, G. Snatzke, *Chem. Ber.* **1980**, *113*, 1855; (j) V. O. Reichsfel'd, B. I. Lein, K. L. Makovetskii, *Dokl. Akad. Nauk. SSSR* **1970**, *190*, 125; (k) C. J. Baddeley, R. M. Ormerod, A. W. Stephenson, R. M. Lambert, *J. Phys. Chem.* **1995**, *99*, 5146.

[8] (a) U. Krüerke, W. Hübel, *Chem. Ber.* **1961**, *94*, 2829; (b) H. Hopff, A. Gati, *Helv. Chim. Acta* **1965**, *48*, 509; (c) W. Reppe, H. Vetter, *Liebigs Ann. Chem.* **1953**, *133*, 585; (d) U. Krüerke, C. Hoogzand, W. Hübel, *Chem. Ber.* **1961**, *94*, 2817.

[9] (a) M. Tsutsui, H. Zeiss, *J. Am. Chem. Soc.* **1959**, *81*, 6090; (b) W. Herwig, W. Metlesics, H. Zeiss, *J. Am. Chem. Soc.* **1959**, *81*, 6203; (c) H. Zeiss, M. Tsutsui, *J. Am. Chem. Soc.* **1959**, *81*, 6255; (d) H. P. Throndesen, W. Metlesics, H. Zeiss, *J. Organomet. Chem.* **1966**, *5*, 176.

[10] G. N. Schrauzer, *Chem. Ber.* **1961**, *94*, 1403.

[11] (a) A. T. Blomquist, P. M. Maitlis, *J. Am. Chem. Soc.* **1962**, *84*, 2329; (b) P. M. Maitlis, D. Pollock, M. L. Games, M. J. Pride, *Can. J. Chem.* **1965**, *43*, 470.

[12] L. B. Luttinger, *J. Org. Chem.* **1962**, *27*, 1591.
[13] (a) F. A. Cotton, W. T. Hall, K. J. Cann, F. J. Karol, *Macromolecules* **1981**, *14*, 233;
     (b) A. Borrini, P. Diversi, G. Ingrosso, A. Lucherini, G. Serra, *J. Mol. Catal.* **1985**,
     30, 181; (c) K. Abdulla, B. L. Booth, C. Stacey, *J. Organomet. Chem.* **1985**, *293*, 103.
[14] E. M. Arnett, J. M. Bollinger, *J. Am. Chem. Soc.* **1964**, *86*, 4729.
[15] A. K. Jhingan, W. F. Maier, *J. Org. Chem.* **1987**, *52*, 1161.
[16] P. Cioni, P. Diversi, G. Ingrosso, A. Lucherini, P. Ronca, *J. Mol. Catal.* **1987**, *40*, 337.
[17] (a) U. Schmidt, U. Zenneck, *J. Organomet. Chem.* **1992**, *440*, 187; (b) D. Böhm, F. Koch,
     S. Kummer, U. Schmidt, U. Zenneck, *Angew. Chem.* **1995**, *107*, 251; *Angew. Chem., Int.
     Ed. Engl.* **1995**, *34*, 198.
[18] Studiengesellschaft Kohle mbH (H. Bönnemann, H. Schenkluhn), DE 2.416.295 (1974).
[19] (a) M. Aresta, M. Rossi, A. Sacco, *Inorg. Chim. Acta* **1969**, *3*, 227; (b) P. Diversi,
     A. Guisti, G. Ingrosso, A. Lucherini, *J. Organomet. Chem.* **1981**, *205*, 239.
[20] G. Vitulli, S. Bertozzi, M. Vignali, R. Lazzaroni, P. Salvadori, *J. Organomet. Chem.*
     1987, *326*, C33.
[21] W. Schultz, H. Pracejus, G. Oehme, *Tetrahedron Lett.* **1989**, *30*, 1229.
[22] (a) D. M. M. Rohe, Ph. D. Thesis, RWTH Aachen, (1979); (b) Studiengesellschaft Kohle
     mbH (H. Bönnemann), DE 3.117.363.2 (1981); (c) Studiengesellschaft Kohle mbH
     (H. Bönnemann), US 4.588.815 (1984).
[23] P. Diversi, L. Ermini, G. Ingrosso, A. Lucherini, *J. Organomet. Chem.* **1993**, *447*, 291.
[24] R. A. Abramovitch, *Chem. Heterocycl. Compd.* **1974**, **1975**, *14*, Suppl. Parts 1–4, Wiley,
     New York, **1975**.
[25] (a) W. Ramsay, *Philos. Mag.* [5] **1876**, *4*, 269; (b) W. Ramsay, *Philos. Mag.* [5] **1877**, *5*,
     24; (c) N. Ljubawin, *J. Russ. Phys.-Chem. Ges.* **1885**, 250; (d) R. Meyer, A. Tanzen, *Ber.
     Dtsch. Chem. Ges.* **1913**, *46*, 3186.
[26] (a) G. Herberich, W. Koch, H. Leuken, *J. Organomet. Chem.* **1978**, *160*, 17; (b) Studien-
     gesellschaft Kohle mbH (H. Bönnemann, B. Bogdanovic), DE Appl. 310.550.1 (1982);
     (c) Studiengesellschaft Kohle mbH (H. Bönnemann, B. Bogdanovic), EP Appl. 83/
     101.246.3 (1983).
[27] (a) Stamicarbon N. V. (J. M. Deumens, S. H. Green) GB 1.304.155 (1973); (b) Stamicar-
     bon N. V. (J. M. Deumens, S. H. Green) US 3.780.082 (1973); (c) *Chem. Mark. Rep.*
     1977.
[28] (a) Y. Kusunoki, H. Okazeku, *Hydrocarbon Process,* **1974**, *53* (11), 129, 131; (b) Y. Ku-
     sunoki, H. Okazaki, *Nippon Kagaku Kaishi* **1981**, *12*, 1969; (c) Y. Kusunoki, H. Oka-
     zaki, *Nippon Kagaku Kaishi* **1981**, *12*, 1971.
[29] (a) Lonza AG (P. Hardt), CH Appl. 76/14.399 (1976); (b) Lonza AG (P. Hardt) DE
     2.751.072 (1978).
[30] S. Goldschmidt, M. Minsinger, DE 952.807 (1956); (b) J. P. Wibaut, C. Hoogzand,
     Chem. Weekblad **1956**, *52*, 357.
[31] M. J. Birchenough, *J. Chem. Soc.* **1951**, 1263.
[32] (a) D. Tatone, Trane Cong Dich, R. Nacco, C. Botteghi, *J. Org. Chem.* **1975**, *40*, 2987;
     (b) G. Cavinato, L. Toniolo, C. Botteghi, S. Gladiali, *J. Organomet. Chem.* **1982**, *229*, 93.
[33] C. Botteghi, private communication, **1975**.
[34] R. Brinkmann, private communication, **1982**.
[35] D. B. Wootton, *Dev. Adhes.* **1977**, *1*, 181.
[36] Schering AG, DE 663.891 (1936).
[37] P. Arnall, N. R. Clark, *Chem. Process. (London)* **1971**, *17*, (10), 9, 11–13, 15.
[38] H. Bönnemann, G. S. Natarajan, *Erdöl, Kohle, Erdgas, Petrochem.* **1980**, *33*, 328.
[39] R. A. Abramovitch, *Chem. Heterocycl. Compd.* **1975**, *14*, Suppl. Part 4, Chapter 15, 189.
[40] (a) E. Shaw, J. Bernstein, K. Losse, W. A. Lott, *J. Am. Chem. Soc.* **1950**, *72*, 4362;
     (b) Olin Mathieson Chemical Corp. (S. Semenoff, M. A. Dolliver), US 2.745.826 (1956).

[41] (a) G. M. Badger, W. H. F. Sasse, *Adv. Heterocycl. Chem.* **1963**, *2*, 179; (b) M. A. E. Hodgson, *Chem. Ind. (London)* **1968**, 49; (c) L. A. Summers, *The Bipyridinium Herbicides*, Academic Press, New York, **1980**.

[42] H. Bönnemann, R. Brinkmann, *Synthesis* **1975**, 600.

[43] A. Naiman, K. P. C. Vollhardt, *Angew. Chem.* **1977**, *89*, 758; *Angew. Chem., Int. Ed. Engl.* **1977**, *16*, 708.

[44] R. E. Geiger, M. Lalonde, H. Stoller, K. Schleich, *Helv. Chim. Acta* **1984**, *67*, 1274.

[45] (a) K. P. C. Vollhardt, J. E. Bercaw, R. G. Bergman, *J. Am. Chem. Soc.* **1974**, *96*, 4996; (b) C. A. Parnell, K. P. C. Vollhardt, *Tetrahedron* **1985**, *41*, 5791; (c) K. P. C. Vollhardt, *Lect. Heterocycl. Chem.* **1987**, *9*, 59.

[46] (a) D. J. Brien, A. Naiman, K. P. C. Vollhardt, *Chem. Soc., Chem. Commun.* **1982**, 133; (b) K. P. C. Vollhardt, *Lect. Heterocycl. Chem.* **1987**, *9*, 60.

[47] K. P. C. Vollhardt, *11th Int. Congr. Heterocycl. Chem.* Heidelberg, **1987**.

[48] H. Bönnemann, X. Chen, *Proc. Swiss Chem. Soc. Autumn Meet, 1987*, Bern, **1987**, 39.

[49] (a) G. S. Sheppard, K. P. C. Vollhardt, *J. Org. Chem.* **1986**, *51*, 5496; (b) R. Boese, H.-J. Knölker, K. P. C. Vollhardt, *Angew. Chem.* **1987**, *99*, 1067; *Angew. Chem., Int. Ed. Engl.* **1987**, *26*, 1035.

[50] (a) R. B. Woodward, M. P. Cava, W. D. Ollis, A. Hunger, H. U. Daeniker, K. Schenker, *Tetrahedron* **1963**, *19*, 247; (b) D. B. Grotjahn, K. P. C. Vollhardt, *J. Am. Chem. Soc.* **1986**, *108*, 2091; (c) K. P. C. Vollhardt, *Lect. Heterocycl. Chem.* **1987**, *9*, 61.

[51] (a) H. Yamazaki, Y. Wakatsuki, *Kagaku Sosetsu* **1981**, *32*, 161; (b) H. Yamazaki, *J. Synth. Org. Chem.* **1987**, *45*, 244.

[52] (a) R. A. Earl, K. P. C. Vollhardt, *J. Am. Chem. Soc.* **1983**, *105*, 6991; (b) P. Diversi, G. Ingrosso, A. Lucherini, S. Malquori, *J. Mol. Catal.* **1987**, *40*, 267.

[53] S. T. Flynn, S. E. Hasso-Henderson, A. W. Parkins, *J. Mol. Catal.* **1985**, *32*, 101.

[54] M. Kajitani, T. Suetsugu, A. Igarashi, T. Akiyama, A. Sugimori, H. Bönnemann, *30th Symp. Organomet. Chem.*, Kyoto, Japan, **1983**, Abstr. A206.

[55] (a) M. Kajitani, T. Suetsugu, A. Igarashi, T. Akiyama, A. Sugimori, *J. Organomet. Cheme.* **1985**, *293,* C15; (b) M. Kajitani, R. Ochiai, N. Kobayashi, T. Akiyama, A. Sugimori, *Chem. Lett.* **1987**, 245.

[56] S. W. Benson, *Thermochemical Kinetics*, Wiley, New York, **1968**.

[57] R. A. Ferrieri, A. P. Wolf, *J. Phys. Chem.* **1984**, *88*, 2256.

[58] P. Caddy, M. Green, E. O'Brien, L. E. Smart, P. Woodward, *J. Chem. Soc., Dalton Trans.* **1980**, 962.

[59] (a) P. M. Maitlis, *Acc. Chem. Res.* **1976**, *9*, 93; (b) P. M. Maitlis, E. A. Kelly, *J. Chem. Soc., Dalton Trans.* **1979**, 167; (c) F. Canziani, C. Allevi, L. Garlaschelli, M. C. Malatesta, A. Albinati, F. Ganazzoli, *J. Chem. Soc., Dalton Trans.* **1984**, 2637.

[60] P. M. Maitlis, *J. Organomet. Chem.* **1980**, *200*, 161.

[61] N. M. Agh-Atabay, L. Carlton, J. L. Davidson, G. Douglas, K. W. Muir, *J. Chem. Soc., Dalton Trans.* **1996**, 999.

[62] (a) W. Hübel, C. Hoogsand, *Chem. Ber.* **1960**, *93*, 103; (b) O. S. Mills, G. Robinson, *Proc. Chem. Soc.* **1964**, 187.

[63] V. Gevorgyan, A. Takeda, Y. Yamamoto, *J. Am. Chem. Soc.* **1997**, *119*, 11.313.

[64] M. Speranza, R. A. Ferrieri, A. P. Wolf, F. Cacace, *J. Labeled Compd. Radiopharm.* **1982**, *19*, 61.

[65] A. Carbonaro, A. Greco, G. Dall'Asta, *Tetrahedron Lett.* **1968**, 5129.

[66] D. M. Singleton, *Tetrahedron Lett.* **1973**, 1245.

[67] A. Chalk, *J. Am. Chem. Soc.* **1972**, *94*, 5928.

[68] L. D. Brown, K. Itoh, H. Suzuki, K. Hirai, J. A. Ibers, *J. Am. Chem. Soc.* **1978** *100*, 8232.

[69] E. Dunach, R. L. Halterman, K. P. C. Vollhardt, *J. Am. Chem. Soc.* **1985**, *107*, 1664.

[70] H. Suzuki, K. Itoh, Y. Ishii, K. Simon, J. A. Ibers, *J. Am. Chem. Soc.* **1976**, *98*, 8494.

[71] M. G. J. Tadic, Ph. D. Thesis, Ruhr-Universität Bochum, **1990**.
[72] (a) W. G. Dauben, M. S. Kellog, *J. Am. Chem. Soc.* **1980**, *102*, 4456; (b) W. G. Dauben, E. G. Olson, *J. Org. Chem.* **1980**, *45*, 3377.
[73] D. W. Macomber, A. G. Verma, *Organometallics* **1988**, *7*, 1241.
[74] K. P. C. Vollhardt, *Pure Appl. Chem.* **1985**, *57*, 1819.
[75] (a) W. D. Sternberg, K. P. C. Vollhardt, *J. Am. Chem. Soc.* **1980**, *102*, 4841; (b) E. D. Sternberg, K. P. C. Vollhardt, *J. Org. Chem.* **1984**, *49*, 1564.
[76] F. Slowinski, C. Aubert, M. Malacria, *Tetrahedron Lett.* **1999**, *40*, 707.
[77] (a) E. D. Sternberg, K. P. C. Vollhardt, *J. Org. Chem.* **1984**, *49*, 1574; (b) H. Butenschön, M. Winkler, K. P. C. Vollhardt, *J. Chem. Soc., Chem. Commun.* **1986**, 388.
[78] T. R. Gadek, K. P. C. Vollhardt, *Angew. Chem., Int. Ed. Engl.* **1981**, *20*, 802.
[79] K. S. Jerome, E. J. Parsons, *Organometallics* **1993**, *12*, 2991.
[80] H. Borwieck, O. Walter, E. Dinjus, J. Rebizant, *J. of Organomet. Chem.* **1998**, *570*, 121.
[81] B. R. Eaton, M. S. Sigmund, A. W. Fatland, *J. Am. Chem. Soc.* **1998**, *120*, 5130.
[82] B. C. Berris, Y.-H. Lai, K. P. C. Vollhardt, *J. Chem. Soc., Chem. Commun.* **1982**, 953.
[83] K. P. C. Vollhardt, *Angew. Chem.* **1984**, *96*, 525; *Angew. Chem., Int. Ed. Engl.* **1984**, *23*, 539.
[84] W. Reppe, O. Schlichting, K. Klager, T. Toepel, *Liebigs Ann. Chem.* **1948**, *560*, 1.
[85] G. Wilke, *Angew. Chem.* **1988**, *100*, 189; *Angew. Chem., Int. Ed. Engl.* **1988**, *27*, 185.
[86] (a) P. Cini, N. Palladino, A. Santambrogio, *J. Chem. Soc. C* **1967**, 835; (b) J. R. Leto, M. F. Leto, *J. Am. Chem. Soc.* **1961**, *83*, 2944.
[87] L. H. Simons, J. J. Lagowski, *Fund. Res. Homogeneous Catal.* **1978**, *2*, 73.

3.3.9   Chemicals from Renewable Resources

Jochen P. Zoller

3.3.9.1   Introduction and General Developments

Sustainable development, responsible care, and process-integrated protection of the environment are the guidelines of chemistry and the chemical industry in the new millennium. A high percentage of industrial processes are based on the catalytic transformation of substrates. The main focus of industrial research is the further development of transformation processes toward sustainability, and the use of renewable resources as substrates.

The use of renewable resources as substrates in chemical processes has recently been reviewed [1]. There are two major topics of current research: (1) oleochemical reactions, and (2) chemical transformation of carbohydrates. Nevertheless, the chemical possibilities of renewable resources as substrates – using homogeneous catalysis – are still very far from being fully exploited.

3.3.9.2   "Oleo Chemistry"

Natural oils and fats comprise the greatest proportion of renewable resources as substrates, since they are produced easily from vegetables and animals. The range of possibilities was summarized in 1988 [2].

The main applications of chemical processes in the field of oleo chemistry involve oxidation and metathesis. Neither is yet involved in industrial processes, but their application moves realistically nearer as the interest in this topic increases.

3.3.9.2.1 The Oxidation of Oils and Fats

As mentioned above, fatty materials as educts for chemical processes are available from vegetables and animals in such purity that they can be used for further chemical conversion.

Unsaturated fatty compounds are the preferred educts in industrial epoxidation. Numerous methods are available to transform then to the corresponding epoxides. Epoxidation with molecular oxygen [3], dioxiranes [4], hydrogen peroxide with methyltrioxorhenium as catalyst [5, 6], the Halcon process [7], or enzymatic reactions [8] are the most important industrial processes (cf. Section 2.4.3).

Due to the excellent stability and activity of oxidized "vegetable oils", these are used as stabilizers in polymerization chemistry. Epoxidation of vegetable oils by the enzyme Novozym 435 [8] (eq. (1)) is an example.

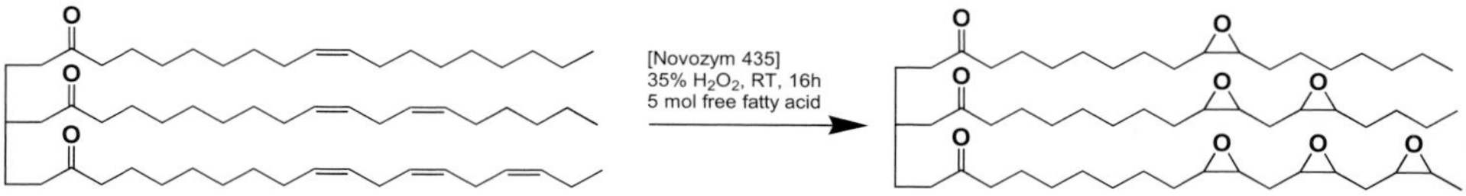

$$\tag{1}$$

Furthermore, oxidation of fatty acids to vicinal diols, as well as their oxidative cleavage, are important industrial applications. Vicinal diols of unsaturated fatty compounds can be prepared by nucleophilic ring opening of the epoxides after epoxidation, but difficult technical conditions are necessary to achieve this ring opening [9]. The use of Re- [10], W- [11], or Mo [1]-based catalysts with hydrogen peroxide can give a *syn*-diol via the epoxide as intermediate (eq. (2)).

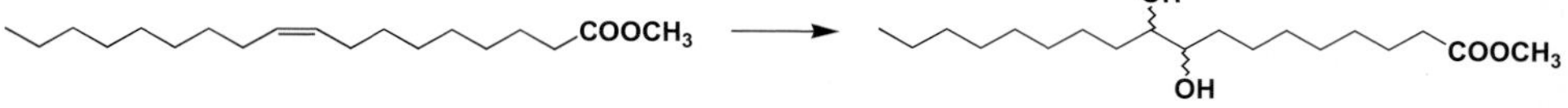

$$\tag{2}$$

The cleavage of oleic acid to azelaic acid and pelargonic acid with ozone as oxidant is one of the important industrial applications of ozonolysis [2]. However, finding a catalytic alternative to this unsuitable and hard-to-handle oxidant is in the interests of research groups all over the world. The cleavage of internal C=C double bonds by use of a Re, Mo, or W catalyst with H_2O_2 as oxidant, or a Ru catalyst with peracetic acid [12], is known.

Oxidative cleavage with hydrogen peroxide as oxidant is more important in oxidation processes of natural products. The use of a three-fold excess of hydrogen peroxide without further additives, except for the catalyst methyltrioxorhenium (MTO), enables the oxidation of certain natural products drawn from styrene

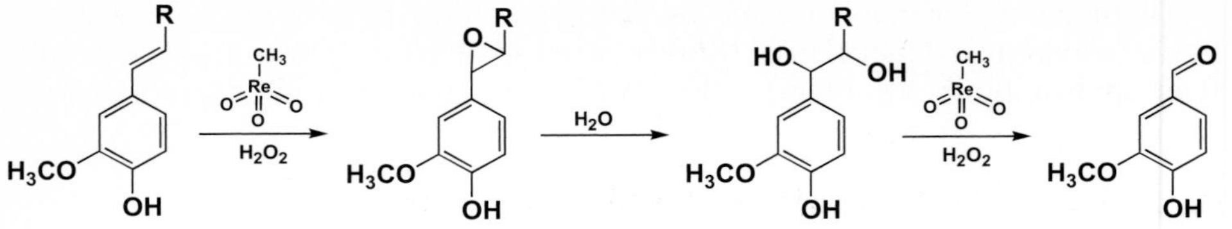

R: -CH$_3$ = **Isoeugenol**
    -COOH = *trans*-**ferulic acid**

Scheme 1

and its derivatives (cf. Section 3.3.13). Thus, a new method for the synthesis of vanillin out of renewable resources such as isoeugenol and *trans*-ferulic acid should be mentioned. Both are derived from renewable resources by the extraction of sawdust (isoeugenol) or agricultural waste (*trans*-ferulic acid) (Scheme 1) [13].

3.3.9.2.2   The Metathesis of Oils and Fats

Metathesis of olefins, through the use of transition metal catalysis, is an important application in the petrochemical as well as in the polymer chemical industry for the production of special olefins and polymers (cf. Section 2.3.3). This chemistry is also applicable to unsaturated fatty acid esters, such as acetic acid methyl ester. However, the high price and unsustainability of the catalysts, compared with the fatty acids as substrates, have made commercial utilization not yet possible; nevertheless they are of great interest for researchers in this field.

Starting from the anchored *heterogeneous* catalysts such as $Re_2O_7 \cdot B_2O_3/Al_2O_3 \cdot SiO_2$, activated by SnR_4, or $CH_3ReO_3 + B_2O_3 \cdot Al_2O_3 \cdot SiO_2$ [14], the use of *homogeneous* catalysts then gained importance. Here, the "Grubbs catalyst" is under investigation [15]. Methyltrioxorhenium is also suitable for the metathesis of unsaturated fatty compounds [16].

From the industrial point of view the co-metathesis of unsaturated fatty compounds, especially methyl oleate, with ethylene to form methyl 9-decenoate and 1-decene is becoming more important (eq. (3)).

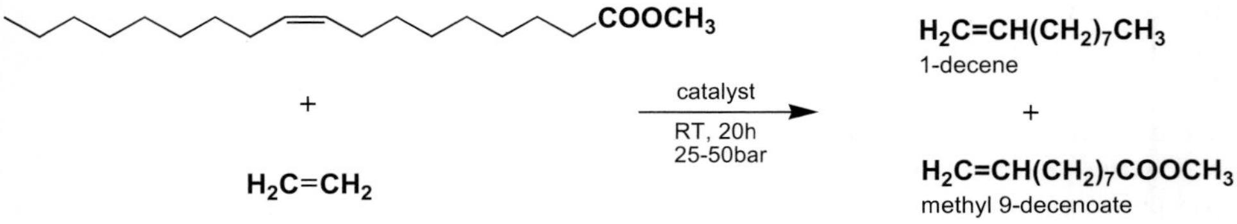

catalyst: $CH_3ReO_3 + B_2O_3 \cdot Al_2O_3 \cdot SiO_2$
    or $Re_2O_7 \cdot B_2O_3 / Al_2O_3 \cdot SiO_2 + SnR_4$

(3)

As generation of oils and fats from of natural products is inexpensive, the investigation of these oils as educts in chemical processes is the focus of current research.

3.3.9.3   The Chemistry of Carbohydrates

Today, metal-catalyzed oxidation with oxygen, being simple, is the most important technology for the conversion of hydrocarbons to bulk industrial chemicals. Such processes involve the use of heterogeneous catalysts. In the manufacture of fine chemicals, the replacement of stoichiometric inorganic oxidants is still the focus of investigations. These catalytic processes employ cheap and environmentally friendly oxidants. Within the same environmental context, carbohydrates are of interest as renewable raw materials in the manufacture of fine chemicals. Oxidation is an obvious method for upgrading carbohydrates, since a mass increase takes place and the character of the carbohydrate compound changes tremendously. Several reviews are available in the field of carbohydrate oxidation [17].

Recently, the oxidative formation of the C_6 group in carbohydrates has became a strong focus of investigation, since when it is applied to the starch molecule, superabsorbing material based on renewable resources can be produced. Therefore, the development of new catalytic methods for the selective oxidation of terminal alcohols C_6 by applying simple and cheap oxidants such as hydrogen peroxide still remains challenging.

3.3.9.4   The Chemistry of Starch

Strong oxidizing agents ($KMnO_4$, MnO_2, SeO_2, etc.) in stoichiometric amounts are necessary to perform the oxidation of alcohols [18]. The *selective* oxidation of primary alcohols, in the presence of secondary ones, is preferred when the stable organic nitroxyl radical 2,2,6,6-tetramethyl-1-piperidinyloxy (TEMPO; Structure **1**) is used as a mediator [19–21]. The nitrosonium ion (**2**) is the intermediate oxidizing species; it becomes reduced to the hydroxylamine **3** during the oxidation process, then it regenerates to **1** during further reaction.

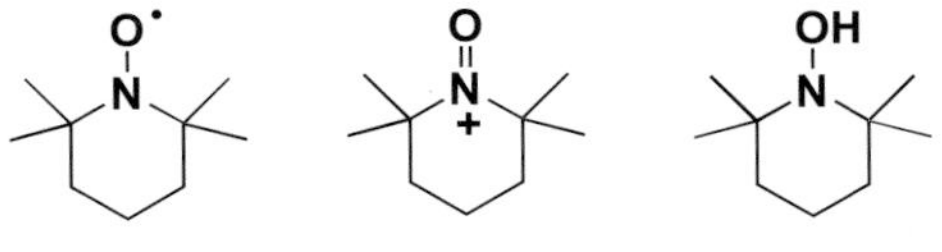

The C_6 primary alcohol group in carbohydrates can be oxidized selectively by the *in situ* generation of the nitrosonium ion **2** using hypochlorite as oxidant and the bromide/hypobromide co-catalyst in water [21, 22]. An analogous system was applied to monosaccharide [22]. Recently it has been reported that MTO also catalyzes the conversion of alcohols to aldehydes in the presence of hydrogen peroxide [23].

From an industrial point of view, the complete oxidation of the C_6-hydroxymethyl group of potato starch yielding the corresponding carboxylic acid would be preferred since such biopolymers can be used in a variety of applications, e. g., as superabsorbing agents. To generate carboxylated starch exclusively by extensive oxidation of the hydroxymethyl groups of the starch molecules to carboxylic acid units, the application of a three-component system (H_2O_2/MTO/ NaBr) without TEMPO is useful; further oxidation of the intermediate aldehyde is expected to occur [23]. The main advantage of this method is the use of aqueous reaction media and the substitution of bleach as an oxidant.

The water-soluble starch used contains 27 % amylose and 73 % amylopectin, and the oxidized starch can be isolated as a gel. The characteristic signal for the carboxylic C_6 group was monitored at $\delta = 176.5$. Signals due to the corresponding aldehyde groups were not seen (Structure 4).

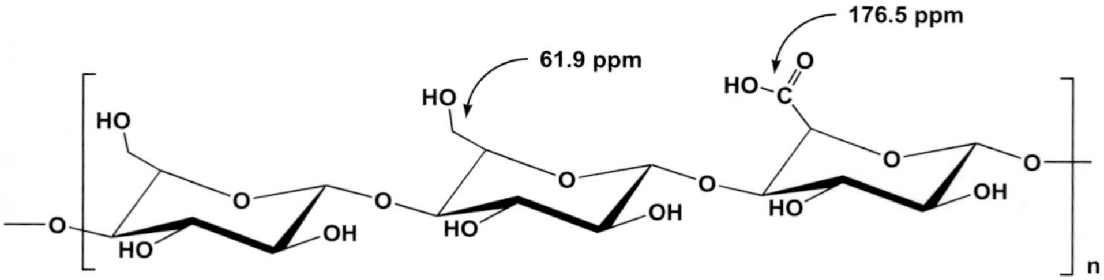

4

The mechanistic proposals for the oxidation of alcohols with MTO/H_2O_2 and NaBr as a co-catalyst are demonstrated in Scheme 2. Further formation of hypobromite in the presence of excess hydrogen peroxide takes place during

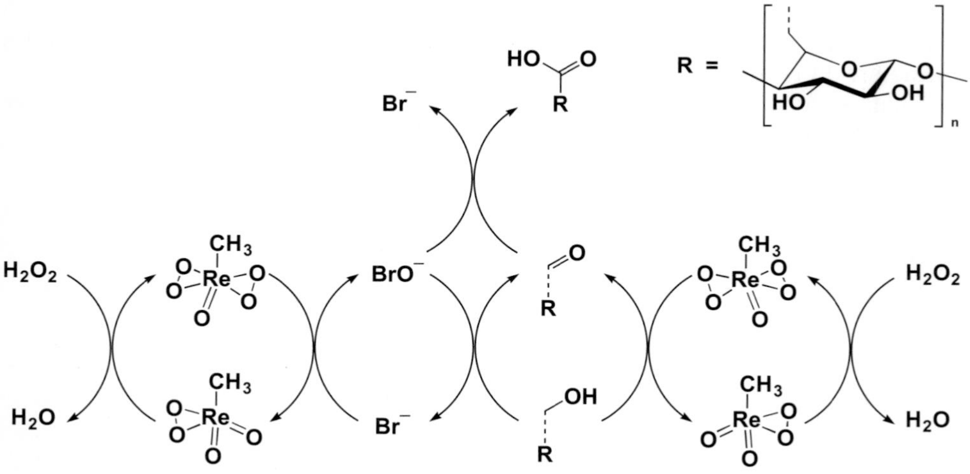

Scheme 2

the oxidation process. This is a reasonable explanation for the observation that no aldehyde is formed, but only the preferred carboxylic acid.

The use of bifunctional cross-linking reagents, such as divinylsulfone, leads to the additional covalent bonding of oxidized starch molecules. This allows a further increase in the water-absorbing properties. Thus, ecologically problematic oxidants such as bleach or NO_2 are becoming obsolete [24].

References

[1] J. O. Metzger et al., *Angew. Chem. Int. Ed.* **2000**, *39*, 2206.
[2] H. Baumann, M. Bühler, H. Fochem, F. Hiersinger, H. Zoebelein, J. Falbe, *Angew. Chem., Int. Ed. Engl.* **1988**, *27*, 41.
[3] M. C. Kuo, C. T. Chou, *Int. Eng. Chem. Res.* **1987**, *26*, 277.
[4] W. Adam, J. Bialas, L. Hadjiarapoglou, *Chem. Ber.* **1991**, *124*, 2377.
[5] W. A. Herrmann, R. W. Fischer, M. U. Rauch, W. Scherer, *J. Mol. Catal. A: Chem.* **1994**, *86*, 245.
[6] W. A. Herrmann, R. W. Fischer, D. W. Marz, *Angew. Chem., Int. Ed. Engl.* **1991**, *30*, 1638.
[7] R. Landau, G. A. Sullivan, D. Brown, *CHEMTECH* **1979**, 602.
[8] M. Rüsch gen. Klass, S. Warwel, *Ind. Crops Prod.* **1999**, *9*, 125.
[9] B. Dahlke, S. Hellbarbt, M. Paetow, W. H. Zech, *J. Am. Oil Chem. Soc.* **1995**, *72*, 349.
[10] W. A. Herrmann, D. W. Marz, J. G. Kuchler, G. Weichselbaumer, R. W. Fischer, DE 3.902.357 A1; *Chem. Abstr.* **1991**, *114*, 143714.
[11] T. M. Luong, H. Schriftmann, D. Swern, *J. Am. Oil Chem. Soc.* **1967**, *44*, 316.
[12] S. Warwel, M. Rüsch gen. Klaas, US 5.321.158 (1994); *Chem. Abstr.* **1996**, *125*, 136578.
[13] W. A. Herrmann, T. Westkamp, J. P. Zoller, R. W. Fischer, *J. Mol. Catal. A: Chem.* **2000**, *153*, 49.
[14] S. Warwel, P. Bavay, M. Rüsch gen. Klaas, B. Wolff, *Perspectiven nachwachsender Rohstoffe in der Chemie,* VCH, Weinheim, **1996**, p. 119.
[15] T. Weskamp, F. J. Kohl, W. Hieringer, D. Gleich, W. A. Herrmann, *Angew. Chem.* **1999**, *111*, 2573.
[16] W. A. Herrmann, W. Wagner, U. N. Flessner, U. Volkhardt, H. Komber, *Angew. Chem., Int. Ed. Engl.* **1991**, *30* ,1636.
[17] A. J. H. F. Arts, E. J. M. Mombarg, H. van Bekkum, R. A. Sheldon, *Synthesis*, **1997**, 597.
[18] J. March, *Advanced Organic Chemistry: Reaction, Mechanisms and Structure,* 4th ed., John Wiley, New York, **1992**.
[19] M. F. Semmelhack, C. S. Chou, D. A. Cortés, *J. Am. Chem. Soc.* **1983**, *105*, 4492.
[20] H. van Bekkum, A. C. Besemer, *Carbohydrates in Europe,* Carbohydrate Research Foundation, **1995**, p. 16.
[21] H. van Bekkum, A. E. J. de Nooy, A. C. Besemer, *Synthesis* **1996**, 1153.
[22] S. L. Flitsch, N. J. Davis, *Tetrahedron Lett.* **1993**, *34*, 1181.
[23] W. A. Herrmann, J. P. Zoller, R. W. Fischer, *J. Organomet. Chem.* **1999**, *579*, 404.
[24] A. Fischbach, Diplomarbeit, Technische Universität München 2001, unpublished results.

3.3.10   Special Reactions in Homogeneous Aqueous Systems

3.3.10.1   Synthesis of Polymers

Bruce M. Novak

3.3.10.1.1 Introduction

Although the synthesis of polymers relies on the same fundamental bond-forming reactions that are commonly used in small molecule synthesis, polymerizations are often complicated by their own distinct issues of high viscosities, low solubilities, and slow molecular diffusion (cf. Sections 2.3.1.1 and 2.3.1.2). Problems emanating from these issues are often amplified, and additional problems introduced, when polymerizations are adapted to aqueous conditions. In spite of these added complications, the use of aqueous media in polymer synthesis is highly desirable. In addition to the more obvious environmental factors, there are a number of kinetic advantages that can be derived by running organic polymerizations under aqueous emulsion conditions.

Aqueous polymerizations can be run homogeneously, heterogeneously, or under emulsion conditions [1]. The standard aqueous-emulsion polymerization system is composed of a water-insoluble (or partially soluble) monomer, an emulsifier, and a water-soluble initiator. One example of a commercial emulsion system is the formation of a styrene–butadiene latex rubber, using sodium dodecylsulfate as a surfactant for the two monomers, with a hydroperoxide–ferrous ion redox system as the initiator [2]. The advantages associated with aqueous-emulsion polymerizations are many and include ease of processing, improved heat transfer [3], reduced viscosities [4], and kinetic advantages that allow for the formation of high-molecular-weight polymer at high polymerization rates [5]. In a standard radical polymerization there is an inverse relationship between the polymerization rate and the polymer's molecular weight (eq. (1)) [6]:

$$v = \frac{k_p^2 [M]^2}{2k_t R_p} \qquad (1)$$

In this equation, v is the kinetic chain length and R_p is the rate of polymerization ($R_p = k_p$ [M]$(fk_d[I]/k_t)^{1/2}$; [M] = monomer concentration; k_p = propagation rate constant; f = frequency factor; k_d = initiator composition constant; k_t = termination rate constant; [I] = initiator concentration). From a practical viewpoint, this inverse relationship makes it difficult to effect large changes in the molecular weight of the polymer. In particular, the difficulty arises in trying to form high-molecular-weight polymer at rapid rates because the high concentration of radicals necessary for fast rates also favors bimolecular termination processes (low-molecular-weight polymer can always be made by the incorporation of chain-transfer agents). The compartmentalization of the reaction within micelles acts to nullify the bimolecular termination by allowing only one propagating chain per micelle.

For obvious reasons, cationic and anionic polymerizations, as well as any other technique that propagates through water-sensitive intermediates, are not applicable to emulsion conditions. Currently, only radical emulsion processes are used commercially.

3.3.10.1.2 Step-Growth Polymerizations

Polymerizations in which chain growth occurs in a stepwise manner are called step-growth polymerizations, or condensation polymerizations [1]. The "step" terminology is used because it accurately reflects the kinetics of the process wherein high-molecular-weight polymers are assembled in a stepwise fashion as monomers react to form dimers. In turn, dimers form tetramers, tetramers become octomers, and so on. This orderly growth scheme is naturally complicated by fragments of any and all sizes reacting with one another (monomer with tetramer, for example). Step-polymers are formed by allowing difunctional monomers with complementary functional groups to react with one another. There are two common types of step-growth processes, involving either the reaction between A–A- and B–B-type monomers or the self-condensation of A–B monomers. In both cases, the A functional groups react exclusively with B groups and B groups exclusively with A groups.

This stepwise method of construction of polymer chains has important consequences for both the molecular weights and the molecular weight distributions of the polymers produced. Probability dictates that the most abundant species tend to co-condense; thus, at the reaction's inception, small chains most likely react with other small chains. This tendency persists to very high conversion, and consequently, high-molecular-weight polymers are not produced until very late in the reaction (i.e., past 99 % conversion) when there is finally a greater probability of larger chains reacting with one another. This statistically determined molecular weight profile means that only very high- yielding reactions can be used to form step-growth polymers. This restriction points to an important distinction between small-molecule organic reactions and step-growth polymerizations. Although a reaction that typically yields 85 % of the desired product is considered "good" in organic synthesis, the same reaction is essentially useless for step-growth polymerizations because high-molecular-weight polymers will never be formed at such low conversions. The successful preparation of high-molecular-weight polymer using step-growth polymerizations is dependent on both the extent of reaction and the stoichiometric match of complementary reacting groups. These relationships can be seen from the Carothers equation (eq. (2)) [7]:

$$X_n = \frac{(1 + r)}{(1 + r + 2rp)} \tag{2}$$

where X_n = the degree of polymerization, r = ratio of reacting functional groups N_a/N_b, and p = fraction of the reactive groups converted. It is clear that incidents that alter the stoichiometric ratio of reactive groups or in any way limit the conversion will drastically reduce the resulting molecular weights.

A long-standing goal in polymer synthesis has been the preparation of soluble poly(p-phenylene) derivatives. Of the various approaches attempted, transition metal-mediated coupling reactions have proved to be the most promising. Unfortunately, most of these coupling approaches are incompatible with aqueous conditions [8]. One method, however, the Suzuki coupling of aryl halides and aryl boronic acids, is carried out in aqueous emulsions using a Pd^0 catalyst (eq. (3)) (cf. Section 2.11) [9].

$$\text{⟨⟩—Br} + \text{⟨⟩—B(OH)}_2 \xrightarrow[\text{C}_6\text{H}_6 \text{ / H}_2\text{O, Ba(OH)}_2]{\text{1 - 5 mol% Pd(PPh}_3)_4} \text{⟨⟩—⟨⟩} \qquad (3)$$

$$> 98\,\%$$

This reaction is significant in that it is one of the few examples of a metal-mediated, carbon–carbon bond-forming reaction that proceeds in the presence of water. This coupling method has been used in the synthesis of a variety of substituted poly(p-phenylenes) which are soluble in organic solvents by virtue of long-chain alkyl substituents [10]. Wallow and Novak [11] showed that this Pd^0-catalyzed cross-coupling can be carried out homogeneously in water [12] by making use of water-soluble phosphine ligands [13]. Specifically, a water-soluble poly(p-phenylene) derivative (Structure **1**) was synthesized by an aqueous adaptation of the Suzuki cross-coupling reaction (eq. (4)).

$$\text{Br—⟨COOH⟩—Br} + \text{(HO)}_2\text{B—⟨⟩—⟨⟩—B(OH)}_2 \xrightarrow[\text{H}_2\text{O}]{\text{PdL}_n} \left[\text{⟨COOH⟩—⟨⟩—⟨⟩}\right]_n$$

L ≡ **TPPMS (cf. Section 3.1.1.1)**

$$\qquad (4)$$

Although side reactions associated with the supporting phospine ligands do not greatly affect small molecule couplings, they can affect polymerizations dramatically. The effect of one of these side reactions will be discussed. A general catalytic cycle for the Suzuki coupling process is shown in Scheme 1.

The coupling pathway in this scheme is complicated by a facile exchange of the aryl groups between the Pd center of Structure **3** and a bound triphenylphosphine ligand [14]. In small-molecule chemistry, this exchange process yields small amounts of unsubstituted biphenyl, but in polymer synthesis, exchange introduces the possibility of incorporating flexible phosphines within the polymer backbone. Specifically, repeated exchanges on the same phosphine result in the formation of new trifunctional monomers that can act as either branch or crosslinking points (Scheme 2) [15].

In the case of bifunctional haloaromatics, this exchange process introduces reactive functionality onto the phosphine ligands. As these catalytic systems have been fine-tuned to insure complete conversion of these functional groups, the probability of incorporating an exchanged phosphine into the polymer structure becomes high. Incorporating the functionalized phosphine after a single

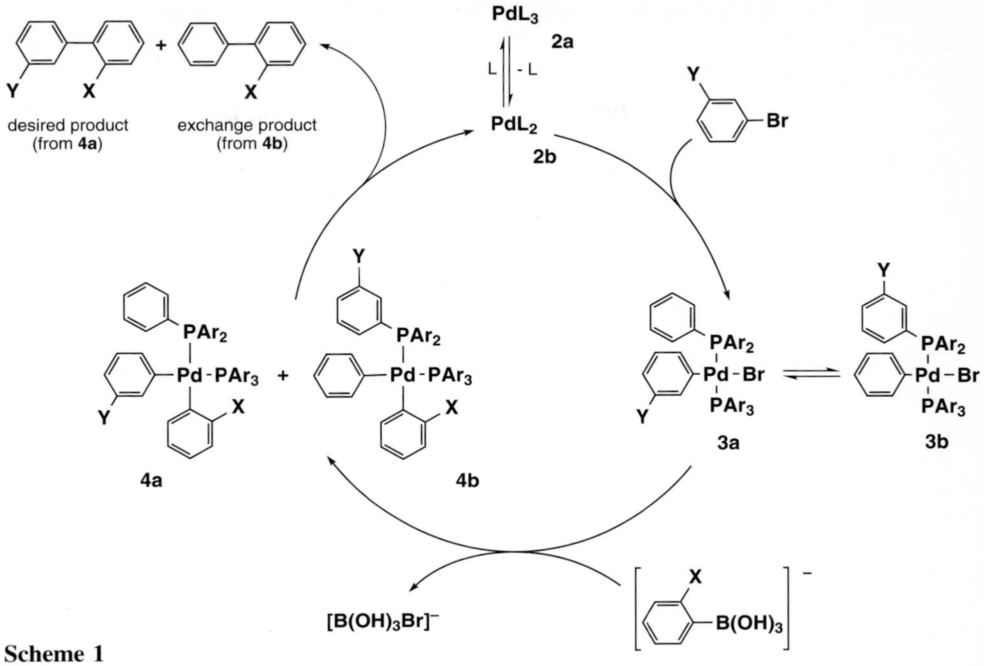

Scheme 1

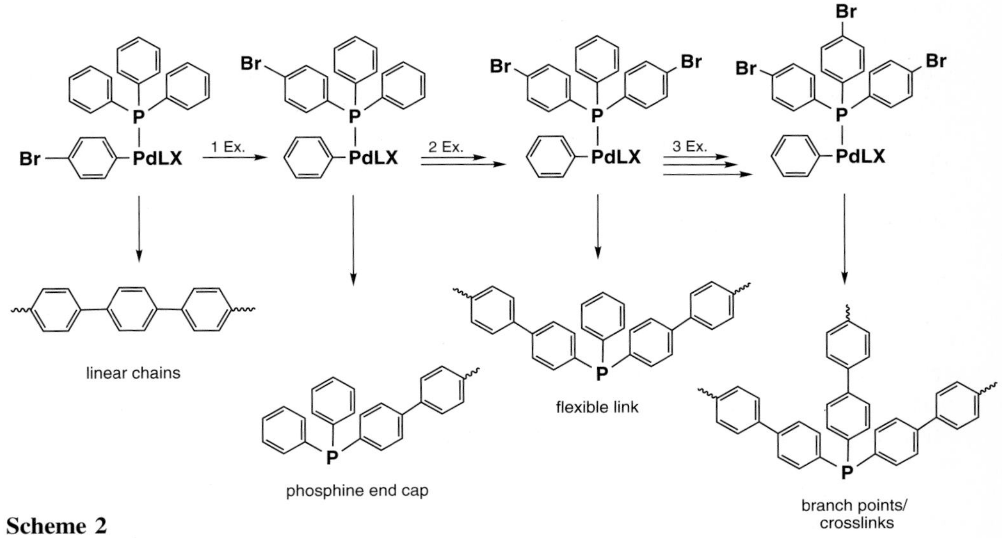

Scheme 2

exchange leads to endcapped chains. Incorporation after a double exchange introduces a flexible link into the polymer chain, and incorporation after a triple exchange leads to branches and/or crosslinks. Hence, polymers synthesized under conditions in which aryl exchange is facile would be expected to be branched, random-coil polymers rather than rigid rods. Phenomenologically, this is observed when comparing the properties of polymers synthesized in the presence or absence (*vide infra*) of phosphine ligands. The sodium and cesium salts of carboxylated poly(phenylenes) synthesized in the presence of phosphines are highly soluble, have very high molecular weights and shape factors consistent with coils rather than rods, and show no evidence of anisotropic behavior in solution. In contrast, sodium salts of carboxylated poly(phenylenes) synthesized in the absence of phosphines are completely insoluble and precipitate during polymerization. The Cs^+ salts of the phosphine-free polymers are sparingly soluble only after prolonged heating. This difference indicates that these materials must be structurally very distinct, although the insolubility is currently acting as an impediment to the full characterization of the polymers from the ligandless systems. Direct evidence for the incorporation of phosphorus into these poly(phenylene) backbones comes from ^{31}P-NMR.

In order to avoid these side reactions, the most obvious modification would be to remove the phosphine ligands altogether [16]. This can be accomplished, although not without penalty, because the reduced palladium centers tend to form colloidal particles that eventually show attenuated activity. The details of this phosphine-free approach have been reported [17]. Convenient sources of Pd^0 include $CpPd(\eta^3-C_3H_5)$ and $Pd_2(dba)_3 \cdot C_6H_5$ (where dba = dibenzylidene acetone). Some of these reactions can be exceedingly fast (99 % yield with 0.02 % $Pd_2(dba)_3 \cdot C_6H_6$ in 45 min).

3.3.10.1.3 Chain-Growth Polymerizations

Chain-growth polymerizations involve sequential addition reactions, either to unsaturated monomers or to monomers possessing other reactive functional groups. Both conceptually and phenomenologically, this approach differs greatly from step-growth processes. As a consequence, chain-growth polymerizations are capable of producing high-molecular-weight polymers relatively early in the reaction, long before all of the monomer is consumed.

The reactive intermediates used in chain-growth polymerizations include radicals, carbanions, carbocations, and organometallic complexes. Of the three common metal catalyzed polymerizations – coordination–insertion, ring-opening metathesis and diene polymerization – the last appears to possess the greatest tolerance toward protic solvents. The polymerization of butadiene in polar solvents was first reported in 1961 using Rh^{3+} salts [18]. It was discovered that these polymerizations could be performed in aqueous solution with an added emulsifier (sodium dodecyl sulfate, for example).

This Rh-catalyzed reaction is selective for the formation of highly crystalline *trans*-1,4-polybutadiene. The activity of the catalyst shows a marked dependence on the nature of the counterions present. Using $RhCl_3$ in 95 % ethanol, no polymer was obtained after 6 h at 80 °C, whereas the nitrate salt displays a polymerization rate of 7 g polymer/g Rh under the same conditions (cf. Section 2.3.2.2).

The following year, the emulsion polymerization of butadiene was reported using a number of transition metal catalysts (Rh^{3+}, Rh^+, Pd^{2+}, Ir^{3+}, Ru^{3+}, and Co^+) in polar solvents [19]. It was found that the microstructure could be varied from all *trans*-1,4 ($> 99.5\%$), to high *cis*-1,4 (88 %), to high 1,2-insertion ($> 98\%$), depending on the metal catalyst employed. Molecular weights varied greatly with choice of catalyst. For example, the Pd^{2+}-catalyzed reactions produce low-molecular-weight oligomers (M_n = 1000–1500), while the Co^+ catalyst produces molecular weights of approximately 300 000 (cf. Section 2.3.2.2).

It was discovered that the addition of 1,3-cyclohexadiene to the Rh^{3+}-catalyzed reactions increased the rate of butadiene polymerization by a factor of over 20 [20]. Considering the reducing properties of 1,3-cyclohexadiene, this effect could be due to the reduction of Rh^{3+} to Rh^+ and stabilization of this low oxidation state by the diene ligands. With neat 1,3-cyclohexadiene, Rh^{3+} is reduced to the metallic state. These emulsion polymerizations are sensitive to the presence of Lewis basic functional groups. A stoichiometric amount of amine (based on Rh) is sufficient to inhibit polymerization completely. It was also discovered that styrene could be polymerized using the Rh^{3+} catalyst. However, the atactic nature of the polymer, along with the kinetic behavior of the reaction, indicated that a free-radical process, rather than a coordination-insertion mechanism, was operative.

There have been suggestions in the literature that the mechanisms of these metal-catalyzed reactions are, in fact, either cationic [21] or free-radical in nature [20]. This assessment, however, is inconsistent with all of the facts. Cationic polymerizations do have a tendency to produce high *trans*-1,4 polymer. For example, using a $TiCl_4/H_2O$ catalyst, polymer containing approximately 75 % *trans*-1,4 units is obtained. However, typical cationic polymerizations are generally carried out at low temperature (–78 °C is common), to reduce the amount of insoluble, crosslinked polymer obtained. In contrast, the Rh^{3+} systems are run between +50 and +80 °C, without appreciable crosslinking occurring. In some polymerizations there does seem to be a competitive free-radical process [22]. This, however, was determined to be due to radical impurities in the surfactants used, and not to a Rh-catalyzed reaction. In fact, Rh^{3+} was found to act as a free-radical *inhibitor* for these reactions. Because of this, the free-radical mechanism was determined to be unimportant at high Rh^{3+} concentrations [23]. In addition, common free-radical inhibitors do not quench the Rh^{3+} polymerizations [19]. All of these facts point strongly to a polymerization mechanism that is substantially different from a classical free-radical or cationic mechanism.

The ring-opening metathesis polymerization (ROMP, cf. Section 2.3.3) of strained-ring cyclic alkenes has attracted considerable attention in recent years due to the discovery that well-characterized metallacyclobutane [24] and metal alkylidene [25] complexes catalyze the living polymerization of monomers such as norbornene. Unfortunately, these catalysts often suffer deactivating side reactions when used with monomers possessing polar functional groups [26]. Significantly greater tolerance to functional groups is observed using catalysts based on later transition metals. For example, 7-oxanorbornene (Structure **5**) derivatives can be polymerized using Group VIII complexes (eq. (5)) [27] that were first reported as ROMP catalysts in the early 1980s [28].

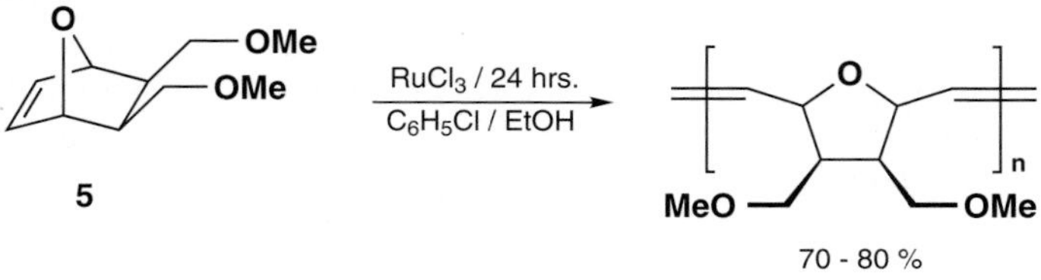

$$\text{(5)}$$

70 - 80 %

The observation that these polymerizations proceeded in alcoholic solution led to the discovery that selected Ru^{3+} and Ru^{2+} complexes catalyze the ROMP of these 7-oxanorbornene derivatives in water alone to provide quantitative yields of the desired ring-opened polymer (eq. (6)) [29].

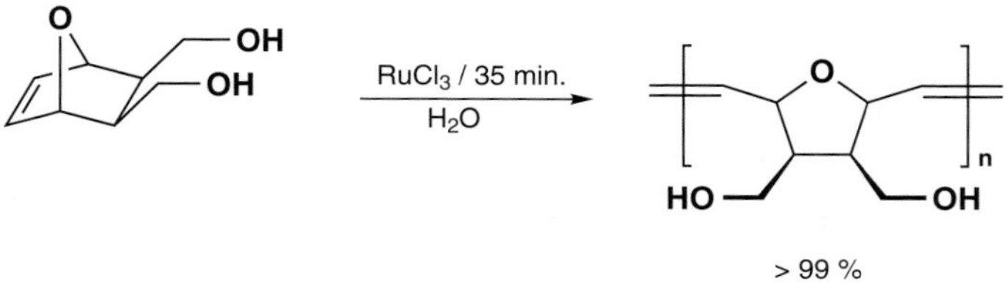

$$\text{(6)}$$

> 99 %

Water appears to co-catalyze this reaction, as evidenced by a decrease in the induction time for the reaction from nearly 24 h in dry organic solvents to 10–15 min in pure water. In addition, conducting the polymerization in protic solution increases the molecular weight by a factor of over 4 (M_w increases from 3.38×10^5 to 1.34×10^6), and decreases the polydispersity from 1.98 to 1.23. The fact that very high-molecular-weight materials form under these aqueous conditions indicates that, if chain-termination or chain-transfer reactions involving the hydrolysis of the carbon–metal bonds are occurring in either the metallacycle or metal alkylidene intermediates, they proceed at a much slower rate (by several orders of magnitude) than the rate of propagation of the polymer. From the average corrected degree of polymerization of the poly-**1** obtained from these aqueous reactions, it can be estimated that approximately 2500–2700 turnovers occur before each termination step.

It was also found that the recycled aqueous catalyst solutions were actually more active than the original solutions. This increase in activity was attributed to the *in situ* formation of ruthenium(II) olefin complexes that were shown, after isolation, to be highly active ROMP catalysts. The mechanism involved in converting these olefin complexes to either metallacyclobutane or metal carbene species remains unknown.

Later, Grubbs and co-workers showed that water-stable ruthenium(II) carbenes can be synthesized by allowing 3,3-diphenylcyclopropene to react with $RuCl_2(PPh_3)_4$ (eq. (7)) [30].

It was found that **6** (a variant of the latter Grubbs catalyst [34]) would catalyze the living polymerization of norbornene in organic solvents. Although not soluble in water, complex **6** will initiate aqueous emulsion polymerizations. To date, all attempts to solubilize **6** in aqueous solution by the incorporation of sulfonated phosphine ligands have failed to yield active catalysts.

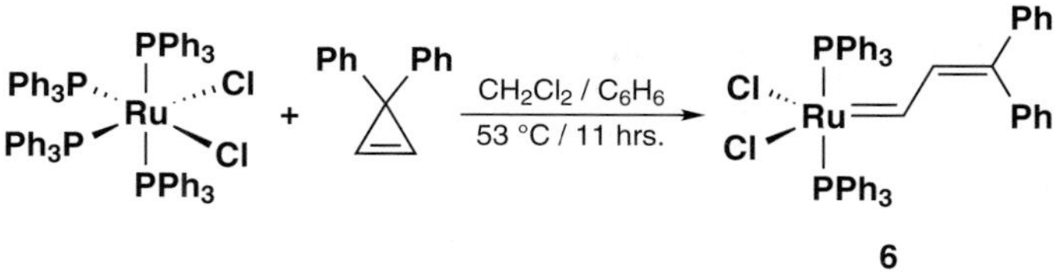

(7)

6

As was shown earlier with the Suzuki coupling reactions, organopalladium intermediates can show good stability to water and other protic sources. This stability has been exploited in the synthesis of polyacetylene under air- and moisture-stable conditions. It was found that simple palladium(II) salts (PdCl$_2$, Pd(CH$_3$CO$_2$)$_2$, etc.) can be used to initiate the 1,2-insertion polymerization of strained cyclic alkenes in water (eq. (8)) [31]. Once formed, poly-**8** can be converted to polyacetylene through a retro-Diels–Alder reaction.

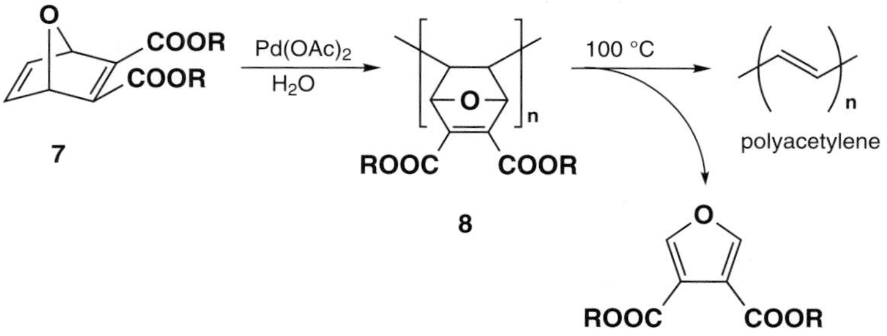

(8)

Using a related reaction, Sen showed that the palladium-catalyzed alternating copolymerization of CO and olefins could be carried out under aqueous conditions by using either the sulfonated chelating phosphine ligand, 1,3-bis(diphenyl-phosphino)propane (dppp), or a sulfonated phenanthroline ligand, Phen-SO$_3$Na (eq. (9)) [32].

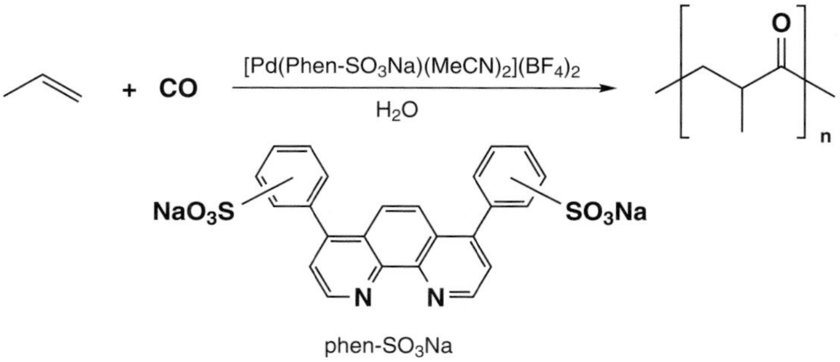

(9)

phen-SO$_3$Na

Finally, there has been at least one report of coordination polymerization of ethylene occurring in aqueous solution. Using $CnRh(CH_3)_n(OTf)_{3-n}$ (where Cn = 1,4,7-trimethyl-1,4,7-triazacyclononane, OTf = triflate or trifluoromethanesulfonate, and n = 1,2), Flood and co-workers reported the very slow oligomerization of ethylene in aqueous solution (90 days, at 24 °C, M_w = 5100) [33]. Although not practical at the present time, this example does help to illustrate the remarkable stability of some of the Group VIII alkyls toward protic solvents.

The stability to water and polar functional groups that characterizes many of the later transition metal organometallics makes this a rich area for new catalyst development. This tolerance extends to a wide range of intermediates and includes metal alkyls, acyls, aryls, and allyls. Coupled with the number of metals that are currently represented, these examples have the potential of acting as the foundation for versatile polymerization catalysts of the future.

References

[1] G. Odian, *Principles of Polymerization*, 2nd ed., Wiley, New York, **1981**.

[2] D. C. Blackley, *Emulsion Polymerization*, Applied Science, London, **1975**.

[3] I. Piirma, J. L. Gardon (Eds.), *Emulsion Polymerizations.* ACS Symp. Series No. 24, American Chemical Society, Washington, DC, **1976**.

[4] D. R. Basset, A. E. Hamielec (Eds.), *Emulsion Polymers and Emulsion Polymerizations.* ACS Symp. Series No. 165, American Chemical Society, Washington, DC, **1981**.

[5] J. L. Gardon, in *Polymerization Processes* (Ed.: C. E. Schildknecht), Wiley-Interscience, New York, **1977**.

[6] Ref. [1] p. 319.

[7] R. B. Seymour, C. E. Carraher, *Polymer Chemistry: An Introduction*, 3rd ed., Marcel Dekker, New York, **1992**, p. 215.

[8] (a) J. K. Stille, *Angew. Chem., Int. Ed. Engl.* **1986**, *25*, 508; (b) I. Colon, D. R. Kelsey, *J. Org. Chem.* **1986**, *51*, 2627.

[9] For a review: A. Suzuki, *Acc. Chem. Res.* **1982**, *15*, 178.

[10] (a) M. Rehahn, A.-D. Schlüter, G. Wegner, W. J. Feast, *Polymer* **1989**, *30*, 1061; (b) M. Rehahn, A.-D. Schlüter, G. Wegner, *Makromol. Chem.* **1990**, *191*, 1991; (c) A.-D. Schlüter, G. Wegner, *Acta Polymer* **1993**, *44*, 59.

[11] T. I. Wallow, B. M. Novak, *J. Am. Chem. Soc.* **1991**, *113*, 7411.

[12] A. L. Casalnuovo, J. C. Calabrese, *J. Am. Chem. Soc.* **1990**, *112*, 4324.

[13] S. Ahrland, J. Chatt, N. R. Davies, A. A. Williams, *J. Am. Chem. Soc.* **1958**, *90*, 276.

[14] K.-C. Kong, C.-H. Cheng, *J. Am. Chem. Soc.* **1991**, *113*, 6313.

[15] (a) T. I. Wallow, T. A. P. Seery, F. E. Goodson, B. M. Novak, *ACS Polymer Prepr.* **1994**, *35*, 710; (b) T. A. P. Seery, T. I. Wallow, B. M. Novak, *ACS Polymer Prepr.* **1993**, *34*, 727.

[16] I. P. Beletskaya, *J. Organomet. Chem.* **1983**, *250*, 551.

[17] T. I. Wallow, B. M. Novak, *J. Org. Chem.* **1994**, *59*, 5034.

[18] R. E. Rinehart, H. P. Smith, H. S. Witt, H. Romeyn, *J. Am. Chem. Soc.* **1961**, *83*, 4864.

[19] A. J. Canale, W. A. Hewett, T. M. Shryne, E. A. Yongman, *Chem. Ind. (London)* **1962**, 1054.

[20] Ph. Teyssie, R. Dauby, *J. Polym. Sci., Polym. Lett.* **1964**, *2*, 413.

[21] D. P. Tate, T. W. Bethea, in *The Encyclopedia of Polymer Science and Technology*, 2nd ed., Wiley-Interscience, New York, **1985**, Vol. 2.

[22] R. E. Rinehart, H. P. Smith, H. S. Witt, H. Romeyn, *J. Am. Chem. Soc.* **1962**, *84*, 4145.

[23] R. Dauby, F. Dawans, Ph. Teyssie, *J. Polyn. Sci. Part C*, **1967**, *16*, 1989.
[24] (a) L. R. Gilliom, R. H. Grubbs, *J. Am. Chem. Soc.* **1986**, *108*, 733; (b) K. C. Wallace, R. R. Schrock, *Macromolecules* **1987**, *20*, 450; (c) K. C. Wallace, A. H. Liu, J. C. Dewan, R. R. Schrock, *J. Am. Chem. Soc.* **1988**, *110*, 4964; (d) L. F. Cannizzo, R. H. Grubbs, *Macromolecules* **1987**, *20*, 1488; (e) L. F. Cannizzo, R. H. Grubbs, *Macromolecules* **1988**, *21*, 1961; (f) W. Risse, R. H. Grubbs, *Macromolecules* **1989**, *22*, 1558.
[25] (a) R. R. Schrock, J. Feldman, L. F. Cannizzo, R. H. Grubbs, *Macromolecules* **1987**, *20*, 1169; (b) R. R. Schrock, R. T. DePue, J. Feldman, C. J. Schaverien, J. C. Dewan, A. H. Liu, *J. Am. Chem. Soc.* **1988**, *110*, 1423; (c) K. Knoll, S. A. Krouse, R. R. Schrock, *J. Am. Chem. Soc.* **1988**, *110*, 4424; (d) R. R. Schrock, *Acc. Chem. Res.* **1990**, *23*, 158.
[26] For references on the metathesis of polar substrates (both cyclic and acyclic) see: (a) J. S. Murdzek, R. R. Schrock, *Macromolecules* **1987**, *20*, 2640; (b) P. D. van Dam, M. C. Mittelmijer, C. Boelhouwer, *J. Chem. Soc., Chem. Commun.* **1972**, 1221; (c) J. C. Mol, *J. Mol. Catal.* **1982**, *15*, 35; (d) S. Matsumoto, K. Komatsu, K. Igarashi, *ACS Polym. Prepr.* **1977**, *18*, 110. (e) S. Matsumoto, K. Komatsu, K. Igarashi, *ACS Polym. Prepr.* **1977**, *18*, 110.
[27] (a) B. M. Novak, R. H. Grubbs, *Proc. Am. Chem. Soc. Div. PMSE* **1987**, *57*, 651; (b) B. M. Novak, R. H. Grubbs, *J. Am. Chem. Soc.* **1988**, *110*, 960.
[28] The list of active Group VIII complexes include a number of Ir^{3+}, Ir^+, Os^{3+}, Ru^{3+}, and Ru^{2+} compounds. See: (a) F. W. Michelotti, W. P. Keaveney, *ACS Polym. Prepr.* **1963**, *4*, 293; (b) F. W. Michelotti, W. P. Keaveney, *J. Polym. Sci.* **1965**, *A-3*, 895; (c) F. W. Michelotti, J. H. Carter, *ACS Polym. Prepr.* **1965**, *6*, 224; (d) H. T. Ho, K. J. Ivin, J. J. Rooney, *J. Mol. Catal.* **1982**, *15*, 245; (e) L. Porri, P. Diversi, A. Lucherini, R. Rossi, *Makromol. Chem.* **1975**, *176*, 3131; (f) L. Porri, R. Rossi, P. Diversi, A. Lucherini, *Markomol. Chem.* **1974**, *175*, 3097.
[29] B. M. Novak, R. H. Grubbs, *J. Am. Chem. Soc.* **1988**, *110*, 7542.
[30] S. T. Nguyen, L. K. Johnson, R. H. Grubbs, *J. Am. Chem. Soc.* **1992**, *114*, 3974.
[31] A. L. Safir, B. M. Novak, *Macromolecules* **1993**, *26*, 4072.
[32] Z. Jiang, A. Sen, *Macromolecules* **1994**, *27*, 7215.
[33] L. Wang, R. S. Lu, R. Bau, T. C. Flood, *J. Am. Chem. Soc.* **1993**, *115*, 6999.
[34] B. Cornils, W. A. Herrmann, R. Schlögl, C.-H. Wong, *Catalysis from A to Z*, Wiley-VCH, Weinheim, **2002**.

3.3.10.2 Homogeneous Catalysis in Living Cells

László Vígh, Ferenc Joó

3.3.10.2.1 Fundamentals

At present, the sole purposeful modification in living cells known to be catalyzed by organometallic complexes is the homogeneous hydrogenation of the polar lipid constituents of cell membranes [1]. The aim of such a modification is the controlled change (modulation) of membrane fluidity which, in turn, is reflected in the functioning of membrane-bound proteins. The latter comprise a wide range of proteins such as – among others – various enzymes, constituents of the light-harvesting complexes in algae and plants, receptors of very highly diverse agents, transport proteins, etc. In such a way, modulation of membrane fluidity triggers a whole array of response reactions in a living cell. Selective *homogeneous* (or heterogeneous) hydrogenation of the unsaturated fatty acyl moieties esterified in polar lipids of the various cell membranes provides a way of confining such modifications to selected regions of the cell (e. g., the plasma membrane) or into selected group of constituents (e. g., a given group of lipid classes). Consequently, the concomitant changes in cell metabolism or in other characteristics (e. g., stress tolerance, viability, etc.) can be directly related to the primary effect of hydrogenation on the fatty acid composition of the membranes.

When discussing synthetic manipulations of living cells it should be always borne in mind that such *living* creatures always strive to maintain whatever chemical composition or physical state is sensed as "optimal" for the cells under the given physiological conditions (temperature, pH, osmotic pressure, etc.). Therefore, the consequences of catalytic hydrogenation should be examined immediately after the cell modification; even then some compensatory changes might have already occurred.

Although the first attempts at biomembrane hydrogenations [2] involved the use of water-insoluble catalysts, such as $RhCl(PPh_3)_3$, modification of living cells was made possible only by the introduction of *water-soluble* hydrogenation catalysts [3, 4]. Complexes of monosulfonated triphenylphosphine (TPPMS, cf. Section 3.1.1.1), such as $RhCl(TPPMS)_3$ and $RuCl_2(TPPMS)_2$, gained limited use [5] and by far the most widely used catalyst is the Pd^{II} complex of alizarin red, $Pd(QS)_2$ (see Structure **1**, [6]). The mechanism of hydrogenations with this catalyst is complex and far from being fully understood. Most probably it involves a heterolytic activation of H_2 (extensive deuteration of substrates occurs in D_2O solutions under H_2 [7]) and participation of the anthraquinone- type ligand in electron transfer to the substrate through formation of semiquinones (as suggested by characteristic ESR spectra of the intermediates [6]).

1

The fatty acid composition of lipids is usually analyzed by gas chromatography following transesterification into methyl esters. Unmodified lipids can be analyzed by HPLC or by soft chemical ionization mass spectrometry. In the course of sample preparation it is often necessary to separate the various membrane fractions (plasma membrane, thylakoid, microsomal, mitochondrial, etc.) by sophisticated gradient centrifugations, as well as the individual lipid classes within a membrane fraction, usually by thin-layer chromatography (TLC).

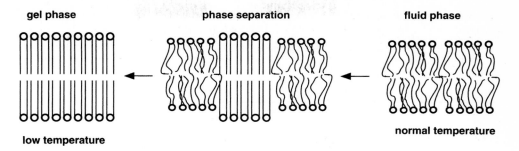

gel phase

phase separation

fluid phase

low temperature

normal temperature

Figure 1. Phase behavior and molecular geometry of membrane phospholipids at normal (physiological) or low temperature.

Membrane fluidity is determined by following anisotropic rotation of fluorescent or spin probes. Liquid-crystalline (or fluid) to gel thermotropic phase transition of lipids (Figure 1) (cf. Section 3.1.11)in liposomes or intact biomembranes can be followed by Fourier transform infrared (FTIR) spectroscopy or differential scanning calorimetry (DSC).

Cells do survive limited hydrogenation [8] and their viability is checked in each and every case when whole-cell characteristics are investigated. The effects of hydrogenation on the physiology or molecular biology of the cells is characterized by the most diverse techniques available, ranging from the very simple (determination of photosynthetic activity by measuring O_2 production with a Clark-type oxygen electrode) to the rather sophisticated (Northern blot hybridization analysis of gene expression).

Fatty acids are usually designated by giving their carbon number:number of usaturated bonds, e. g., 16:0 is palmitic acid, 18:1 oleic acid, etc.

3.3.10.2.2 Compensatory Response of Cells

Protoplasts prepared from tobacco (*Nicotiana plumbaginifolia*) leaf were hydrogenated using $RuCl_2(TPPMS)_2$ as catalyst [9]. Surprisingly, in the first 30 min of the rather slow reaction the general unsaturation of lipids *increased* (Table 1) and this was reflected in fluidization of the membranes as shown by ESR measurements. By pulse labeling the protoplasts with [1-^{14}C] oleate, precursor of *de novo* fatty acid synthesis, it could be unambiguously demonstrated that hydrogenation of the original unsaturated lipid pool of the protoplast membranes triggered an immediate and fast synthesis of fatty acids replacing the hydrogenated ones on the glycerin backbone. Radioactivity was highest in the 18:3 (linolenate) content of phosphatidylcholine, showing that *de novo* fatty acid synthesis was accompanied by vigorous desaturase action. Limited resources of the cells (lack of the O_2 required by desaturase enzymes) resulted in gradual loss of unsaturated acids upon prolonged hydrogenation (Table 1, data for 60 min).

Table 1. Changes in the fatty acid composition of living protoplasts from *N. plumbaginifolia* during hydrogenation with $RuCl_2(TPPMS)_2$ (adapted from [9]).[a]

Fatty acid	Fatty acid composition [wt. %][b]				
	0 min	30 min		60 min	
	C	C	H	C	H
16:0	15.6	15.8	16.0	16.1	20.5
16:1	0.7	0.7	tr	0.5	tr
18:0	3.2	3.1	3.4	3.2	12.2
18:1	13.0	13.1	10.1	13.5	21.5
18:2	11.9	11.7	6.8	12.8	15.0
18:3	55.6	55.3	63.8	53.9	30.8

[a] Conditions: 2×10^5 protoplasts/mL, 0.5 mg/mL catalyst, 0.7 MPa H_2, 30 °C, in 20 mL of W–6 medium.
[b] C = control; H = hydrogenated; tr = trace.

This example strikingly demonstrates the activity of living systems and their ability to maintain the optimal state of their membranes, and is a clear example of so-called homeoviscous adaptation. A definite advantage of homogeneous hydrogenation as compared with other methods is that isothermal conditions could be used.

3.3.10.2.3 Temperature Perception and Cold Acclimatization of Plants and Algae

Frost resistance or sensitivity of culture plants is a problem of highest economic importance. Furthermore, low (or high) temperature is only one of the several types of environmental shock (draught, high or low salt concentration, etc.) and there are indications that plants use a fairly common strategy to cope with the different kinds of such stress. It has long been debated whether temperature acclimatization is achieved through the intrinsic temperature dependence of the activity of certain enzymes or through specific modification of the lipid composition of the membranes, and consequently their physical state, which in turn leads to altered genetic and enzymic activity. This complex question could be addressed by selective hydrogenation of blue–green algae, well characterized photosynthetic organisms which are often regarded as simple models of higher plants.

Anacystis nidulans has a rather simple membrane structure (mainly plasma membrane and thylakoid); furthermore, these membranes have fairly simple lipid compositions. The latter changes distinctly with the growth temperature, the lipids being more unsaturated (more fluid) when the algae are grown at lower temperatures (Table 2, columns 1 and 3). By careful hydrogenation of

Table 2. Fatty acid composition of total lipids of *A. nidulans* grown at 28 °C (1), grown and hydrogenated at 28 °C (2) and grown at 38 °C (3) (adapted from [10]).[a]

Fatty acid	Fatty acid composition [mol %]		
	1	2	3
14:0	2.1	3.7	1.5
14:1	4.3	2.5	1.1
16:0	44.9	50.5	48.5
16:1	43.5	38.2	37.8
18:0	1.8	2.2	4.8
18:1	3.4	2.5	6.3

[a] Conditions of hydrogenation: 0.1 mM catalyst, 0.7 MPa H_2, 28 °C, in 20 mL Kratz and Myers medium C, cell density 10 mg chlorophyll/L.

the cells acclimatized at low temperature (28 °C) [10], their lipids could be saturated to the point when the overall composition resembled very closely that of the membranes of cells grown at 38 °C (Table 2, column 2).

When such algae cells are gradually cooled down from their growth temperature, the fluidity of their membranes decreases and – depending on the lipid composition – finally a gel-like state is reached. In parallel with these phase transitions, the photosynthetic oxygen evolution activity drops considerably, the greatest change being observed around a well-defined transition temperature (Figure 2).

It is seen from Figure 2 that, independently of their actual growth temperature, the cells with similar lipid composition behaved very similarly, the transition temperatures being 15 °C and 12 °C, as opposed to 4 °C in case of nonhydrogenated

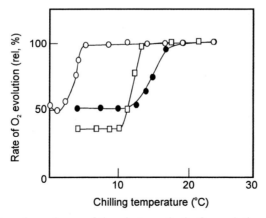

Figure 2. Temperature dependence of the photosynthetic O_2 evolution in *Anacystis nidulans* cells grown at 28 °C (○), grown and hydrogenated [Pd(QS)$_2$ catalyst] at 28 °C (●) and grown at 38 °C (□). (Adapted from [1]).

algae adapted at 28 °C. In further studies [11] it could also be demonstrated that selective hydrogenation of plasma membrane in *A. nidulans* resulted in the same changes of photosynthetic activity despite the fact that the photosynthetic apparatus is located exclusively in the thylakoid (inner) membrane, which remained untouched during short-term hydrogenations. In a broader context it can be concluded that changes in the plasma membrane fluidity are the primary signals of temperature (stress) for the cells.

The latter point was unambiguously proven with an other species of blue–green algae, *Synechocystis PCC6803*. It was observed [12] that lowering the temperature resulted in increased production of mRNA on the *desA* desaturase gene (Figure 3 A). Mild isothermal hydrogenation [13] of permeaplasts of *Synechocystis PCC6803* at the growth temperature of the algae led to the same level of

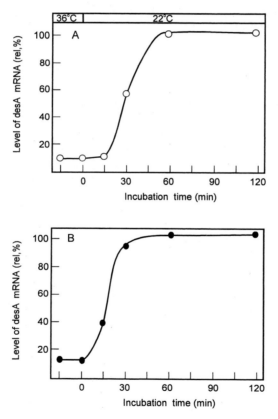

Figure 3. Changes in the *desA* transcript level in *Synechocystis* permeaplasts upon a shift of temperature from 36 °C to 22 °C (A) and upon 4 min hydrogenation at 36 °C followed by incubation at 36 °C (B). In both cases the transcript level is expressed in units relative to the level determined at 60 min incubation time. (Adapted from [12] and [13]).

gene transcript (Figure 3 B). All this evidence allows the conclusion that the primary signal for biological perception of temperature in algae and plants is the change of fluidity of cell plasma membranes [14].

3.3.10.2.4 Other Uses of Organometallic Catalysis in Living Cells

The usefulness of hydrogenating living cells has been demonstrated here by describing rather simple examples. However, the scope of such catalytic manipulations is much wider, including investigations on surface expression of antigens (receptors) in tumor cells [15, 16]. Other, related reactions are also practiced (deuteration [17] and isomerization [18] of lipid fatty acids) or considered (selective catalytic oxidation). Heterogeneous hydrogenation catalysts find their use in this field, too [19, 20].

References

[1] P. J. Quinn, F. Joó, L. Vígh, *Prog. Biophys. Molec. Biol.* **1989**, *53*, 71.

[2] D. Chapman, P. J. Quinn, *Chem. Phys. Lipids* **1976**, *17*, 363.

[3] P. A. Chaloner, M. A. Esteruelas, F. Joó, L. A. Oro, *Homogeneous Hydrogenation (Catalysis by Metal Complexes)*, Kluwer, Dordrecht, **1994**, pp. 183–233.

[4] W. A. Herrmann, C. W. Kohlpaintner, *Angew. Chem.* **1993**, *105*, 1588; *Angew. Chem., Int. Ed. Engl.* **1993**, *32*, 1524.

[5] L. Vígh, F. Joó, P. R. van Hasselt, P. J. C. Kuiper, *J. Mol. Catal.* **1983**, *22*, 15.

[6] F. Joó, N. Balogh, L. I. Horváth, G. Filep, I. Horváth, L. Vígh, *Anal. Biochem.* **1991**, *194*, 34.

[7] F. Joó, L. Vígh, in *Handbook of Nonmedical Applications of Liposomes*, Vol. III, (Eds.: Y. Barenholz, D. Lasic), CRC Press, Orlando, FL, USA, **1995**, pp. 257–271.

[8] L. Vígh, I. Horváth, G. A. Thompson, Jr., *Biochim. Biophys. Acta* **1988**, *937*, 42.

[9] L. Vígh, F. Joó, A. Cséplõ, *Eur. J. Biochem.* **1985**, *146*, 241.

[10] L. Vígh, F. Joó, *FEBS Lett.* **1983**, *162*, 423.

[11] L. Vígh, Z. Gombos, F. Joó, *FEBS Lett.* **1985**, *191*, 200.

[12] D. Los, I. Horváth, L. Vígh, N. Murata, *FEBS Lett.* **1993**, *318*, 57.

[13] L. Vígh, D. A. Los, I. Horváth, N. Murata, *Proc. Natl. Acad. Sci. USA* **1993**, *90*, 9090.

[14] B. Maresca, A. R. Cossins, *Nature (London)* **1993**, *365*, 606.

[15] S. Benkö, H. Hilkmann, L. Vígh, W. J. van Blitterswijk, *Biochim. Biophys. Acta* **1987**, *896*, 129.

[16] E. Duda, S. Benkö, I. Horváth, E. Galiba, T. Páli, F. Joó, L. Vígh, in *Advances in Psychoneuroimmunology* (Eds.: I. Berczi, J. Szélenyi), Plenum Press, New York, **1994**, pp. 181–190.

[17] Z. Török, B. Szalontai, F. Joó, C. Wistrom, L. Vígh, *Biochem. Biophys. Res. Commun.* **1993**, *192*, 518.

[18] Y. Pak, F. Joó, L. Vígh, Á. Kathó, G. A. Thompson, Jr., *Biochim. Biophys. Acta* **1990**, *1023*, 230.

[19] F. Joó, S. Benkö, I. Horváth, Z. Török, L. Nádaski, L. Vígh, *React. Kinet. Catal. Lett.* **1992**, *48*, 619.

[20] F. Joó, F. Chevy, O. Colard, C. Wolf, *Biochim. Biophys. Acta* **1993**, *1149*, 231.

3.3.11   Cyclic Hydrocarbons from Diazoalkanes

Wolfgang A. Herrmann, Horst Schneider

3.3.11.1   Introduction

Diazoalkanes **1** form a versatile class of functionalized organic compounds [1]. Their undisputed significance in organic synthesis is manifested in a number of organometallic and other metal-induced reactions [2], some of which have entered catalytic applications. Cyclopropanation is one of them (cf. Section 3.1.7) but *intramolecular carbon–hydrogen insertion* appears of much potential in synthesis, too. This type of reaction relates to the easily available, normally nonexplosive *α*-diazocarbonyl compounds (*α*-diazoketones, Structure **2**).

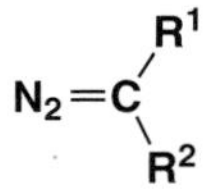

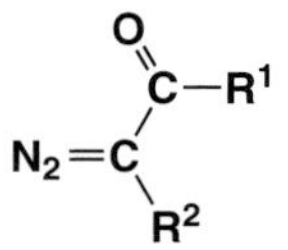

3.3.11.2   Scope and Definition

The remote functionalization of carbon–hydrogen bonds by *α*-diazoketones according to the general eq. (1) is efficiently catalyzed by rhodium(II) complexes and yields cyclopentanones, lactams, and lactones, depending on the substituent Y [3, 4]. Typical reaction conditions are boiling methylene chloride or boiling benzene.

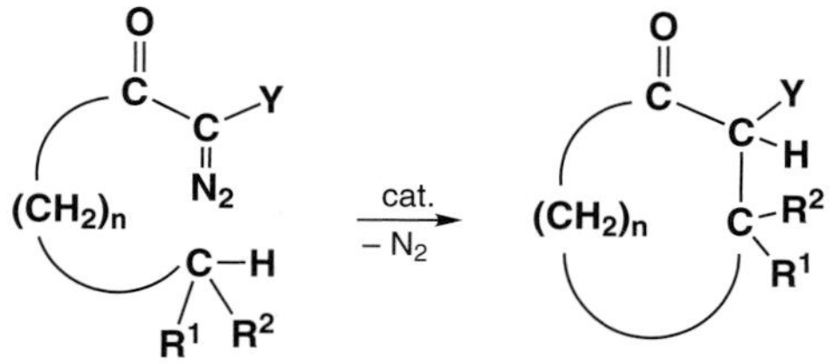

$$(1)$$

Rhodium(II) acetate (Structure **3**), a dinuclear molecule of D_4 symmetry and vacant coordination sites ($\rightarrow$) at each metal atom, is the most commonly employed catalyst for this reaction. Copper catalysts are no longer used because they are inferior in terms of both activity and selectivity. The diazoalkane constitution in eq. (1) includes compounds with

$$Y = H, COCH_3, COOR$$

In addition, an amide group can be attached to the carbonyl function. A possible side reaction is carbene dimerization [1], which can be suppressed by slowly adding the diazoalkane to the catalyst at a temperature appropriate for smooth N_2 elimination. The precise conditions depend on the nature of the *a*-diazoketone and vary broadly in terms of stability and reactivity [1, 2].

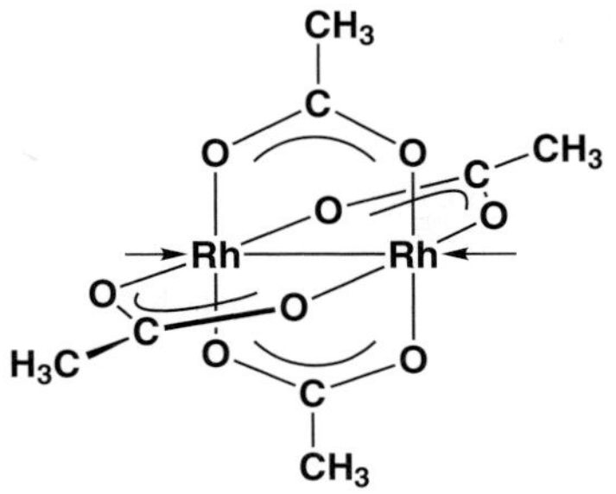

3

3.3.11.3 Mechanistic Considerations

It is commonly accepted that rhodium–carbene intermediates are the active species preceding the C–C bond-forming insertion step (cf. Scheme 1 with M = Rh). The *in situ* generation of carbenes is in line with the characteristics of diazoalkane reactivity [1, 2]. However, neither has the "carbenoid" primary adduct **B** been observed nor is there any spectroscopic evidence of the metal–carbene species **C**. It is likely that the electrophilic addition of the "active catalyst" **A** (e. g., sol-

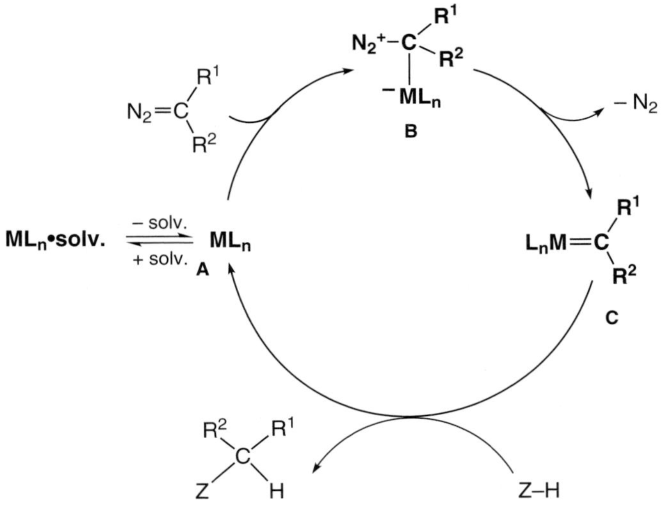

Scheme 1

vent-free rhodium(II) acetate, **3**) is the rate-determining step; N_2 elimination takes place around -20 °C in case of ethyl diazoacetate.

It remains unclear whether the catalyst retains the quadruply bridged structure throughout the catalytic cycle. If this is the case, the insertion step of the carbene CR_2 into the Z–H hydrocarbon would proceed in a sterically rather congested environment (basically a square-planar RhO_4 unit!). Since C–H insertion is an electrophilic process, the metal seems to stabilize the carbene in the carbocation form.

3.3.11.4   Catalytic Cyclization

The literature on catalytic cyclization of α-diazoketones has a rather recent history, with the majority of papers originating from the 1980s. The copper catalysts originally used (e. g., $CuSO_4$) suffer from an unspecific product spectrum [5] and have largely been replaced by rhodium catalysts, mainly through the work of Doyle and colleagues [3].

3.3.11.4.1 Cyclopentanones

A broad spectrum of α-diazo-β-ketoesters (e. g., **4**), -sulfones, and -phosphonates (e. g., **6**) have been converted in one-step procedures and in decent yields into cyclopentanones such as **5** and **7**, respectively (eqs. (2) and (3)).

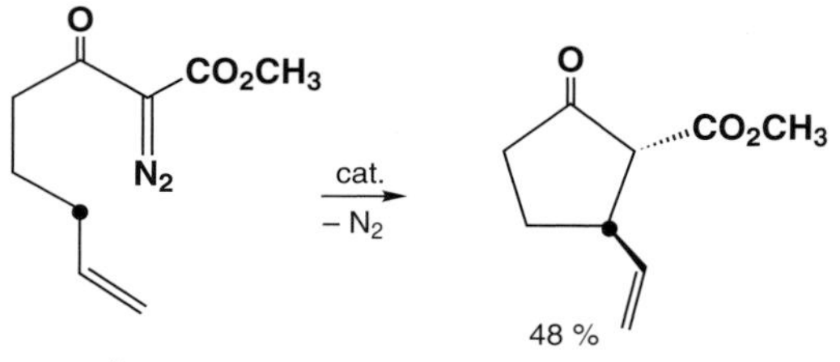

$$(2)$$

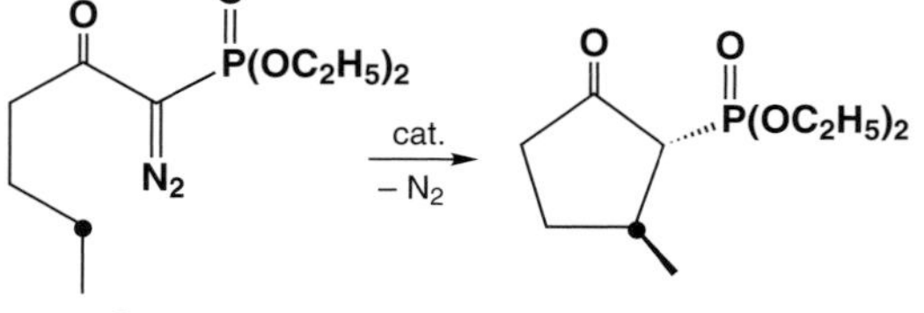

$$(3)$$

Outstanding regioselectivities have been reported when rhodium(II) acetate was employed as a catalyst [6–11]. The reactivity of the hydrocarbon component decreases in the order tertiary > secondary > primary C–H [8]. While the α,δ-insertion yields the preferred cyclopentanones, the α,γ-mode has occasionally been observed, too: both the four-membered spirocycle (Structure **10**) and the bicyclic (five-membered) product **9** result in a 2:3 ratio from the α-diazoketone **8** at 83 % conversion (eq. (4)) [12].

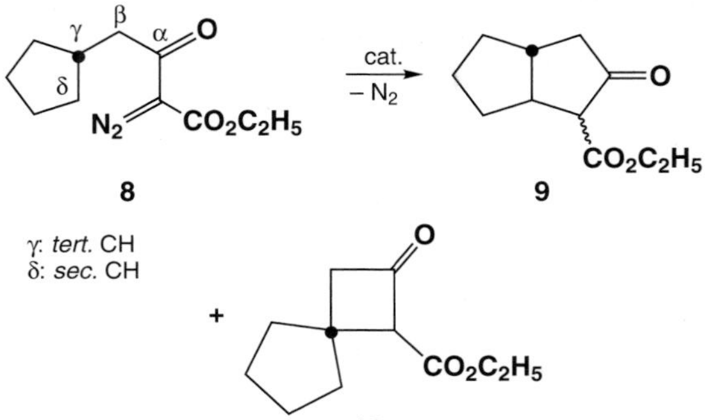

There is supporting evidence from numerous other examples that the regioselectivity is not simply explainable from electronic factors; the (unknown) geometries and energies of the transition states seem to govern the final result in a particularly subtle way [1].

3.3.11.4.2 Lactams

α-Diazoacetamides undergo cyclization to β-lactams. Rhodium(II) acetate is once again much more efficient than copper catalysts. For example, the β-lactam **12** is obtained in 75 % yield (Rh) vs. 25 % yield (Cu) from the α-ketodiazoacetamide **11** according to eq. (5) [13].

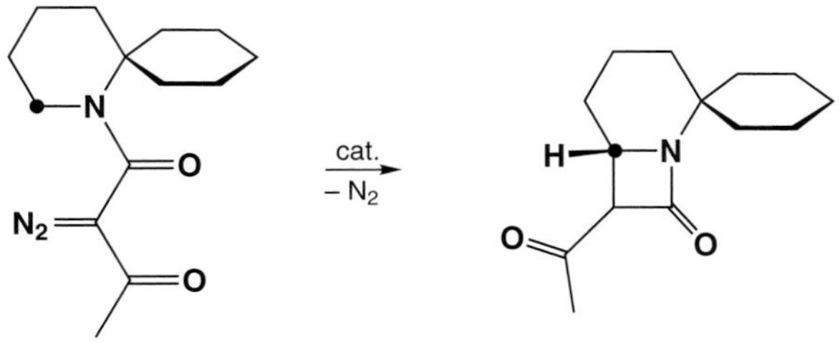

β-Lactam formation (eqs. (5)–(7)) can result in either *cis* or *trans* configuration; the stereochemistry is not yet easy to predict but seems to depend on the type of bridging ligand on the $Rh_2(O_2CR)_4$ catalysts. For example, the diazoacetamide **13** gives exclusively the *trans* isomer **14** in 96 % isolated yield if R = CH$_3$, while structure **15** gives in 89 % yield the *cis* isomer **16** if R = CF$_3$(CF$_2$)$_2$CF$_2$ (cf. eqs. (6) and (7)) [14]. Recent literature lends support to the generality of this lactam synthesis [15].

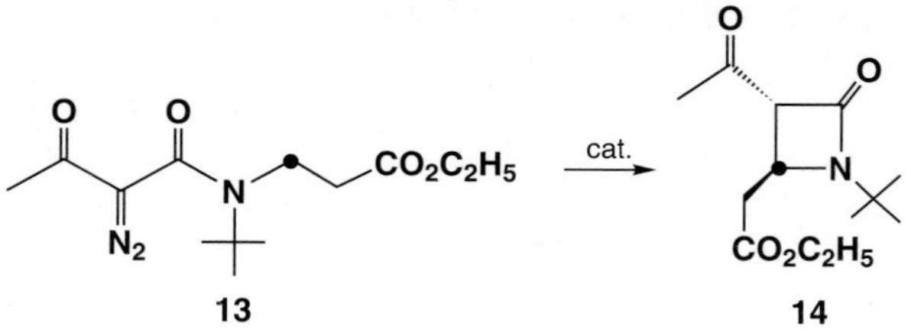

(6)

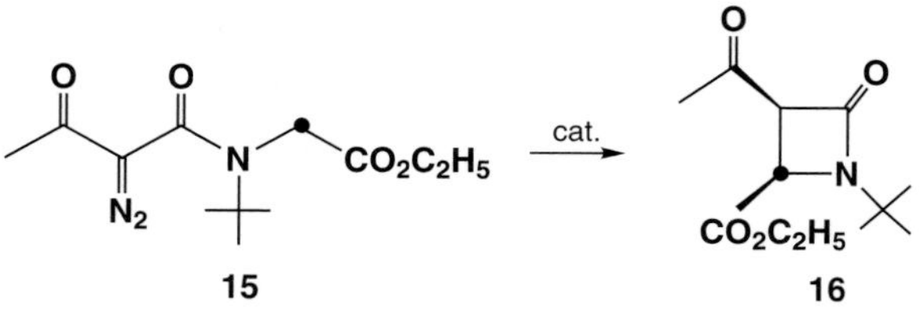

13   **14**

(7)

15   **16**

3.3.11.4.3 Lactones

The first case of an *intra*molecular C–H carbenoid insertion was reported by Cane and Thomas in 1984 [12], with the special diazoacetate **17** forming the spirocyclic δ-lactone **18** in 45 % yield according to eq. (8). Doyle et al. recognized that this is a general methodology for the synthesis of γ-butyrolactones [16]. The reactivity of the C–H bond toward carbene insertion is increased in the vicinity of an ether functionality. Thus, the 3(2*H*)-furanone **20**, as a useful building block in the total synthesis of (+)-muscarine, results in 40 % yield from the diazo precursor compound **19** [17].

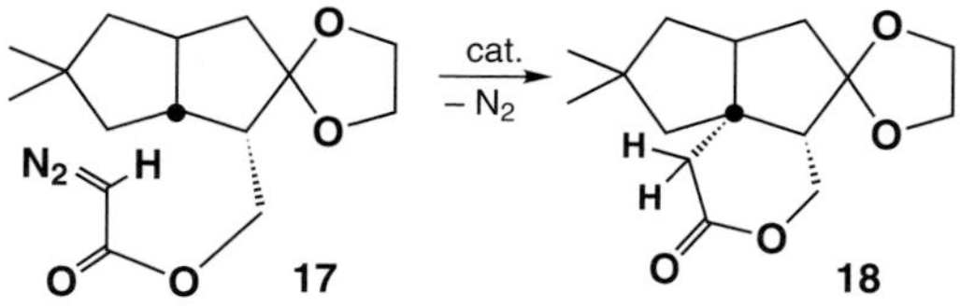

(8)

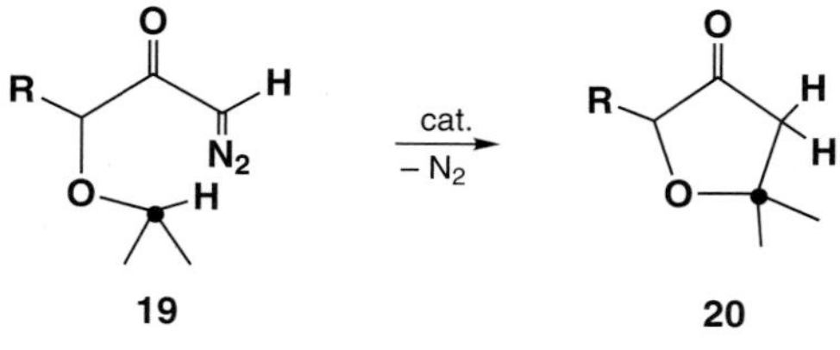

17   **18**

(9)

19   **20**

R = isobutyl

3.3.11.5 Enantioselective Cyclization

Chiral catalysts with structures related to rhodium(II) acetate should principally afford optically pure enantiomeric γ-lactones from diazoacetates of type **21**. As a matter of fact, Doyle et al. have obtained alkoxy-substituted γ-lactones **22** in 85–90 % *ee* (eq. (10)) upon using a Rh_2X_4-catalyst derived from chiral 2-pyrrolidinones [18]. Related results suggest that the catalyst has a rigid stereochemistry throughout the catalytic cycle [19], which conclusion had already been drawn for enantioselective cyclopropanation [20] (cf. Section 3.1.7).

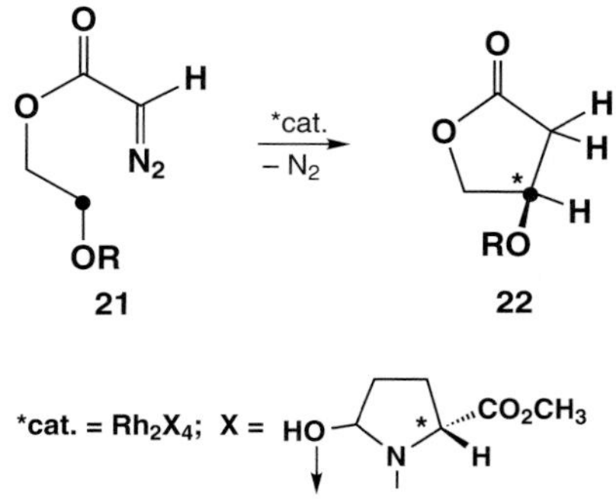

$$\text{(10)}$$

Related results suggest that the catalyst has a rigid stereochemistry throughout the catalytic cycle [19, 20], a conclusion which had already been drawn for enatioselective cyclopropanation [21] (cf. Section 3.1.7). In some cases even β-lactones could be obtained as major products when using this catalyst [22]. In general, acyclic diazoacetates give higher yields of β-lactones than cyclic ones [23].

3.3.11.6 Perspectives

The intramolecular cyclization according to eq. (1) has great potential in the synthesis of four-, five-, and six-membered carbo- and heterocycles. The mechanistic knowledge of this reaction is still rudimental, however, and for this reason even crude rules of how to direct regio- and stereoselectivity are lacking. We suggest the catalyst structure to be modified beyond the bridging ligands. The most significant progress is expected from chiral catalysts; enantioselective formation of carbo- and heterocyclic compounds should soon enter the methodological arsenal of natural product synthesis, especially since the required diazo precursor compounds are normally easy to synthesize by standard techniques [1]. A prerequisite of mechanistic knowledge is further establishment of the coordination chemistry of diazoalkanes, of which only a few general lines are yet visible [2, 24].

References

[1] Monograph: M. Regitz, *Diazoalkanes*, Thieme, Stuttgart, **1977**.

[2] Review: W. A. Herrmann, *Angew. Chem.* **1978**, *90*, 855; *Angew. Chem., Int. Ed. Engl.* **1978**, *17*, 800.

[3] Reviews: (a) M. P. Doyle, *Acc. Chem. Res.* **1986**, *19*, 348; (b) M. P. Doyle, *Chem. Rev.* **1986**, *86*, 919.

[4] G. Maas, *Top. Curr. Chem.* **1987**, *137*, 75.

[5] S. D. Burke, P. A. Grieco, *Org. React.* **1979**, *26*, 361.

[6] D. F. Taber, E. H. Petty, *J. Org. Chem.* **1982**, *47*, 4808.

[7] D. F. Taber, E. H. Petty, K. J. Raman, *J. Am. Chem. Soc.* **1985**, *107*, 196.

[8] D. F. Taber, R. E. Ruckle jr., *J. Am. Chem. Soc.* **1986**, *108*, 7686.

[9] H. J. Monteiro, *Tetrahedron Lett.* **1987**, *28*, 3459.

[10] B. Corbel, D. Hernot, J.-P. Haelters, G. Sturtz, *Tetrahedron Lett.* **1987**, *28*, 6605.

[11] D. F. Taber, S. A. Salch, R. W. Korsmeyer, *J. Org. Chem.* **1980**, *45*, 4699.

[12] D. E. Cane, P. J. Thomas, *J. Am. Chem. Soc.* **1984**, *106*, 5295.

[13] R. J. Ponsford, R. Southgate, *J. Chem. Soc., Chem. Commun.* **1979**, 846.

[14] M. P. Doyle, J. Taunton, H. Q. Pho, *Tetrahedron Lett.* **1989**, *30*, 5397.

[15] (a) M. P. Doyle, M. N. Protopopova, W. R. Winchester, K. L. Daniel, *Tetrahedron Lett.* **1992**, *33*, 7819; (b) M. P. Doyle, L. J. Westrum, N. E. W. Wolthuis, M. M. See, W. P. Boone, V. Bagheri, M. M. Pearson, *J. Am. Chem. Soc.* **1993**, *115*, 958.

[16] (a) M. P. Doyle, V. Bagheri, M. M. Pearson, J. D. Edwards, *Tetrahedron Lett.* **1989**, *30*, 7001; (b) M. P. Doyle, A. B. Dyatkin, *J. Org. Chem.* **1995**, *60*, 3035.

[17] J. Adams, M.-A. Poupart, L. Grenier, *Tetrahedron Lett.* **1989**, *80*, 1749.

[18] M. P. Doyle in *Homogeneous Transition Metal Catalyzed Reactions* (Eds.: W. R. Moser, D. W. Slocum), Adv. Chem. Ser., Vol. 230, American Chemical Society, Washington DC, **1992**, pp. 443–461.

[19] (a) M. P. Doyle, Q.-L. Zhou, C. E. Raab, G. H. P. Roos, *Tetrahedron Lett.* **1995**, *36*, 4745; (b) M. P. Doyle, A. B. Dyatkin, S. Jason, *ibid.* **1994**, *35*, 3853; (c) M. P. Doyle, A. B. Dyatkin, G. H. P. Roos, F. Canas, D. A. Pierson, A. von Basten, P. Mueller, P. Polleux, *J. Am. Chem. Soc.* **1994**, *116*, 4507; (d) N. McCarthy, M. A. McKervey, T. Ye, M. McCann, E. Murphy, M. P. Doyle, *Tetrahedron Lett.* **1992**, *33*, 5983.

[20] (a) M. P. Doyle, D. G. Ene, D. C. Forbes, T. H. Pillow, *Chemcomm*, **1999**, 1691; (b) M. P. Doyle, J. S. Tedrow, A. B. Dyatkin, C. J. Spaans, D. G. Ene, *J. Org. Chem.* **1999**, *64*, 8907.

[21] M. P. Doyle, R. J. Pieters, S. F. Martin, R. E. Austin, C. J. Oalmann, P. Müller, *J. Am. Chem. Soc.* **1991**, *113*, 1423.

[22] M. P. Doyle, A. V. Kalinin, D. G. Ene, *J. Am. Chem. Soc.* **1996**, *118*, 8837.

[23] H. W. Yang, D. Romo, *Tetrahedron*, **1999**, *55*, 6403.

[24] W. A. Herrmann, *Adv. Organomet. Chem.* **1981**, *20*, 159.

3.3.12   Acrolein and Acrylonitrile from Propene

Wolfgang A. Herrmann

3.3.12.1   Introduction

Propene, as one of the most powerful petrochemical feedstocks, depends to a large extent upon metal-containing catalysts for its further "refinement" [1]. While hydroformylation (Rh) is the prototype of homogeneous catalysis, and the Ziegler-type polymerization (Ti, Zr) has at least molecular mechanistic features (but is normally microheterogeneous), the *oxidation of propene* is based on heterogeneous catalysts [2]. Of key importance in industry is a group of reactions leading to the allylic oxidation products *acrolein* and *acrylonitrile* (eqs. (1) and (2)), commonly referred to as SOHIO (*S*tandard *O*il of *O*hio) oxidations [3]. The major follow-up product is acrylic acid, resulting from acidic hydrolysis of acrylonitrile (eq. (3)). Alternative routes to acrylic acid, including oxidative carbonylation of ethylene (homogeneous Pd catalysis; Union Oil process), have been discussed in Chapter 1.

$$H_2C=CH-CH_3 + O_2 \xrightarrow{\text{cat.}} H_2C=CH-C{\overset{O}{\underset{H}{\diagup}}} + H_2O + 368\ \text{kJ/mol} \tag{1}$$

$$H_2C=CH-CH_3 + {}^3/_2\,O_2 + NH_3 \xrightarrow{\text{cat.}} H_2C=CH-C{\equiv}N + 3\,H_2O + 502\ \text{kJ/mol} \tag{2}$$

$$H_2C=CH-C{\equiv}N + {}^1/_2\,H_2SO_4 \xrightarrow[-\ {}^1/_2\,(NH_4)_2SO_4]{2\,H_2O} H_2C=CH-COOH \tag{3}$$

The allylic oxidation of propene typifies the so-called "bimetallic heterogeneous catalysis" [4], a *terminus technicus* to emphasize cooperative effects in catalytic conversions (for multicomponent homogeneous catalysis, see Section 3.1.5). Nevertheless, the SOHIO-type oxidation is included in this book because one can imagine a number of mechanistic implications on a *molecular* platform, too. Studies on organometallic model compounds and reactions are available in ref. [2].

3.3.12.2   Scope and Technological Features

The oxidation of propene to acrolein has been applied in industry since 1958, when Shell introduced a gas-phase oxidation based on a $Cu_2O/SiC/I_2$ catalyst system. This process made acrolein a commodity product. A more efficient technology, still state-of-the-art, was subsequently developed by Standard Oil of Ohio (from 1957 onward), using bismuth molybdate and bismuth phosphatecatalysts in a fixed-bed tube reactor to handle the strongly exothermic oxidation process. Typical side products (from over-oxidation) are acetaldehyde, acrylic acid, and carbon dioxide.

The SOHIO "ammoxidation" to make acrylonitrile is a modification of the simple allylic oxidation. It converts an activated methyl group into a carbonitrile functionality (eq. (2)). Equimolar amounts of propene and ammonia are reacted in a fluidized-bed reactor at ca. 450 °C/0.03–0.2 MPa with oxygen from air. After the product has been washed with water, the acrylonitrile is refined by multistep distillation to > 99 % purity, as is mandatory for the production of fibers. The product selectivity is ≥ 70 %. The side products are acetonitrile which is normally burned. Hydrogen cyanide which, at a production of ca. 15 wt. % relative to the propene conversion, contributes significantly to the capacities of this base chemical. It is interesting to note that the directed synthesis of hydrogen cyanide is also based on an ammoxidation-type reaction, namely the direct conversion of methane in the Andrussow process according to eq. (4). However, a Pt/Rh catalyst is used in this particular case since a π-allyl intermediate cannot be traversed (see Section 3.3.12.3). An additional technology is applied by Degussa AG in the so-called "BMA process" (*B*lausäure–*M*ethan–*A*mmoniak) (eq. (5)). Heterogeneous Pt, Ru, or Al catalysts are being used for this dehydrogenation reaction at 1250 °C with methane conversions of approx. 90 % [20]. In contrast to the ammonoxidation, this reaction of hydrocarbons with ammonia is called ammondehydrogenation (ammonolysis + dehydrogenation).

$$CH_4 + {}^3/_2 O_2 + NH_3 \xrightarrow{\text{cat.}} HC{\equiv}N + 3 H_2O + 480 \text{ kJ/mol} \qquad (4)$$

$$CH_4 + NH_3 \xrightarrow{\text{cat.}} HC{\equiv}N + 3 H_2 \qquad (5)$$

Other applications of the ammoxidation include the reactions of isobutene ($\rightarrow$ α-methacrylonitrile), α-methylstyrene ($\rightarrow$ atropanitrile), β-picoline ($\rightarrow$ nicotine nitrile and nicotinamide), toluene ($\rightarrow$ benzonitrile), and xylenes ($\rightarrow$ phthalonitrile, terephthalonitrile, and isophthalonitrile on the way to fiber- grade diamines).

3.3.12.3   Catalyst Principles and Mechanism

Most efficient in the ammoxidation of propene are catalysts containing simultaneously

(1) multivalent Main-Group elements – preferably bismuth, antimony, or tellurium,
(2) oxidic molybdenum, and
(3) a redox-active component: $Fe^{2+/3+}$, $Ce^{3+/4+}$, or $U^{5+/6+}$,

in solid-state matrix [5]. The standard catalyst could in an utterly simplistic way be formulated as $Bi_2O_3 \cdot nMoO_3$. The first SOHIO patent on this type of catalyst was filed in 1957. Ammoxidation is a six-electron reaction, indicating that a number of mechanistic steps must be traversed.

According to common opinion, both the SOHIO oxidation ($\rightarrow$ acrolein) and the propene ammoxidation ($\rightarrow$ acrylonitrile) receive their unexpected selectivites (albeit far off 100 %) from a specific type of crystal-lattice oxygen as the actual reagent, quite typically exemplifying the *Mars/van Krevelen* mechanism. In support of this view, bismuth molybdate is reduced by propene and can be reoxidized by air or oxygen yielding the original valence state; this was shown by $^{18}O_2$ labeling experiments [6]. The catalysts have to fulfill the following demands:

(1) strong oxidative power with regard to the hydrocarbon to be converted,
(2) susceptibility to regeneration by *elemental* oxygen,
(3) activation of ammonia (in the case of ammoxidation).

In the presence of ammonia, some oxidic molybdenum sites (Mo=O) are likely to be replaced by imino (Mo=NH) or diimino functions (Mo($=NH)_2$) which then couple with the allyl group. The final product, acrylonitrile, is obtained after dehydrogenation and the catalyst is reoxidized with air.

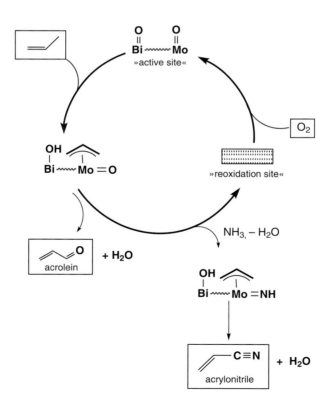

Scheme 1

At the same time, degradation and side reactions should be suppressed (see above).

The ammoxidation mechanism was investigated intensively by Grasselli and his team in the Standard Oil of Ohio laboratories. Despite supporting experimental evidence, the textbook mechanism remains speculative to a certain extent. It can be taken for granted, however, that the commonly employed Bi/Mo catalysts exercise their double-site activation such that the propene undergoes α-H abstraction at a BiO functionality while the remaining allyl group coordinates to the high-valent molybdenum (Scheme 1). The bond-making step to follow in case of the acrolein synthesis is little defined.

The key allyl species in the Grasselli mechanism [5 a] remains unspecified: does it coordinate to the high-valent molybdenum in a σ/π fashion, or does it occur as radicals? What is the role of free allyl radicals? Also, the product-forming C–O and C–N connection steps, respectively, leave open questions. For example, are there intramolecular rearrangements like those shown in Scheme 2 responsible for the formation of the new bonds? How are the product precursors detached from the metal(s)?

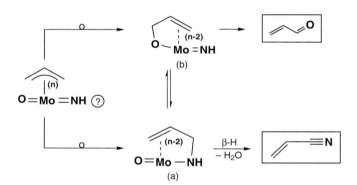

Scheme 2

3.3.12.4   Organometallic Models

Allyl coordination to high-valent metal oxides was unknown until very recently [7, 8]. Strinkingly, this ligand coordinates at heptavalent rhenium in a σ rather than in a π mode, and the same is true for the related indenyl derivative (cf. Structures **1** and **2**) although the ligand sphere would not be sterically congested in the case of π coordination (cf. Structure **3**). The σ-allyl complex undergoes a radical-path decomposition, yielding predominantly 1,5-hexadiene [7 c, 8]. The rhenium–oxygen bond is probably too strong for the formation of oxygenated hydrocarbons to occur, even if there was a Re $\rightarrow$ O rearrangement of the allyl group. Whether lower metal oxidation states favor π-bonding, as in the tungsten(IV) complex **4** [9], is not likely to be so simple a motive in coordination chemistry. The π-donating effects of the oxo ligands in **1** and **2** partly explain the structural details (Structure **3** contains the electron-rich ligand C_5Me_5).

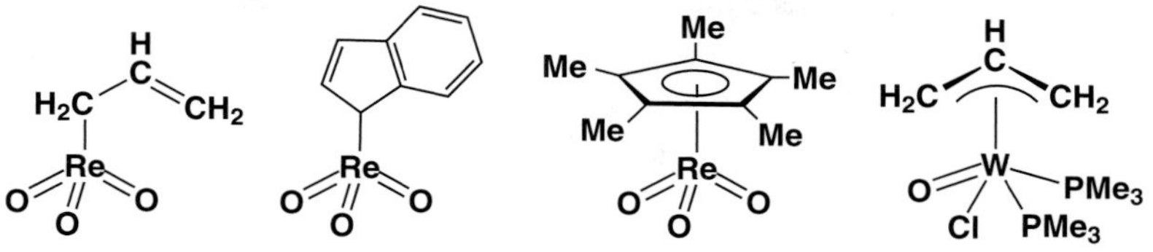

While an allyl-to-oxo coupling step is not available from any molecular model system, a rearrangement of allyl groups from one to another oxo group has been detected for the oxomolybdenum(VI) complex **5a** $\rightleftharpoons$ **5b** [10–12]. The intramolecular isomerization is thought to involve a cationic allylic intermediate via a [3,3]-sigmatropic shift similar to a Claisen rearrangement [11]. Further, an oxygen-to-nitrogen allyl migration is known from the chemistry of allyloxo imido complexes of tungsten(VI); cf. eq. (7) for Structure **6** [12]. A plethora of oxo-, imido-, and nitridomolybdenum complexes is known in the meantime; they offer stoichiometric model reactions in the context of the SOHIO oxidation [13, 14].

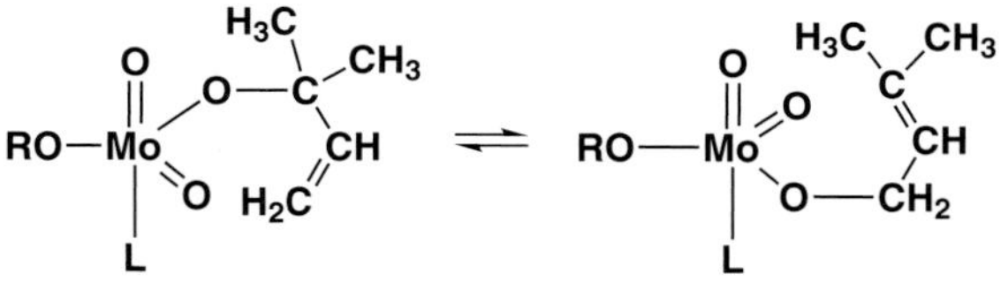

$$(6)$$

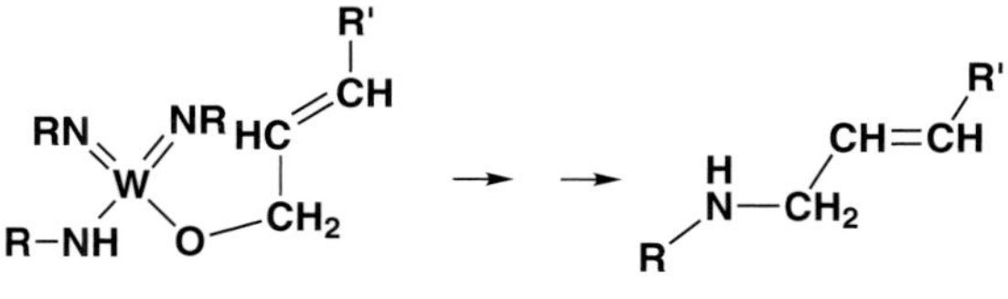

$$(7)$$

3.3.12.5 The "Amm(on)dehydrogenation"

There are recent literature reports according to which alcohols and aldehydes dehydrogenate in the presence of ammonia to form nitriles. Process are operated by Rohm & Haas, Ruhrchemie, and Bayer [21–23]. Molybdenum nitrides were found most efficient as heterogeneous catalysts [15]; cf. eqs. (8) and (9) (R = n-C_3H_7). They also effect dehydrogenation of amines, as demonstrated for n-butylamine in eq. (10).

$$R\text{-}CH_2OH \; + \; NH_3 \quad \xrightarrow{\text{cat.}} \quad R\text{-}C\equiv N \; + \; H_2O \; + \; 2\,H_2 \qquad\qquad (8)$$

$$R\text{-}C\!\!\overset{O}{\underset{H}{\diagdown}} \; + \; NH_3 \quad \xrightarrow{\text{cat.}} \quad R\text{-}C\equiv N \; + \; H_2O \; + \; H_2 \qquad\qquad (9)$$

$$R\text{-}CH_2NH_2 \quad \xrightarrow{\text{cat.}} \quad R\text{-}C\equiv N \; + \; 2\,H_2 \qquad\qquad (10)$$

Using supported γ-Mo$_2$N as a surface-mediated catalyst, a new proposal for the manufacture of aromatic nitriles and dinitriles from the relatively cheap aldehydes was made [16, 17]. Thus, isophthalic dinitrile can be produced in a gas-phase "amm(on)dehydrogenation" (cf. Section 3.3.12.2) at 400 °C according to eq. (11) with 80 % conversion when a 50-fold excess of ammonia is used [16]. It is obvious that imine intermediates are formed (from the aldehyde and ammonia) which then dehydrogenate under the reaction conditions.

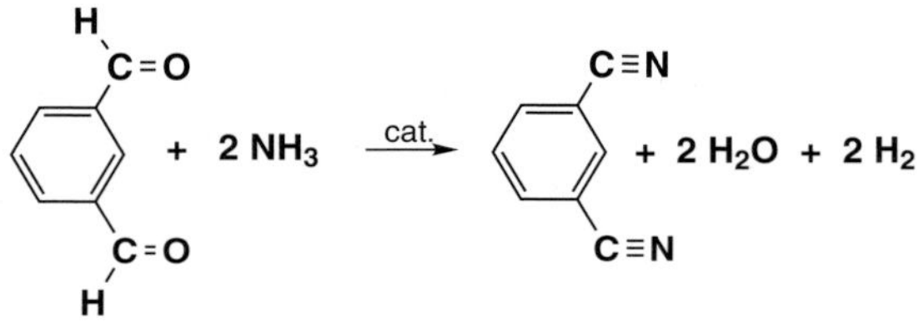

$$+ \; 2\,NH_3 \quad \xrightarrow{\text{cat.}} \quad + \; 2\,H_2O \; + \; 2\,H_2 \qquad\qquad (11)$$

This type of reaction also works with complex catalysts resulting from the decomposition of molecular molybdenum nitrides such as **7** [14, 17]. The thermolysis product (Structure **8**) is mostly δ-MoN (eq. (12)), and the incorporated oxygen is likely to originate from air and moisture. In the absence of ammonia, a product with lower nitrogen content (Mo/N = 1:0.68) is obtained [17].

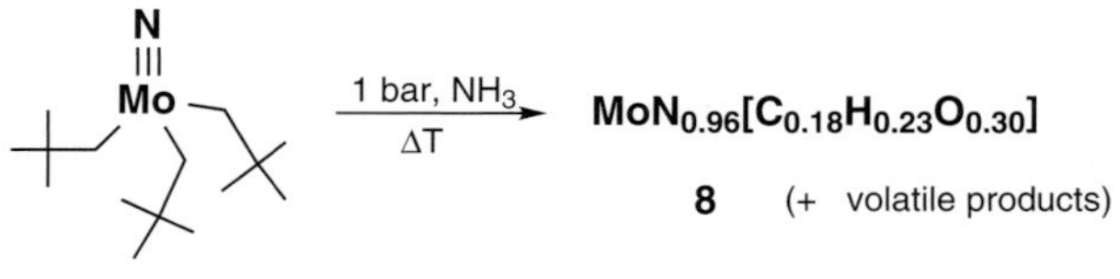

$$\xrightarrow[\Delta T]{1\ \text{bar, NH}_3} \quad MoN_{0.96}[C_{0.18}H_{0.23}O_{0.30}] \qquad\qquad (12)$$

8   (+ volatile products)

7

Since little is known about the mechanism other than that ammonia reacts first to form an imine ($-CH=NH$, [18]), studies related to the chemistry of $Mo\equiv N$ structures (e. g., ^{15}N labeling experiments) seem necessary in order to find out at which step the surface nitrogen enters the reaction. It would also be interesting to see which surface sites (N vs. Mo) are taken by the hydrogen resulting from dehydrogenation of the intermediate imine. Surface grafting of molecular precursors such as **7** [19] may be a new technique to generate molecularly defined catalyst species.

3.3.12.6 Perspectives

More than 5 million tons of acrylonitrile are made annually. It is synthesized industrially by the gas-phase heterogeneous ammoxidation of propene. New catalysts based on Bi–Mo or V–Sb oxides may lead to manufacture of this important compound from propane. Although the new process has a considerably lower selectivity to acrylonitrile, the lower cost of the alkane makes it economically interesting (the propane method can cut at least 20 % from the production costs of the propene route). Nevertheless, increases in the selectivity of the catalysts, especially at higher conversions, will be necessary for this process to compete with the usual process of acrylonitrile synthesis [24].

Propene oxidation by means of oxidic bimetallic catalysts is a unique example of selectivity synergism in heterogeneous catalysis. The secret seems to be the specific reactivity of certain lattice-oxygen atoms (ions?) upon the surface-generated allyl intermediates. Now that this basic mechanistic feature has strong support, organometallic molecular chemistry should search deliberately for the electronic prerequisites of well-defined metal–oxo species to engage in allyl–oxygen coupling reactions. Complexes of type $H_2C=CH-CH_2-MO_x$ may be screened for their reactivity pattern, e. g., under electrochemical conditions, to learn more about the circumstances under which an MO_x fragment is ready to undergo O–C coupling (this question, by the way, does not apply only to allyl species). In this context, a synthetic study related to oxygen-bridged bismuth–molybdenum model compounds is worth the laborious efforts to be expected on this uncharted sea: 40 years after the first SOHIO patent on this topic!

References

[1] K. Weissermel, H.-J. Arpe, *Industrial Organic Chemistry,* 3rd ed., VCH Publishers, New York, **1988**.
[2] Summary: W. A. Herrmann, *Kontakte* (Merck, Darmstadt), No. 3, **1991**, pp. 29–52.
[3] (a) P. N. Rylander in *Catalysis – Science and Technology* (Eds.: J. R. Anderson, M. Boudart), Springer, New York, **1983**, Vol. 4, p. 27; H. Heinemann, in *ibid.* **1981**, Vol. 1, p. 30; (b) R. K. Grasselli, *J. Chem. Educ.* **1986**, *63*, 216; (c) H. Schaefer, *Chem.-Tech.* **1978**, *7*, 231.
[4] J. H. Sinfelt, *Bimetallic Catalysts* (Exxon Monograph), John Wiley, New York, **1983**.
[5] Mechanistic studies: (a) R. K. Grasselli, J. D. Burrington, *Adv. Catal.* **1981**, *30*, 133; (b) G. W. Keulks, L. D. Krenzke, T. M. Notermann, *ibid.* **1978**, *27*, 183; (c) J. D. Burrington, C. T. Kartisek, R. K. Grasselli, *J. Catal.* **1984**, *87*, 363; (d) R. K. Grasselli, J. D. Burrington, J. F. Brazdil, *Faraday Discuss. Chem. Soc.* **1982**, *72*, 203; (e) L. C. Glaeser, J. F. Brazdil, M. A. Hazle, M. Mehicic, R. K. Grasselli, *J. Chem. Soc., Faraday Trans. 1* **1985**, *81*, 2903.
[6] (a) G. W. Keulks, *J. Catal.* **1970**, *19*, 232; (b) C. C. McCain, G. Gough, G. W. Godin, *Nature (London)* **1963**, *198*, 989.
[7] (a) W. A. Herrmann, F. E. Kühn, C. C. Romao, H. Tran Huy, *J. Organomet. Chem.* **1994**, *481*, 227; (b) W. A. Herrmann, F. E. Kühn, *ibid.* **1995**, *495*, 209; (c) F. E. Kühn, Ph.D. Thesis, Technische Universität München, **1994**.
[8] Review: W. A. Herrmann, *J. Organomet. Chem.* **1995**, *500*, 149.
[9] L. M. Atagi, S. C. Critchlow, J. M. Mayer, *J. Am. Chem. Soc.* **1992**, *144*, 1483.

[10] J. Belgacem, J. Kress, J. A. Osborn, *J. Chem. Soc., Chem. Commun.* **1993**, 1125.

[11] J. Belgacem, J. Kress, J. A. Osborn, *J. Am. Chem. Soc.* **1992**, *114*, 1501.

[12] J. Belgacem, J. Kress, J. A. Osborn, *J. Mol. Catal.* **1984**, *86*, 267.

[13] W. A. Nugent, J. M. Mayer, *Metal–Ligand Multiple Bonds,* Wiley–Interscience, New York, **1988**, and references cited therein.

[14] W. A. Herrmann, S. Bogdanoviè, R. Poli, T. Priermeier, *J. Am. Chem. Soc.* **1994**, *116*, 4989.

[15] See, for example: H. Abe, A. T. Bell, *J. Catal.* **1993**, *142*, 430.

[16] SKW Trostberg AG (J. Graefe, K. Wernthaler, H.-G. Erben), DE Patent appl. P 195.18.398.3 (**1995**).

[17] N. Hansen, Diploma Thesis, Technische Universität München, **1995**.

[18] Ruhrchemie AG (H. Goethel, B. Cornils, H. Feichtinger, H. Tummes, J. Falbe), DE 2.048.750 (**1970**).

[19] W. A. Herrmann, A. W. Stumpf, Th. Priermeier, J.-M. Basset, *Angew. Chem. Int. Ed.* **1996**, *35*, 2803.

[20] F. Endter, *Chem. Ing. Tech.* **1958**, *30*, 305; F. Endter, *Dechema Monogr.* **1959**, *33*, 28.

[21] Rohm & Haas (L. R. U. Spence, E. Park, D. J. Butterbaugh, F. W. Robinson, US 2.337.421 (1941) and US 2.337.422 (**1941**).

[22] Ruhrchemie AG (G. Horn, D. Fröhlich, H. Liebern), EP 0.038.507 A1 (**1981**).

[23] Bayer AG (F. Hagedorn et al.), DE 3.216.382 (**1983**).

[24] (a) G. Centi, S. Perathoner, *Chemtech* **1998**, *28*, 13; (b) G. Centi, S. Perathoner, *J. Chem. Soc., Faraday Trans.* **1997**, *93*; (c) G. Senti, S. Perathoner, *C&EN* **1997**, *75*, 15.

3.3.13   Chemistry of Methyltrioxorhenium (MTO)

3.3.13.1   Fine Chemicals via Methyltrioxorhenium as Catalyst

Fritz E. Kühn, Michelle Groarke

3.3.13.1.1 Introduction

The title compound has been known for less then 25 years [1]. For a considerable part of this time, it has been widely regarded as a mere curiosity. This picture changed dramatically during the last decade. Today, not only is an amazing wealth of derivatives and reaction products known and easily accessible, several of these compounds, most notably methyltrioxorhenium(VII) itself, have found numerous very interesting applications in both catalysis and material sciences.

3.3.13.1.2 Synthesis of Methyltrioxorhenium

Methyltrioxorhenium(VII) (**1**), usually abbreviated to MTO, was first reported in 1979 by Beattie and Jones [2]. Due to the difficult synthesis and the low yields of **1** that could be obtained, the compound was not examined further. The breakthrough came in 1987, when Herrmann and co-workers presented the first efficient synthetic route, starting from dirhenium heptoxide and tetramethyltin (eq. (1)) [2].

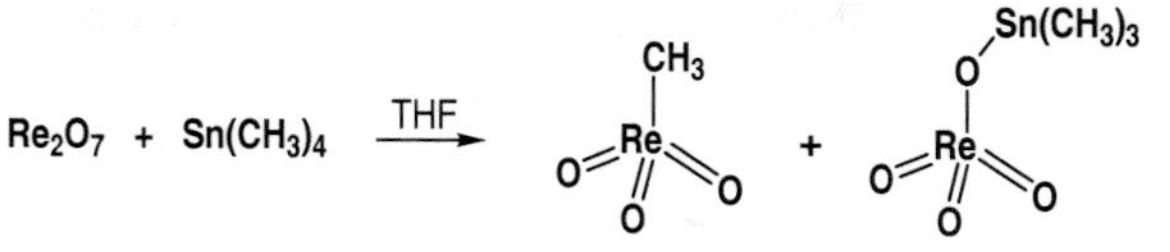

$$Re_2O_7 + Sn(CH_3)_4 \xrightarrow{THF} \quad \mathbf{1 \ (MTO)} \quad + \quad \tag{1}$$

This preparation of **1** has been improved considerably since then [2]. Since 1993, MTO has been commercially available (cf. Section 3.3.13.1.3). In 1997, the synthetic method starting from trimethylstannyl perrhenate has been extended to other perrhenates so that the moisture-sensitive Re_2O_7 can be replaced by more conveniently handled starting materials [2]. Nowadays **1** can be synthesized directly from rhenium powder in amounts of several kilograms. This synthetic progress was accompanied by the discovery of a plethora of derivatives and catalytic applications of organorhenium(VII) oxides. The use of these complexes in catalysis, however, is still strongly dominated by **1** itself [2, 3].

At room temperature **1** is a colorless solid, crystallizing in colorless needles. It is readily sublimed and soluble in all common solvents. MTO decomposes only above 300 °C. Heated in water to ca. 70 °C for several hours it forms a golden polymer of empirical formula $\{H_{0.5}[(CH_3)_{0.92}ReO_3]\}$ [4].

3.3.13.1.3 Applications of Methyltrioxorhenium(VII) in Catalysis

Oxidation Catalysis

The catalytic activity of **1** and some of its derivatives is known [5]. However, the breakthrough in the understanding of the role of MTO in oxidation catalysis was the isolation and characterisation of the reaction product of MTO with excess H_2O_2; it is a bis(peroxo) complex of stoichiometry $(CH_3)Re(O_2)_2O$ (**2**) [6 a].

In the solid state, it is isolated as an adduct with a donor ligand L (L = H_2O, **2a**; L = O=P(N(CH$_3$)$_2$)$_3$, **2b**) [2, 3, 6 a], which is lost in the gas phase. The structures of **2** (electron diffraction) [3], **2a** and **2b** (X-ray diffraction) were determined; the structure of ligand-free complex **2** is known from the gas phase.

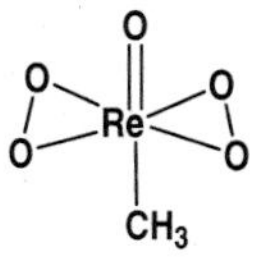

2 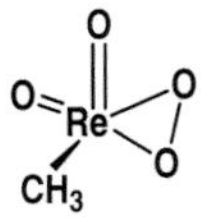　　　**3** 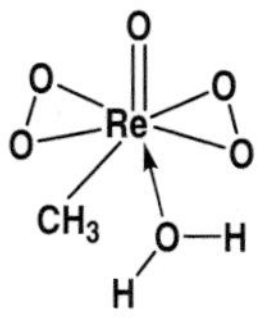　　　　　**2a** 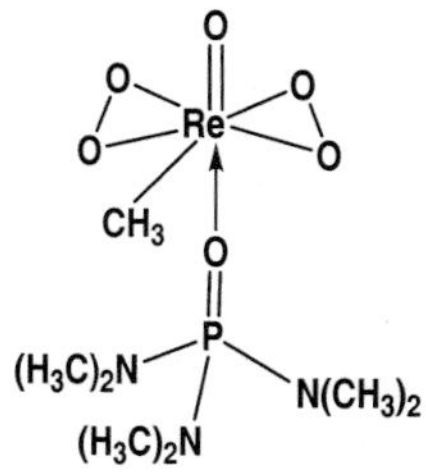　　　　　**2b**

^{17}O-NMR experiments showed that **2a, b** exchange their ligand L rapidly in solution, especially in coordinating solvents [2, 3, 6 a]. Only the terminal oxygen atom is involved in an oxygen exchange with water [6].

Experiments with the isolated bis(peroxo)complex **2a** have shown that it is an active species in oxidation catalysis, e. g., in the oxidation of olefins [6 a]. *In situ* experiments show that the reaction of MTO with one equivalent of H_2O_2 leads to a monoperoxo complex **3** [2–4]. Complex **3** has never been isolated and exists only in equilibrium with MTO and **2** [3, 6 b].

The decomposition of **1**, **2**, and **3** in solution was also examined [3, 7]. The full kinetic pH profile for the base-promoted decomposition of complex **1** to CH_4 and $[ReO_4]^-$ has been examined. In the presence of hydrogen peroxide, complexes **2** and **3** decompose to methanol and perrhenate with a rate that is dependent on $[H_2O_2]$ and $[H_3O]^+$ [3].

The activation parameters for the coordination of H_2O_2 to MTO indicate a mechanism involving nucleophilic attack (see also [8]).

Oxidation of Alkenes

One of the most intensively examined catalytic processes using organorhenium (VII) oxides, in particular MTO, as catalysts is in the epoxidation of alkenes [2, 3, 9, 10]. Usually, less than 85 wt. % of H_2O_2 is employed and MTO is typically used in concentrations of 0.2–1.0 mol %. Turnover numbers of up to 2000 [mol/mol catalyst] and turnover frequencies of ca. 1200 [mol/mol catalyst per hour] can be achieved.

The catalytic MTO/H_2O_2 system is already active below room temperature, e. g., at –30 °C. The reaction between **2** and alkenes are ca. one order of magnitude faster in semi-aqueous solvents (e. g., 85 % H_2O_2) than in methanol. The rate constants for the reaction of **2** with aliphatic alkenes correlate closely with the number of alkyl groups on the alkene carbons. Theoretical calculations support these results [10 a, b]. The reactions become significantly slower when electron-withdrawing groups, such as –OH, –CO, –Cl, and –CN are present.

Two catalytic pathways for the alkene epoxidation may be described, corresponding to the concentration of the hydrogen peroxide used. If 85 % hydrogen peroxide is used, only **2** appears to be responsible for the epoxidation activity (Scheme 1, cycle I).

When a solution of 30 wt. % or less of H_2O_2 is used the monoperoxo complex **3** is also responsible for the epoxidation process and a second catalytic cycle is involved, as shown in Scheme 1, cycle II.

For both cycles, a concerted mechanism is suggested in which the electron-rich double bond of the alkene attacks a peroxidic oxygen of **2**. It has been inferred, from experimental data, that the system may involve a spiro arrangement [3, 5 a]. The selectivity toward epoxides can be enhanced by the addition of Lewis *O*- or *N*-bases such as quinuclidine, pyridine, pyrazole or 2,2'-bipyridine to the system [3, 6 d, 10 g–k]. Lewis acids catalyze ring-opening reactions and diol formation. These reactions are suppressed after the addition of Lewis bases. An

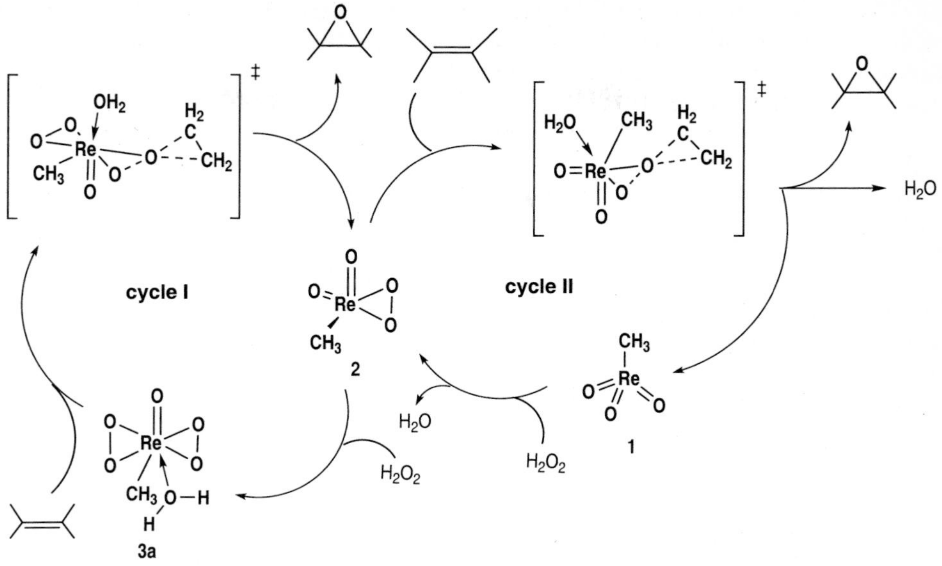

Scheme 1

excess of aromatic *N*-base ligands, especially under two-phase conditions, leads to accelerated reactions [3, 10 a, b, g–k]. Functionalized epoxides can be prepared in high yields by this method [2, 3]. The activity of the peroxo complexes in these cases depends on the Lewis bases, the redox stability of the ligands, and the excess of the ligands. Nonaromatic nitrogen bases were found to reduce the catalytic performance. It is known that polar noncoordinating solvents increase the rate of the epoxidation reaction [2, 3]. The epoxidation of alkenes in the presence of 0.1 mol % MTO using H_2O_2 (60 %) has also been successfully achieved in trifluoroethanol with only 5 mol % of pyrazole present [10 c].

In comparison with the standard system for epoxidation which uses *m*-chloroperoxybenzoic acid as oxidizing agent, the MTO/H_2O_2/aromatic Lewis base system displays several advantages: (a) it is safer but equal in price; (b) due to the suppression of epoxide ring opening, it has a much broader scope; (c) its selectivity is higher, and (d) it is more reactive, requires less solvent, the product work-up is easier, and the only by-product formed is water.

Another possibility for enhancing the selectivity toward epoxides is use of the urea–H_2O_2 (UHP) adduct. This enables the oxidation to be carried out in water-free solutions, thus avoiding formation of any diols and other side reactions. In the case of the oxidation of chiral allylic alcohols (see below) high diastereoselectivities have been achieved [3]. The ability to transform olefins to epoxides diastereoselectively seems to indicate that the reaction proceeds through a peracid-like transition state. However, a drawback of the urea–H_2O_2 system is the insolubility of the polymeric complex.

The zeolite NaY acts as an efficient heterogeneous host for MTO-catalyzed selective epoxidation of alkenes with H_2O_2 (85 %) [10 d]. The supernatant liquid was found to be catalytically inactive, even in the presence of pyridine. The Re catalyst is located inside 12 Å supercages of the zeolite. This inhibits the Lewis acid assisted hydrolysis of the epoxide to the diol by steric means. This catalytic system is comparable with the MTO/UHP and MTO/H_2O_2 (30 %)/pyridine systems although conversions may be improved by the addition of catalytic amounts of pyridine.

By tailoring the reaction conditions, alkenes may be cleaved oxidatively to yield aldehydes and carboxylic acids [10 f]. The presence of a Brønsted acid, e. g., HBF_4 or $HClO_4$, required to accelerate the oxidation of the olefin with respect to the deactivation of the catalyst, results in the synthesis of the aldehyde without overoxidation to the acid. Significant quantities of the diol are also present, however. A three-fold excess of H_2O_2 is required to oxidize the alkene to the acid. However, this also results in the simultaneous oxidation of the methyl *tert*-butyl ether solvent to the peroxide.

Oxidation of Conjugated Dienes

Conjugated dienes are oxidized to epoxides (or diols, if water is present) with the MTO/H_2O_2 system [11]. Urea/H_2O_2 avoids the subsequent epoxide ring opening. Electron-rich and conjugated dienes are more easily oxidized than electron-poor dienes and dienes with isolated double bonds. According to kinetic measurements complex **3** plays no important role as catalyst in this case. Compound **2** is an active species [11a].

The biphasic system MTO/H_2O_2/CH_2Cl_2 oxidizes 1,4-polybutadiene efficiently. The oxidation system is highly efficient and the extension of epoxidation (10–50 %) can be modulated by the amount of oxidant added, without significant change in the molecular weight of the polymer [11b] (cf. also Section 3.1.1.1.2).

Epoxidation of Allylic Alcohols and 1,3-Transposition of Allylic Alcohols

Allylic alcohols are epoxidized to the epoxy alcohols by hydrogen peroxide in the presence of MTO [3, 12]. Provided acid is not added, the product is mostly epoxide, accompanied by small amounts of triol that results from acid-catalyzed ring opening of the epoxide. With added acid, only the triol is obtained. According to kinetic data only the bisperoxo complex **2** and not the monoperoxo complex **3** acts as a catalyst.

Furthermore, MTO catalyzes the 1,3-transposition of allylic alcohols (eq. (2)) [12 c]. This reaction does not require the presence of peroxides or peroxo complexes. Theoretical investigations on the allylic rearrangement have also been performed [12 d]. Recently it has been reported that allylic alcohols as well as alkenes can be oxidized in an ambient-temperature ionic liquid using MTO and the urea hydrogen peroxide [12 e].

$$\text{(2)}$$

Oxidation of Aromatic Compounds

Arenes are oxidized to *para*-benzoquinones by hydrogen peroxide with MTO as catalyst [2, 3, 13]. The high regioselectivity achieved is noteworthy, particularly in the industrially interesting synthesis of vitamin K_3 (see eq. (3)). Since water is an inhibitor, concentrated (85 wt. %) H_2O_2 is preferred. Alternatively, commercially available 35 % H_2O_2 in acetic anhydride can be employed; a considerable regio-selectivity is obtained with this system. The conversion is higher for electron-rich arenes (nearly 100 %) and selectivities of more than 85 % have been reached [13 a]. Biphenylene can be oxidized with the MTO system in chloroform affording an *o*-quinone product (83 % conversion) [13 b]. Hydroxy-substituted arenes can be oxidized by aqueous hydrogen peroxide (85 wt. %) in acetic acid to afford the corresponding *p*-quinones in isolated yields of up to 80 % [3]. It has been shown that using a mixture of acetic acid and acetic anhydride further improves the product yield [13 d, e]. Instead of acetic acid, HBF_4 in EtOH may also be used [2, 3]. Anisol was also found to undergo selective oxidation with the MTO/H_2O_2 system to yield *o*- and *p*-methoxyphenols. There is no need to use a solvent in this case. For the mechanism of the arene oxidation see ref. [13 e].

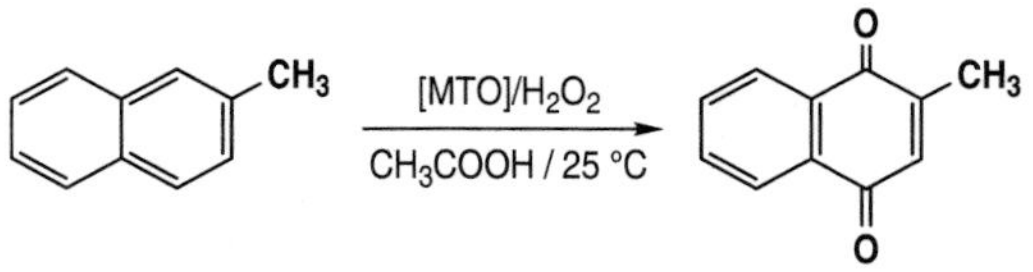

$$\text{(3)}$$

Benzaldehydes with hydroxyl or methoxy substituents in *ortho* or *para* positions are oxidized to the corresponding phenols (carboxylic acids are formed as by-products) in good yields [2, 3]. The yield is temperature- and solvent-dependent.

Baeyer–Villiger Oxidation and Dakin Reaction

γ-Butyrolactones are obtained in good yields and high regioselectivity from the corresponding cyclobutanones on oxidation with H_2O_2 catalyzed by MTO. Lactonization was found to be chemoselective in the presence of double bonds, aromatic rings, and chlorine substitutents [3, 14 a]. It has been shown that **2** also acts as an active species in the Baeyer–Villiger oxidation of ketones (eq. (4)) and in the Dakin reaction [14].

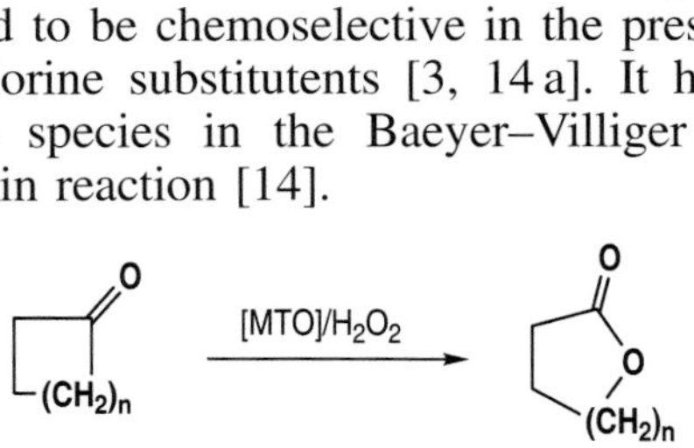

$$\text{(4)}$$

It is somewhat surprising that the MTO/H_2O_2 system presents this activity since these oxidations involve nucleophilic attack at the carbonyl group which is in contrast to all the preceding examples where the substrates attacked the electrophilic

Re–peroxo complexes, e. g., in olefin epoxidation. Nevertheless, **2** reacts stoichiometrically with cyclobutanone in the absence of H_2O_2. This reversed behavior may be due to substrate binding to rhenium. The unsymmetrical geometry of compound **2a**, displaying a polarity within the peroxo ligands, may also be responsible for the observed behavior.

Using thiantrene-5-oxide as a mechanistic probe for oxygen transfer reactions, the Baeyer–Villiger reaction was found to be strongly solvent-dependent [14 b, 15]. Donor solvents such as acetonitrile seem to enhance the nucleophilicity of the peroxo groups. Low H_2O_2 concentrations are sufficient and no H_2O_2 decomposition is observed at temperatures up to 70 °C. This is an advantage of the catalytic MTO/H_2O_2 system over the known transition-metal Baeyer–Villiger catalysts containing V, Mo, Mn, or Os. However, the nucleophilic character of the peroxidic atoms in **2** is not as pronounced as in Pt or Ir peroxo complexes that react with CO_2 or SO_2 to give isolable cycloaddition products [2, 3]. In the case of **2** TOFs of 18 000 [mol/mol catalyst per hour] are obtained for cyclobutanone, but in other cases TONs of up to 100 are usual [3]. Cycloketones can be converted into lactones even below room temperature (15 °C) by dilute hydrogen peroxide (10 wt. %).

Oxidation of Sulfur Compounds

Organic sulfides can be oxidized to the corresponding sulfoxides by hydrogen peroxide in the presence of MTO (eq. (5)) [2, 3, 16]. Both complexes **2** and **3** appear to be active in this reaction but kinetic results indicate that **3** might be more active than **2**. Using ethanol as solvent, the MTO/H_2O_2 system can be used to oxidize dialkyl, diaryl, and alkyl aryl sulfides to sulfoxides (R_2S/H_2O_2 = 1:1.1) or sulfones (R_2S/H_2O_2 = 1:2.2) with excellent yields and selectivity even in the presence of oxidatively sensitive functions on the sulfide side chain [3, 16]. The rate constants for the oxidation of sulfoxides to sulfones are significantly smaller than for the oxidation of sulfides to sulfoxides and the reaction rate is negligible without a catalyst. MTO can also be used in the oxidation of sulfides with the water-free urea/hydrogen peroxide system as oxidant in acetonitrile.

The results of a kinetic study are consistent with a nucleophilic attack of the sulfur atom on the peroxide oxygen group. While the bisperoxo complex **2** is an active catalyst, the monoperoxo complex **3** shows no activity [16 a].

$$ R{-}S{-}R \xrightarrow[\text{[MTO]}]{H_2O_2} R{-}\overset{\overset{O}{\|}}{S}{-}R \xrightarrow{[MTO/H_2O_2]} \overset{O}{\underset{O}{\|}}S\overset{R}{\underset{R}{\cdots}} \tag{5}$$

Sulfoxides are also readily accessible by the oxidation of thioether Fischer carbenes using 4 % MTO and H_2O_2 in a methanol/methylene chloride solvent [3].

Aryl disulfides are oxidized to thiosulfinates using MTO-catalyzed H_2O_2. The use of an excess of H_2O_2 eventually affords the sulfonic acids through thiosulfinate intermediates [16 b]. Disulfides themselves may be oxidized by H_2O_2 in the absence of MTO but this requires a large excess of oxidant, and elevated temperatures.

Thioketones are sequentially oxidized through sulfines (thioketone-*S*-oxide) to sulfur monoxide and the parent ketone on treatment with the relevant amount of H_2O_2 catalyzed by MTO. A 1:1 ratio of the R_2CS/H_2O_2 generates the sulfine without further oxidation or side product synthesis. A second equivalent of H_2O_2 results in the oxidative cleavage of the sulfine to release SO and the corresponding ketone [16 c, d]. However, this step occurs much more slowly. An application of this approach is in the synthesis of bridged SO species [16 e].

Oxidation of Phosphines, Arsines and Stibines

Tertiary phosphines, triarylarsines, and triarylstibines are converted to their oxides, R_3EO (E = P, As, Sb) by MTO/H_2O_2. Kinetic studies lead to the assumption that **2** and **3** have similar catalytic activities in all cases. The kinetic data support a mechanism involving nucleophilic attack of the substrate at the rhenium peroxides.

In the absence of peroxides, MTO also catalyzes the oxidation of tertiary phosphines to phosphine oxides [3, 17].

Oxidation of Anilines and Amines

A broad range of aromatic and aliphatic amines are readily oxidized to their corresponding *N*-oxides using the MTO/H_2O_2 catalytic system. The oxidation of aryl amines proceeds ca. 50 times faster than without catalyst [3, 18]. Nitrosobenzene is obtained by the oxidation of aniline (eq. (6)) while the oxidation of 4-substituted *N,N*-dimethylanilines yields the *N*-oxide as the only product. It has been found that electron-withdrawing substituents present on the substrate inhibit the reaction.

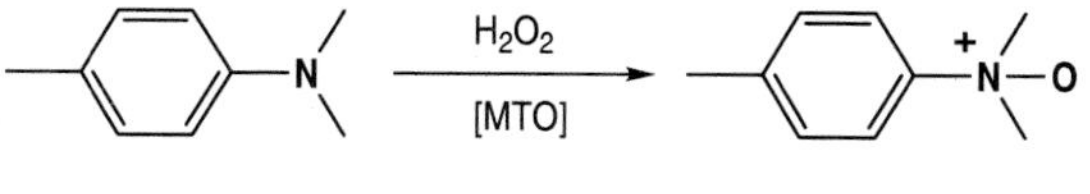

$$ (6) $$

Kinetic results suggest that both compounds **2** and **3** are involved in the oxidation process [2, 3]. It is proposed that the rate-determining step is the nucleophilic attack of the nitrogen lone-pair electrons of the aromatic amines on a peroxidic oxygen of the catalyst. Electron-donating groups attached to the nitrogen atom of aniline increase the rate constant. In general, the reactions are facile and high yielding at or below room temperature [3]. Furthermore, a broad variety of aromatic and aliphatic secondary amines are oxidized to the corresponding *N*-oxides [3]. The amines are converted to the corresponding hydroxylamines before transformation to the nitrones in very good yields. The hydroxylamine formation is rate-determining [2, 3]. Both H_2O_2 and the urea–hydrogen peroxide complex can be used together with MTO. Benzylamines are selectively oxidized to oximes. Primary amines are oxidized to nitro compounds by MTO catalysis (eq. (7)) [3].

$$RNH_2 \xrightarrow[\text{[MTO]}]{H_2O_2} RNO_2 \qquad (7)$$

Oxidative Cleavage of *N,N*-Dimethylhydrazones

N,N-Dimethylhydrazones of aldehydes react with hydrogen peroxide in the presence of catalytic amounts of MTO to give the corresponding nitriles in high yield [19]. The reaction commences with the oxidation of the *N,N*-dimethyl-hydrazone and presumably proceeds through an intermediate which can undergo a Cope-type elimination to yield the nitrile (eq. (8)) [3, 19 b].

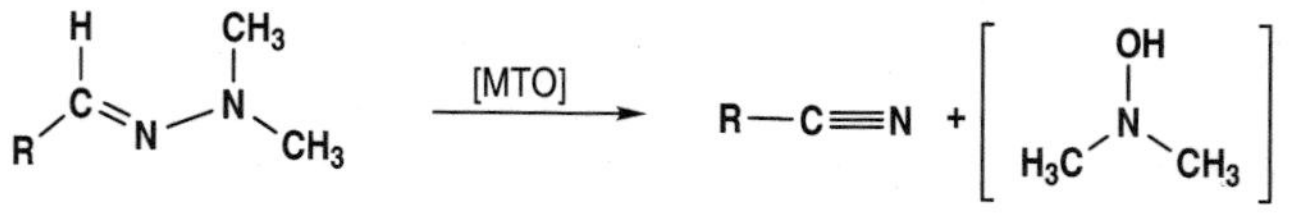

$$\underset{R}{\overset{H}{\underset{}{C}}}=N-\underset{CH_3}{\overset{CH_3}{N}} \xrightarrow{\text{[MTO]}} R-C\equiv N \; + \; \left[\underset{H_3C}{\overset{OH}{\underset{}{N}}} _{CH_3} \right] \qquad (8)$$

A more recent development shows that *N,N*-dimethylhydrazones derived from ketones may be cleaved oxidatively to revert back to the parent carbonyl compound in excellent yield [19 c].

Oxidation of Halide Ions

Another application of the MTO/H$_2$O$_2$ system is the catalytic oxidation of chloride and bromide ions in acidic aqueous solutions. The chloride oxidation steps are three to four orders of magnitude slower than the corresponding bromine oxidation steps. Both compounds **2** and **3** have been shown to be active catalysts in these processes. In both cases the catalyzed reactions were about 10^5 times faster than the uncatalyzed ones under similar conditions. In a first step HOX is formed, then HOX reacts with X$^-$ to form X$_2$. When H$_2$O$_2$ is used in excess the reaction yields O$_2$ [2, 3, 20].

Oxidation of C–H and Si–H Bonds

MTO/H$_2$O$_2$ also catalyzes the insertion of oxygen into a variety of activated and nonactivated C–H bonds. Alcohols or ketones are formed as shown in eq. (9). In the case of tertiary substrates alcohols are obtained as products. Suitable substrates proved that the reaction is stereospecific with retention of the configuration [2, 3]. The reaction can be accelerated by the addition of pyrazine-2-carboxylic acid, which also increases the total yield [3].

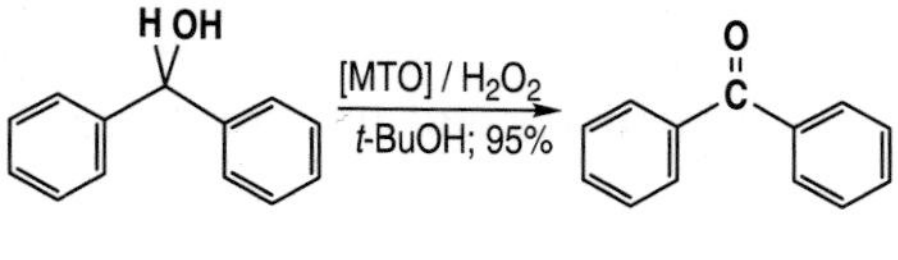

$$\xrightarrow[\text{t-BuOH; 95\%}]{\text{[MTO] / H_2O_2}} \qquad (9)$$

The catalyst system MTO/H_2O_2 also catalyzes oxygen atom insertion into Si–H bonds. Silanols and disiloxanes are formed as products, with the latter being the major ones [3, 21 a]. When UHP is used as an oxygen source instead of aqueous H_2O_2, **1** catalyzes the oxidation of silanes to silanols in high conversions and excellent selectivities in favor of the silanol (eq. (10)).

$$
\begin{array}{c}
R^1 \\
| \\
R^2\!-\!Si\!-\!H \\
| \\
R^3
\end{array}
\xrightarrow[\text{[MTO]}]{\text{H}_2\text{O}_2\,/\,\text{CH}_2\text{Cl}_2}
\begin{array}{c}
R^1 \\
| \\
R^2\!-\!Si\!-\!OH \\
| \\
R^3
\end{array}
\; + \;
\left[
\begin{array}{c}
R^1 \\
| \\
R^2\!-\!Si\!-\!O \\
| \\
R^3
\end{array}
\right]_2
\qquad (10)
$$

major product

More recently, it has been found that MTO absorbed in NaY zeolite allows the selective oxidation of silanes to silanols in the presence of H_2O_2 (85 %) in excellent yield [10 e]. No oxidation was found in the absence of MTO. In most cases studied, yields of disiloxanes were low (< 6 %).

In the presence of catalytic amounts of **1**, methyl trimethylsilyl ketene acetals are oxidized with urea hydrogen peroxide to afford α-hydroxy and α-siloxy esters [3, 21 b].

Oxidation of Other Compounds

Internal alkynes yield carboxylic acids and α-diketones when oxidized with the MTO/H_2O_2 system [22]. Rearrangement products are observed only for aliphatic alkynes. Terminal alkynes give carboxylic acids, derivatives thereof and α-keto acids as the major products. The yields of these products vary with the solvent used [22].

Primary and secondary alcohols are oxidized using the MTO/H_2O_2 catalyst system to aldehydes and ketones, respectively [23]. The dominant and reactive form of the catalyst is compound **2** [10 j]. The addition of a catalytic quantity of bromide ions, such as HBr or NaBr, significantly enhances the reaction rate [23 b]. The bromide is oxidized to the hypobromide ion, BrO^- which combines with additional bromide to give bromine. Bromine oxidizes the alcohols to aldehydes and ketones. The system MTO/H_2O_2/HBr/TEMPO (TEMPO = 2,2,6,6-tetramethyl-1-piperidinyloxy) catalyzes the selective oxidation of terminal alcohols to the corresponding aldehydes with excellent selectivities and yields [23 c]. The system allows the oxidation of alcohols either selectively to aldehydes or to the corresponding acids, depending on the reaction parameters. This technique is especially applicable to the oxidation of carbohydrates [23 c].

The MTO/hydrogen peroxide system oxidatively cleaves furans in yields usually > 70 % to enediones. Substituted pyranones are obtained from furans with hydroxymethyl groups at the 2-position. The yields in this case are > 75 %. Acetonitrile as solvent leads to the fastest reactions although the work-up is reported to be easier in CH_2Cl_2 [23 d]. Silyl enol ethers are oxidized to α-hydroxy ketones by MTO/H_2O_2 with subsequent desilylation with KF (eq. (13)) [23 e]. Yields are usually > 90 %. In the case of conjugated systems, the yields are significantly lower.

The MTO/H_2O_2 system furthermore catalyzes the oxidation of cyclic β-diketones to carboxylic acids (eq. (11)) [15]. Conversions are usually above 85 %; the product selectivity is nearly quantitative. It has been assumed that enolic forms which exist in solution are initially epoxidized. After a rearrangement step the C–C bond is cleaved and an oxygen inserted. Then an a-diketone intermediate forms which is finally oxidized to the carboxylic acid [15].

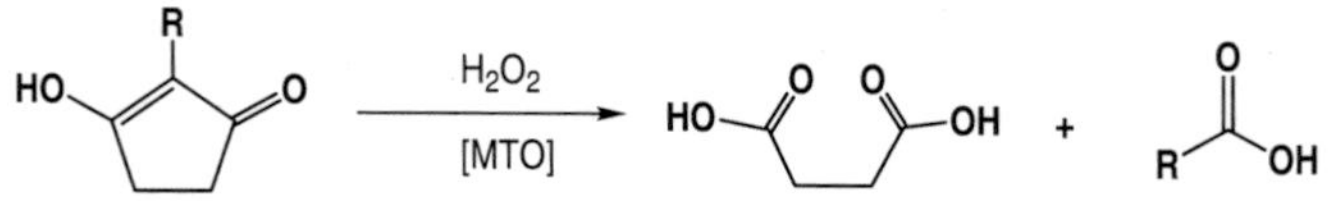

$$\text{(11)}$$

Oxidation of Metal Carbonyls

MTO catalyzes the oxidation of metal carbonyls to metal oxides with H_2O_2 (eq. (12)) [24 a–c]. These reactions proceed at room temperature and yields of up to 90 % are obtained. However, only organometal carbonyls with oxidation-resistant organic groups can be oxidized, e. g., (pentamethylcyclopentadienyl)tricarbonylrhenium(I) [24 a]. In all other cases, the organic ligand is also oxidized, leading to decomposition of the product complex [24 c].

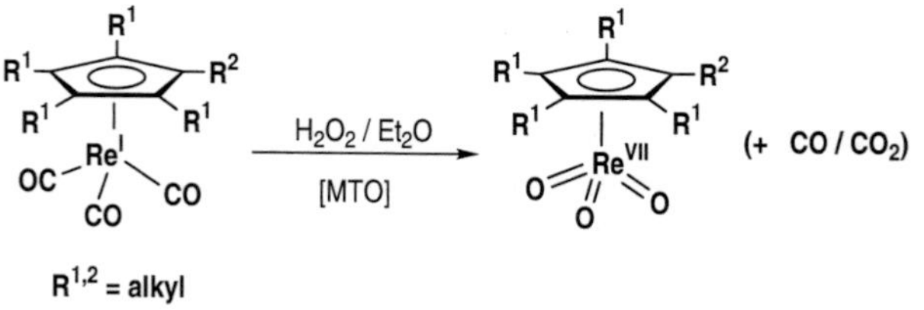

$$\text{(12)}$$

$$R^{1,2} = \text{alkyl}$$

3.3.13.1.4 Aldehyde Olefination and Related Reactions

Aldehydes or strained cycloketones, treated with aliphatic diazoalkanes in the presence of an equimolar amount of a tertiary phosphine and **1** as catalyst, afford an olefinic coupling product in good yields already at room temperature according to eq. (13) (cf. also Section 3.2.10) [2, 3, 25].

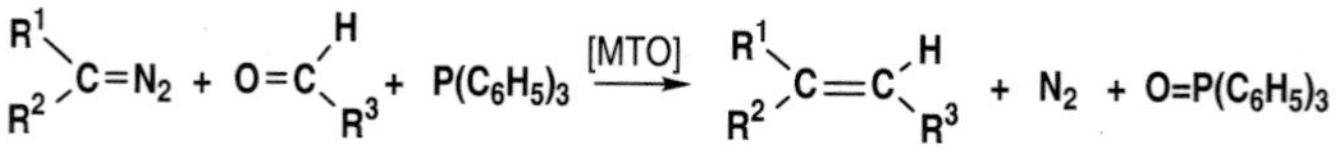

$$\text{(13)}$$

The *trans* selectivity is between 60 and 95 %, depending on the substrate, and the yields are around 85 %. The advantage of this method over Tebbe-Grubbs coupling is that it does not require the use of a stoichiometric amount of an organometallic coupling reagent [3, 25].

The deoxygenation of epoxides, sulfoxides, N-oxides, and triphenylarsine and triphenylstibine oxides at room temperature is also catalyzed by MTO with PPh$_3$ as oxygen acceptor [26]. Again, a Re^V intermediate, containing the (ligand-stabilized) methyldioxorhenium, seems to be involved. A catalytic amount of MTO also allows the stereospecific desulfurization of thiiranes (episulfides) by Ph_3P [25 b]. It is proposed that the Re^V again is the active catalyst in this reaction. When MTO is initially treated with H_2S the reaction rate is significantly enhanced. It is not entirely clear what the active species in the MTO/H_2S system is.

3.3.13.1.5 Olefin Metathesis

The system Re_2O_7/Al_2O_3 is an effective heterogeneous catalyst for carrying out olefin metathesis under mild conditions and its activity can be further increased by the addition of tetraalkyl tin compounds (cf. Section 2.3.3) [3, 26].

Since tin-containing co-catalysts are essential for the metathesis of functionalized olefins [26], it was soon discovered that **1** supported on acidic metal oxides forms metathesis catalysts that are active without additives even for functionalized olefins [26]. Standard supports are Al_2O_3–SiO_2, or Nb_2O_5 and the activity is related to the surface acidity [2, 3, 26]. A high metathesis activity is observed when MTO is chemisorbed on the surface. No evidence for a surface carbene species was obtained, but there appears to be a correlation between the catalytic activity and the presence of an alkyl fragment on the surface [26 a–c].

It was also possible to encapsulate **1** in zeolite, maintaining its metathesis activity. IR and EXAFS data indicate that the structure of **1** remains unchanged and that it is anchored by hydrogen bridges to the zeolite oxygens [2, 3]. Adsorption of water causes the de-aggregation of the guest molecules. Thermal treatment around 120 °C is found to yield methane and water together with the formation of an intrazeolite cluster species containing Re–Re bonds [2, 3].

The MTO supported on Al_2O_3–SiO_2 catalyzes in particular the self-metathesis of allyl aldehydes, ethers, silanes, and unsaturated carboxylates and nitriles, but also the ethenolysis of olefins with internal double bonds [26]. The catalyst system is also suitable for the metathesis of simple open-chain and cyclic olefins. Otherwise, frequent side reactions such as double-bond isomerization and olefin dimerization are insignificant. Ring-opening polymerization is catalyzed by the homogeneous catalyst MTO/R_nAlCl_{3-n} (R = CH_3, C_2H_5; n = 1, 2). As in the case of the heterogeneous olefin metathesis, the reaction can be performed at room temperature [26a]. Several functionalized diolefins cyclize to hydroazulenes via olefin metathesis in the presence of **1** [26 d].

3.3.13.1.6 Diels–Alder Reaction

MTO enhances the Diels–Alder reactivity of unsaturated C=C compounds, the standard case of which is given in eq. (14) [2, 3, 27].

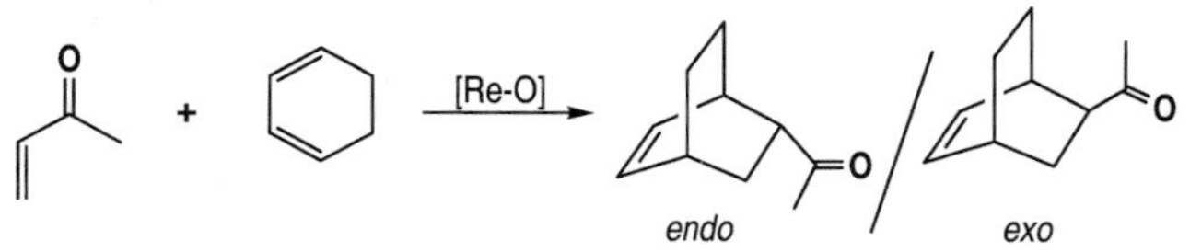

(14)

MTO proves to be an efficient and effective catalyst in this reaction when the dienophile is an *α,β*-unsaturated ketone or aldehyde. It is especially active in water, usually with isolated yields $> 90\%$. Kinetic studies show that the reaction rate is proportional to the catalyst concentration. The desirability of **1** as a Diels–Alder catalyst stems from a combination of favorable properties: the tolerance for many substrates, the inertness to air and oxygen, the use of aqueous medium, and the absence of product inhibition. The initial step appears to be the coordination of the carbonyl oxygen to the rhenium center. Steric crowding around rhenium inhibits reactions of the larger dienophiles [27].

3.3.13.1.7 Other Reactions

In the presence of **1** the catalytic alkoxylation of cyclohexene oxide with secondary and tertiary alcohols can be performed. This catalyst is known to cause disproportionation of epoxides, yielding olefins and diols. FT-IR spectroscopy indicated the formation of an active intermediate composed of **1** and epoxide. The carbocationic intermediate species is highly reactive with respect to nucleophilic compounds [28].

Ethyl diazoacetate (EDA) decomposes in the presence of **1**, thereby allowing access to a wide range of products. In the absence of other reactants, this decomposition results in the formation of both fumarate and diethyl maleate with the azine ($EtO_2CC=N-N=CCO_2Et$) also being formed [29 a]. Excellent yields of alkoxy and phenoxy esters are achieved from the OH insertion of low molecular weight primary and phenyl alcohols into EDA in the presence of MTO. SH and NH insertion reactions are accessible by treatment of EDA with thiols and amines in the presence of MTO to give excellent yields of thio esters and glycine esters, respectively [29 a]. The EDA decomposition in the presence of MTO allows the formal addition of a carbene to unsaturated systems. One such addition is the formation of epoxides from aldehydes and ketones [50]. Carbene addition to aromatic imines yields aziridines in excellent yields whereas the addition to alkenes furnishes cyclopropanes [29 b].

MTO has also been claimed to be the first transition metal complex to catalyze the direct, solvent-independent formation of ethers from alcohols [30]. Aromatic alcohols give better yields than aliphatic ones and reactions between different alcohols have been used to prepare asymmetric ethers. Also catalyzed by **1** is the dehydration of alcohols to form olefins at room temperature. When primary or secondary amines, respectively, are used as the limiting reagents, direct amination of alcohols gives the expected secondary or tertiary amines in yields of ca. 95%. Disproportionation of alcohols to carbonyl compounds and alkanes is also observed for aromatic alcohols in the presence of MTO as catalyst.

MTO has found application in the cyclotrimerization of aldehydes to yield 1,3,5-trioxanes in excellent yield [31]; 1 mol% of the catalyst is employed and water was found to inhibit the reaction. No other products were observed.

However, the introduction of bulky or electron-withdrawing substituents at the α-position limits the rate of formation of the trioxane.

References

[1] G. Rouschias, *Chem. Rev.* **1974**, *74*, 531.
[2] Recent reviews on organorhenium oxides: (a) W. A. Herrmann, F. E. Kühn, *Acc. Chem. Res.* **1997**, *30*, 169; (b) C. C. Romão, F. E. Kühn, W. A. Herrmann, *Chem. Rev.* **1997**, *97*, 3197.
[3] Recent reviews dealing with catalytic applications of methyltrioxorhenium: (a) J. H. Espenson, M. M. Abu-Omar, *ACS Adv. Chem.* **1997**, *253*, 3507; (b) G. S. Owens, J. Arias, M. M. Abu-Omar, *Catal. Today*, **2000**, *55*, 317; (c) W. Adam, C. M. Mitchell, C. R. Saha-Möller, O. Weichold, in *Structure and Bonding* (Ed.: B. Meunier), Springer Verlag, Berlin, **2000**, *Vol. 97*, p. 237.
[4] (a) W. A. Herrmann, W. Scherer, R. W. Fischer, J. Blümel, M. Kleine, W. Mertin, R. Gruehn, J. Mink, H. Boyson, C. C. Wilson, R. M. Iberson, L. Bachmann, M. R. Mattner, *J. Am. Chem. Soc.* **1995**, *117*, 3231.
[5] (a) W. A. Herrmann, R. W. Fischer, D. W. Marz, *Angew. Chem., Int. Ed. Engl.* **1991**, *30*, 1638.
[6] (a) W. A. Herrmann, R. W. Fischer, W. Scherer, M. U. Rauch, *Angew. Chem., Int. Ed. Engl.* **1993**, *32*, 1157; (b) P. Gisdakis, S. Antonczak, S. Köstlmeier, W. A. Herrmann, N. Rösch, *Angew. Chem., Int. Ed. Engl.* **1998**, *37*, 2211.
[7] (a) J. H. Espenson, H. Tan, S. Mollah, R. S. Houk, M. D. Eager, *Inorg. Chem.* **1998**, *37*, 4621; (b) K. A. Brittingham, J. H. Espenson, *Inorg. Chem.* **1999**, *38*, 744.
[8] O. Pestovsky, R. vanEldik, P. Huston, J. H. Espenson, *J. Chem. Soc., Dalton. Trans.* **1995**, 133.
[9] S. Yamazaki, J. H. Espenson, P. Huston, *Inorg. Chem.* **1993**, *32*, 4683.
[10] (a) F. E. Kühn, A. M. Santos, P. W. Roesky, E. Herdtweck, W. Scherer, P. Gisdakis, I. B. Yudanov, C. DiValentin, N. Rösch, *Chem. Eur. J.* **1999**, *5*, 3603; (b) M. C. A. VanVliet, I. W. C. E. Arends, R. A. Sheldon, *J. Chem. Soc., Chem. Commun.* **1999** 821; (d) W. Adam, C. R. Saha-Möller, O. Weichold, *J. Org. Chem.* **2000**, *65*, 2897; (e) W. Adam, C. R. Saha-Möller, O. Weichold, *J. Org. Chem.* **2000**, *65*, 5001; (f) W. A. Herrmann, T. Weskamp, J. P. Zoller, R. W. Fisher, *J. Mol. Catal. A*, **2000**, *153*, 49; (g) M. Nakajima, Y. Sasaki, H. Iwamoto, S. Hashimoto, *Tetrahedron Lett.* **1998**, *39*, 87; (h) W. A. Herrmann, R. M. Kratzer, H. Ding, W. R. Thiel, H. Glas, *J. Organomet. Chem.* **1998**, *555*, 293; (i) H. Rudler, J. R. Gregorio, B. Denise, J. M. Brégeault, A. Deloffre, *J. Mol. Catal. A* **1998**, *133*, 255; (j) W. D. Wang, J. H. Espenson, *J. Am. Chem. Soc.* **1998**, *120*, 11335; (k) A. L. Villa de P., D. E. DeVos, C. C. deMontes, P. A. Jacobs, *Tetrahedron Lett.* **1998**, *39*, 8521.
[11] (a) H. Tan, J. H. Espenson, *Inorg. Chem.* **1998**, *37*, 467; (b) J. R. Gregrio, A. E. Gerbase, M. Martinelli, M. A. M. Jacobi, L. de Luca Freitas, M. L. A. v. Holleben, P. D. Marcico, *Macromol. Rapid Commun.* **2000**, *21*, 401.
[12] (a) H. R. Tetzlaff, J. H. Espenson, *Inorg. Chem.* **1999**, *38*, 881; (b) W. Adam, C. M. Mitchell, C. R. Saha-Möller, *Eur. J. Org. Chem.* **1999**, 785; (c) J. Jacob, J. H. Espenson, J. H. Jensen, M. S. Gordon, *Organometallics,* **1998**, *17*, 1835; (d) S. Bellemin-Laponnaz, J. P. LeNy, A. Dedieu, *Chem. Eur. J.* **1999**, *5*, 57; (e) G. S. Owens, M. M Abu-Omar, *J. Chem. Soc., Chem. Commun.* **2000**, 1165.
[13] (a) W. Adam, W. A. Herrmann, J. Lin, C. R. Saha-Möller, R. W. Fischer, J. D. G. Correia, *Angew. Chem., Int. Ed. Engl.* **1994**, *33*, 2475; (b) W. Adam, M. Balci, H. Kilic, *J. Org.*

Chem. **1998**, *63*, 8544; (c) F. E. Kühn, J. J. Haider, E. Herdtweck, W. A. Herrmann, A. D. Lopes, M. Pillinger, C. C. Romão, *Inorg. Chim. Acta* **1998**, *279*, 44; (d) W. A. Herrmann, J. J. Haider, R. W. Fischer, *J. Mol. Catal. A*, **1999**, *138*, 115; (e) J. Jacob, J. H. Espenson, *Inorg. Chim. Acta* **1998**, *270*, 55.

[14] (a) A. M. F. Phillips, C. Romão, *Eur. J. Org. Chem.* **1999**, 1767; (b) W. A. Herrmann, R. W. Fischer, J. D. J. Correia, *J. Mol. Catal. A* **1994**, *94*, 213.

[15] M. M. Abu-Omar, J. H. Espenson, *Organometallics* **1996**, *15*, 3543.

[16] (a) D. W. Lahti, J. H. Espenson, *Inorg. Chem.* **2000**, *39*, 2164; (b) Y. Wang, J. H. Espenson, *J. Org. Chem.* **2000**, *65*, 104; (c) R. Huang, J. H. Espenson, *J. Org. Chem.* **1999**, *64*, 6935; (d) R. Huang, J. H. Espenson, *J. Org. Chem.* **1999**, *64,* 6374; (e) R. Huang, I. A. Guzei, J. H. Espenson, *Organometallics* **1999**, *18*, 5420; (f) H. N. Q. Gunaratne, M. A. McKervey, S. Feutren, J. Finlay, J. Boyd, *Tetrahedron Lett.* **1998**, *39*, 5655.

[17] M. D. Eager, J. H. Espenson, *Inorg. Chem.* **1999**, *38*, 2533.

[18] (a) Z. Zhu, J. H. Espenson, *Synthesis* **1998**, 417.

[19] (a) H. Rudler, B. Denise, *J. Chem. Soc., Chem. Commun.* **1998**, 2145; (b) S. Stankovic, J. H. Espenson, *J. Chem. Soc., Chem. Commun.* **1998**, 1579; (c) S. Stankovic, J. H. Espenson, *J. Org. Chem.* **2000**, *65*, 2218.

[20] (a) J. H. Espenson, O. Pestovsky, P. Huston, S. Staudt, *J. Am. Chem. Soc.* **1994**, *116*, 2869; (b) P. J. Hansen, J. H. Espenson, *Inorg. Chem.* **1995**, *34*, 5389.

[21] (a) W. Adam, C. M. Mitchell, C. R. Saha-Möller, O. Weichold, *J. Am. Chem. Soc.* **1999**, *121*, 2097; (b) S. Stankovic, J. J. Espenson, *J. Org. Chem.* **2000**, *65*, 5528.

[22] Z. Zhu, J. H. Espenson, *J. Org. Chem.* **1995**, *60,* 7728.

[23] (a) T. H. Zauche, J. H. Espenson, *Inorg. Chem.* **1998**, *37*, 6827; (b) J. H. Espenson, Z. Zhu, T. H. Zauche, *J. Org. Chem.* **1999**, *64*, 1191; (c) W. A. Herrmann, J. P. Zoller, R. W. Fischer, *J. Organomet. Chem.* **1999**, *581*, 404; (d) J. Finlay, M. A. McKervey, H. N. Q. Gunaratne, *Tetrahedron Lett.* **1998**, 5651; (f) S. Stancovic, J. H. Espenson, *J. Org. Chem.* **1998**, *63*, 4129.

[24] (a) W. A. Herrmann, J. D. G. Correia, F. E. Kühn, G. R. J. Artus, C. C. Romão, *Chem. J. Eur.* **1996**, *2*, 168; (b) W. R. Thiel, R. W. Fischer, W. A. Herrmann, *J. Organomet. Chem.* **1993**, *459*, C9; (c) W. A. Herrmann, M. R. Geisberger, F. E. Kühn, G. R. J. Artus, E. Herdtweck, *Z. Anorg. Allg. Chem.* **1997**, *623*, 1229.

[25] (a) W. A. Herrmann, M. Wang, *Angew. Chem., Int. Ed. Engl.* **1991**, *30*, 1641; (b) J. Jacob, J. H. Espenson, *J. Chem. Soc., Chem. Commun.* **1999**, 1003.

[26] (a) W. A. Herrmann, W. Wagner, U. N. Flessner, U. Volkhardt, H. Komber, *Angew. Chem., Int. Ed. Engl.* **1991**, *30*, 1636; (b) R. Buffon, A. Auroux, F. Lefebvre, et al., *J. Mol. Catal. A* **1992**, *76*, 287; (c) R. Buffon, A. Choplin, M. Leconte, et al., *J. Mol. Catal. A* **1992**, *72*, L7; (d) T. M. Mathews, J. A. K. duPlessis, J. J. Prinsloo, *J. Mol. Catal. A* **1999**, *148*, 157.

[27] Z. Zhu, J. H. Espenson, *J. Am. Chem. Soc.* **1997**, *119*, 3501.

[28] A. B. Kholopov, A. V. Nikitin, V. L. Rubailo, *Kinet. Katal.* **1995**, *36*, 101.

[29] (a) Z. Zhu, J. H. Espenson, *J. Org. Chem.* **1995**, *60*, 7090; (b) Z. Zhu, J. H. Espenson, *J. Org. Chem.* **1995**, *60*, 7728.

[30] Z. Zhu, J. H. Espenson, *J. Am. Chem. Soc.* **1995**, *118*, 9901.

[31] Z. Zhu, J. H. Espenson, *Synthesis* **1998** 417.

3.3.13.2   Pilot-Plant Synthesis of MTO

Wolfgang A. Herrmann

3.3.13.2.1 Introduction

The simple organorhenium(VII) compound methyltrioxorhenium (Structure **1** in Scheme 1) – called MTO – has developed a plethora of applications in catalytic processes [1]. This rapid development occurred in the decade of 1990–2000. The epoxidation of olefins (cf. Section 2.4.3) became attractive to industrial applications. There is sound evidence that MTO represents the most efficient catalyst for this process, being active even for highly dilute solutions of hydrogen peroxide. The latter oxidant is not decomposed by MTO, as opposed to many other metal complexes (cf. Section 3.3.13.1).

Due to the increasing industrial demand, a laboratory pilot-plant synthesis of the catalyst was developed. In the period of 1993–2000, a total of 189 papers and patents on MTO applications have appeared, showing the great interest in both academia and industry.

3.3.13.2.2 Principle of Synthesis

MTO is generated by methylation of oxidic Re^{VII} precursor compounds under nonreducing conditions. Thus, dimethylzinc and methyl Grignard compounds are not very well suited for this purpose. The first commonly applied synthesis started from dirhenium heptoxide using the toxic tetramethyltin as methylating reagent (eq. (1)). The major drawback in this otherwise excellent approach is the loss of half of the Re due to formation of the low-reactivity trimethylstannyl perrhenate [2].

$$Re_2O_7 + Sn(CH_3)_4 \longrightarrow CH_3ReO_3 + (CH_3)_3SnOReO_3 \qquad (1)$$
$$\mathbf{1}$$

An improvement was the use of the mixed ester of perrhenic and trifluoroacetic acid, avoiding the chemical loss of rhenium [3, 4]. At the same time, the much less toxic tris(n-butyl)methyltin was used for the selective methylation in eq. (2). This route reached the laboratory pilot-plant stage in 1999.

$$\tfrac{1}{2} Re_2O_7 + \tfrac{1}{2} [CF_3\text{-}C(=O)]_2O \longrightarrow [CF_3\text{-}C(=O)O]ReO_3 \qquad (2a)$$

$$[CF_3\text{-}C(=O)O]ReO_3 + (^nBu)_3SnCH_3 \longrightarrow CH_3ReO_3 + (^nBu)_3Sn\text{-}OC(=O)CF_3 \qquad (2b)$$
$$\mathbf{1}$$

A second industrially feasible synthesis has particularly focused on cheap starting compounds, abandoning both the trifluoric acetic anhydride and the (hygroscopic) dirhenium heptoxide. Instead, easily available perrhenates $M^+[ReO_4]^-$ are used ($M^+ = Ag^+$, $[NH_4]^+$, Na^+, K^+). They result from easy H_2O_2 oxidation of elemental Re according to eq. (3).

$$Re \xrightarrow{\quad H_2O_2 \quad} H[ReO_4] \xrightarrow[\;-HX\;]{\;+MX\;} M^+[ReO_4]^- \tag{3}$$

According to Scheme 1, the perrhenate is converted into the silylperrhenate, which undergoes subsequent stepwise transformation to the reactive chlorotrioxo-rhenium [5–7]. Half of the rhenium from the chlorination process of Re_2O_7 is recycled to the trimethyl stannyl perrhenate. The final methylation of $ClReO_3$ is achieved by tetramethyltin in near quantitative yields.

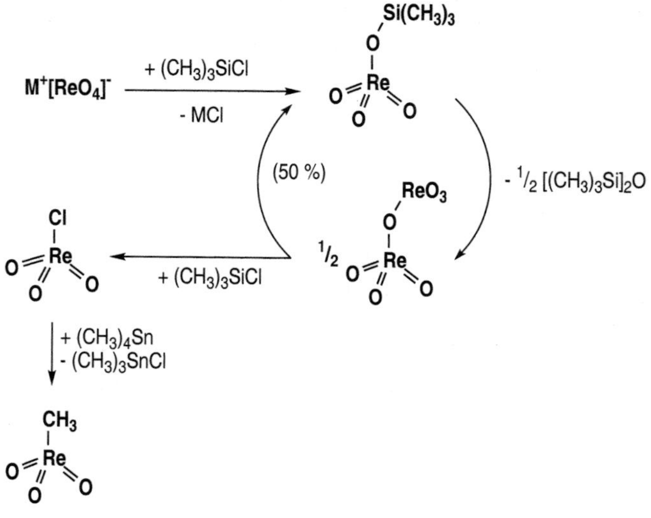

Scheme 1. Synthesis of MTO **1** from perrhenates.

The net reaction follows eq. (4). The overall yields range between 50 and 85 % based on rhenium. This methodology has the advantage that cheap precursor compounds can be used and that a basically unlimited scale-up is possible. Amounts up to 500 g of pure MTO **1** have thus been made.

$$M^+[ReO_4]^- + 2\,(CH_3)_3SiCl + Sn(CH_3)_4 \longrightarrow$$

$$CH_3ReO_3 + [(CH_3)_3Si]_2O + (CH_3)_3SnCl + MCl \tag{4}$$

The silver perrhenate, especially, forms reproducible, analytically pure, off-white MTO [8]. Selected catalytic applications of MTO are treated in Section 3.3.13.1.

3.3.13.2.3 Synthetic Procedures

The Mixed-Anhydride Route

Trifluoroacetic anhydride (105.1 g, 70.7 mL, 0.5 mol) is dissolved in 1500 mL of anhydrous acetonitrile. Freshly sublimed dirhenium heptoxide (Re_2O_7) (243.2 g, 0.5 mol) is added with vigorous stirring to avoid aggregation of Re_2O_7. After the Re_2O_7 has completely dissolved, the solution turns slightly green. Tris(n-butyl)-methylstannane (305.1 g, 1 mol) is added, whereby the color of the solution changes to dark brown. The mixture is stirred overnight (ca. 12 h); longer reaction times have no negative influence on yield and purity. The solvent is then removed at room temperature in an oil-pump vacuum (10^{-2} Torr) until the residue forms a paste. The product sublimes at 80 °C/10^{-2} Torr as colorless needles. If necessary the sublimate is washed with cold n-pentane to remove the last impurities. The off-white product was then dried *in vacuo*. Yield: 400–450 g (80–90 %).

The Perrhenate Route

Method A

Sodium perrhenate [$NaReO_4$] (100 g, 0.366 mol) is suspended in 1.5 L of acetonitrile. After addition of 102 mL (0.8 mol) of trimethylchlorosilane [$ClSi(CH_3)_3$] and 56 mL (0.4 mol) of tetramethyltin, the reaction mixture is heated for 10 h under reflux. At ambient temperature, the yellow-orange mother liquor is separated from the insoluble residue by filtration; the solvent is evaporated under reduced pressure. After the trimethylchlorotin has been removed as the first fraction of the sublimation, MTO is separated from the remaining residue by sublimation at 50 °C and 10^{-2} mbar. The MTO obtained may contain small amounts of trimethylchlorotin (detectable by its characteristic smell; **caution: trimethylchlorotin is highly poisonous!**). In this case, either a second careful sublimation (in several fractions) is recommended, or the MTO is stored on a filter paper in a flask under a slight vacuum until the undesired, very volatile by-product has gone. However, care has to be taken not to lose the MTO, too. Yield: 60–70 g (65–75 %).

Method B

The following reagents and respective amounts are needed: 1.5 L of acetonitrile, 1400 g (39 mol) of silver perrhenate [$AgReO_4$], 1080 mL (85 mol) of trimethylchlorosilane [$ClSi(CH_3)_3$], 600 mL (4.3 mol) of tetramethyltin.The quick formation of the reactive intermediate species is indicated by the spontaneous precipitation of silver chloride after the silver perrhenate and the trimethylchlorosilane are mixed together. Yield: 780–880 g (80–90 %).

The product thus obtained is of higher purity than MTO made in the same way from sodium perrhenate.

The perrhenates $M^+[ReO_4]^-$ ($M^+ = Ag^+$, K^+) are made from rhenium powder according to [9].

3.3.13.2.4 Properties

Methylrhenium trioxide MTO can be stored at room temperature without decomposition. The compound forms pale yellow needles, m. p. 112 °C. Direct exposure to light should be avoided.

IR (KBr): $v = 1002$ (vs), 950 cm^{-1}(vs. br. Re=O).
1H NMR (CDCl$_3$, 28 °C): $\delta = 2.61$ (s, CH$_3$).
^{13}C NMR (CDCl$_3$, 28 °C): $\delta = 19.03$ [d, ^1J(C, H) $=138$ Hz, CH$_3$].
^{17}O NMR (CDCl$_3$, 28 °C): $\delta = 829$.
EI-MS: $m/z = 248/250$ (molecular ion peak with correct isotope distribution ^{185}Re/^{187}Re for CH$_3$ReO$_3$, 249.21).

Solubility in water: 50 g/L (0.20 mol/L)
pKs(H$_2$O, 25 °C): 7.5 (sat. solution, pH $\approx$ 4)
$\rho = 4.103$ g cm^{-3}; $\mu = 2.6$ D (C$_6$H$_6$, 25 °C); magnetic susceptibility $\chi = -55 \cdot 10^{-6}$ cm^{-3} mol^{-1}; ionization potential $I_1 = 11.80$ eV; dissociation energy D(CH$_3$-Re) $= 319$ kJ $\cdot$ mol^{-1} (calc.).

Solid-state crystal structure (X-ray): d(Re–C) $= 204(3)$ pm, d(Re–O) $= 168(2)$ pm.

Small amounts of MTO are also available commercially from (a) Aldrich: 41,291–0 (100 mg, 500 mg) and (b) Fluka: 69489 (50 mg, 250 mg).

References

[1] Reviews: (a) W. A. Herrmann, F. E. Kühn, *Acc. Chem. Res.* **1998**, *30*, 169; (b) C. C. Romão, F. E. Kühn, W. A. Herrmann, *Chem. Rev.* **1997**, *97*, 3197; (c) *Aqueous-Phase Organometallic Catalysis-Concepts and Applications* (Eds.: B. Cornils, W. A. Herrmann), **1998**, pp. 529–538; (d) *Transition Metal Catalyzed Reactions* (Eds.: W. A. Herrmann, F. E. Kühn, in: S. I. Murahashi, S. G. Davies), IUPAC Series for the 21st Century Monographs, Blackwell Science, **1999**, pp. 375–390; (e) *Structure and Bonding* (Eds.: F. E. Kühn, W. A. Herrmann, in: B. Meunier), **2000**, *97*, pp. 211–234. Recent applications: W. A. Herrmann, J. P. Zoller, R. W. Fischer, *J. Organomet. Chem.* **1999**, *579*, 404; W. A. Herrmann, T. Weskamp, J. P. Zoller, R. W. Fischer, *J. Mol. Catal. A: Chemical*, **2000**, *153*, 49.
[2] W. A. Herrmann, J. G. Kuchler, J. K. Felixberger, E. Herdtweck, W. Wagner, *Angew. Chem.* **1988**, *100*, 420; *Angew. Chem., Int. Ed. Engl.* **1988**, *27*, 394.

[3] W. A. Herrmann, F. E. Kühn, R. W. Fischer, W. R. Thiel, C. C. Romão, *Inorg. Chem.* **1992**, *31*, 4431; W. A. Herrmann, R. W. Fischer, M. U. Rauch, W. Scherer, *J. Mol. Catal.* **1994**, *86*, 243.

[4] R. W. Fischer, Ph.D. Thesis, Technische Universität München, 1994.

[5] W. A. Herrmann, R. M. Kratzer, R. W. Fischer, *Angew. Chem.* **1997**, *109*, 2767; *Angew. Chem., Int. Ed.* **1997**, *36*, 2652.

[6] R. Kratzer, Ph.D. Thesis, Technische Universität München, 1998.

[7] F. E. Kühn, R. W. Fischer, W. A. Herrmann, *Chem. Unserer Zeit*, **1999**, *33*, 192.

[8] Further information on MTO and its uses are available at the author's homepage under the following Internet address: http://aci.anorg.chemie.tu-muenchen.de

[9] W. A. Herrmann, R. W. Fischer, M. Groarke, F. E. Kühn, in: *Synthetic Methods of Organometallic and Inorganic Chemistry* (Ed.: W. A. Herrmann), Vol. 10, Enke Verlag, Stuttgart, **2001**.

3.3.14 Acetoxylations and Other Palladium-Promoted or Palladium-Catalyzed Reactions

Reinhard Jira

3.3.14.1 Historical and Economic Background

Acetoxylations (oxyacylations) have to be seen in context with olefin oxidation to carbonyl compounds (Wacker process, Section 2.4.1). With the lowest olefin, ethylene, acetaldehyde is formed. In water-free acetic acid no reaction takes place. Only in the presence of alkali acetates – the acetate ion shows higher nucleophilicity than acetic acid – ethylene reacts with palladium salts (eq. (1)) to give vinyl acetate, the expected product, as first reported by Moiseev et al. [1]. Stern and Spector [2] independently used $[HPO_4]^{2-}$ as base in a mixture of iso-octane and acetic acid. This reaction could be exploited for a commercial process to produce vinyl acetate and closed the last gap replacing acetylene by the cheaper ethylene, a petrochemical feed material, for the production of large-tonnage chemical intermediates.

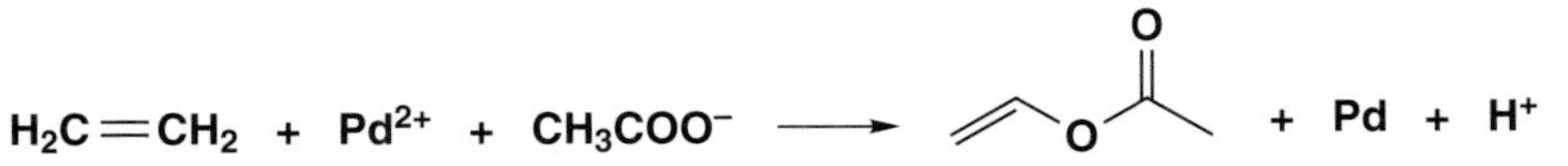

$$H_2C{=}CH_2 \ + \ Pd^{2+} \ + \ CH_3COO^- \ \longrightarrow \ \text{(vinyl acetate)} \ + \ Pd \ + \ H^+ \qquad (1)$$

The industrial and scientific success of these two reactions initiated a boom in palladium chemistry which is still continuing.

3.3.14.2 Chemical Background

Acetoxylation of olefins according to eq. (1) is an oxidative reaction which can be widely applied. However, it does not occur in such a distinct manner as olefin oxidation in aqueous solution. Depending on reaction conditions, diverse primary and secondary by-products arise. Thus, a main by-product in vinyl acetate synthesis is ethylidene diacetate; this has been shown to be a primary by-product, since with CH_3COOD as a reactant the product does not contain any deuterium [3] . This would not be the case if deuterated acetic acid adds to vinyl acetate in a secondary reaction. Glycol mono- and diacetates are found in the presence of a large amount of lithium nitrate [4] and nitric acid [5].

With higher olefins the product distribution becomes more variable. Not only the expected enol acetates but also allylic acetates are formed. Thus, propene forms isopropenyl acetate along with some *n*-propenyl acetate and allyl acetate [6, 7]. Higher and cyclic olefins react to form mainly allylic esters [8–18]; moreover pre-isomerization of the olefins give rise to an even broader spectrum of products. The results published differ from each other, probably because of different reaction conditions and composition of reaction mixtures.

Quite analogously to the olefin oxidation in aqueous medium, acetoxylation of olefins can also be carried out catalytically by addition of oxidants such as benzoquinone [1] , cupric chloride, and cupric acetate (a survey of the patent literature has been given by Krekeler and Schmitz [19] and Miller [20]) which oxidize the metallic palladium to the active oxidation state Pd^{II} (eq. (2)). Cuprous chloride is reoxidized by oxygen (eq. (3)) and the overall reaction according to eq. (4) becomes catalytic.

$$Pd + 2\,CuCl_2 \longrightarrow PdCl_2 + 2\,CuCl \qquad (2)$$

$$2\,CuCl + 2\,HCl + {}^1\!/_2\,O_2 \longrightarrow 2\,CuCl_2 + H_2O \qquad (3)$$

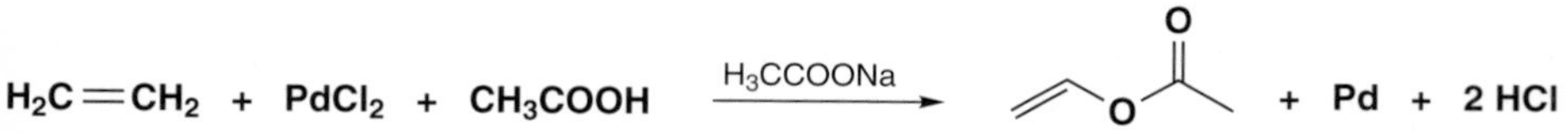

$$H_2C{=}CH_2 + PdCl_2 + CH_3COOH \xrightarrow{\;H_3CCOONa\;} \quad + \; Pd + 2\,HCl \qquad (1a)$$

$$H_2C{=}CH_2 + CH_3COOH + {}^1\!/_2\,O_2 \xrightarrow{\;PdCl_2,\,CuCl_2,\,H_3CCOONa\;} \quad + \; H_2O$$

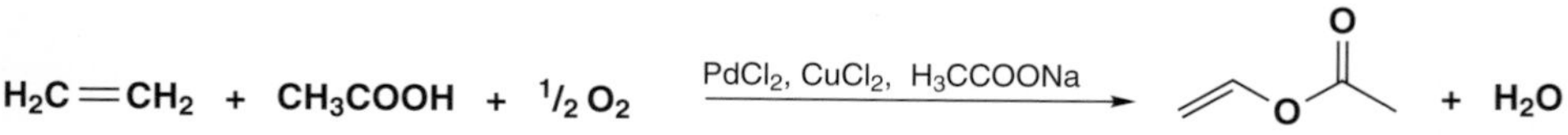

$$(4) = (1a) + (2) + (3)$$

Since water is formed in this reaction, acetaldehyde is also a by-product. It can arise (1) directly according to olefin oxidaton (Wacker process), (2) through hydrolysis of vinyl acetate which occurs very easily in the presence of $PdCl_2$, or – rather unlikely – (3) through an interaction of vinyl acetate with acetic acid in the presence of palladium chloride, a reaction according to eq. (5) and published by Clement and Selwitz [21] forming acetaldehyde and acetic anhydride.

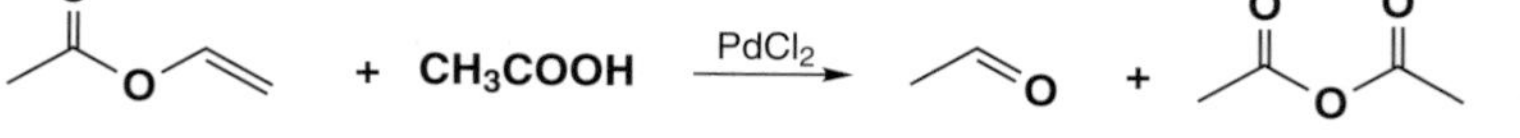

$$+ \; CH_3COOH \xrightarrow{\;PdCl_2\;} \quad + \qquad (5)$$

In vinyl acetate synthesis in the presence of cupric chloride according to eq. (4), other by-products are mono- and diacetates of glycol and β-chloroethyl acetate which under certain conditions become the predominant products [22]. In order to avoid side reactions, chloride-free catalyst systems such as $Pd^{II}/H_9PMo_6V_6O_{40}$ have been described [23].

For the commercial production of vinyl acetate, a procedure with a heterogeneous fixed-bed catalyst is exclusively applied today. The catalysts usually consist of palladium salts, mostly the acetate, or palladium metal together with alkali acetate supported on a carrier such as alumina, silica, or carbon without any additional oxidant. This process avoids the formation of larger amounts of by-products. Thus, from ethylene vinyl acetate and from propene, allyl acetate is obtained exclusively.

3.3.14.3 Kinetics and Mechanism

3.3.14.3.1 Homogeneous Reaction

Kinetic investigations have been carried out in the presence and absence of chloride. They seem to give a somewhat confusing picture of the mechanistic features of this reaction. However, some details also show certain similarities with the kinetics of the Wacker reaction in aqueous medium.

If eq. (6), derived by Ninomiya et al. [24] by studying the reaction of ethylene with palladium chloride in a mixture of acetic acid and *p*-xylene, is transformed by replacing [NaOAc] with $1/[H^+]$, it adopts the form of eq. (13) of Section 2.4.1 [25], showing an activating or an inhibiting effect of H^+ ions at low or higher H^+ concentrations, respectively. A similar behavior of Cl^- ions, also shown in the equation mentioned, was observed by van Helden et al. [26]. Clark et al. [27] published a rate equation (eq. (7)) quite similar to that of the Wacker reaction in aqueous system (see eq. (9) in Section 2.4.1).

$$\frac{-d\,[C_2H_4]}{d\,t} = \frac{k\,[PdCl_2]\,[HOAc]^2\,[NaOAc]\,[C_2H_4]}{1 + K\,[NaOAc]^2} \qquad (6)$$

$$\frac{-d\,[C_2H_4]}{d\,t} = \frac{k\,[Pd(II)]\,[LiOAc]\,[C_2H_4]}{[LiCl]^{1\text{-}2}} \qquad (7)$$

Other authors found other rate expressions. Thus, Moiseev et al. [28] found at high sodium acetate concentration a dependence according to eq. (8): while sodium acetate accelerates the reaction at low concentration (also found by Grover et al. [29]). They interpreted the activation by sodium acetate in terms of the formation of a mononuclear Pd complex from polynuclear palladium chloride according to eqs. (9) and (10).

$$\frac{-d\,[C_2H_4]}{d\,t} = \frac{k\,K\,[Na_2Pd(OAc)_4]\,[C_2H_4]}{[NaOAc]^2} \qquad (8)$$

$$Pd_nCl_{2n} + n\ NaOAc \rightleftharpoons Na_2Pd_2Cl_4(OAc)_2 \qquad (9)$$

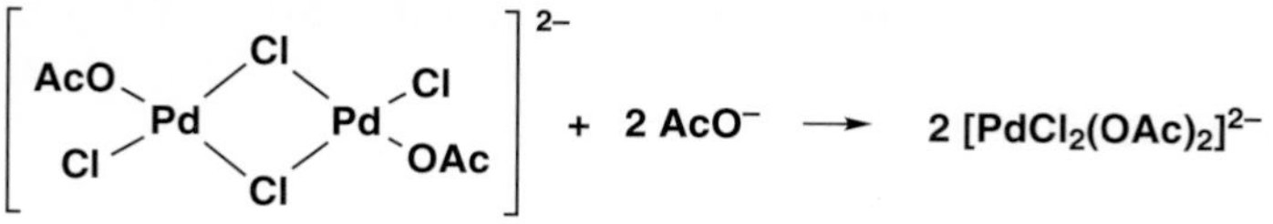

$$\left[\begin{array}{c} AcO \diagdown \quad \diagup Cl \diagdown \quad \diagup Cl \\ Pd \qquad Pd \\ Cl \diagup \quad \diagdown Cl \diagup \quad \diagdown OAc \end{array}\right]^{2-} + 2\ AcO^- \longrightarrow 2\ [PdCl_2(OAc)_2]^{2-} \qquad (10)$$

For a possible mechanism most authors assume an acetoxypalladation step after complexing of the olefin analogous to the reaction in aqueous medium (eq. (11)) with X = Cl^-, OAc^-, solvent).

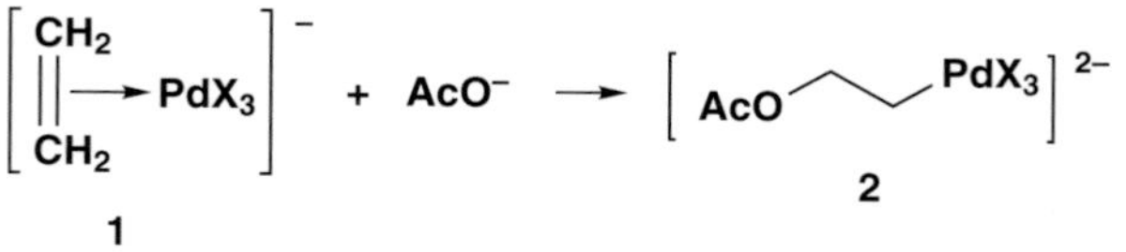

$$\left[\begin{array}{c} CH_2 \\ \| \longrightarrow PdX_3 \\ CH_2 \end{array}\right]^- + AcO^- \longrightarrow \left[AcO \diagup\diagdown\diagup PdX_3 \right]^{2-} \qquad (11)$$

$$\mathbf{1} \qquad\qquad\qquad \mathbf{2}$$

A *trans* attack, i.e., an attack of the acetoxy anion from the solution, is assumed [30] but a *cis* attack, i.e., a ligand insertion, cannot be excluded since *trans* attack has been proven with a cyclic olefin consisting of a rigid skeleton; but this is not typical for linear olefins, and the above interpretation of the activation by sodium acetate [27] would make some sense with two coordinated acetate ligands of which one would be in a *cis* position relative to a coordinated olefin.

Vinyl acetate is formed by a β-elimination of a hydridopalladium moiety (Structure **3**), which was the first step of the hydride shift in the acetaldehyde mechanism (eq. (12)) (Section 2.4.1).

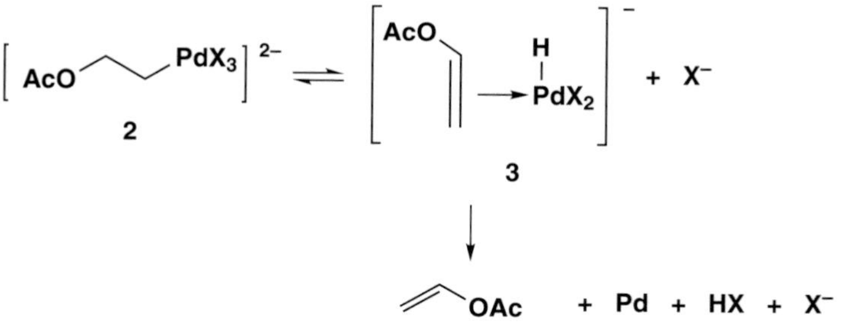

$$\left[AcO \diagup\diagdown\diagup PdX_3 \right]^{2-} \rightleftharpoons \left[\begin{array}{c} AcO \diagdown \qquad H \\ \| \quad | \\ \longrightarrow PdX_2 \end{array}\right]^- + X^-$$

$$\mathbf{2} \qquad\qquad\qquad\qquad \mathbf{3} \qquad (12)$$

$$\downarrow$$

$$\diagup\diagdown OAc \quad + Pd + HX + X^-$$

Extraction of a hydridopalladium moiety by β-elimination is a common step in many palladium-catalyzed sequences of reactions.

Hydride shift, as in olefin oxidation in aqueous medium, forming carbonyl compounds (see eq. (20) in Section 2.4.1) is completed under conditions in which, instead of vinyl compounds, ethylidene diacetate or acetals are formed, since using deuterated acids or alcohols, e. g., AcOD or ROD, the respective products do not contain any deuterium [3] . According to eqs. (13) and (14) with R = OAc^-, O-alkyl$^-$, the step leading to these products can be interpreted as reductive elimination.

$$\left[\begin{array}{c} RO \\ \| \longrightarrow \overset{\overset{\displaystyle H}{|}}{PdX_2} \end{array} \right]^{-} + RO^{-} \quad \rightleftharpoons \quad \left[(RO)X_2Pd \overset{\displaystyle }{\diagdown} OR \right]^{2-} \qquad (13)$$

$$\underset{\mathbf{3a}}{} \qquad\qquad\qquad\qquad\qquad\qquad \underset{\mathbf{4}}{}$$

$$\left[RO \overset{\displaystyle }{\diagup} PdX_2(OR) \right]^{2-} \longrightarrow \quad RO \overset{\displaystyle }{\diagdown} OR \; + \; Pd \; + \; 2\,X^{-} \qquad (14)$$

$$\underset{\mathbf{4}}{}$$

For the formation of glycol derivatives such as glycol mono- and diacetates and 1-acetoxy-2-chloroethane, the initial acetoxypalladation complex **2** might be the key intermediate. The diacetate may arise out of this complex through a β-elimination together with a coordinated acetate. As this reaction preferably occurs in the presence of nitrates and nitrites, coordinated nitro groups may assist [5] . For the monoacetate a less simple route has to be assumed as, surprisingly, with an ^{18}O-labeled nitrocomplex the ^{18}O appears exclusively in the acetate group of the monoester [31]. An acetyl group shift via the nitro ligand has been proposed. 1-Acetoxy-2-chloroethane is formed in the presence of a high excess of cupric chloride [22]. A bi- or oligonuclear Pd–Cu cluster may be responsible (see also the formation of 2-chloroethanol in aqueous medium described in Section 2.4.1.5.1).

3.3.14.3.2 Heterogeneous Reaction

Commercial production of vinyl acetate is nowadays carried out in the gaseous phase with a fixed-bed catalyst. A mixture of acetic acid, ethylene and oxygen is led over a catalyst consisting of palladium acetate or paladium-metal and alkali, mostly potassium acetate, and occasionally some activating metals such as gold or others on a carrier such as silica, alumina, or active carbon. The question is whether the reaction takes place in a pure heterogeneous phase or as a homogenous catalysis in a heterogenized liquid phase on the surface or in the pores of the carrier. Some authors assume the first case and give proof for this assumption [32–34]. Thus Davidson et al. [33] reached this conclusion from the fact that with higher olefins allylic esters are formed exclusively. From an initial-rate kinetic study [34] a mechanism with hydrogen abstraction from absorbed ethylene and acetic acid followed by combination of the radicals and of the absorbed hydrogen with activated oxygen has been proposed. Carbon dioxide, the main by-product, should be formed mainly by oxidation of acetic acid. At least this fact contradicts the findings with the commercial process, where a yield of 99 % with respect to acetic acid has been obtained while CO_2 formation is 7–14 % with respect to the carbon feed (see Section 3.3.14.4).

Others [35–38] assume a principally liquid-phase reaction on the carrier, although their kinetic investigations are less helpful to prove this assumption.

Some experimental hints, however, give more information. Thus, finely divided palladium metal is readily oxidized under the reaction conditions by oxygen in the presence of acetic acid, even in the absence of any additional oxidant (eq. (15)) [27, 39].

$$Pd + 2\,AcOH + {}^1\!/_2\,O_2 \longrightarrow Pd(OAc)_2 + H_2O \tag{15}$$

Accordingly, whether the original catalyst is metallic or a bivalent salt, it will adopt the same configuration after some time. The effect of additional activators such as gold is to prevent agglomeration of the palladium metal, which would gradually deactivate the catalyst. Even chloro compounds, which to a great extent inhibit the gaseous-phase reaction completely, activate in trace amounts [40], and it can be assumed that they facilitate the oxidation of palladium into the bivalent state. A study on the selectivity of vinyl acetate formation [38] shows that for the formation of vinyl acetate and carbon dioxide, the main by-product, two different active centers of the catalyst are responsible. Only a recent investigation on the role of acetic acid in this reaction [41] showed the likelihood that Pd^{II} is the active species of the catalyst.

If for this gas-phase reaction the contemporary presence of a palladium^{II} species, acetic acid, alkali acetate, ethylene, and oxygen is necessary, a classical heterogeneous catalysis seems to be rather unlikely; preferably a sequence of single reactions, as in the homogeneous phase, has to be assumed. This could occur within the acetic acid film adsorbed on the carrier. Thus vinyl acetate formation in the gas-phase might occur according to eq. (1b) (M = Li, Na, K) and the overall reaction follows eq. (16).

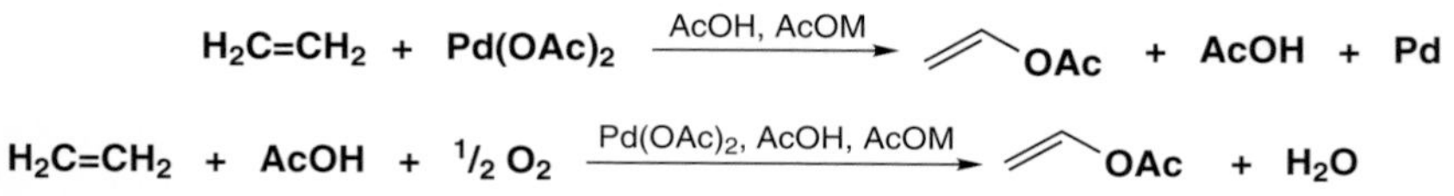

$$H_2C{=}CH_2 + Pd(OAc)_2 \xrightarrow{AcOH,\ AcOM} \quad \diagup\!\!\!\diagdown OAc + AcOH + Pd \tag{1b}$$

$$H_2C{=}CH_2 + AcOH + {}^1\!/_2\,O_2 \xrightarrow{Pd(OAc)_2,\ AcOH,\ AcOM} \quad \diagup\!\!\!\diagdown OAc + H_2O$$

$$(16) = (1b) + (15)$$

3.3.14.3.3 Allylic Oxidation

This reaction describes the entrance of a nucleophile into the allylic position of an olefin. In aqueous medium this reaction is of minor importance but in nonaqueous medium, particularly under the conditions of acetoxylation, it attracts broad interest. As already mentioned above and outlined later (see Section 3.3.14.6), higher and cyclic olefins give exclusively allylic esters. Two mechanisms have been proposed. One possibility is according to eq. (17) hydride abstraction through the palladium of an oxypalladation moiety by β-elimination from the adjacent C-atom which had not been added to the nucleophile [9].

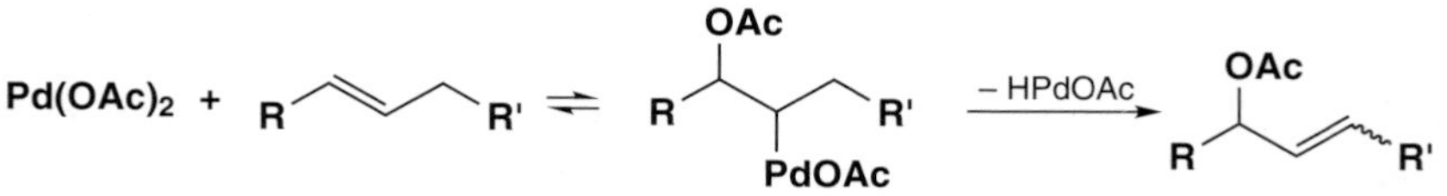

The other route includes a π-allylpalladium intermediate according to eq. (18). π-Allyl complexes can be obtained from olefins, preferably branched at the double bond. Their formation is often supported by a base for proton abstraction [42–44].

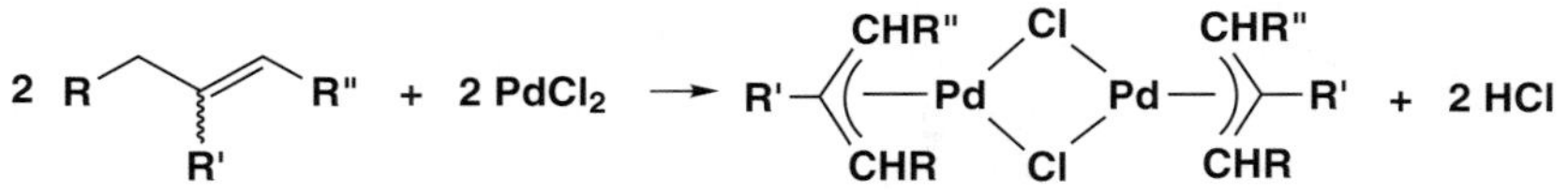

$$2\ R\diagup\!\!\!\diagdown\!\!\diagup\!\!\!\diagdown R'' + 2\ PdCl_2 \longrightarrow R'-\!\!\Big\langle\!\!\Big(\!\!\underset{\text{CHR}}{\overset{\text{CHR''}}{-}}Pd\underset{\text{Cl}}{\overset{\text{Cl}}{\diagup\diagdown}}Pd\!-\!\!\Big)\!\!\Big\rangle\!\!-\!R' + 2\ HCl \quad (18)$$

π-Allyl complexes are also formed from 1,3-diolefins [45], from allyl halides [46–48], and from allyl alcohol [44, 49]. By the latter reaction Hafner and coworkers [44] obtained the first π-allyl complex of palladium and firstly described the π-allylic bond as a three-hapto, four-electron bond if the palladium is assumed to be in the oxidation state Pd^{II}.

π-Allyl palladium complexes react with nucleophiles to give allylic compounds, e. g., according to eq. (19):

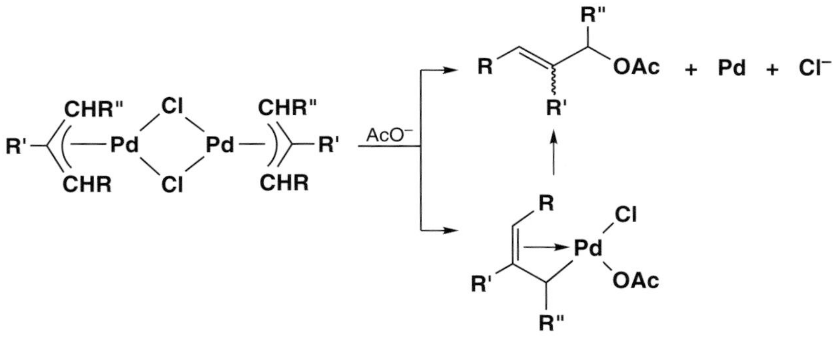

$$(19)$$

This route shows a possible intermediate from which the allylic product is obtained by reductive elimination. Also, for the formation of allyl acetate in the palladium-catalyzed gas-phase oxidation of propene, a π-allylic species on the catalyst surface has been proposed [35]. π-Allyl complexes of various metals have been synthesized in recent years. For many catalytic reactions they are assumed to be intermediates [50–53].

3.3.14.4 Commercial Processes

3.3.14.4.1 Vinyl Acetate

Vinyl acetate is the basis of a broad spectrum of polymers and copolymers mainly used as emulsions for adhesives, paints, concrete admixtures, coatings, and binding agents for paper, textiles, etc.

Thus, immediately after the publication by Moiseev et al. [1] (cf. Section 2.4.2), many industrial laboratories focused their research activities on development of a commercial vinyl acetate manufacturing process quite analogous to the liquid-phase Wacker acetaldehyde process. Commercial plants have been operated by lCl (UK), Celanese (USA), and Tokuyama Petrochemical (Japan) [54], but they have been shut down already. Not only the large number of by-products – acetaldehyde is the main one because of the water formed during the catalytic reaction according to eq. (4) – but evidently enormous corrosion problems might have been the reasons for this, since titanium (the construction material in the Wacker process) is not stable against the corrosive cupric chloride in non-aqueous media.

More than 80 % of the world's capacity of vinyl acetate is now produced by gas-phase technology. Hoechst/Bayer [55–57] and National Distillers Products Corporation [58] have developed their respective processes.

Although recent literature gives evidence for a homogeneous reaction on the catalyst surface (cf. Section 3.3.5.3.2) the commercial processes show typical features of heterogeneous catalysis and do not comply with the definition of homogeneous catalysis given in the preface. Therefore a description of the manufacturing process is not included here but it is dealt with in [59].

3.3.14.4.2 Allyl Acetate

Allyl Acetate can be a starting material for polymers as a comonomer, for glycidyl acetate as a component of epoxy resins, for glycerol, for allyl alcohol by hydrolysis, and for the synthesis of allyl esters of higher acids or other compounds by transesterification or transallylation. Processes for the commercial production of allyl acetate have been developed by Bayer and Hoechst [60], and some Japanese companies. Such processes work in the gas-phase under similar conditions to those used in vinyl acetate manufacture and are currently operated by Showa Denko and Daicel [61].

3.3.14.4.3 1,4-Diacetoxy-2-Butene

Butene-1,4-diol is obtained by hydrolysis of 1,4-diacetoxybutene and is an intermediate for the production of tetrahydrofuran, pharmaceuticals, terpenes, polyesters, etc. [62, 63]. According to eq. (20), 1,4-diacetoxybutene is formed by acetoxylation of butadiene over a palladium catalyst.

$$\text{\hspace{2cm} + 2 AcOH + } \tfrac{1}{2}\, O_2 \xrightarrow[-\,H_2O]{Pd^{II},\,Pd} \text{AcO} \diagdown\hspace{-0.2cm}\diagup\hspace{-0.2cm}\diagdown \text{OAc} \hspace{2cm} (20)$$

This reaction has been commercialized by Misubishi Kasei Corporation [61, 62, 64] using a Pd-on-carbon catalyst activated by tellurium in the liquid phase.

3.3.14.5 Transvinylation

Vinyl compounds undergo acetoxylation or generally oxyacylation with replacement of the nucleophile of the vinylic compound by the entering carboxylic group (eqs. (21) and (22)). Thus, vinyl chloride gives vinyl acetate with acetic acid in the presence of palladium chloride [65, 66].

$$
\diagup\!\!\diagup\text{Cl} + \text{AcO}^- \xrightarrow[\text{AcOH}]{\text{PdCl}_2} \diagup\!\!\diagup\text{OAc} + \text{Cl}^- \tag{21}
$$

If vinyl esters react with carboxylic acids, an equilibrium between the two vinyl esters is of course obtained [67]:

$$
\diagup\!\!\diagup\text{O}\underset{\text{O}}{\overset{\text{O}}{\diagdown}}\text{R} + \text{R'COOH} \xrightleftharpoons[]{\text{PdCl}_2, \text{AcO}^-} \diagup\!\!\diagup\text{O}\underset{\text{O}}{\overset{\text{O}}{\diagdown}}\text{R'} + \text{RCOOH} \tag{22}
$$

It has been shown that it is the vinyl group that is transferred in this reaction, not an alcoholic one as in acid- or alkali-catalyzed transesterification of esters with saturated alcoholic components. With ^{18}O-labeled acetic acid in the transvinylation of vinyl-propionate, the labeled oxygen remained completely in the acetate group [68].

This reaction gained technical importance for the synthesis of vinyl esters of higher carboxylic acids [69–72]. Thus, divinyl adipate is now produced by Wacker-Chemie from adipic acid and vinyl acetate.

Alkyl-substituted vinyl groups such as iso- and *n*-propenyl groups can be transferred in the presence of palladium salts [68, 73]. The new group enters exclusively at that carbon atom to which the leaving group is bound.

Transvinylation occurs stereospecifically, but a difference between vinylic esters and chlorides has been observed. While the transvinylation of *n*-propenyl esters is accompanied by an inversion of the geometric configuration (i.e., the *cis* starting ester results in a *trans* product and vice versa [68]), in the case of the corresponding chlorides the resulting esters retain their configuration [50, 74]. This can be explained in the first case by a *cis* addition (*cis*-carboxy-palladation) and *cis* elimination or *trans* addition and *trans* elimination, and in the second case by a *trans* addition and *cis* elimination or vice versa. It cannot yet be decided which of the two versions would in fact occur. But if it is assumed that *cis*-elimination is the most likely route in both cases the acetoxypalladation (addition) must be *trans (Anti)* in vinyl transfer from vinyl chloride and *cis (syn)* in that from vinyl esters. *syn*-acetoxypalladation was also found in addition to chiral allylic alcohols [75]. These findings prove that in Wacker-type reactions both *cis (syn)* and *trans (anti)* addition reactions are possible, depending on substrates and reaction conditions.

The vinyl group can also be transferred to alcohols to form vinyl ethers starting from vinyl chloride [66], vinyl esters [76], and vinyl ethers [77]. The transfer to water corresponds to a hydrolysis which readily takes place with all vinylic compounds in the presence of palladium salts to form acetaldehyde [78].

3.3.14.6   Acetoxylation in Organic Synthesis

Acetoxylation is a valuable method for the introduction of an OH group into organic compounds, which can be used for further syntheses. In Section 3.3.14.2 it has been mentioned that acetoxylation of higher and cyclic olefins with palladium salts, or catalyzed by palladium salts or metal, mostly leads to allylic derivatives. This also takes place in the catalytic acetoxylation of terpenic olefins [79, 80].

In the presence of nitrates [4, 81], palladium nitro complexes [5, 31], and with different oxidants, glycol derivatives are produced according to eq. (23):

$$CH_2=CH_2 \ + \ \text{oxidant} \ + \ \text{AcOH} \ \xrightarrow{\ Pd^{II}\ } \ X\diagdown\diagup\diagdown_{OAc} \tag{23}$$

A survey of this reaction is given by Henry [82]. The oxidants include $K_2Cr_2O_7$, $NaNO_2$, $CuBr_2$, MnO_2, $Pb(OAc)_4$, etc. Ferric and molybdic salts and others are inactive; they are more suitable as oxidants in monoacetoxylation of olefins, e. g., to form vinyl acetate. Although 1-acetoxybutadiene was found in the reaction of butadiene with $PdCl_2$ [2], in the presence of oxidants generally linear [83–86] and cyclic 1,3-diolefins [87] undergo diacetoxylation to form 1,4-diacetoxy-2-enes with homogeneous and heterogeneous palladium^{II} and metal catalysts. This reaction occurs regio- and stereospecifically, as could be proven with cyclic 1,3-dienes [88].

In the presence of LiCl, palladium acetate catalyzes the regio- and stereospecific 1,4-acetoxychlorination of linear and cyclic 1,3-dienes. The oxidant is benzoquinone (eq. (24)) [89].

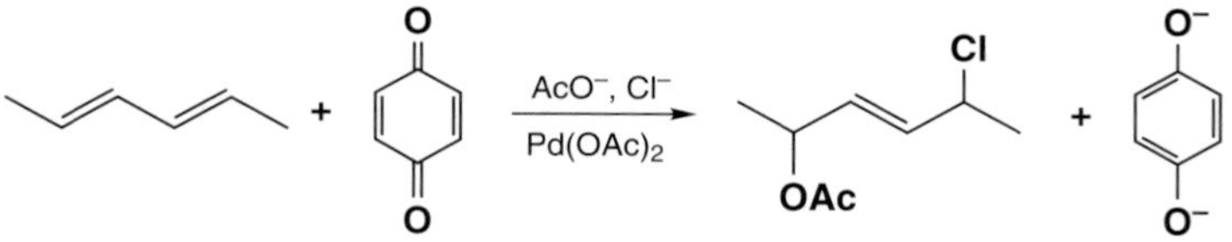

$$\tag{24}$$

Examples of the use of 1,4-addition products in organic synthesis have been given by Bäckvall [90, 91].

Acetoxylation of aromatics (eq. (25)) was first carried out by Davidson and Triggs [92] to produce phenyl acetate, which could give rise to the development of a new phenol synthesis. The formation of phenyl acetate is accompanied by the formation of biphenyl (see Section 3.3.14.7.2). In the presence of oxidants such as $Pb(OAc)_4$, $NaNO_2$, $NaNO_3$, $KMnO_4$, K_2CrO_7 [93] and P–Mo–V heteropolyacids [94], phenyl acetate is the main product. The favored ring acetoxylation with high-oxidation-state Pd catalyst over dimerization with low-oxidation-state catalyst has recently been confirmed [95]. With toluene, probably in an allylic-like oxidation, benzyl acetate is obtained [92].

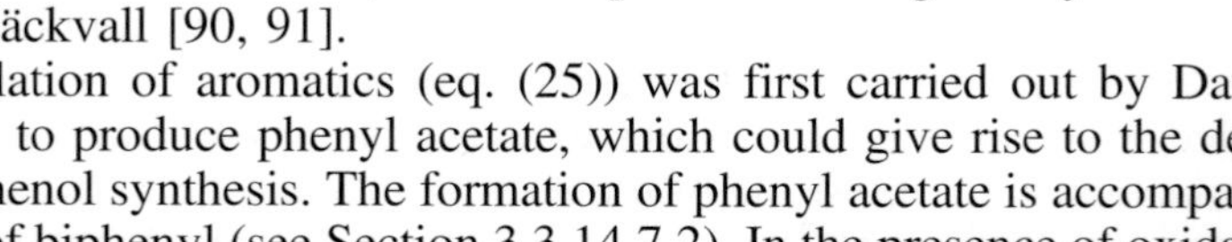

$$\text{(ring)} \ + \ \text{Pd(OAc)}_2 \ \xrightarrow{\ \text{AcOH, AcONa}\ } \ \text{(ring)} \ + \ \text{Pd} \ + \ \text{AcOH} \tag{25}$$
$$\text{OAc}$$

3.3.14.7 Other Palladium-Promoted or Palladium-Catalyzed Reactions

3.3.14.7.1 Alkoxylations

Reactions of olefins with alcoholic solutions of palladium salts lead mainly to acetals or ketals [93] (eq. (26) with R = H, alkyl; R' = alkyl).

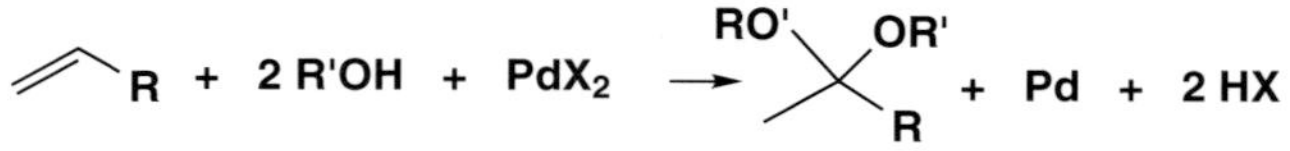

$$\overset{}{\diagup}R \ + \ 2\,R'OH \ + \ PdX_2 \ \longrightarrow \ \overset{RO'\quad OR'}{\underset{R}{\times}} \ + \ Pd \ + \ 2\,HX \qquad (26)$$

Corresponding vinyl ethers are formed only to a minor extent. Their formation can, however, be favored by diluting the reaction mixture with an inert solvent [2]. With glycols, cyclic acetals are obtained, e. g., dioxolanes from 1,2-glycols and 1,3-dioxanes from 1,3-glycols [97]. This reaction has been claimed as an alternative to the Wacker route producing acetaldehyde when it is carried out in the presence of suitable oxidants.

Acetals or ketals respectively are primary products since, quite analogously to the formation of ethylidene diacetate in the acetoxylation of ethylene with ROD, only deuterium-free acetals or ketals are obtained [3]. The mechanism ought to be analogous, too (cf. Section 3.3.14.3).

3.3.14.7.2 Oxidative Coupling

C–C coupling of aromatics is catalyzed by $Pd(OAc)_2$ in the presence of sodium acetate (eq. (27)) [98].

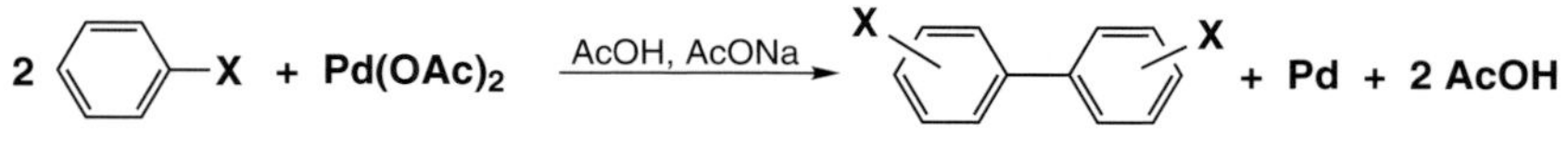

$$2 \ \langle\!\!\!\bigcirc\!\!\!\rangle\!\!-X \ + \ Pd(OAc)_2 \ \xrightarrow{\text{AcOH, AcONa}} \ \overset{X}{\langle\!\!\!\bigcirc\!\!\!\rangle}\!\!-\!\!\overset{X}{\langle\!\!\!\bigcirc\!\!\!\rangle} \ + \ Pd \ + \ 2\,AcOH \qquad (27)$$

Biphenyls are also by-products of acetoxylation of aromatics [92]. Their formation is favored with a palladium metal catalyst in the absence of oxidants [93–95]. Vinyl acetate undergoes oxidative coupling under similar conditions to form 1,4-diacetoxy-1,3-butadiene [99], and aromatics and heterocycles can substitute an olefinic H-atom [100] according to eq. (28) (with X = H, CN, AcO, EtO) [100–102].

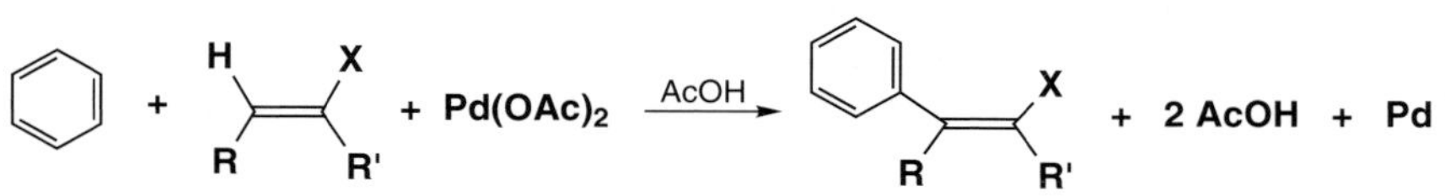

$$\langle\!\!\!\bigcirc\!\!\!\rangle \ + \ \overset{H\quad X}{\underset{R\quad R'}{\diagdown\!\!\!\diagup}} \ + \ Pd(OAc)_2 \ \xrightarrow{\text{AcOH}} \ \overset{\langle\!\!\!\bigcirc\!\!\!\rangle}{\underset{R\quad R'}{\diagdown\!\!\!\diagup}}\!\!\!X \ + \ 2\,AcOH \ + \ Pd \qquad (28)$$

(see also Section 3.1.9 and Heck reaction, Section 3.1.6).

$Pd(OAc)_2$ also activates aromatic hydrogen atoms for *non*-oxidative addition to alkenes and alkynes [103, 104].

3.3.14.7.3 Catalytic Coupling Using Organometallics

Main-group organometallics such as those of Li, Mg, Sn, Pb, Tl, Hg, etc., react with olefinic compounds and palladium(II) compounds to give coupling products according to the general scheme in eq. (29):

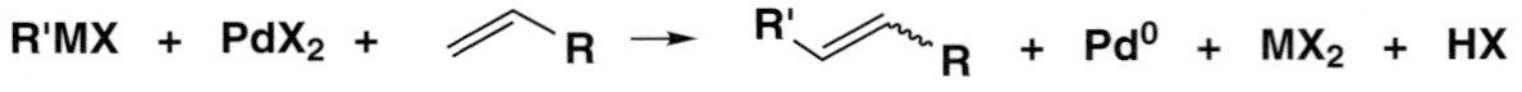

$$\text{R'MX} + \text{PdX}_2 + \;\diagup\!\!\!\diagdown_R \;\longrightarrow\; {}^{R'}\!\diagup\!\!\diagdown\!\!\diagup_R + \text{Pd}^0 + \text{MX}_2 + \text{HX} \qquad (29)$$

The organometallics can be derived from aliphatics and aromatics, even functionalized ones; the olefinics also show a broad spectrum in their applicability. Heck [105–107] studied this reaction extensively with M = Hg. With oxidants like $CuCl_2$ the reaction can be carried out catalytically. It is now known as the "Heck reaction". Special applications are referenced in [108–112]; see also Sections 2.1.2 and 3.1.6.

3.3.14.7.4 Oxidative Carbonylation (Carboxylation)

This reaction was first published by Tsuji et al. [113], who reacted ethylene with carbon monoxide in the presence of palladium chloride (eq. (30)).

$$\text{CH}_2{=}\text{CH}_2 + \text{CO} + \text{PdCl}_2 \;\xrightarrow{\text{benzene}}\; \text{Cl}\diagup\!\!\diagdown\!\!\diagup\!\!\overset{\displaystyle O}{\underset{\displaystyle}{\diagdown}}_{\text{Cl}} + \text{Pd} \qquad (30)$$

In acetic acid as solvent, acrylic acid is produced (eq. (31)), accompanied by some acetoxypropionic acid [114].

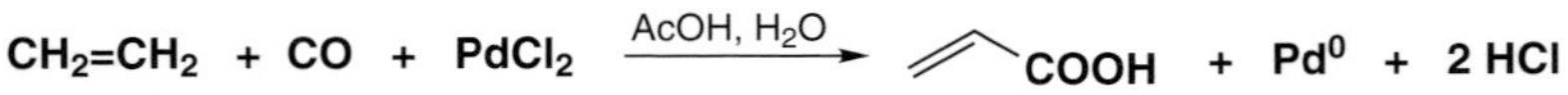

$$\text{CH}_2{=}\text{CH}_2 + \text{CO} + \text{PdCl}_2 \;\xrightarrow{\text{AcOH, H}_2\text{O}}\; \diagup\!\!\!\diagdown_{\text{COOH}} + \text{Pd}^0 + 2\,\text{HCl} \qquad (31)$$

With alcohols, esters are formed. Oxidants like $CuCl_2$ render the reaction catalytic, too (cf. Section 2.1.2.5).

Similarly, aromatic [115–117] and even aliphatic [118, 119] compounds are carboxylated catalytically to give aromatic and aliphatic carboxylic acids respectively with carbon monoxide and oxidants such as $K_2S_2O_8$, *t*-BuOOH or O_2 in the presence of palladium acetate, copper acetate and trifluoroacetic acid. Acetic acid has also been obtained in high yield by carboxylation of methane with CO in the presence of VO(acetylacetonato)$_2$ [120] and $CaCl_2 \cdot H_2O$ [121] as catalysts, and oxidants like $K_2S_2O_8$ and trifluoroacetic acid. Excitingly, aromatic [122] and aliphatic compounds such as methane [119] react under similar conditions with carbon dioxide also to give the respective carboxylic acids. These reactions are evidently promoted by a simultaneous aromatic and aliphatic C–H bond activation through the transition metal compound.

3.3.14.7.5 Dimerization

The dimerization of olefins catalyzed by noble metal compounds is a non-oxidative reaction (eq. (32)). It takes place in the absence of any other reactant and often accompanies other reactions [78].

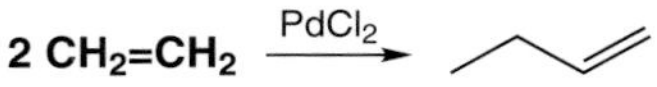

$$2 \; CH_2=CH_2 \xrightarrow{PdCl_2} \diagup\diagdown\diagup\diagdown \tag{32}$$

Cramer studied the mechanism using Rh compounds [123, 124]. The linear dimerization of acrylonitrile in the presence of hydrogen and a ruthenium catalyst to form 1,4-dicyano-1-butene [125–127] received much interest in industrial laboratories, mostly in Japan, so a large number of patents have been filed, in some of which palladium catalysts have also been claimed. The above compound can be hydrogenated to give the industrially important adipodinitrile and hexamethylene diamine.

3.3.14.7.6 Telomerization

Telomerization is the functionalization of olefins simultaneously with oligomerization, e. g., eq. (33) (cf. Section 2.3.5) [128, 129].

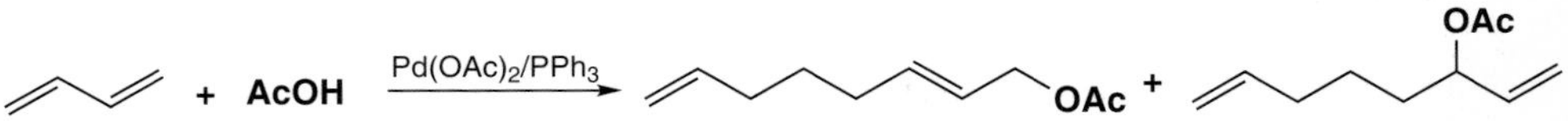

$$\diagup\diagdown\diagdown + \; AcOH \xrightarrow{Pd(OAc)_2/PPh_3} \text{\small (products)} \; OAc \; + \; \text{\small (products)} \; OAc \tag{33}$$

Telomerization with other nucleophiles, such as OH^-, or through carbonylation has also been described.

3.3.14.7.7 Double-Bond Shift (Isomerization)

Non-oxidative isomerizations often occur when olefinic compounds react with noble metal compounds, e. g., in Wacker oxidation of higher olefins. An example is found in the oxidation of 1-octene where octane-2-, 3-, and 4-ones are formed, in this example with an immobilized Pd^{II} catalyst [130]. A plausible mechanism with a hydridorhodium species as catalytically active moiety has been described by Cramer [131].

In another example substituted allyl alcohols undergo non-oxidative double-bond shift if there is no labile coordination site available for the hydride transfer. Isomerization is explained by a reversible hydroxypalladation reaction [132].

3.3.14.7.8 Skeletal Rearrangement

Skeletal rearrangement takes place, for instance, with vinylcycloalkanes which react under ring extension taking the vinyl group into the ring to form the respective enlarged cycloalkenes, e. g., vinylcyclopropane gives cyclopentene [133]. Further examples are listed in [50, pp. 143–146, Table XIII].

3.3.14.7.9 Addition Reactions

Other non-oxidative reactions catalyzed by palladium compounds are addition reactions e. g., eq. (34). Thus, alcohols easily add to double bonds of olefinic ketones [134].

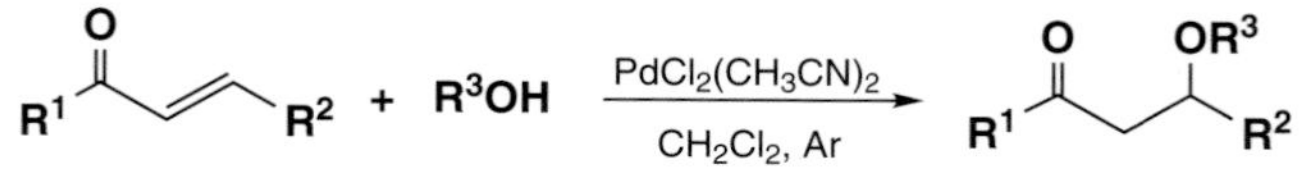

$$
\underset{R^1}{\overset{O}{\|}} \diagdown \diagup \diagdown {}_{R^2} \ + \ R^3OH \ \xrightarrow[\text{CH}_2\text{Cl}_2, \ \text{Ar}]{\text{PdCl}_2(\text{CH}_3\text{CN})_2} \ \underset{R^1}{\overset{O}{\|}} \diagdown \diagup \underset{R^2}{\overset{OR^3}{|}} \tag{34}
$$

Alcohols and amines add also to acetylenic compounds in the presence of palladium compounds to form intermediates for syntheses of heterocycles, e. g., furans and pyrroles [135].

3.3.14.8   Conclusions

This brief listing of reactions, not including oxyacylations, ought to demonstrate the broad applicability of palladium reagents in synthetic organic chemistry. These reactions are used in multistep syntheses to produce pharmaceuticals, fragrances, natural products, etc., and are documented in recent publications. Surveys are given in many review articles and books [50–53, 136–148].

References

[1] I. I. Moiseev, M. N. Vargaftik, Y. K. Syrkin, *Dokl. Akad. Nauk SSSR* **1960**, *133*, 377.
[2] E. W. Stern, M. L. Spector, *Proc. Chem. Soc. (London)* **1961**, 370.
[3] I. I. Moiseev, M. N. Vargaftik, *Izv. Akad. Nauk SSSR, Ser. Khim.* **1965**, 759.
[4] M. Tamura, T. Yasui, *Chem. Commun.* **1968**, 1209.
[5] M. G. Volkhonskii, V. A. Likholobov, Yu. I. Ermakov, *Kinet. Katal.* **1983**, 24, 347, 578; *Kinet. Katal. Engl. Transl.* **1983**, 24, 289, 488.
[6] A. P. Belov, G. Yu. Pek, I. I. Moiseev, *Izv. Akad. Nauk SSSR, Ser. Khim.* **1965**, 2204.
[7] I. I. Moiseev, A. P. Belov, Y. K. Syrkin, *Izv. Akad. Nauk SSSR, Ser. Khim.* **1963**, 1527.
[8] A. P. Belov, I. I. Moiseev, *Izv. Akad. Nauk SSSR, Ser. Khim.* **1966**, 139.
[9] W. Kitching, Z. Rappoport, S. Winstein, W. G. Young, *J. Am. Chem. Soc.* **1966**, 88, 2054.
[10] M. N. Vargaftik, I. I. Moiseev, Y. K. Syrkin, V. V. Yakshin, *Izv. Akad. Nauk SSSR, Otd. Khim. Nauk* **1962**, 930.
[11] C. B. Anderson, S. Winstein, *J. Org. Chem.* **1963**, *28*, 605.

[12] S. Hansson, A. Heumann, T. Rein, B. Åkermark, *J. Org. Chem.* **1990**, *55*, 975.

[13] St. E. Byström, E. M. Larsson, B. Åkermark, *J. Org. Chem.* **1990**, *55*, 5674.

[14] B. Åkermark, S. Hansson, T. Rein, J. Vågberg, A. Heumann, J.-E. Bäckvall, *J. Organomet. Chem.* **1989**, *369*, 433.

[15] A. Heumann, B. Åkermark, S. Hansson, T. Rein, *Org. Synth.* **1990**, *68*, 109.

[16] S. Uemura, Sh. Fukuzawa, A. Toshimitsu, M. Okano, *Tetrahedron Lett.* **1982**, *23*, 87.

[17] M. Green, R. N. Haszeldine, J. Lindley, *J. Organomet. Chem.* **1966**, *6*, 107.

[18] Y. Odaira, T. Yoshida, S. Tsutsumi, *Technol. Report, Osaka Univ.* **1966**, *16*, 737; *Chem. Abstr.* **1967**, *67*, 90334.

[19] H. Krekeler, H. Schmitz, *Chem.-Ing.-Tech.* **1968**, *40*, 785.

[20] S. A. Miller in *Ethylene and its Industrial Derivatives* (Ed.: S. A. Miller, Ernest Benn), London, **1969**, p. 946.

[21] W. H. Clement, C. M. Selwitz, *Tetrahedron Lett.* **1962**, 1081.

[22] P. M. Henry, *J. Org. Chem.* **1967**, *32*, 2575.

[23] I. V. Kozhevnikov, V. E. Taraban'ko, K. I. Matveev, V. D. Vardanyan, *React. Kinet. Catal. Lett.* **1977**, *7*, 297.

[24] R. Ninomiya, M. Sato, T. Shiba, *Bull. Jpn. Petr. Inst.* **1965**, *7*, 31.

[25] R. Jira, J. Sedlmeier, J. Smidt, *Liebigs Ann. Chem.* **1966**, *693*, 99.

[26] R. van Helden, C. F. Kohll, D. Medema, G. Verberg, T. Jonkhoff, *Rec. Trav. Chim. Pays-Bas* **1968**, *87*, 961.

[27] D. Clark, P. Hayden, R. D. Smith, *Disc. Faraday Soc.* **1967**, *46*, 98.

[28] I. I. Moiseev, A. P. Belov, V. A. Igoshchin, Y. K. Syrkin, *Dokl. Akad. Nauk SSSR* **1967**, *173*, 863.

[29] G. S. Grover, R. V. Chaudhari, *Chem. Eng. J.* **1986**, *32*, 93.

[30] O. S. Andell, J.-E. Bäckvall, *J. Organomet. Chem.* **1983**, *244*, 401.

[31] F. Mares, St. E. Diamond, F. J. Regina, J. P. Solar, *J. Am. Chem. Soc.* **1985**, *107*, 3545.

[32] S. Nakamura, T. Yasui, *J. Catal.* **1971**, *23*, 315.

[33] J. M. Davidson, P. C. Mitchell, N. S. Raghavan, *Front. Chem. React. Eng.* **1984**, *1*, 300.

[34] H. Debellefontaine, J. Besombes-Vailhé, *J. Chim. Phys. Phys.-Chim. Biol.* **1978**, *75*, 801.

[35] T. Kunugi, H. Arai, K. Fujimoto, *Bull. Jpn. Petr. Inst.* **1970**, *12*, 97.

[36] S. A. H. Zaidi, *Appl. Catal.* **1988**, *38*, 353.

[37] B. Samanos, P. Boutry, R. Montarnal, *J. Catal.* **1971**, *23*, 19.

[38] A. Rabl, A. Renken, *Chem.-Ing.-Tech.* **1986**, *58*, 434.

[39] R. Jira, unpublished results.

[40] Wacker-Chemie GmbH, Hoechst AG (D. Dempf, L. Schmidhammer, G. Dummer, G. Roscher, K.-H. Schmidt, E. Selbertinger, R. Strasser), EP 48.946 (1982).

[41] St. M. Augustine, J. P. Blitz, *J. Catal.* **1993**, *142*, 312.

[42] R. Hüttel, H. Christ, *Chem. Ber.* **1963**, *96*, 3101.

[43] R. Hüttel, H. Christ, *Chem. Ber.* **1964**, *97*, 1439.

[44] J. Smidt, W. Hafner, *Angew. Chem.* **1959**, *71*, 284.

[45] B. L. Shaw, *Chem. Ind. (London)* **1962**, 1190.

[46] R. Hüttel, J. Kratzer, *Angew. Chem.* **1959**, *71*, 456.

[47] R. Hüttel, J. Kratzer, M. Bechter, *Chem. Ber.* **1961**, *94*, 766.

[48] R. Jira, J. Sedlmeier, *Tetrahedron Lett.* **1971**, 1227.

[49] W. Hafner, H. Prigge, J. Smidt, *Liebigs Ann. Chem.* **1966**, *693*, 109.

[50] R. Jira, W. Freiesleben in *Organometallic Reactions*, Vol. 3 (Eds.: E. Becker, M. Tsutsui), John Wiley, New York, **1972**.

[51] P. M. Maitlis, *The Organic Chemistry of Palladium*, Vol. 2, Academic Press, New York, **1971**.

[52] P. M. Henry in *Palladium Catalyzed Oxidation of Hydrocarbons*, D. Reidel, Dordrecht, **1980**, pp. 41–223.

[53] J. Tsuji, *Organic Synthesis with Palladium Compounds*, Springer, Berlin, **1980**.

[54] Anon., *Eur. Chem. News* **1973**, 21, 22.

[55] Bayer AG, BE 627.888 (1963).

[56] Knapsack AG, DE 1.244.766 (1965).

[57] Hoechst AG, DE 1.296.138 (1967).

[58] Nat. Distillers, US 3.190. 912 (1962), GB 976.613 (1963).

[59] G. Roscher, E. Hofmann, K. A. Adey, W. Jeblink, H.-J. Klimisch, H. Kieczka, *Ullmann's Encycl. Techn. Chem. 4th ed.* **1983**, Vol. 23, pp. 601–604.

[60] J. Grolig, *Ullmann's Encycl. Techn. Chem. 5th ed.* **1985**, Vol. A1, pp. 434–436.

[61] I am indebted to Dr. Akio Mitsutani, Nippon, Chemtec Consulting Inc., 665–0022 Takarazuka City, Nogami 3-chome, 11–10, Japan for leaving to me respective chapters of the evaluation reports of the company.

[62] Mitsubishi Kasei Corp., *Chemtech* **1988**, 759.

[63] H. Pommer, A. Nurrenbach, *Pure Appl. Chem.* **1975**, *43*, 527.

[64] Y. Tanabe, *Hydrocarbon Proc.* **1981**, *60*, 187.

[65] C. F. Kohll, R. van Helden, *Rec. Trav. Chim. Pays-Bas* **1968**, *87*, 481.

[66] E. W. Stern, *J. Catal.* **1966**, *6*, 152.

[67] J. Smidt, W. Hafner, R. Jira, R. Sieber, J. Sedlmeier, A. Sabel, *Angew. Chem.* **1962**, *74*, 93; *Angew. Chem. Int. Ed. Engl.* **1962**, *1*, 80.

[68] A. Sabel, J. Smidt, R. Jira, H. Prigge, *Chem. Ber.* **1969**, *102*, 2939.

[69] Consortium für Elektrochem. Ind. (J. Smidt, A. Sabel), DE 1.127.888 (1962).

[70] Wacker-Chemie GmbH (K. Blum, R. Strasser), DE-OS 3.047.347 (1982).

[71] A. A. Ketterling, A. S. Lisitsyn, A. V. Nosow, V. A. Likholobov, *Appl. Catal.* **1990**, *66*, 123.

[72] F. J. Waller, *Chem. Ind. (Dekker)* **1994**, *53*, 397.

[73] Consortium für Elektrochem. Ind. (J. Smidt, A. Sabel), DE 1.277.246 (1968).

[74] E. W. Stern, *Catal. Rev.* **1967**, *1*, 73, 125.

[75] O. Hamed, P. M. Henry, *Organometallics* **1997**, *16*, 4903.

[76] Imperial Chemical Industries (D. Clark, P. Hayden), NL Appl. 6.703.724, DE 1.273.525 (1968).

[77] Imperial Chemical Industries (D. Clark, P. Hayden), DE 1.275.532 (1968).

[78] J. Smidt, W. Hafner, R. Jira, J. Sedlmeier, R. Sieber, R. Rüttinger, H. Kojer, *Angew. Chem.* **1959**, *71*, 176.

[79] L. El Firdoussi, A. Benharref, S. Allaoud, A. Karim, Y. Castanet, A. Mortreux, F. Petit, *J. Mol. Catal.* **1992**, *72*, L1.

[80] L. El Firdoussi, A. Bagga, S. Allaoud, B. Ait Allal, A. Karim, Y. Castanet, A. Mortreux, *J. Mol. Catal. A:* **1998**, *135*, 11.

[81] E. J. Mistrik, A. Mateides, *Chem. Technol.* **1983**, *35*, 90.

[82] P. M. Henry, *J. Org. Chem.* **1973**, *38*, 1681.

[83] Kuraray (T. Shimizu, T. Yasui, S. Nakamura), GB 1.368.505 (1974); *Chem. Abstr.* **1975**, *83*, 9234a.

[84] Mitsubishi Chem. Ind. (T. Onoda, A. Yamura, J. Toriya, I. Kasahara, M. Sato, N. Ishizaki), JP 74.101.322 (1974); *Chem. Abstr.* **1975**, *82*, 86089p.

[85] Mitsubishi Chemical (T. Onoda, J. Haji), DE-OS 2.217.452 (1972), *Chem. Abstr.* **1973**, *78*, 57786a.

[86] BASF AG (H. M. Weitz, J. Hartwig), DE-OS 2.421.408 (1975); *Chem. Abstr.* **1976**, *84*, 58665w.

[87] R. G. Brown, J. M. Davidson, *J. Chem. Soc. (A)* **1971**, 1321.

[88] J.-E. Bäckvall, St. E. Byström, R. E. Nordberg, *J. Org. Chem.* **1984**, *49*, 4619.

[89] J.-E. Bäckvall, R. E. Nordberg, J.-E. Nyström, *Tetrahedron Lett.* **1982**, *23*, 1617.

[90] J.-E. Bäckvall, *Pure Appl. Chem.* **1983**, *55*, 1669.

[91] J.-E. Bäckvall, *Pure Appl. Chem.* **1992**, *64*, 429.

[92] J. M. Davidson, C. Triggs, *J. Chem. Soc. A* **1968**, 1331.

[93] P. M. Henry, *J. Org. Chem.* **1971**, *36*, 1886.

[94] L. Pachkovskaya, K. I. Matveev, G. N. Il'inich, N. K. Eremenko, *Kinet. Katal.* **1977**, *18*, 1040; *Kinet. Katal. Engl. Transl.* **1977**, *18*, 854.

[95] J. E. Lyons, G. Suld, C. Y. Hsu, *Chem. Ind. (Dekker) (Catal. Org. React.)* **1988**, *33*, 1.

[96] I. I. Moiseev, M. N. Vargaftik, *Dokl. Akad. Nauk SSSR* **1966**, *166*, 370.

[97] W. G. Lloyd, B. J. Luberoff, *J. Org. Chem.* **1969**, *34*, 3949.

[98] R. van Helden, G. Verberg, *Rec. Trav. Chim. Pays-Bas* **1965**, *84*, 1263.

[99] C. F. Kohll, R. van Helden, *Rec. Trav. Chim. Pays-Bas* **1967**, *86*, 193.

[100] S. Danno, I. Moritani, Y. Fujiwara, *Tetrahedron* **1969**, *25*, 4819.

[101] C. Jia, W. Lu, T. Kitamura, J. Fujiwara, *Org. Lett.* **1999**, *1*, 2097.

[102] K. Hirota, H. Kuki, Y. Maki, *Heterocycles* **1994**, *37*, 563.

[103] C. Jia, D. Piao, J. Oyamada, W. Lu, T. Kitamura, Y. Fujiwara, *Science* **2000**, *287*, 1992.

[104] C. Jia, W. Lu, J. Oyamada, T. Kitamura, K. Matsuda, M. Irie, Y. Fujiwara, *J. Am. Chem. Soc.* **2000**, *122*, 7252.

[105] R. F. Heck, *J. Am. Chem. Soc.* **1968**, *90*, 5518.

[106] R. F. Heck, *J. Am. Chem. Soc.* **1968**, *90*, 5535.

[107] R. F. Heck, *J. Am. Chem. Soc.* **1968**, *90*, 5538.

[108] J. K. Stille, *Pure Appl. Chem.* **1985**, *57*, 1771.

[109] R. C. Larock, *Pure Appl. Chem.* **1990**, *62*, 653.

[110] S. Cacchi, *Pure Appl. Chem.* **1990**, *62*, 713.

[111] E. Negishi, T. Takahashi, K. Akiyoshi, *J. Organomet. Chem.* **1987**, *334*, 181.

[112] A. Suzuki, *Pure Appl. Chem.* **1985**, *57*, 1749.

[113] J. Tsuji, M. Morikawa, J. Kiji, *Tetrahedron Lett.* **1963**, 1061.

[114] D. M. Fenton, K. L. Olivier, *Chem. Technol.* **1972**, 220.

[115] T. Jintoku, Y. Fujiwara, I. Kawata, T. Kawauchi, H. Taniguchi, *J. Organomet. Chem.* **1990**, *385*, 297.

[116] Y. Taniguchi, Y. Yamaoka, K. Nakata, K. Takaki, Y. Fujiwara, *Chem. Lett.* **1995**, 345.

[117] W. Lu, Y. Yamaoka, Y. Taniguchi, T. Kitamura, K. Takaki, Y. Fujiwara, *J. Organomet. Chem.* **1999**, *580*, 290.

[118] K. Nakata, T. Miyata, T. Jintoku, A. Kitani, Y. Taniguchi, K. Takaki, Y. Fujiwara, *Bull. Chem. Soc. Jpn.* **1993**, *66*, 3755.

[119] M. Kurioka, K. Nakata, T. Jintoku, Y. Taniguchi, K. Takaki, Y. Fujiwara, *Chem. Lett.* **1995**, 244.

[120] Y. Taniguchi, T. Hayashida, H. Shibasaki, D. Piao, T. Kitamura, T. Yamaji, Y. Fujiwara, *Org. Lett.* **1994**, *1*, 557.

[121] M. Asadullah, T. Kitamura, Y. Fujiwara, *Angew. Chem., Int. Ed. Engl.* **2000**, *39*, 2475.

[122] H. Sugimoto, I. Kawata, H. Taniguchi, Y. Fujiwara, *J. Organomet. Chem.* **1984**, *266*, C44.

[123] R. Cramer, *J. Am. Chem. Soc.* **1965**, *87*, 4717.

[124] R. Cramer, *Acc. Chem. Res.* **1968**, *1*, 186.

[125] A. Misono, Y. Uchida, M. Hidai, H. Kanai, *Chem. Commun.* **1967**, 357.

[126] W. Strohmeier, A. Kaiser, *J. Organomet. Chem.* **1976**, *114*, 273.

[127] D. T. Tsou, J. D. Burrington, E. A. Maher, R. K. Grasselli, *J. Mol. Catal.* **1985**, *30*, 219.

[128] J. Tsuji, *Acc. Chem. Res.* **1973**, *6*, 8.

[129] J. Tsuji, *Adv. Organomet. Chem.* **1979**, *17*, 141.

[130] H. G. Tang, D. C. Sherrington, *J. Mol. Catal.* **1994**, *94*, 7.

[131] R. Cramer, *J. Am. Chem. Soc.* **1966**, *88*, 2272.

[132] J. W. Francis, P. M. Henry, *J. Mol. Catal. A: Chem.* **1996**, *112*, 317 and literature cited therein.

[133] K. Hiroi, Y. Arinaga, *Tetrahedrin Lett.* **1994**, *35*, 153.

[134] T. Hosokawa, T. Shinohara, Y. Ooka, S. Murahashi, *Chem. Lett.* **1989**, 2001.
[135] K. Utimoto, *Pure Appl. Chem.* **1983**, *55*, 1845.
[136] J. Tsuji, *Pure Appl. Chem.* **1981**, *53*, 2371.
[137] R. F. Heck, *Pure Appl. Chem.* **1981**, *53*, 2323.
[138] R. A. Sheldon, J. A. Kochi, *Metal-Catalyzed Oxidations of Organic Compounds*, Academic Press, New York, **1981**.
[139] R. F. Heck, *Palladium Reagents in Organic Synthesis*, Academic Press, New York, **1985**.
[140] R. G. Schultz, D. E. Gross, *ACS Adv. Chem. Ser.* **1968**, *70*, 97.
[141] Review on oxidative coupling: E. Negishi, *Acc. Chem. Res.* **1982**, *15*, 340.
[142] J. Tsuji, *Synthesis* **1990**, 739.
[143] J. Tsuji, Palladium Reagents, in *Innovation in Organic Synthesis*, John Wiley, New York, **1996**.
[144] A. Yamamoto, T. Yamamoto, F. Ozawa, *Pure Appl. Chem.* **1985**, *57*, 1799.
[145] M. Catellani, G. P. Chiusoli, M. Costa, *Pure Appl. Chem.* **1990**, *62*, 623.
[146] S. Cacchi, *Pure Appl. Chem.* **1990**, *62*, 713.
[147] T. Hayashi, A. Kubo, F. Ozawa, *Pure Appl. Chem.* **1992**, *64*, 421.
[148] J. Tsuji, *Transition Metal Reagents and Catalists*, in: Innovation in Organic Synthesis, Wiley, New York 2000.

4

Epilogue

4.1   Homogeneous Catalysis – *Quo vadis?**

Wolfgang A. Herrmann, Boy Cornils

> "The discovery of truly new reactions is likely to be limited
> to the realm of transition-metal organic chemistry, which will
> almost certainly provide us with additional 'miracle reagents'
> in the years to come."
>
> D. Seebach
> *Angew. Chem., Int. Ed. Engl.* **1990**, *29*, 1320

The impact of *homogeneous* catalysis on industrial process technology has grown since 1960 to the same extent as organometallic chemistry has developed as a science (cf. Figure 1 in Chapter 1). As a matter of fact, new knowledge regarding the structure and reactivity of organometallic compounds has, with a certain time delay, created new catalytic processes in industry or has re-established old catalytic features under improved conditions. Important examples include the replacement of cobalt by rhodium in a number of processes that are otherwise not feasible industrially (or not as elegantly or economically, e. g., methanol carbonylation and hydroformylation), which during the 1964–1976 period caused much scepticism directed at the recycling of the precious noble metal. Nowadays, ironically, it is often the price of a more or less sophisticated ligand that dictates the economics of a new process, while the metals themselves – even rhodium, palladium, or platinum – are irrelevant in this respect; cf. Table 3 in Chapter 1.

Homogeneous catalysis and organometallic chemistry have stimulated and supported each other since the early days of hydroformylation (1938), olefin polymerization (1953), and acetaldehyde synthesis (1959), to name just the "battleships" of organometallic catalysis [1]. While it is certainly true that many epoch-making discoveries came along by serendipity [2], a typical such case being Karl Ziegler's "*Metallorganische Mischkatalysatoren*" [3], the majority of homogeneous catalytic applications has resulted from continued step-by-step development of industrial laboratory and plant-site research alongside the scientific growth of organometallic chemistry.

This brings to mind the first comprehensive monograph on *Die Chemie der Metallorganischen Verbindungen* (The Chemistry of Organometallic Compounds), written by E. Krause and A. von Grosse. It appeared in 1937 [4], shortly before hydroformylation was discovered. In this 900-page book only a little is reported on the present-day champions of homogeneous catalysis (e. g., platinum: three pages and six references), and there is no mention of cobalt, rhodium, and palladium. What a contrast, when the monumental work entitled *Comprehensive*

---

* Homogeneous Catalysis – where now?

Organometallic Chemistry – nine volumes, 8750 printed pages – appeared 45 years later, with the second edition comprising 14 volumes already in print by 1996 [5]! A total of approx. 30 500 organometallic compounds has been registered by the *Chemical Abstracts Service* up to the time of writing [6, 7]. The *Journal of Organometallic Chemistry* was chartered in 1963 and, since then, has seen over 600 volumes with a total of 25 000 papers on 230 000 journal pages [8]. Two more key journals – the *Journal of Molecular Catalysis* and *Organometallics* – broadly cover catalytic applications of organometallic compounds [9]. This was why the Preface of this book referred to homogeneous catalysis as a success story of organometallic chemistry. It is indeed the overwhelming diversity of structures and reactivities which forms the basis of so many applications mentioned in this book. Early monographs and reviews have recognized this interdependence [10]. No chemistry student leaves university today without having been introduced to the basic principles of homogeneous catalysis. Nevertheless, *heterogeneous* catalysis strongly dominates the industrial scene, where approx. 80 % of all known catalytic processes being heterogeneous would probably be a good estimate.

Homogeneous catalysis and *industry* have their own "love affair": wherever it appeared economically and technically useful, the advantages of homogeneous catalysis have been exploited: more than nine million tons each of oxo products [132] and terephthalates, nearly three million tons of acetic acid and anhydride, and hundreds of thousands of tons of organic feedstocks (acetic aldehyde, carboxylic acids, ω,ω'-dinitriles, di- and oligoolefins, etc.) are convincing proof of this. On the way are products of which the potential has been recognized but markets are still small: they come from partial hydrogenation of unsaturated aldehydes, amidocarbonylations, copolymerizations of carbon monoxide, hydrodimerizations, and the growing number of stereoselective catalytic syntheses.

Of specific potential are several processes which presently experience (or even surpass) the pilot-plant stage: vinyl acetate from syngas, precursors of polymers such as polycarbonate and polyurethanes via reductive or oxidative carbonylation, methyl methacrylates and adipic acid through alternative routes, polypropene and COCs (cf. Section 4.1.14) by means of metallocenes (cf. Section 2.3.1.5) – new routes have been opened in all these cases. The last-named example emphasizes in an almost classical way the principle of tailor-making novel, optimized, homogeneous catalysts. Chapter 3 should again be consulted for details.

Numerous scientific and industrial problems of catalysis in general, and homogeneous catalysis in particular, remain to be solved. Depending on their viewpoint, objectives, and position, readers will have identified pending problems – these of a general and those of a specific nature – in every section of this book. Organometallic catalysis is far from being a mature scientific field. In the Editors' opinion (which is shared by a great number of colleagues in industry and academics), the following key problems of catalysis warrant intensive research [137].

4.1.1 Immobilization of Homogeneous Catalysts

It has been proven in this book over and over again that the strength of organo-metallic catalysts – *activity, chemo-, regio-,* and *stereoselectivity* – result from this simple concept: specific ligands keep the catalytic metal in a low-nuclearity, normally mononuclear state of molecularly defined stereochemistry; at the same time, these metal complexes participate in dissociation equilibria, thus promoting the reactivity of the metal center by making appropriate coordination sites available. While solubility in common organic solvents is an additional advantage in terms of site availability, it constitutes a severe methodological drawback in terms of catalyst recycling (cf. the Introduction and Figure 1 in Chapter 1). The relationships between heterogeneous and homogeneous catalysis are sketched in Scheme 1, which shows the link between surface organometallic catalysis and phase transfer catalysis (cf. Sections 3.1.1.1, 3.1.1.4, and 3.2.4).

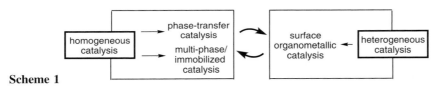

Scheme 1

The royal road (*Königsweg*) to overcome the problem is liquid/liquid two-phase catalysis (cf. Section 3.1.1.1). Here, the catalyst retains its beneficial mole-cularity, but nevertheless becomes immobilized as such (by forming a second liquid phase as a "liquid support"). Another approach to facilitating catalyst/product separation has long been the attachment of the catalyst to a polymeric resin [11], be it organic (e. g., polystyrene) or inorganic (polysiloxanes). However, no resin-bound catalyst of industrial – e. g., economic – importance is known: catalyst "leaching" remains the central problem indeed (cf. Section 3.1.1.3). Potent catalysts such as osmium tetraoxide (for stereoselective *cis*-hydroxylation of ole-fins) will not see an industrial plant before a reliable immobilization technique is available; product contamination, toxicity, and the costs related to insufficient metal recycling must be mentioned in this context.

The leaching normally results from the dissociation of the metal from one of the anchored ligands (eq. (1)), thus liberating the (active) molecular catalyst. Leaching can also originate from structural changes with concomitant weakening of certain bonds during the catalytic cycle, where the coordination sphere of a metal under-goes continuous changes [132].

For this reason, the specific type of bonding between catalyst and support ap-pears crucial. The disadvantage of leaching has hitherto only been discussed with regard to the metal centers of catalysts; it is noted, however, that bleeding losses can also involve expensive ligands, especially those necessary in stereoselective catalysis. Nevertheless, every possible class of catalysis-supporting ligand should be considered further for methods of immobilization. *N*-Heterocyclic carbenes (eq. (2)) promise a successful approach to the problem: they effect certain catalytic C–C coupling reactions [12] but, unlike phosphines, they do not engage in dissociation from the catalytically active metals. Inorganic backbones are preferred over organic polymers as catalyst supports because of their much higher thermal and mechanical stability. However, they may participate in the metal- bonding themselves and thus influence the elementary catalytic steps. Diffusion problems related to the polymers surrounding the catalyst must also be taken into consideration.

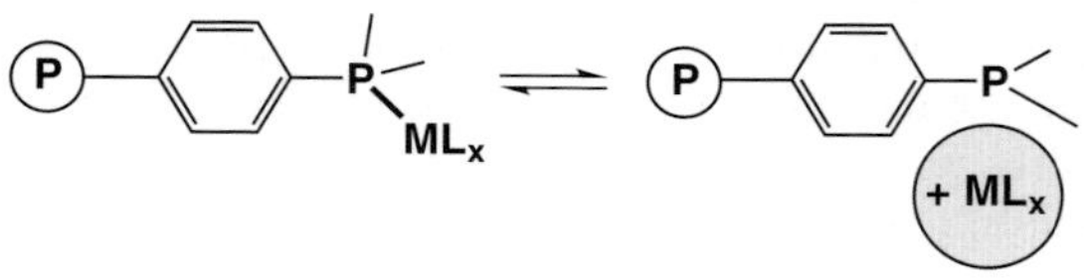

$$
\text{(image of equation 1)} \tag{1}
$$

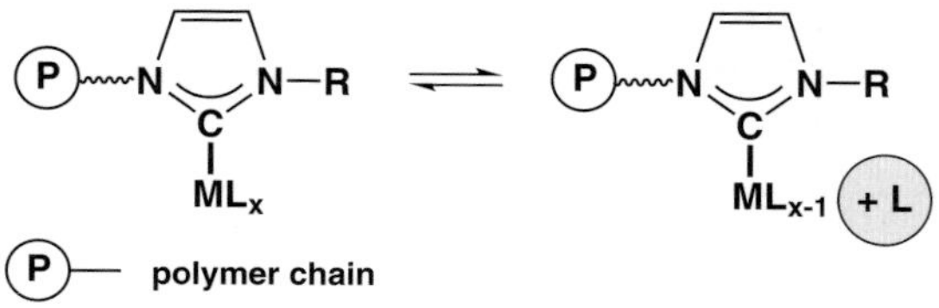

$$
\text{(image of equation 2)} \tag{2}
$$

(P)— polymer chain

A little-explored field concerns zeolites and polyoxometallates as active support materials, at least where well-defined anchoring of organometallic catalysts is concerned. The landmark discovery of the titanium effect in the Enichem TS-1 catalyst tells the story of how dramatically a transition metal in an ordered environment (e. g., silicalite) can influence the catalytic performance [13]. An ostensibly promising concept is *surface organometallic chemistry* (Section 3.1.1.4): molecular catalyst precursors are reacted with surface groups of (normally inorganic) supports, for example silanol groups of silica, thus yielding a specific surface attachment of organometallics as quasimolecular domains [14]. The Zr-catalyzed degradation of C–C polymers (e. g., polypropene) to oligomeric fragments and monomers may open horizons for this technique (cf. Section 4.1.9) [15].

This area also includes the introduction of additional catalytic sites in zeolites by organometallic precursor compounds. Both the adjustment of Brønsted/Lewis acidity and the directed fixation of active transition metal centers inside zeolite cages and channels seem possible, but the size and structures of these precursors will be crucial. The "molecular modification" appears all the more important as many catalytic processes will remain heterogeneous in nature, especially when the shape-selectivity of sodalites, zeolites, pentasils, etc. is exploited. Once again, the possible versatility of organolanthanoid complexes deserves emphasis, since these metals verify to the above-mentioned acidity and catalysis effects [92].

"Heterogeneous modifications" of homogeneous catalysts comprise the supramolecular concepts of sol/gel entrapment [16, 17], template- and host/guest interactions [18, 76, 82, 84, 85, 89], metal colloids [24, 83, 85], hybrid organic-inorganic zeolite analogues [133], and the "ship-in-the-bottle" principle in zeolite catalysis [19, 20]. These belong to "crossing the border from homogeneous to heterogeneous catalysis" [87] and reflect the still-unsolved question posed by Heinemann: "Homogeneous and heterogeneous catalysis – common frontier or common territory?" [88].

In view of the growing industrial successes, it is suffice to say that two-phase catalysis (liquid/liquid) has not yet reached its culmination point [93]. Novel (water-soluble) ligands pose a challenge to the synthetic chemist. New reaction media will be required, too, but it seems questionable whether the recently announced perfluorinated hydrocarbons (cf. Section 3.1.1.2.1), supercritical fluids (cf. Section 3.1.13), or non-aqueous ionic liquids (cf. Section 3.1.1.2.2) indicate a true breakthrough in this direction [21]. These studies have shown, however, that temperature-dependent miscibility of the different reactants and solvents could be exploited as a simple means of verifying the "single-phase catalysis/ two-phase separation" concept without additional solubilizers. It is to be seen how the elegant possibility of "re-immobilization" (cf. Section 3.1.1.6) can be exploited for industrial syntheses. There could be a bright future in the combination of such catalysts with the avantgarde membrane reactors [22].

4.1.2   Colloidal Organometallic Catalysts

Metal colloids – links between homogeneous and heterogeneous catalysts [24, 83, 85] – represent a fairly underestimated group of catalysts, in spite of the fact that their pre-stage organometallic clusters enjoyed strong scientific interest for some time (cf. Section 3.1.1.5) [23]. Although oligo- and multinuclear metal aggregates are subject to degradation (denucleation) under a number of catalytic conditions (e. g., high-pressure carbonylation) or undergo conglomeration to yield the bulk metal (e. g., temperature effects, depletion of stabilizing ligands), there is some evidence that colloidal systems are engaged even in catalysis that is normally considered "homogeneous" [24, 25]. A good example is the long-known hydrosilylation where low-valent colloidal platinum seems to work best [25 d]. Little is known as to how conventional ligands and other ingredients stabilize metal colloids during a given "homogeneous" process. Such knowledge would set the scene for the directed application of "colloid ligands". Both the theory of colloids (formation, structure, stability) and their catalytic performance will clearly form a center of gravity of future research.

Recent progress has been reported by the groups of Bönnemann, Toshima, and Schmid [24, 25]. Nanometal colloids were generated from metal halides by reduction with (surfactant) hydridoboranates (e. g., $[N(octyl)_4][HB(C_2H_5)_3]$). Applied on supports, the resulting catalysts are claimed to be of much higher activity than the currently used (heterogeneous) catalysts. Chiral cinchonidine-"stabilized" platinum colloids effect the hydrogenation of ketones to alcohols with conversions of 50–100 % and 69–85 % *ee*, depending on the particle size [25e]. Chiral protective shells may induce stereoselective reactions at metal colloid catalysts and may act, at the same time, as inhibitors of catalyst poisoning.

4.1.3   Multicomponent and Multifunctional Catalysis

Bimetallic multi*component* and multi*functional* catalysis is a common feature in heterogeneous catalytic processes [26, 27]. There is hardly a reaction that just employs a "pure metal" catalyst. Usually, doping additives exhibit electronic or steric effects (e. g., Ag–Re/Cs in the Shell oxirane process, a heterogeneous catalysis), and for many conversions intermetallic phases have been developed. It would once again be attractive to implement this as a concept on the molecular level into homogeneous catalysis. To become catalytic, numerous reactions require the activation of at least two substrates by different metals. On the other hand, a single-site (molecular) catalyst could take advantage of a second metal nearby (e. g., as an electronic reservoir) to become re-activated. This is not an easy synthetic problem, since the coordination chemistry of each of the metals to be coupled for such purposes is normally quite different (e. g., Pd–Cu in the Wacker–Hoechst oxidation). Beyond that, molecular seats of Lewis and Brønsted acidity – like those so crucial for surface-type catalysts (SiOH, Al^{3+}) – could be combined with a catalytic metal site. This topic implies high-risk research from both the synthetic and the catalytic–mechanistic sides, but is of great methodological importance to the future "refinement" of homogeneous catalysts. Literature does not yet discriminate clearly enough between multicomponent and multifunctional catalyst systems. *Multicomponent catalysts* contain two or several metals that enforce the catalytic effect upon a given reaction (e. g., Pt/Sn in hydroformylation). *Multifunctional catalysts* effect different reactions (e. g., simultaneous or subsequent oligomerization combined with hydroformylation; cf. Section 3.1.5).

4.1.4   Stereoselective Catalysis

It is an undisputed fact that the homogeneous catalysis of the future will have its greatest strength of versatility in *stereoselective synthesis*. The chiral-drug market of the world was estimated for 1994 to comprise a sales total amounting to US$ 45.2 billion, which corresponds to an increase of 27 % in one single year. An annual growth of ca. 9 % has been predicted for the near future, only including drugs that have already been approved by drug authorities worldwide [28]. Among these products, the leading anti-inflammatory drug (*S*)-(+)-ibuprofen alone is expected to have a sales potential of US$ 1 billion per year as an over-the-counter drug. It is remarkable to see that only 11 % of the chiral pharmaceuticals introduced before 1983 were produced in the single stereoisomeric form; this fraction increased to 26 % for synthetic drugs introduced from 1983 to 1987, and there is a conservative estimate that now 80 % of all chiral pharmaceuticals will be produced optically pure [29, 96].

This implies an enormous challenge to stereoselective synthesis as a major branch of "applied homogeneous catalysis". At the same time, the organic synthesis of tomorrow will be governed by the "chiral selection principle" [30]. Optically stable ligands – in addition to phosphines and others – will continue to be targets for the synthetic organic and organometallic chemist [31]. Once again, immobilized and surface-anchored catalysts are in high demand, not only for toxicity reasons. Chiral modification of (metal) surfaces seems to bring about stereoselectivity [32], again an area of hitherto little emphasis. It looks as if bidentate ligands of axial chirality (atropisomers) will dominate future development, at least in the area of the ubiquitous organophosphines. The prototype BINAP (Structure **1**) [33 a] is being succeeded by a number of related systems, of which enantiopure BINAS (**2**) [33 b] and BINAPHOS (**3**) [33 c–g] seem to solve even the notoriously difficult problem of enantioselective hydroformylation [95].

The area of (chiral) *N/O*-oligodentate ligand synthesis – see, for example, the known Structures **4–6** – also merits future attention. Smart ideas are required because these ligands become more and more sophisticated [34]. Logically, *N/O*-ligands are more resistant to oxidation than phosphines, one of the reasons

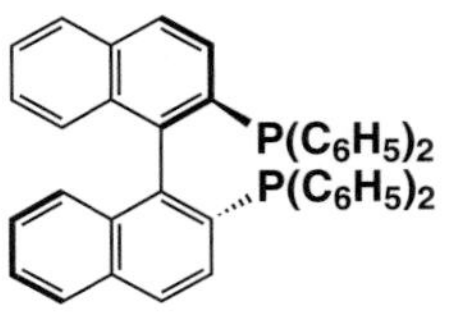

(-)-BINAP
1

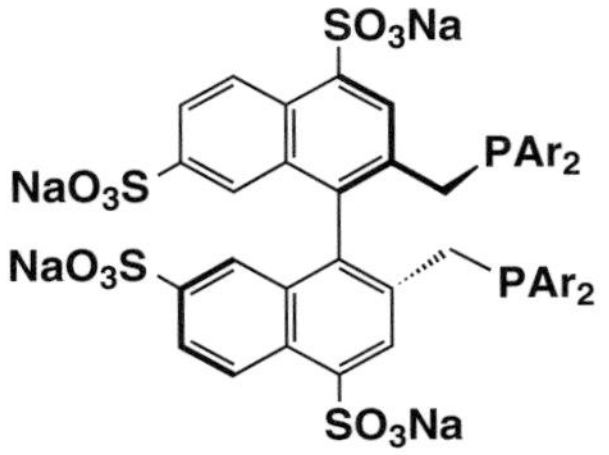

(-)-BINAS
2

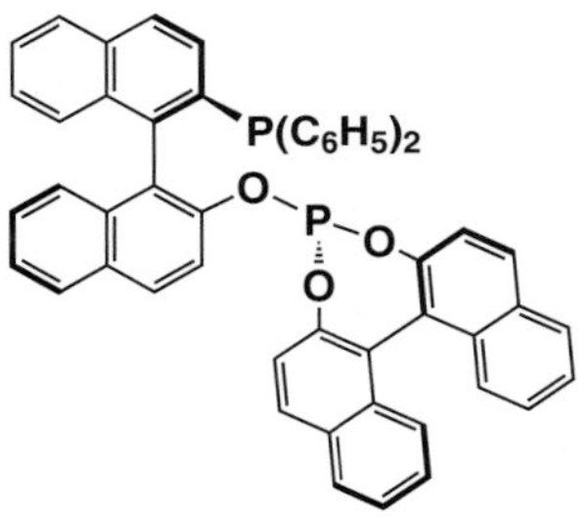

R,S-BINAPHOS
3

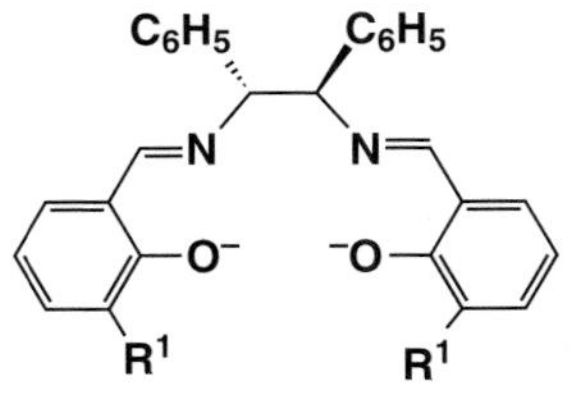

Ar = C$_6$H$_4$-*m*-SO$_3$Na

R^1 = Bu^t
4

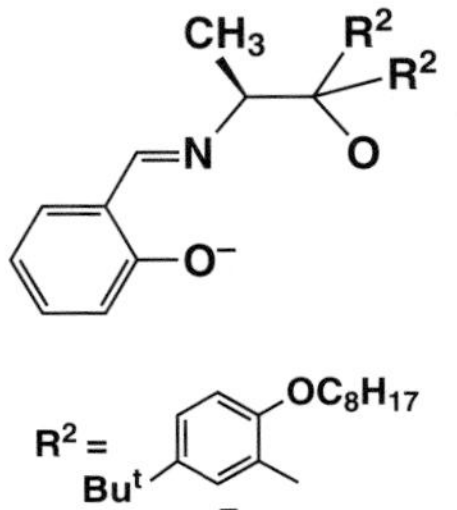

R^2 =
Bu^t
5

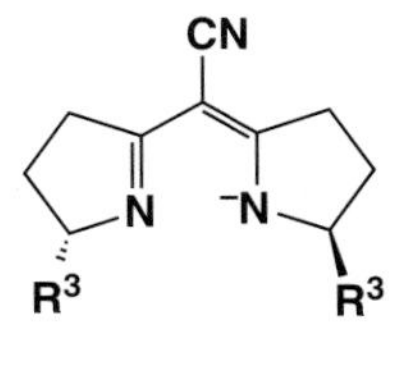

R^3 = -C(CH$_3$)$_2$OH
6

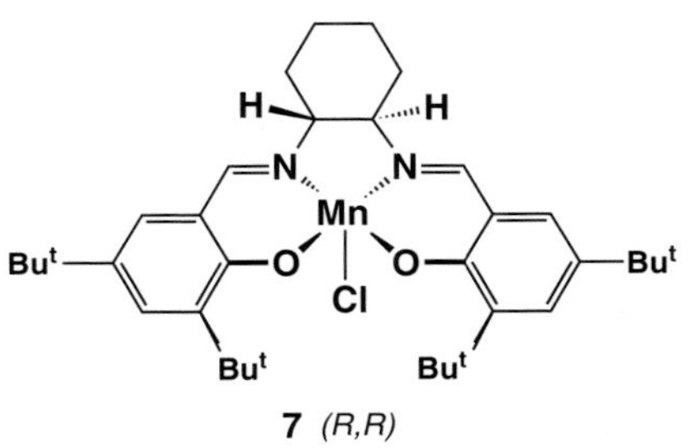

7 *(R,R)*

why tetradentate salicylaldimine derivatives like **7** prove so successful in stereo-selective epoxidation [35, 36] (cf. Sections 2.4.3 and 2.9). High optical inductions are also seen for Cu^{II}- or Co^{II}-attached semicorrin and Schiff-base ligands in olefin cyclopropanation reactions with diazoalkanes [37] (cf. Section 3.1.7). This work has opened new access to insecticides (permethrin) and drugs (Cilastatin® by Sumitomo). It seems that rigid ligands in an equatorial arrangement around the metal are particularly efficient in chirality transfer.

Ligands from the "chiral pool" have previously been underrepresented inboth coordination chemistry and stereoselective catalysis, e. g., sugars, cyclodextrins, and amino and nucleic acids. Furthermore, almost nothing is known about organometallic chemistry in biological systems such as cells.

Of particular interest is the phenomenon of so-called *chirality amplification*, where a chiral ligand of *low* optical purity yields products of *high* enantioselectivity. For example, a 95 % *ee* alcohol forms from the achiral substrates benzaldehyde and diethylzinc when some (–)-DAIB ((–)-3-*exo*-(dimethylamino)borneol) as all-organic catalyst of only 15 % optical purity is present (eq. (3)) [37 a, b].

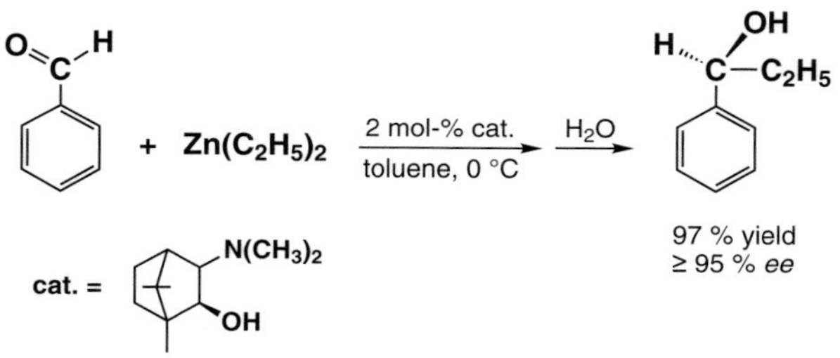

(3)

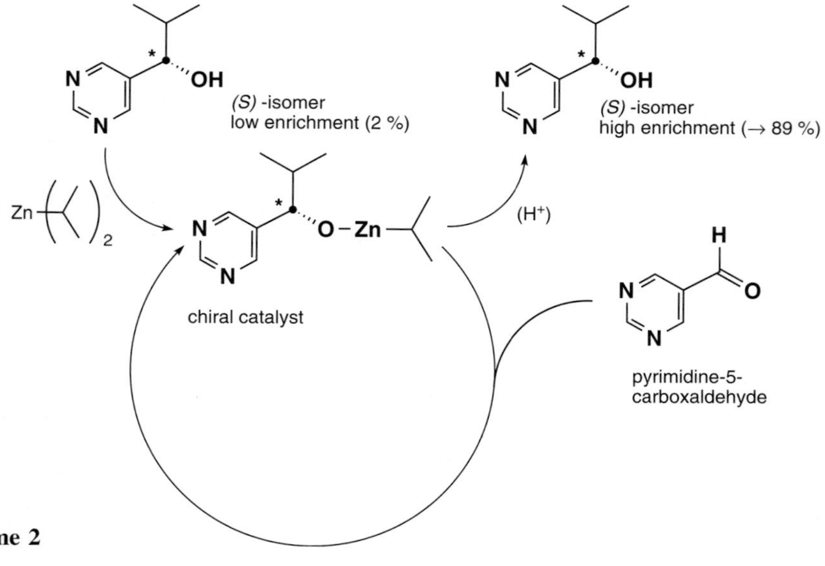

Scheme 2

This is a case where chiral and achiral metal complexes compete with each other ($> 600:1$ by rate). *Asymmetric autocatalysis* is certainly an attractive, yet still largely unexplored field. Two-phase processes for stereoselective syntheses are under investigation [38], and phase transfer catalysis must be mentioned in this context.

A recent convincing case of chiral amplification opens a wide horizon in stereoselective catalysis [37 c]: thus, when a 5-pyrimidyl alcohol with a small (2 %) enantiomeric excess is treated with diisopropylzinc and pyrimidine-5-carboxaldehyde, it undergoes an autocatalytic reaction to generate more of the alcohol. The chiral catalyst is formed from the initial alcohol. The *ee* result of the (S)-isomer was successively increased in the series 2 % → 10 % → 57 % → 81 % → 89 %. Amplification factors of up to ca. 1700 were recorded with the catalytic system of Scheme 2 [37 c]. This is the first case in which the enantiomeric excess of the product is greater than that of the chiral catalyst [97].

4.1.5   Metals from Stoichiometric Reactivity to Catalytic Efficiency

Almost every *new* synthetic procedure in organic chemistry includes a metal-mediated step. However, most of these metal-containing reagents suffer from only working stoichiometrically, so that a "true catalysis chemist" would not count them in his area of research. One typical example is the cobalt-mediated Pauson–Khand reaction which makes cyclopentenones as useful synthetic building blocks available from cheap precursors (Section 3.3.7). Metals that were previously unheard of in organic synthesis now effect synthetic steps with good yields and perfect selectivities: niobium(III), for example, N–C-couples aldehydes with imines or alkynes [39]; samarium(II) iodide is a selective one-electron reductant that C–C-couples keto compounds [40, 41]. It is a challenge to surround these metals with ligand spheres so as to make their individual reactions catalytic (i.e., $\ll 1$ mol % of catalyst). The titanium-mediated McMurry reaction to synthesize olefins from keto compounds needs further improvement in terms of catalytic applications (Section 3.2.12). Immobilization techniques such as sol/gel entrapment [16, 17], use of templates [17, 76] and host/guest relationships [18, 84, 85], and molecular recognitions [18, 19, 77, 85] promise success in this area, too.

Great potential is seen in the relatively little-explored organic chemistry of the rare earth metals (cf. Section 3.2.5), where a tuning of reactivity seems most feasible because of the gradation of properties within the lanthanoids. Normally, the "tuning" in organometallic catalysis works through the ligands, while in the lanthanoid series the metal properties (size, Lewis acidity) come to the fore. Superacidic soluble and immobilized catalysts may well have their future in the chemistry of the lanthanoids, including their "smaller brothers" scandium and yttrium [92].

It is obvious that toxicity/price reasons (cf. Table 3 in Chapter 1) must push all these marvelous reagents toward their catalytic application and then, just one step beyond, to their perfect separation from the product(s) [41]. "Miracle reagents" (Seebach) must become "miracle catalysts" to enter the industrial scene success-fully. Thus, for any newly discovered reagent one always should think of a proper "ligand outfit" to (re-)enter a catalytic cycle. Tremendous efforts in the coordina-tion and organometallic chemistry of these ligands are to be made, especially with regard to synthesis and structure.

4.1.6   Mechanistic Knowledge and Theory – Keys to Catalyst Design

The declared goal of "going catalytic" with metal-containing reagents will depend on a firm foundation of mechanistic studies, including thermodynamic considera-tion. This is not a new concept, but should always be kept in mind [10, 42, 43]. Carefully determined conversion/time diagrams, *in situ* spectroscopic studies, and, if possible, kinetic time laws are among the fundamentals of catalysis research. Such data will pave the way toward a mechanistic understanding. It is frequently seen here that certain discoveries are not taken up quickly enough in an interdis-ciplinary effort. The versatile Heck reaction, for example, has long remained an "organic chemistry domain," where it has its synthetic benefits. It was only recently that through an accurate mechanistic study *new* catalysts – phospha-pal-ladacycles (**8**) – were found to be superior to the conventional ones (cf. Section 3.1.6) [44 a]. These new catalysts mediate other C–C and N–C coupling reactions, too [44 b]. It is certainly not a matter of chance that the latest synthesis of the famous antitumor reagent taxol (**9**) depends on an intramolecular palladium catalyzed Heck coupling [44 c, d].

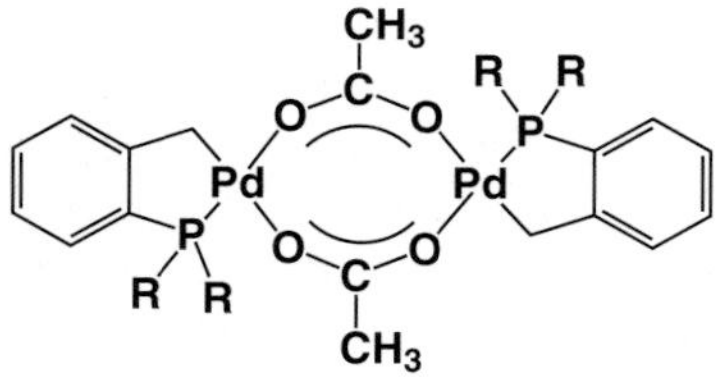

R = *o*-tolyl, mesityl
phospha-palladacycles

8

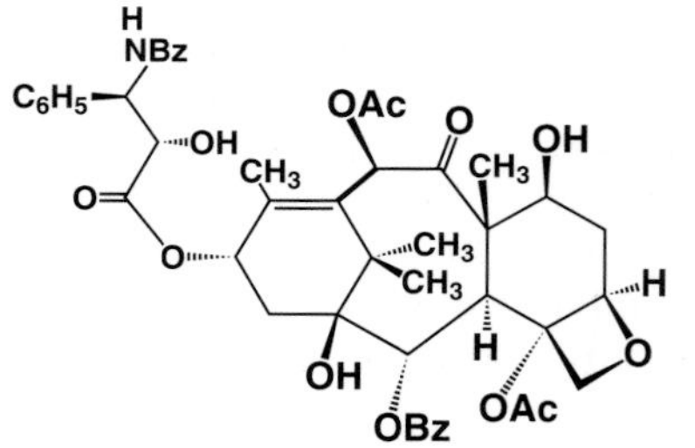

taxol

9

Homogeneous catalysis faces a number of obviously trivial but still unresolved problems that depend primarily on mechanistic insight. For example, why do some metal oxides (V_2O_5, MoO_3, CH_3ReO_3) catalyze olefin expoxidation while others (OsO_4) catalyze olefin hydroxylation? What are the stereoelectronic prerequisites for this fundamentally different reactivity? If these differences become known, how can a "switchable catalyst" be synthesized?

Unfortunately, only a few reliable thermodynamic data of catalytically relevant bonding situations are available in the literature, for which reason this area of physico-organometallic chemistry should receive greater emphasis. In this context, theoretical approaches should be integrated in mechanistic studies much better than previously. Quantum chemical calculations, most notably the density functionality theory (DFT) taking care of relativistic effects of the (heavier) metals, should quickly become established in homogeneous catalysis research [45]. Furthermore, molecular modeling has reached a predictive state, at least when series of similar catalysts (e. g., the ligand periphery of a given metal) are to be compared with each other (cf. Section 3.1.2). Spectroscopic techniques are beginning to enter the scene to support theoretical approaches [46]. There are cases in which older mechanistic schemes or kinetic data are superseded by new and more detailed proposals due to improved theoretical methodology [47, 48]. It remains to be seen whether this improvement entails a new catalyst and process design or even new techniques.

4.1.7 Catalyst Performance/New Techniques to Generate and Activate Catalysts

An ideal situation is reached when the catalytic intermediates and at least the rate-determining step(s) are established. This body of knowledge normally takes a long time to attain and will depend strongly on the above-mentioned kinetic studies. In addition, however, only the overall catalyst performance finally decides the industrial feasibility: what is the stability of the catalyst, under which process conditions, and how can the catalyst lifetime be improved? From experience, continuous micro(laboratory) plants are of pivotal importance to predict the performance of a new catalyst. *In situ* spectroscopy (mainly IR and NMR techniques, e. g., parahydrogen labeling) are to be applied here as a means of detecting structural features of the catalytic species involved. Chemical (model) reactions with the (pre-)catalyst *outside* the reactor, and either in the absence of one component or in the real reaction conditions, will always give only part of the truth with regard to the catalytic reaction. Mass and energy transport features are worth of consideration at the beginning of a catalytic project.

Catalyst performance has of course been a permanent theme in industry. For example, the catalytic *activity* of oxo catalysts (in hydroformylation) has improved in the past 50 years by a factor of 10 000: change from diadic and triadic process technology to continuous plant operation, replacement of cobalt by rhodium, tailoring of the ligand sphere (phosphines), change of phase application (from mono- to two-phase processes). At the same time, an improvement of *selectivity* has been achieved, apart from the ease of product/catalyst separation [132]. A similar development seems to occur in the Monsanto acetic acid process [49].

Homogeneous catalysts applied in typical captive-use situations normally imply a high degree of specific in-house expertise, for which reason not many details are known in the "open" literature. Often generated *in situ*, these may remain secret upon licensing, so any independent industrial catalysis research to some extent risks the duplication of known science.

Increasing desire for easy-to-make, storable, organometallic catalysts is a common trend. Their stability to air, water, and temperature is decisive in their success. In many cases, only the precatalysts fully share these requirements. Therefore, techniques to generate the active catalyst species are desirable. Beyond chemical *in situ* techniques, some physical methods have recently come to the fore. For example, the microwave technique [99] was used to generate a catalyst system for allylic alkylations with high regio- and stereoselectivities (cf. eq. (4)) [100]. The reaction times are remarkably short. Due to the high temperatures attained by microwave heating (up to 220 °C), the active catalyst species, however, must be thermally robust.

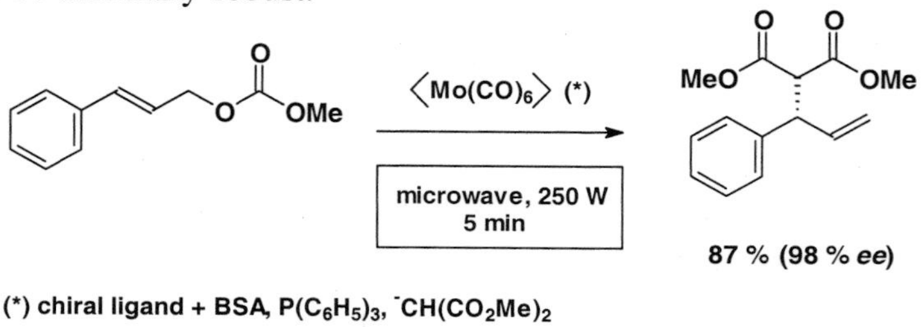

$$(4)$$

87 % (98 % *ee*)

(*) chiral ligand + BSA, P(C$_6$H$_5$)$_3$, ⁻CH(CO$_2$Me)$_2$

Ultrasound is involved in another technique of catalyst activation. Known from (stoichiometric) ligand substitution, catalytic applications have now become available.

A methodological breakthrough in the elucidation of catalytic mechanisms comes from the ultrafast electron diffraction (UED) technique. Even though only the most simple models are accessible as yet, it is possible in principle to view "hot" reaction intermediates on a multi-picosecond (and femtosecond [101]) time-scale after their formation, as shown for CO elimination from Fe(CO)$_5$ [101].

The characterization of catalysts has become the object of high-throughput screening. Experimental arrays employing microsystems techniques are now available. The area is promising, at least to identify *relative* data of catalyst activity and selectivity [102]. Combinatorial catalysis, as the field is called, will not replace the innovative chemical idea, as amply shown in this book (cf. Section 3.1.3).

4.1.8   Organometallic Electrocatalysis and Biomimetic Catalysis

For a long time, electrocatalysis (cf. Section 3.2.8) has not been considered important in organometallic chemistry although numerous known processes depend on redox reactions (e. g., Wacker-type catalysis, oxidative olefin carbonylation, metal activation of carbon dioxide). It seems, however, that a number of catalytic problems will resist solution unless an electrochemical step is included. To this end, our knowledge about electron-transfer and radical processes has to be improved [50]. In numerous metal-mediated reactions, the work-up and reactivation of the used metal (compound) appears possible only by means of electric current – not the worst solution, from a number of considerations. In this context, radical-type catalysis should be given emphasis; two relevant areas are the activation of halohydrocarbons [51] and the gas-phase metal chemistry of hydrocarbons in general. In the latter field, sophisticated mass-spectroscopic methodology can be exploited to learn more about the elementary steps of organometallic catalysis [52, 53].

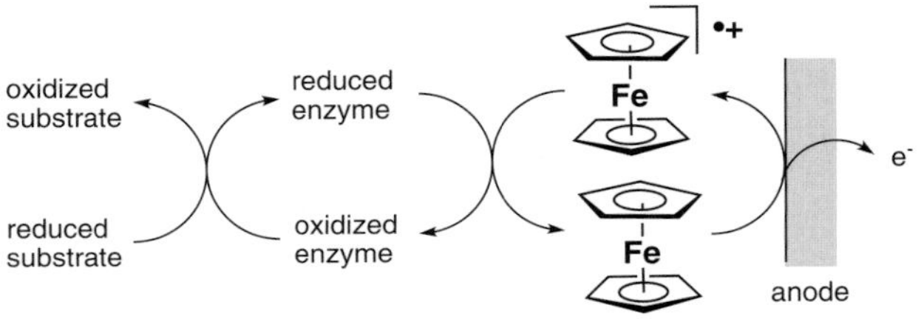

Scheme 3

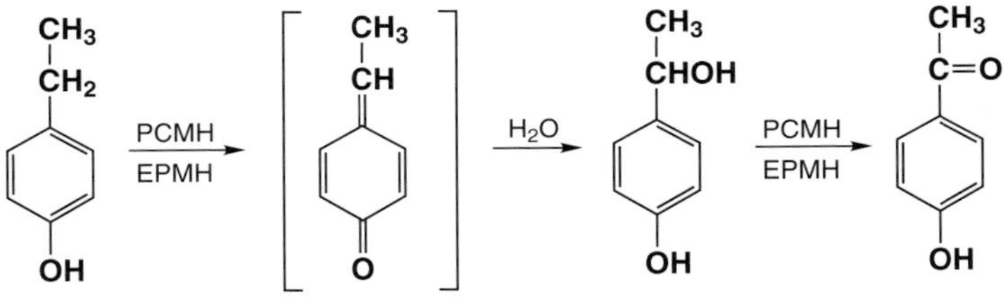

Scheme 4   **PCMH, EPMH cf. [53]**

Electroenzymatic processes in enzyme membrane reactors may be considered as bridges toward biocatalytic reaction systems, with the ferricinium cation acting as a clean one-electron oxidant (Scheme 3). To give an example, 4-ethylphenol can be oxidized to 4-hydroxyacetophenone according to Scheme 4 [53].

Biocatalytic, enzyme-analogous, and antibody processes are under intensive investigation (cf. Section 3.2.1; see, for example, [17, 78, 82]).

One recent example should suffice to demonstrate the future of the biomimetic catalysis approach [103]: An amazingly selective NAD(P)H regeneration was achieved with the organorhodium complex **11**. It was coupled with an isolated monooxygenase (HbpA, E.C. 1.14.13.44) from *P. azelaica*, which is required by the NADH as coenzyme, eq. (5). The selective *orthohydroxylation* of various α-substituted phenols is thus achieved according to Scheme 5. The reducing agent **11** was generated electrochemically (−750 mV vs. Ag/AgCl$_{sat}$) from the corresponding $(Rh^{IV})^{2+}$ precursor complex **10**. The reduction **10** → **11** is also possible chemically, e.g., with formate $[HCO_2]^-$. The (non-optimized) conversions are at ca. 200 mg L^{-1} h^{-1}, which is approx. 50 % of the (optimized) fermentative process (390 mg L^{-1} h^{-1}).

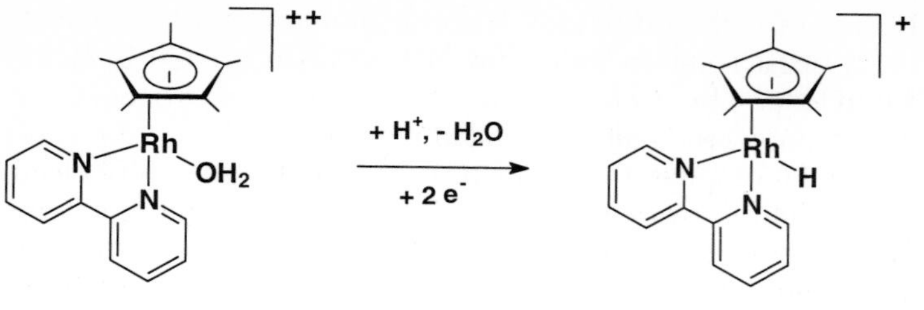

$$+ \text{H}^+, - \text{H}_2\text{O}$$
$$+ 2\,\text{e}^-$$

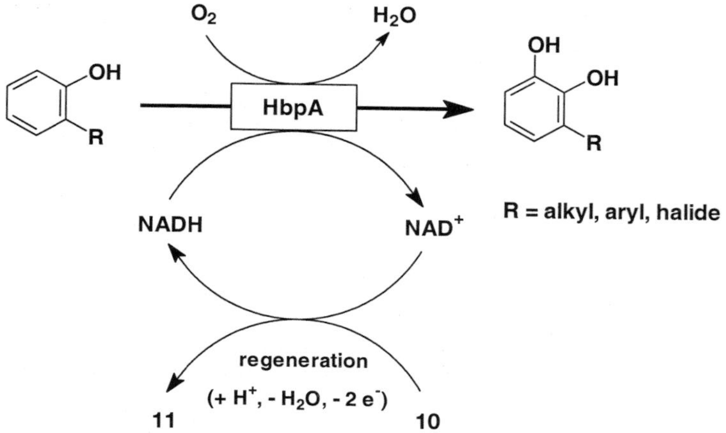

(5)

10   **11**

R = alkyl, aryl, halide

Scheme 5   **11**   regeneration ($+\text{H}^+$, $-\text{H}_2\text{O}$, $-2\,\text{e}^-$)   **10**

4.1.9   New Chemical Feedstocks for Homogeneous Catalysis and Renewable Resources

Organometallic catalysis has largely been based on feedstock chemicals such as alkynes, alkenes, dienes, carbon monoxide, and hydrogen. The classical mechanistic principles – σ/π-complexation, oxidative addition, insertion/migration, reductive elimination – suffice to explain the majority of reactions thus comprising catalytic cycles. Unfortunately, only reluctant reactivity has been seen for alternative, particularly cheap, feedstock chemicals. Methane, ethane, propane, and carbon dioxide deserve specific mention because of their general and abundant availability. Saturated C_{1-3} building blocks suffer from the failure that their metal chemistry is not really catalytic as yet [54–56]. Nevertheless, possible advantages are clear to see:

(1) *Methane* as an abundant natural resource chemical could partially substitute for both olefins (refinery ethylene and propene) and carbon monoxide.

(2) *Propane* is the most prominent refinery gas (av. 60 %), followed by *butane* (ca. 30 %), so their functionalization has been a serious topic in heterogeneous catalysis research for a long time.

(3) *Carbon dioxide* could at least be engaged in the synthesis of carboxylic compounds and certain heterocycles, quite apart from the consideration that it may eventually substitute for the more expensive carbon monoxide in specific applications. For reductive processes, however, a CO_2-based feedstock stituation could prove more expensive since an extra equivalent of a reductant is required (cf. Sections 3.2.11 and 3.3.4).

The selective *and* catalytic C–C and C–H bond cleavage of (unreactive, saturated) hydrocarbons, like ethane and methane, is still another "Holy Grail" in chemistry. It is obvious then, that the catalytic activation of such molecules under mild (e. g., nonphotolytic) conditions would revolutionize catalysis. This capability would contrast the classical high-temperature reforming processes of the petrochemical industry, which constitute the largest-scale industrial catalyses. No solution to the problem is available as yet. However, a number of recent approaches regarding C–H, C–C, and C–F bonds appear promising (cf. Sections 2.8, 3.1.6, or 3.3.6).

Carbon–Hydrogen Bonds

Methane is broken up at +10 °C upon treatment with the iridium complex $(\eta^5$-$C_5Me_5)IrH(PMe_3) \cdot CH_2Cl_2$ to give the oxidative addition product $(\eta^5$-$C_5Me_5)$ $Ir(H)_2(PMe_3)CH_3$ [57a]. This reaction matches with earlier findings according to which the less dissociation-labile carbonyl complex $(\eta^5$-$C_5H_5)Ir(CO)_2$ affords the hydrido-methyl derivative $(\eta^5$-$C_5H_5)Ir(CO)H(CH_3)$ under photochemical conditions [54 c]. The reader is further reminded of Watson's elegant work concerning $^{13}CH_3$ vs. $^{12}CH_3$ exchange in the "metal-acidic" lutetium(III) complex $(\eta^5$-$C_5Me_5)_2$ $Lu(CH_3)$ by means of methane [57 b] and to a novel approach of methane activation by *N*-heterocyclic carbenes [135].

Carbon–Carbon Bonds

Milstein et al. demonstrated (Scheme 6) that a methyl group is first C–H-activated before it eliminates as methane via hydrogenolysis from the intermediate **10** [57 c]. The prior C–H bond activation occurs readily, and is expectedly much more rapid than the subsequent C–C-cleavage reaction. It is certainly true that the formation of the new rhodium–phenyl bond of **13**, in addition to the very special type of chelating *P,C,P*-coordination, is the driving force here, but nevertheless the cleavage of a low-reactivity bond between sp^2- and sp^3-hybridized carbon atoms has been achieved. The more difficult problem to cope with is the cleavage of purely aliphatic C–C bonds because their directed sp^3-orbitals along the bond axis are inaccessible to metals and are thus much less reactive.

Transfer of a C–H-activated methyl group after consecutive C–C bond cleavage as a methylene unit to other substrates is an interesting alternative. Starting from

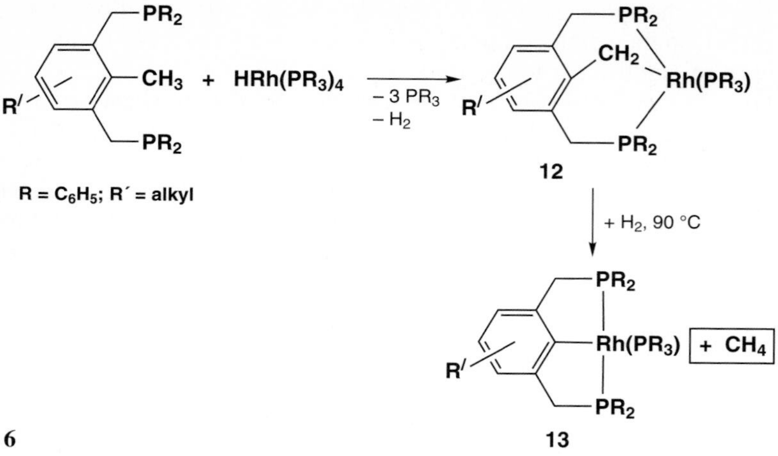

Scheme 6

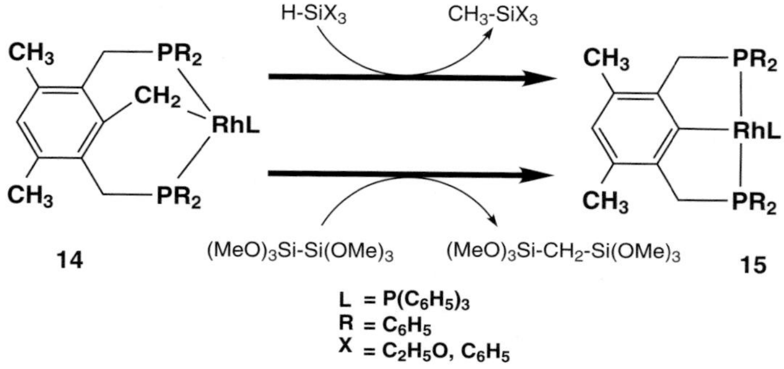

Scheme 7

the strained rhodacycle **14**, for example, the insertion reactions of Scheme 7 were achieved, at least in stoichiometric reactions [57 d, e].

It must be noted that the cleavage of strong C–C bonds by transition metal insertion under mild conditions depends thermodynamically on an auxiliary chemical reaction coupled to the cleavage process (e. g., hydrogenation). The work of Milstein shows that an oxidative addition process according to the simplifying eq. (6) is highly dependent on the electron density of the metal center, but can be thermodynamically more favorable than the competing C–H activation process. C–C bond cleavage may start with a sterically less hindered C–H bond activation.

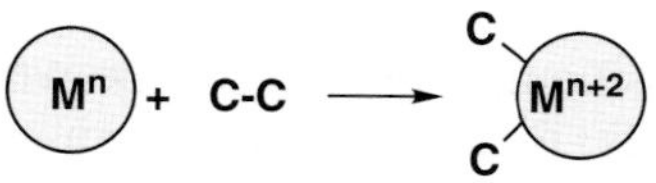

$$\boxed{M^n} + C\text{-}C \longrightarrow \begin{smallmatrix}C\\ \\C\end{smallmatrix}\boxed{M^{n+2}} \tag{6}$$

A challenging example of a paraffinic C–C bond metathesis is the recently discovered conversion of ethane into methane and propane under mild conditions

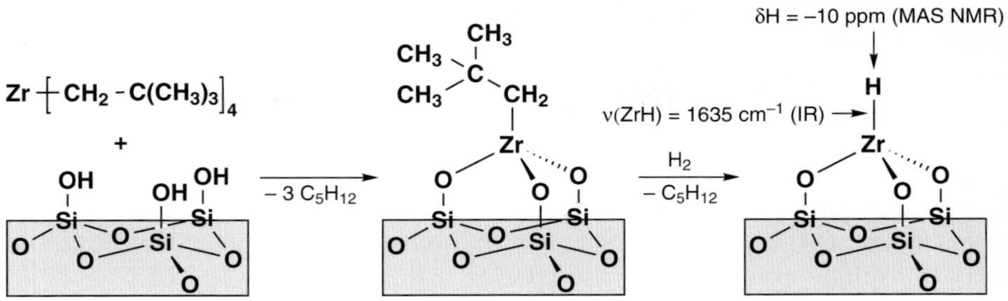

Scheme 8

using a surface-attached but molecularly defined electron-deficient catalyst. This unprecedented hydrocarbon "dismutation" of eq. (7) once again starts with a C–H bond activation, thus involving conventional alkylmetal intermediates [57 i]. This result compares with a hydridozirconium-catalyzed hydrogenolysis of simple alkanes at 25 °C on the same type of catalyst, which is generated from tetrakis(neopentyl)zirconium and partially dehydroxylated silica according to Scheme 8; the final products are methane and ethane, which is not cleaved by the catalyst under a hydrogen atmosphere [57 k].

$$2 \ CH_3{-}CH_3 \ \xrightarrow{\text{cat.}} \ CH_4 \ + \ CH_3CH_2CH_3 \tag{7}$$

$$\left.\begin{array}{c} \textbf{polyethylene} \\ or \\ \textbf{polypropylene} \end{array}\right\} \ \xrightarrow[H_2]{\text{cat.}} \ 2 \ C_nH_{2n+2} \tag{8}$$

$$n \geq 1 \ (\text{mainly } CH_4 \ \& \ C_2H_6)$$

This principle can be applied to the exhaustive degradation of polyethylene and polypropylene to low molecular weight oligomers, ethane, and methane when the zirconium catalyst of Scheme 7 is administered (cf. eq. (8)) [57 l]. It will be interesting to see how polymer wastes containing chlorinated polymers (e. g., PVC) behave against this type of highly electron-deficient catalyst. In any case, Basset's results are the only "puristic" cases concerning catalytic cleavage of *saturated* hydrocarbons so far.

Carbon–Fluorine Bonds

A specific challenge remains for the metal-induced cleavage of the strongest bond that carbon can form, the C–F bond [57 f]. The functionalization of polyfluorinated organic compounds – important building blocks of chemical, pharmaceutical, and advanced materials industries – would promise a diverse market for such a process. First success in this area is based on the stoichiometric reaction of eq. (9), the thermodynamics of which clearly relies on the strength of the silicon–fluorine bond. The perfluorophenyl–rhodium bond of complex **16** also favors this kind of C–F bond activation [57 g]. The related rhodium–phosphine complex was found to allow the catalytic hydrogenolysis of hexafluorobenzene: when this fluorocarbon is used as a solvent, it eliminates fluorine upon heating with the catalyst (PMe$_3$)$_4$RhH in the presence of a base (e. g., NEt$_3$ or NEt$_3$/K$_2$CO$_3$) and hydrogen; the released hydrogen fluoride is captured by the base. The proposed catalytic cycle for the hydrogenolysis (Scheme 9) involves electron-rich hydridorhodium(I)–phosphine complexes as the species that induce cleavage of C–F bonds. This implies that other complexes which can serve as a good source of such species are also likely to be active. The synthetic potential of carbon–fluorine activation is enormous, provided that efficient catalysts can be found. So far, turnover numbers up to only 114 have been recorded [57 h]. Heterogeneously catalyzed hydrogenolysis of carbon–fluorine bonds is known; however, it requires very high temperatures and is nonselective, once again showing the future potential of homogeneous organometallic catalysis (cf. also [136]).

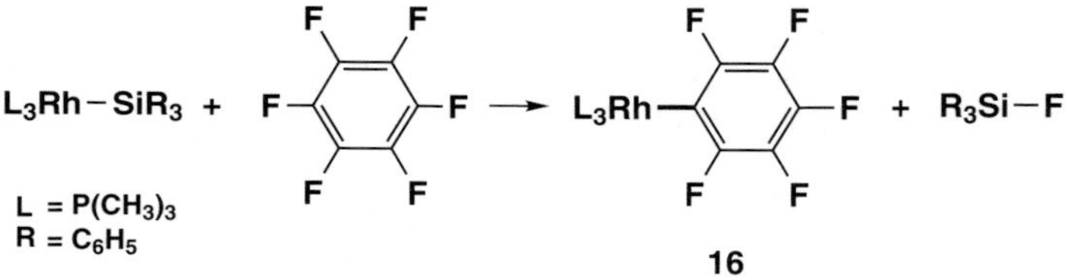

$$(9)$$

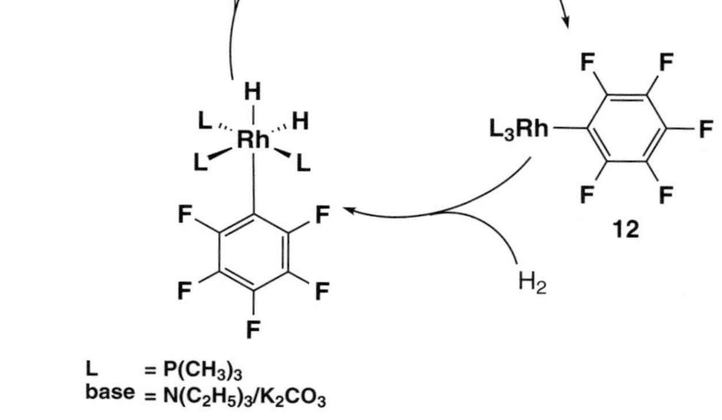

Scheme 9

In this context, the catalytic degradation of polymers as a possible new basic feedstock for organic chemicals seems worthy of reinforced efforts [15, 57 i–l]. Catalytic activation of methane and other hydrocarbons remains of central interest in catalysis research in general [54]. To this end, any stoichiometric chemistry of this molecule has a significance of its own.

The wealth of present-day chemistry is based on petroleum feedstock and inorganic minerals, such as rock salt, pyrite, dolomite, and bauxite. As far as the enormous variety of chemicals and their refined products (e. g., pharmaceuticals) are concerned, it is quite obvious that, in two or three generations from now, a new feedstock basis is inevitable. Even if large-scale recycling technologies are available by then (e. g., in the plastics area), there will not be enough easily available oil to cover the entire demand of the chemical industry and energy generation. Instead, we shall depend more and more on renewable resources.

However, little attention is as yet given to this chemistry. First of all, new brands of plant materials are necessary that grow fast enough, especially under bad climate and soil conditions, and exploit a specific property. Plant genetics clearly is the way to produce new "industrial" plant species that, for example, contain just the desired chemical and not a mixture of many products (which natural plants normally do).

Again, chemical catalysis must take on the duty of performing the necessary transformations. The main problem is normally the complex chemical composition of natural products, and the presence of numerous functional groups, e. g., NH_2, CH_2OH, $COOH$, SH. This requires highly developed catalysts that are both selective and resistant to functional groups.

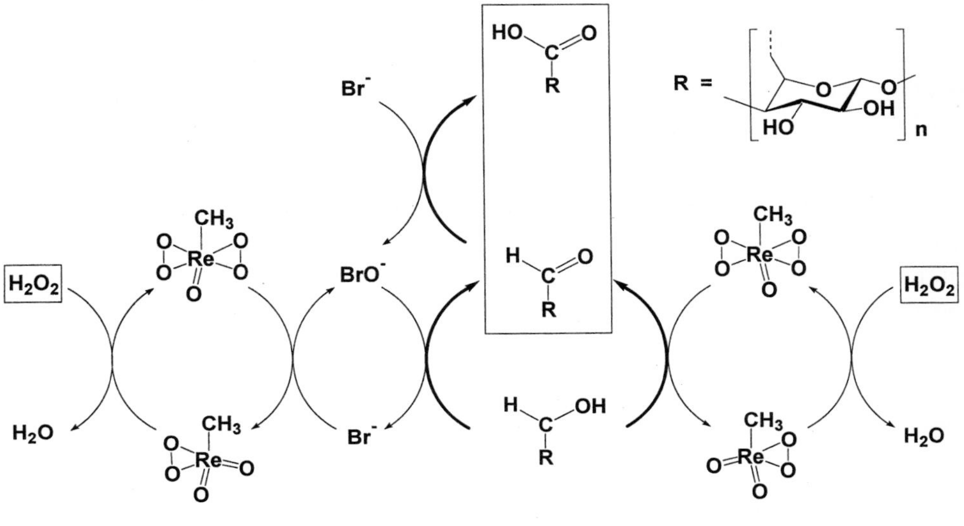

Scheme 10

Slow but significant progress is visible in this area. For example, potato starch (containing 27 % amylase and 73 % amylopectin) can be oxidized to superabsorbing biopolymers. The three-component system $H_2O_2/HBr/CH_3ReO_3$ works in the formation of carboxylated starch according to the mechanism proposed in Scheme 10 [104].

Isoeugenol, a product from sawdust, and the agricultural waste product *trans*-ferulic acid can be converted into vanillin by consecutive oxidation steps, again employing CH_3ReO_3 as a catalyst and hydrogen peroxide as a "green", yet still expensive, oxidant [105]. Biological wastes are thus shown to form highly value-added products by virtue of organometallic catalysis.

4.1.10   Catalysis Under Supercritical Conditions and Supported by Ionic Liquids

Earlier, no emphasis was given to homogeneous catalysis in supercritical fluids, albeit the technique *per se* is now well established in extraction technology (e. g., for coffee, tea, hops, spices, and natural flavors). Incorporated in homogeneous catalyst processes [58–60], supercritical conditions can dramatically change the solubility profile of solvents and the reactivity of certain chemicals.

A comprehensive review (1999) highlighted the enormous opportunities afforded by supercritical fluids (SCF) in organometallic catalysis (cf. Section 3.1.13) [106].

Of particular interest is carbon dioxide ($scCO_2$) because of its relative chemical inertness and the convenient critical data. In contrast, water (21.8 MPa, 374 °C) is a rather bad candidate for both the critical data and the reactivity. Generally, catalytic reactions in SCFs are limited to narrow temperature ranges. The T_c has to be little below or within the desired range, otherwise too much pressure is required. It is relevant to the kinetics of the catalytic reactions that the dielectric constant, viscosity, and other physical data of an SCF medium stongly depend on the pressure (via the density function). Since many reactions have polarized reactants, products, or transition states, or exhibit a polar solvent effect, the pressure dependence can be exploited to optimize a process. Therefore, mechanistic studies can benefit from polarity effects tunable in supercritical media. Since supercritical carbon dioxide has a high degree of compressibility, it provides the opportunity to explore density as an additional reaction parameter. The schematic phase diagram of carbon dioxide is seen in Figure 1 of Section 3.1.13.

With regard to organometallic reactivity, little is known on the various SCFs. Even carbon dioxide can be reactive if it inserts in M–H, M–R, M–OR, or M–NR$_2$ bonds. In this case, $scCO_2$ can be used simultaneously as C_1 building block.

An impressive example of selectivity effects comes from olefin metathesis. A certain Ru–carbene catalyst strictly yields the ring-closure metathesis product in supercritical carbon dioxide at densities above 0.65 g mL^{-1} (eq. (10)), while at lower densities acyclic oligomers are formed [107]. The separation of the muscarine fragrant exhibiting Structure **17** is so easy that olefin metathesis should gain a new profile of application with the SCF-OMC technique (cf. [108]).

$$
\begin{array}{c}
\text{Ru cat (RCM)} \\
\xrightarrow{\hspace{2cm}} \\
\text{scCO}_2 \\
\text{d > 0,65 g mL}^{-1}
\end{array}
\qquad
\textbf{17, RCM product}
\qquad + \; \text{C}_2\text{H}_4
\tag{10}
$$

$$
\begin{array}{c}
\text{Ru cat (ADMET)} \\
\xrightarrow{\hspace{2cm}} \\
\text{scCO}_2 \\
\text{d < 0,65 g mL}^{-1}
\end{array}
\qquad
\boxed{\text{oligomers}}
$$

Convincing results also became available for CO_2 hydrogenation using the catalyst $RuH_2[P(CH_3)_3]_4$ at 50 °C/8.5 MPa H_2: while in standard solvents, e. g., $N(C_2H_5)_3$, H_2O, or THF, initial TOFs < 100 h^{-1} were recorded, activities far above 4000 h^{-1} could be achieved in scCO$_2$/CH$_3$OH and scCO$_2$/DMSO [109]. Furthermore, a strictly alternating polyketone was made in scCO$_2$ from C_2H_4 and CO in the presence of the Ni^{II} catalyst; 11 kg of the polyketone per g of Ni was obtained, which represents the best data reported as yet [110].

The asymmetric hydrovinylation (cf. Section 3.3.3) of styrene with excellent chemo-, regio-, and stereoselectivity was achieved in scCO$_2$ using the known Ni^{II} catalyst but – instead of the flammable co-catalyst $(C_2H_5)_3Al_2Cl_3$ – the boranate ("BARF"). This was possible because all components are soluble in scCO$_2$ [111] (cf. eq. (6) of Section 3.1.13).

Beyond these effects, carbon dioxide is an "environmentally responsible" solvent and deserves investigation of its technical uses for this reason, too.

Asymmetric hydrogenation combined with *catalyst recycling* using ionic liquids and scCO$_2$ highlights the potential of supercritical media [112].

Ionic liquids are salts that are liquid at low temperature, at least below 100 °C. They form biphasic systems with many organic compounds and product mixtures (cf. Section 3.1.1.2.2).

Within a given class of ionic liquids, the melting ranges and viscosities can be adjusted greatly by changing the substituents, but the anions also have a strong influence. The density changes mainly according to the bulkiness of the groups R.

Advantages of ionic liquids in homogeneous catalysis are as follows [113, 114]:

(1) They are nonmolecular, ionic solvents.
(2) Product separation is facile, due to negligible vapor pressure.
(3) Ionic liquids have good solubility for organometallic compounds.
(4) Their melting ranges, viscosities, densities, solubility characteristics, acidity, and coordination ability are easily adjustable.
(5) They are available commercially.
(6) Recovery and clean-up are easy due to biphasic process technology.

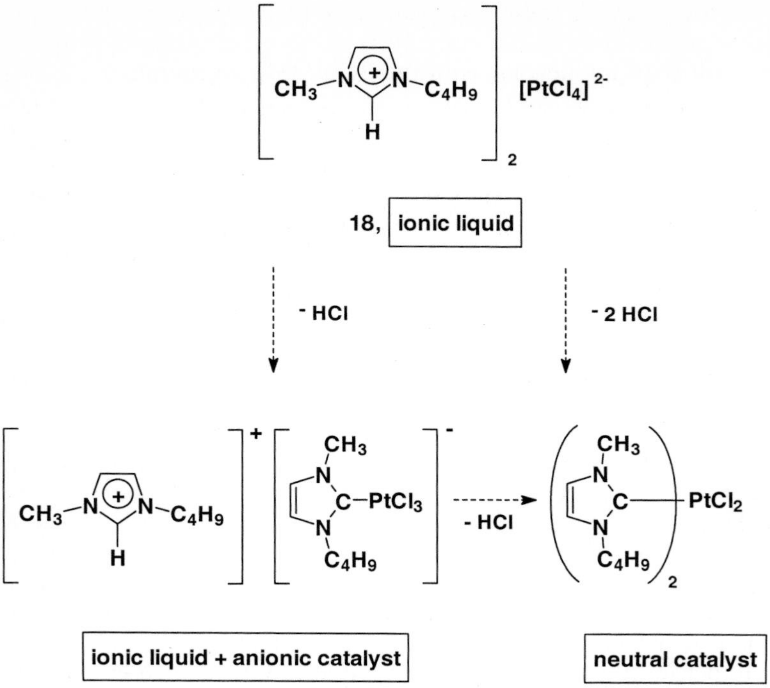

Scheme 11

A main question – not yet really considered – concerns the inertness of ionic liquids. Not only are the anions potential ligands, especially for neutral and cationic metal complexes; one has also to take into consideration what is known for cations like (imid)azolium: formation of carbene complexes via deprotonation is a rather facile process especially if ligands of sufficient basicity are present, e. g., –OR, –NR$_2$. Therefore, several of the impressive catalytic results [113] deserve mechanistic investigation to find out whether they are really limited to the ionic liquid effects. For example, solvent and complexation effects are likely to enhance one another in the Heck coupling reactions that were run in the presence of Structure **18**, Scheme 11 [115].

Promising results have been reported by various laboratories since 1990 on catalysis in molten salts, notably for catalytic hydrogenation, hydroformylation, oxidation, alkoxycarbonylation, hydrodimerization/telomerization, oligomerization, and Trost–Tsuji coupling [113]. A continuous-flow application to the linear dimerization of 1-butene on an ionic-liquid nickel catalyst system reached activities with TON $> 18\,000$ [116].

4.1.11 New Reactions, Improved Catalysts

Oxygen, water, and *ammonia* are preeminent when inorganic reagents of applied homogeneous catalysis are under discussion. As a matter of fact, oxidative processes comprise the greatest share among all homogeneous catalytic processes if metal-mediated gas-phase oxidations (e. g., terephthalic acid; Section 2.8.1.2) are included [61]. A specific opportunity for homogeneous catalysis can be seen in the knowledge that oxidation processes are limited in their selectivities (ca. 85 %) when they operate on the basis of heterogeneous catalysis. Nevertheless, selective activation of elemental oxygen is difficult to achieve, and in a number of cases coupled processes, such as the combination of a secondary alcohol with oxygen, are required [62]. Ironically, the Wacker-type reactions – prototypes of organometallic oxidations – exploit the catalyst metal to oxidize the ethylene, while the oxygen only reactivates the palladium (cf. Section 2.4.1). No other oxygen reaction using organometallics as activators is close to being given any application. Here is an ostensibly rich field for high-oxidation state organometallic chemistry [63].

Hydrogen peroxide is a more reactive but more expensive substitute for oxygen. It has a broad and relatively well-investigated metal coordination chemistry [62 a]. While it normally does not meet the tight economic requirements for the oxidative production of industrial bulk chemicals, the priority list

$$\text{oxygen} > \text{hydrogen peroxide} > t\text{-butyl peroxide} > \text{other oxidants}$$

is generally accepted [61 a–c]. It is sure enough that stoichiometric oxidants such as "chromic acid" will be excluded in future times from technical-scale applications for environmental reasons, even if higher-price chemicals such as vitamin K_3 and others are concerned [61 d–f]. Catalytic synthesis is often the only reasonable alternative (cf. Section 3.3.13). It is questionable, even for stereoselective oxidations, whether oxidants yielding appreciable amounts of salt (e. g., NaOCl bleach) [63] will be able to access large-scale applications. It is thus by force of demand that the old topic of *oxygen activation* enters the high-priority list of future research in both coordination/organometallic chemistry and homogeneous catalysis. It has to be noted that the apparently primitive question of oxygen transfer from peroxometal intermediates to olefins is not undisputed in terms of the mechanism. While the OsO_4-mediated dihydroxylation of olefins with hydrogen peroxide has long been known, *atmospheric oxygen* can now be employed for the same purpose. It may be of strong industrial relevance that transformations following eq. (11) are effected by catalytic amounts (0.5 mol %) of $K_2[OsO_2(OH)_4]$, with convincing evidence for stereoselective varieties, as tested with 1-octene and a-methylstyrene [117]. Relatively low pressures (0.3–0.9 MPa) at low catalyst loadings (cat/substrate 1:4000) are promising features of this elegant reaction. It seems that the oxygen regenerates the active Os^{VIII} species from the reduced form $[OsO_2(OH)_4]^{2-}$ of hexavalent osmium. This is reminiscent of the Wacker-type oxidation of ethylene where the oxygen also serves to reoxidize the catalyst metal ($Pd^0 \rightarrow Pd^{II}$) (cf. also Section 2.4.1).

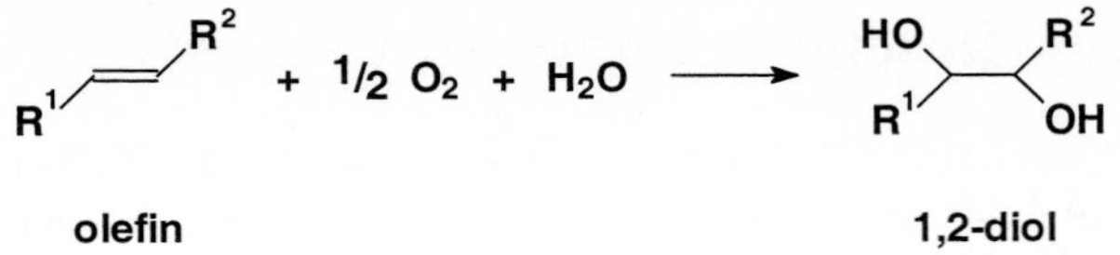

$$(11)$$

olefin   1,2-diol

Water is another ideal, environmentally sound reagent with exciting prospects, but numerous questions are still open. Could water be added to olefins in the anti-Markownikov mode? Primary alcohols would thus be cheaply available (eq. (12)). Similarly, what is the appropriate catalyst to add ammonia across a double bond (hydroamination, eq. (13), Section 2.7) to yield organic amines as starting materials for a number of fine chemicals?

$$R\text{-}CH{=}CH_2 \; + \; H_2O \xrightarrow{\;cat.\;} R\text{-}CH_2\text{-}CH_2OH \qquad (12)$$

$$R\text{-}CH{=}CH_2 \; + \; NH_3 \xrightarrow{\;cat.\;} R\text{-}CH_2\text{-}CH_2NH_2 \qquad (13)$$

$$n\,CH_2{=}CH_2 \; + \; n\,CO \xrightarrow[(Pd)]{\;cat.\;} \left(\!CH_2\text{-}CH_2\text{-}\!\!\underset{\underset{O}{\|}}{C}\!\right)_{\!n} \qquad (14)$$

polyketones

On the other hand, several well-established reactions are missing certain speciality applications. For example, what would be an efficient catalyst for the hydroformylation of (per)fluoroalkenes, of which the products are of broad use as pharmaceuticals [64]? How can functionalized olefins enter industrial applications, based upon a recent development employing special tungsten–carbene complexes **19** for the metathesis even of C–C-unsaturated thioethers [65]? Is the directive ethylene/carbon monoxide coupling of Drent et al. according to eq. (14) (cf. Section 2.3.4) [66 a–c] and the structural principle **20** (intermediate) of general use (e. g., to obtain functionalized olefins such as fluoroolefins, or to use isonitriles in place of carbon monoxide)? What are the structural prerequisites for multiple carbonylation reactions? A highly enantioselective alternating co-polymerization of propene or styrene and carbon monoxide with a chiral phosphine-phosphite (BINAPHOS; **3**) palladium catalyst was achieved (eq. (15)). The optically active polymer had a molecular weight of about 10^5 and M_w/M_n = 1.6 when the cationic catalyst $[CH_3Pd(R,S\text{-}BINAPHOS)(N{\equiv}C\text{-}CH_3)]^+$ was applied [66 d].

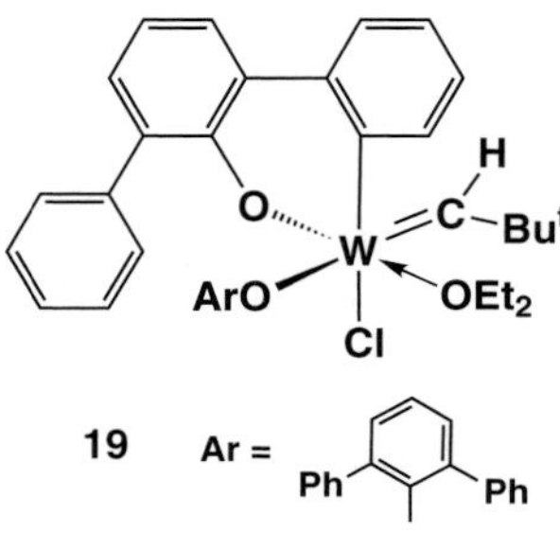

19   Ar =

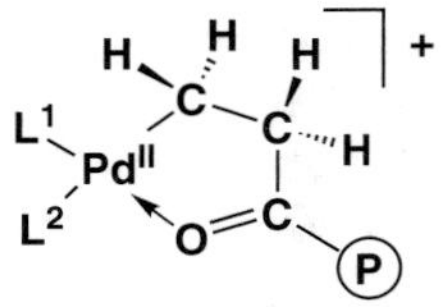

20 Ⓟ = polymer chain

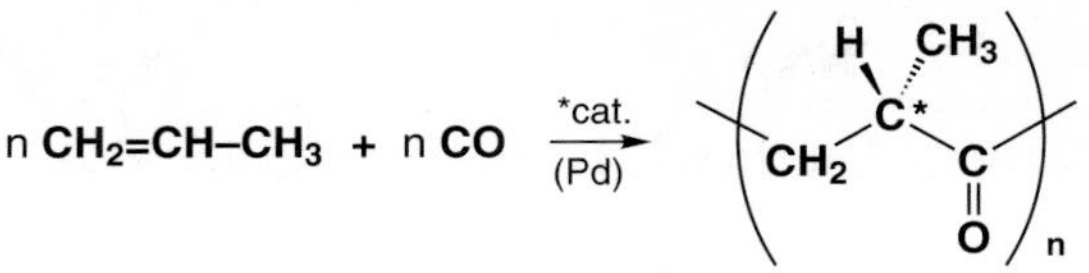

$$n \ CH_2=CH-CH_3 \ + \ n \ CO \ \xrightarrow[\text{(Pd)}]{\text{*cat.}} \ \left(\begin{array}{c} H \quad CH_3 \\ \overset{|}{\underset{|}{C}}* \\ CH_2 \quad C \\ \parallel \\ O \end{array} \right)_n \tag{15}$$

A new generation of polymers, e. g., Shell's Carilon, was developed from the discovery of the perfect CO/olefin alternating principle within the short time span of less than ten years. Systematic mechanistic work in this area has yielded a highly efficient carbonylation of propene (TON $= 4 \times 10^4$) in the presence of palladium(II) catalysts to methyl methacrylate (cf. Section 2.3.2.3) [66 e].

It is noteworthy that the attractive homologation reaction – a formal methylene (CH$_2$) insertion (cf. Section 3.2.7) – according to eq. (16) has received little attention as yet [67]. This synthetic principle looks promising for homologous compounds of which only one certain derivative is easily available.

$$R\text{-}Y \ + \ CO \ + \ 2\,H_2 \ \longrightarrow \ R\text{-}CH_2\text{-}Y \ + \ H_2O$$

$$Y = OH, \quad C \overset{\displaystyle O}{\underset{\displaystyle OH}{\diagup}} \tag{16}$$

Novel C–C coupling reactions are about to enter the scope of metal complex catalysis. Examples are asymmetric aldolizations (Mukaiyama [79]), Diels–Alder reactions [80], and indium- or zinc-mediated alkylations [86].

The topic of molecular recognition should gain increased attention in catalyst design. For example, specific structural interactions of higher olefins (e. g., 1-decene) with chemically modified β-cyclodextrins allow efficient hydroformylation in a two-phase aqueous system even though the olefin is completely insoluble in water; at the same time, olefin isomerization at the rhodium catalyst is hampered [81].

Efficient chiral molybdenum catalysts (Structure **21**) which are, at the same time, easy to handle were generated *in situ* and used without further purification in asymmetric olefin metathesis. For example, the RCM following eq. (17) yields $> 80\%$ of the desired product at $> 88\%$ stereoselectivity [118].

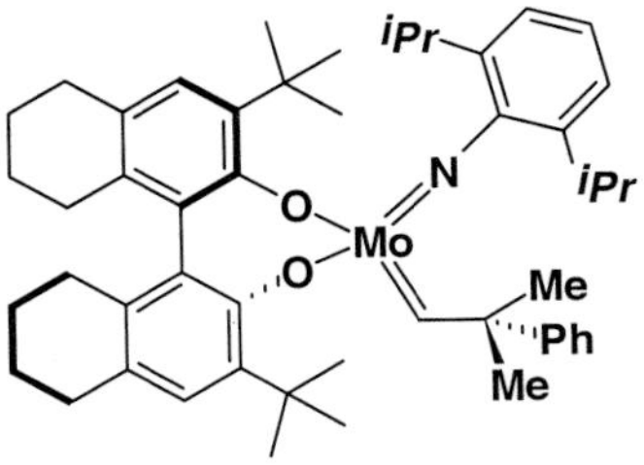

21

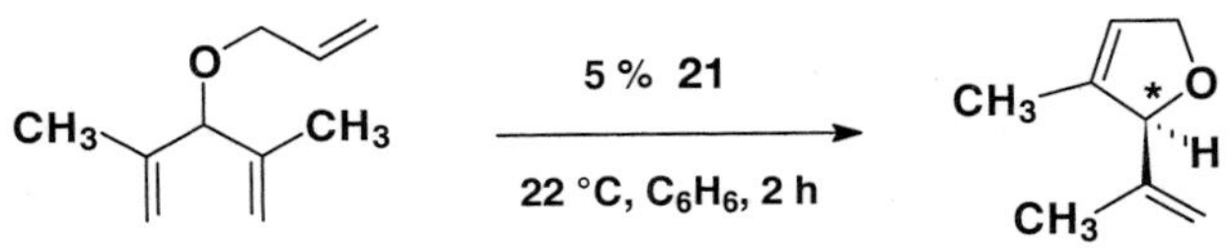

$$\xrightarrow[\text{22 °C, C}_6\text{H}_6\text{, 2 h}]{5\% \ \mathbf{21}} \tag{17}$$

Significant progress is being made in catalytic N–C bond formation. Thus, the *stereoselective* hydroamination of styrene derivatives [119] and norbornene [120] was achieved with BINAP catalysts (Pd and Ir, respectively) (cf. eq. (18)).

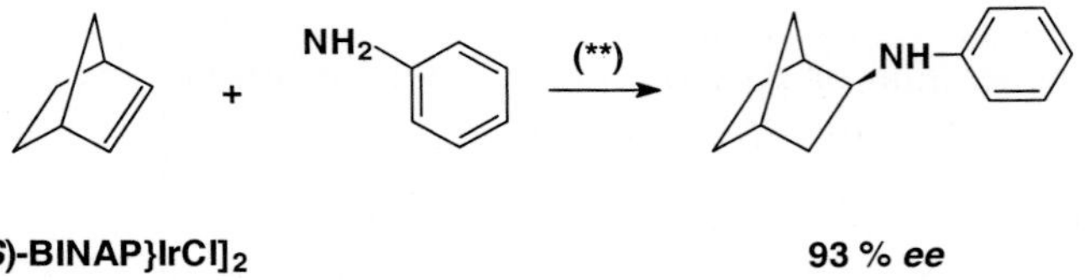

$$[\{(S)\text{-BINAP}\}\text{IrCl}]_2 \qquad\qquad\qquad 93\;\%\;ee \tag{18}$$

4.1.12   A New Generation of Catalyst Ligands

Since the first edition of this book went into print (1996), many new catalyst ligands have been discovered. They very closely resemble one another in terms of their structural features and their coordination behavior. However, there has also come along a completely new generation of ligands that in part substitutes and in part supplements the ubiquitous organophosphanes (Structure **22**): *N*-heterocyclic carbenes **23** (cf. also Section 3.1.10) [121].

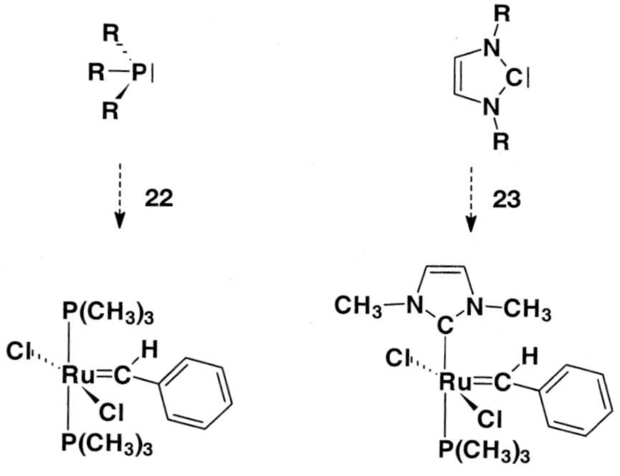

Scheme 12   **24**   **25**

Whereas phosphanes have a conical shape, much of which is decisive in stereoselective catalysis, the *N*-heterocyclic carbenes (NHCs) exhibit flat core structures. More importantly, phosphanes dissociate from metal centers in common catalysts, whereas the *C*-coordination of the new ligands is much more stable under catalysis conditions, thus preventing the catalytically active metal from aggregating and precipitating. Theoretical studies (DFT; cf. Section 3.1.2) showed that the phosphanes in the Ru metathesis catalyst **24** dissociate much more easily (ca. 27 kcal/mol) from the metal than the carbene in **25** (ca. 21 kcal/mol); the remaining phosphane of the latter is labilized by the *trans*-carbene (dissociation of Ru–PMe$_3$: ca. 25 kcal/mol). The NHC ligands strengthen the π-olefin bonding to initiate their metathesis transformations at the catalytic site. This result is in accord with the experimental results concerning olefin metathesis [122]. However, a strong discrimination in terms of ligand-to-metal bonding is generated by steric effects: bulky groups like adamantyl (C$_{10}$H$_{15}$) in Structure **26** facilitate dissociation as indicated by the ΔH_f data, bringing them again close to the class of phosphanes **22**.

26

N-Heterocyclic carbenes are compatible with metals in quite different oxidation states and structural environments. They are easily accessible, easy to handle, thermally fairly robust, structurally variable, and cheap. Functionalized, chelating, water-soluble, chiral, and immobilized derivatives are now available [121]. The latter are all the more important as catalyst leaching seems not to occur due to the strong metal–ligand bonding. Numerous applications in important catalytic processes have proven successful, particularly as compared with related metal–phosphane catalysts. Examples are olefin metathesis, *Heck–Suzuki* and *Stille* coupling, *Grignard* cross-coupling (*Kumada* reaction), alkyne coupling, hydroformylation, olefin hydrogenation, and hydrosilylation, as well as several cyclization reactions [121, 122, 136].

A further advantage is the possible *in situ* generation of catalysts from simple metal salts or complexes (e. g., Ni(OR)$_2$, PdCl$_2$) and azolium salts. The nickel-catalyzed *Grignard* cross-coupling of aryl chlorides at room temperature [123] and the activation of aryl fluorides [136] are convincing examples.

4.1.13 Rare Earth Catalysts

Rare earth organometallic chemistry and catalysis have been a *"Sleeping Beauty"* for many decades (cf. Section 3.2.5). Only recently, ways of molecular activation were discovered that set the scene for systematic studies in homogeneous catalysis. In particular, the synthetic accessibility was greatly improved by virtue of the "silylamide route" of Scheme 13 [124]. In contrast to the standard salt metathesis reactions, this method merits the advantages that (1) noncoordinating solvents can be used due to the precursors' excellent solubility; (2) mild reaction conditions can be applied (e. g., room temp.); (3) halide contaminations are excluded and redox side reactions (at Ln) are rare; (4) product purification is easy (b. p. HN(SiMe$_3$)$_2$: 125 °C; b. p. HN(SiHMe$_2$)$_2$: 93–99 °C); (5) base-free products of Ln^{II} and Ln^{III} are obtained; (6) quantitative yields are obtained in many if not most cases; and (7) mono- and bimetallic lanthanoid precursor compounds are

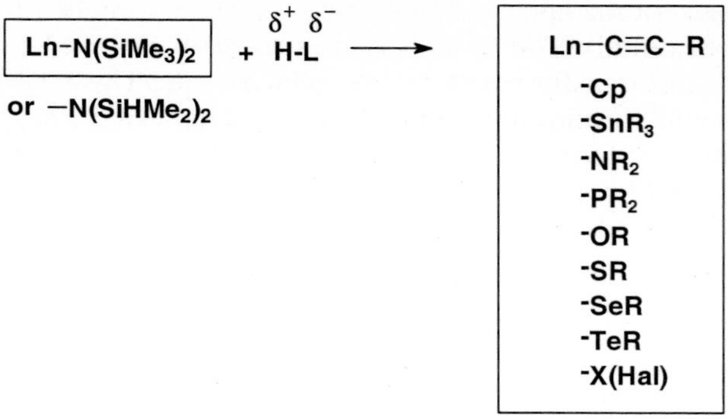

Scheme 13

easily available and can be tuned in terms of their reactivity by the right choice of amide ligands (e. g., –N(SiMe$_3$)$_2$ vs. –N(SiHMe$_2$)$_2$). Using this technique, catalytically relevant rare earth complexes with salen (Structure **27**), (substituted) linked-indenyl (**28**), and sulfonamide ligands (**29**) have been made [124, 125].

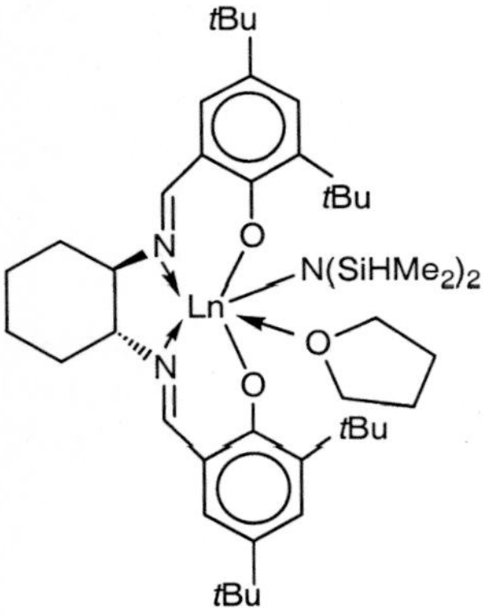

27	**28**	**29**

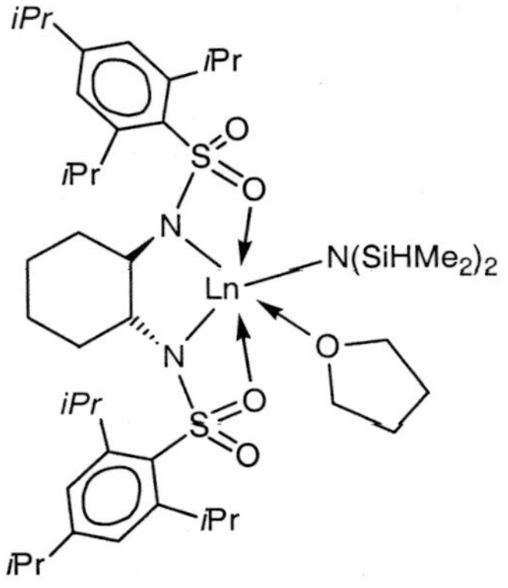

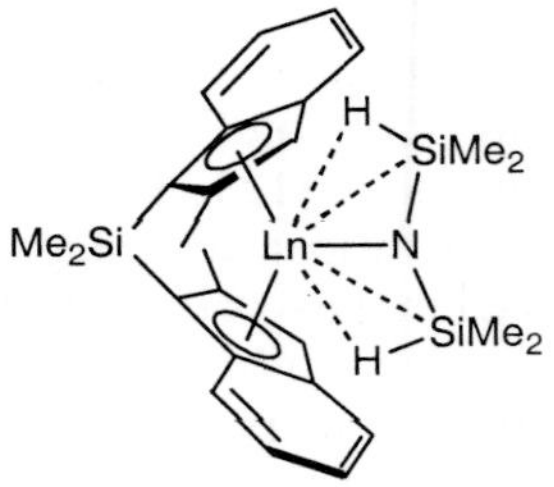

In spite of countless applications of rare earth activation in industrial heterogeneous catalysis, most soluble complexes have long been limited to more or less stoichiometric reactions. An early example is the *Kagan* C–C coupling mediated by samarium(II) iodide [126]. Meanwhile, true catalytic reactions have become available. Highlights are considered the organolanthanide-catalyzed hydroamination of olefins [127], the living polymerization of polar and nonpolar monomers [128], and particularly the polymerization of methyl methacrylate [129]. In the first case, lanthanocene catalysts of type **27** are employed [127].

High molecular weight polymers with a very narrow molecular weight distribution obtained via living polymerization were generated from methyl methacrylate catalyzed by organolanthanoid alkyls and hydrides (Ln = Lu, Sm, Y), with excellent stereotacticity being another striking feature [128]. The product data are fascinating to polymer chemists: $M_n > 500 \times 10^3$, $M_w/M_n < 10^5$, syndiotacticity $> 95\%$, with typical catalysts being the lanthanocenes **30** (syndiotactic polymers) and the alkylytterbium(II) complexes **31** (isotactic polymers, $> 97\%$); $M_n > 200 \times 10^3$, $M_w/M_n = 1.1$). High polymer yields are obtained throughout [128].

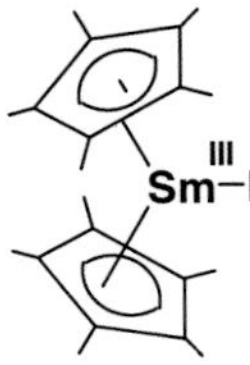

30 R = H, CH₃   **31**

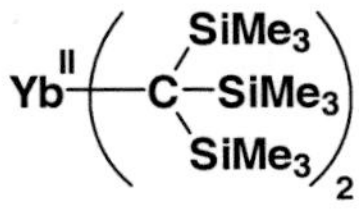

The major advantage of the Periodic Table's "footnotes" originates from (1) the "lanthanoid" contraction effect which makes the chemistry "tunable" according to size and related properties (e.g., metal Lewis acidity, coordination number, steric bulk); (2) their pronounced oxophilicity, "hardness", and (tunable) size. Their chemistry is ruled by simple principles such as ionic binding and HSAB theory. For this reason, combinatorial chemistry could prove an avant-garde tool for ligand fine-tuning. Supermolecular aspects such as dendrimer chemistry [129], immobilization [130], and stacking host–guest interactions [131] are at the top of the synthetic chemist's agenda. Organolanthanoid catalysis, however, is still badly underestimated with regard to its potential.

4.1.14 Organometallic Catalysts for Polymers

In polymer chemistry, new C–C coupling products, such as COC materials (*cycloolefin copolymers*) with special properties like high transparency and hardness, appear possible through tailored organometallic catalysts (cf. Section 2.3.1). The novel polyketones with their high melting points ($> 220\ ^\circ$C) have already been mentioned above [66]. As a matter of fact, Ziegler–Natta catalysis remained for a long time the only true organometallic catalysis in macromolecular chemistry. Once again, the area was revolutionized by the discovery that certain zirconocene derivatives (**32**) of stereorigid, C_2-symmetrical structures catalyze the strictly isotactic polymerization of propene. It must be emphasized that it was an interdisciplinary effort that wrote this success story: Sinn observed the methylalumoxane effect [68], Brintzinger designed and made the chiral zirconocene complexes [69], and Kaminsky discovered the polymerization characteristics related to these structures [69, 70]. At the same time, a strong industrial interest has pushed forward a worldwide development in numerous laboratories since 1984. New perspectives came along with novel structures such as **33** and **34** originating from a chemistry that had not been popular previously. Special copolymers are at present the first-priority goals in this area of catalysis. The first polymers based upon the new generation of metallocene-type catalysts have appeared on the market: Hostacen®

(Hoechst), Exact® (Exxon), and Affinity® (Dow). Some but not all expectations have been fulfilled since the first edition of our book appeared (cf. Section 2.3.1.5).

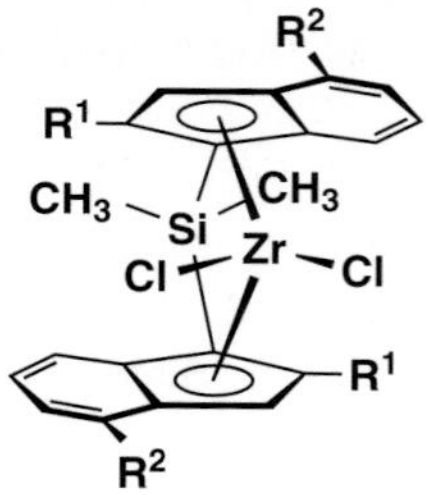

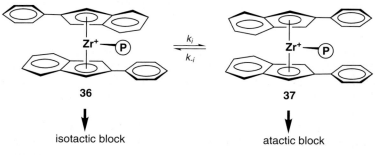

X = CI, N(CH₃)₂

32

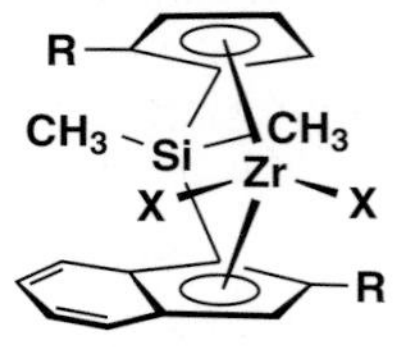

X = CI, N(CH₃)₂
33

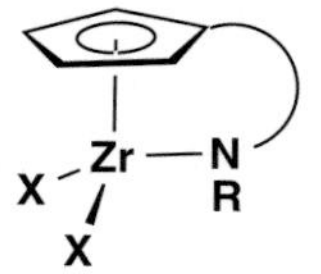

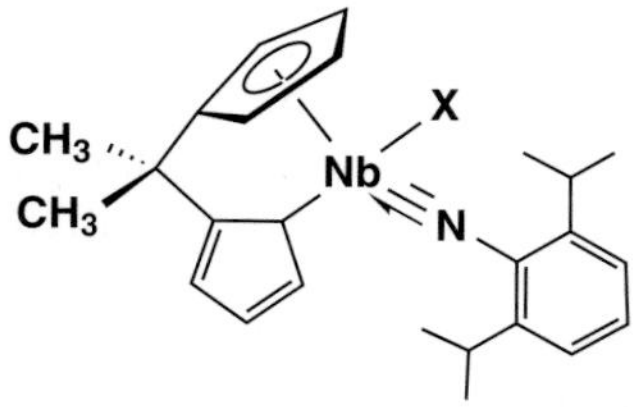

X = N(CH₃)₂
34

X = CI, N(CH₃)₂
35

Of particular "intelligence" are Waymouth's unbridged zirconocene variants **36** and **37**: They are "switchable" from isotactic to atactic block polymerization due to their specific conformations. The oscillation between chiral and achiral coordination geometries is the molecular basis for making novel thermoplastic elastomeric polypropene, with the isotacticity depending on the temperature and propene pressure (isotactic product content 6–28 %) [71 a]. This is an excellent example of how a "gray-hair" type of organometallics fertilizes an applied area of technology.

In spite of all the recent success in the metallocene area, the purely synthetic part of the game has narrowed down to zirconium and hafnium. Little is known

36

37

isotactic block

atactic block

Ⓟ = growing polymer chain

on *ansa*-metallocenes of related metals like niobium and tantalum, in which fact a loss of "synthetic culture" is seen. The imido complex **35** is one of the few such examples [71 b].

In the near future, it is believed that a major part of macromolecular chemistry will receive a strong impact from homogeneous organometallic catalysis. Thus, poly-coupling reactions with redox-sensitive precursor compounds to give polymers like **38** were discovered [72].

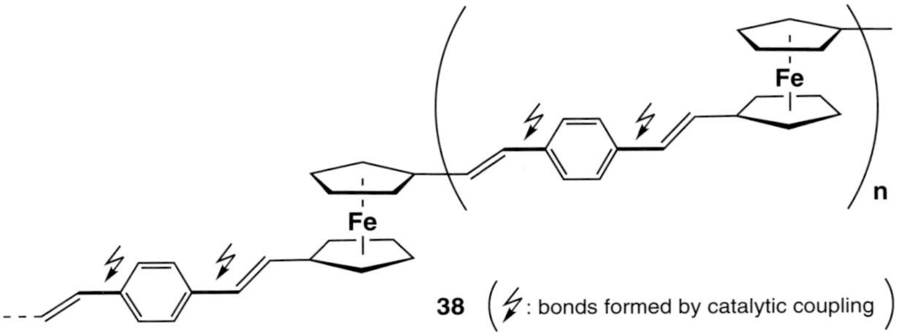

38 (⚡ : bonds formed by catalytic coupling)

The coupling of alkynes, an old field of polymer chemistry, is on new tracks resulting from defined metal–alkyl (**39**) and metal–carbene (**40**) catalysts [73, 74]; previously simple metal halides (e. g. $NbCl_5$) of unspecified active catalyst structures were often employed. The polymer stereochemistry can be switched from 100% *trans* (R = *t*-Bu) to 100% *cis* (R = $C(CF_3)_2CH_3$) in the case of the imido(carbene) complex **40** upon making poly(2,3-bistrifluoromethyl)norbornadiene [74].

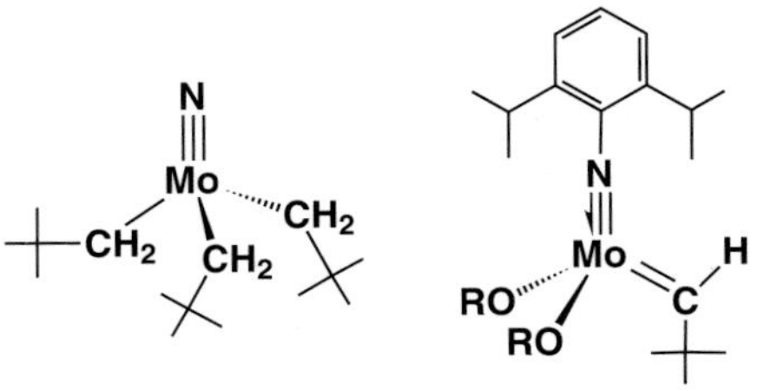

39   **40**

R = $C(CF_3)_2CH_3$ or $C(CH_3)_3$

As well as certain monomer syntheses, defined polymers can be made through homogeneous catalysis, e. g., olefin metathesis (CdF Norsorex®, Hüls Vestenamer®, Hercules DCP®), special ring-opening metathesis (cf. Section 2.3.3), CO/C_2H_4 copolymerization (cf. Section 2.3.4), and other reactions (cf. Section 3.3.10.1).

Yet another era of organometallic polymer chemistry appears to arise from new cationic nickel(II)- and palladium(II) complexes **41** of *N,N*-chelate (diimine) ligands (cf. eq. (19)). According to Brookhardt et al., both the homo- and copolymerization of *α*-olefins proceed with activities that compare in the case of nickel with those commonly seen with metallocene catalysts. For the first time, simple variation of pressure, temperature, and ligand substituents yields an ethylene homopolymer whose structure varies from a highly branched, completely amorphous to a linear, semi-crystalline, high-density material with a degree of branching from 1 to 300 branches per 1000 carbons. As an alternative to **41**, the catalyst system (diimine)NiBr$_2$ + methylalumoxane is equally suitable for the polymerization of ethylene and higher α-olefins [90 a].

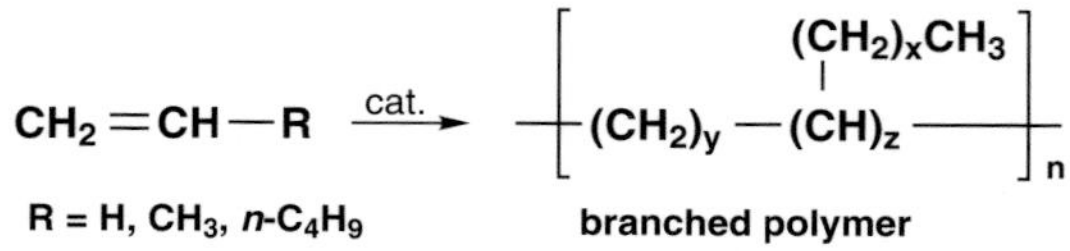

$$CH_2=CH-R \xrightarrow{cat.} \left[(CH_2)_y - (CH)_z \underset{|}{\overset{(CH_2)_xCH_3}{}} \right]_n \tag{19}$$

R = H, CH$_3$, *n*-C$_4$H$_9$ branched polymer

cat. =

M = Ni (a)
Pd (b)

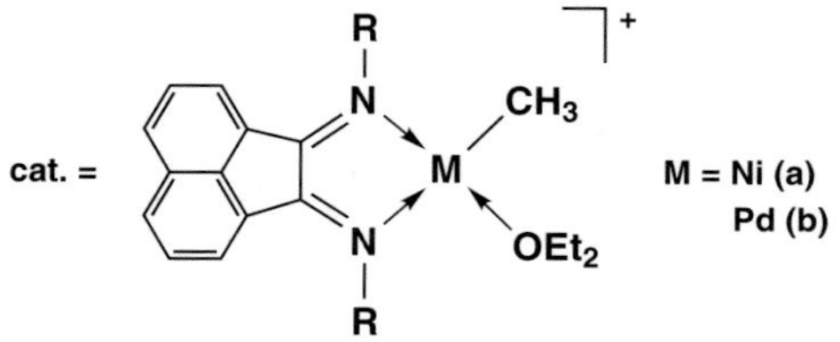

41 a, b

The special brightness of the new catalysts comes from their ability to include functionalized vinyl monomers which normally terminate polymerizations at the oxophilic early transition-metal catalysts. On the other hand, late transition metals most often dimerize or oligomerize olefins, especially nickel, due to the preference of β-hydride elimination. For this reason, ethylene–acrylate and ethylene–vinyl acetate copolymers are exclusively manufactured by radical-type processes, which often enough require high-pressure conditions. Palladium complexes of type **41** allow the formation of high molecular weight random copolymers; the acrylate co-monomer is equally distributed over all molecular weights of the monomodal distribution. The branching amounts to ca. 100 branches/1000 carbon atoms, with the ester groups being predominantly located at the ends of branches (eq. (20)). The slightly modified catalyst **42** adds the olefins in a reversible manner, while the intermediate **43** allows for the chain growth, e. g., consecutive insertion of the ethylene into the Pd-C-alkyl bond (cf. eq. (21)) [90 b]. Brookhart's work has opened a new possibility of organometallic catalysis in macromolecular chemistry. DuPont has filed patents in this area and expects commercialization [91].

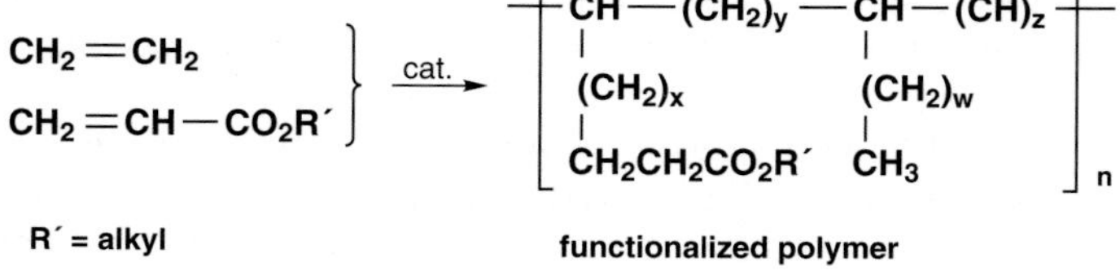

$$\left. \begin{array}{l} CH_2=CH_2 \\ CH_2=CH-CO_2R' \end{array} \right\} \xrightarrow{cat.} \left[\begin{array}{cc} CH-(CH_2)_y-CH-(CH_2)_z \\ | & | \\ (CH_2)_x & (CH_2)_w \\ | & | \\ CH_2CH_2CO_2R' & CH_3 \end{array} \right]_n \tag{20}$$

R′ = alkyl   functionalized polymer

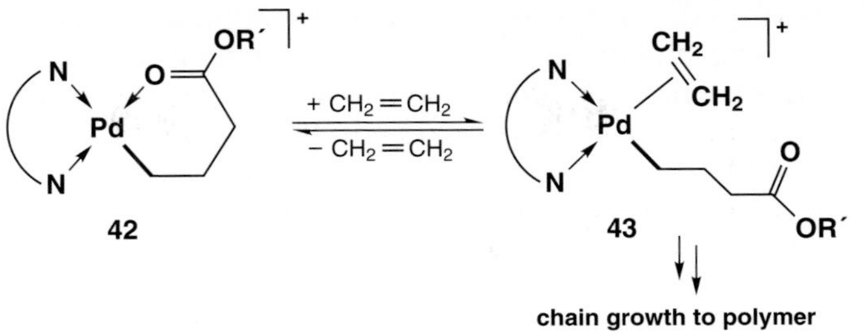

$$(21)$$

chain growth to polymer

4.1.15   Catalyst Reactivation, Process, and Reactor Technology

Only a small minority of organometallic reactions have cleared the hurdle to become catalytic reality; in other words, catalyst reactivation under process conditions is a relatively rare case. As a matter of fact, the famous Wacker/Hoechst ethylene oxidation achieved verification as an industrial process only because the problem of palladium reactivation, $Pd^0 \rightarrow Pd^{II}$, could be solved (cf. Section 2.4.1). Academic research has payed relatively little attention to this pivotal aspect of catalysis. However, a number of useful metal-mediated reactions wind up in thermodynamically stable bonding situations which are difficult to reactivate. Examples are the "early transition metals" when they extrude oxygen from ketones to form C–C-coupled products and stable metal oxides; cf. the McMurry (Ti) and the Kagan (Sm) coupling reactions. Only co-reactants of similar oxophilicity (and price!) are suitable to establish catalytic cycles (cf. Section 3.2.12). In difficult cases, electrochemical procedures should receive more attention because expensive chemicals could thus be avoided. Without going into details here, it is the basic, often inorganic, chemistry of a catalytic metal, its redox and coordination chemistry, that warrant detailed study to help achieve catalytic versions.

4.1.16   Final Closure

Many questions raised in this "Epilogue" will certainly remain open problems for the third edition of this book. On the other hand, there has never been such a strong emphasis of organometallic chemistry on catalytic applications than at present. More than ever before will collaborations "between the disciplines" prove crucial for success – the problems left are, as ever, the more difficult ones. However, a much improved methodology is available these days, and the importance of theory should be mentioned here. At the same time, large-scale sophisticated organometallic preparations – for example, of metallocenes – are becoming standard in industry. It is now clear to many workers in the field that the philosophy of the "roaring sixties" – to make each compound for its own sake and for the fun it gives – applies only for rare segments these days. It is all the more important to intensify the collaboration of homogeneous catalysis with coordination and organometallic chemistry. Beyond that, theoretical chemistry, chemical kinetics, and chemical engineering must be integrated and intensified in current research, not least for the sake of an early assessment regarding the industrial feasibility of a certain reagent/reaction/mechanism/process combination.

We note that the problem of the much talked-about "gap" within heterogeneous catalysis – namely how to perform structural investigations under realistic conditions and to derive reliable conclusions therefrom for the working catalyst – will remain central. Heterogeneous catalysis still has empirical status. In contrast, homogeneous catalysis has its greatest potential in step-by-step improvements, based on the possibility of examining (and understanding!) the molecular details of mechanism(s) under true catalytic conditions. Unlike in heterogeneous catalysis, an encyclopedic collection of catalysts and their efficiency (so-called "expert systems" [94]) is thus not required to choose a homogeneous catalyst for a special purpose.

It is generally observed that chemical companies include homogeneous catalysis in their research and production. For example, Ciba–Geigy commercialized their first organometallic homogeneous catalytic process (100 tons per year), the synthesis of the herbicide Prosulfuron® via the $Pd(dba)_2$-catalyzed Matsuda reaction of 3,3,3-trifluoropropene with an aryldiazonium salt [98].

Organometallic chemistry has become so central an interdisciplinary science that the opportunities for it to serve in catalysis are a daily exciting challenge. Let us hope that homogeneous and heterogeneous catalysis, as a modified, less apodictic version of C. P. Snow's "two cultures" [75], come to unification. The recent Nobel Prize (2001) to Knowles, Sharpless, and Noyori underlines the successes of molecular organometallic catalysis in a convincing way. A 35-author team [137] of experts supports much of what has said in this book when they considered the future catalysis research needs of relevance to carbon management. The chemical feedstock situation will greatly govern the catalytic sciences in the near future, for sure.

References and Notes

[1] To be historically correct, there were earlier examples of metal-mediated homogeneous catalysis. For example, the Hg^{2+}-catalyzed hydration of acetylene to acetaldehyde became an industrial process in 1912. There is an intermediate π-acetylene complex to activate the substrate. The "lead chamber process" to make sulfuric acid (NO_x catalysis) is even older but does not involve metals or metal complexes as catalysts [134].

[2] The word "serendipity" was coined by Horace Walpole in a letter to Sir Horace Mann in 1754 and is based on the fairy tale about the adventures of "The Three Princes of Serendip" (or Serendib, an ancient name for Ceylon, now known as Sri Lanka): R. M. Roberts, *Serendipity – Accidental Discoveries in Science*, John Wiley, New York, **1989**. The term is now used for fortunate, totally unexpected discoveries – discoveries by accident and sagacity.

[3] "Organometallic mixed catalysts" for ethylene polymerization, discovered (by serendipity) when nickel-contaminated autoclaves were used to carry out an *Aufbaureaktion* (reaction of $Al(C_2H_5)_3$ with ethylene). The "nickel effect" lead to the zirconium-catalyzed ethylene polymerization in Ziegler's laboratory on October 26, 1953, see: F. M. MacMillan, *The Chain Straighteners*, The MacMillan Press Ltd., p. 62f., London, **1979**.

[4] E. Krause, A. von Grosse, *Die Chemie der Metallorganischen Verbindungen*, Gebrüder Borntraeger, Berlin, **1937**; reprint by Dr. Martin Sändig oHG, Wiesbaden, **1965**. Specific treatments of organomagnesium (Grignard) and organoalkaline metal compounds are older, for example: W. Runge, *Organometallverbindungen*, 1st ed., Wissenschaftliche Verlagsgesellschaft mbH, Stuttgart, **1931**; 2nd ed. **1944**.

[5] E. W. Abel, G. Wilkinson, F. G. A. Stone (Eds.), *Comprehensive Organometallic Chemistry*, 1st ed., Pergamon, Oxford, **1982**; 2nd ed., **1995-1996**.

[6] (a) J. Buckingham (Ed.), *Dictionary of Organometallic Compounds*, Vols. 1–3, Chapman and Hall, London, **1984**; (b) B. J. Aylett, M. F. Lappert, P. L. Pauson (Eds.), *Dictionary of Organometallic Compounds*, 2nd ed., Vols. 1–5, Chapman and Hall, London, **1995**.

[7] *Chemical Abstracts Service*, December **2001**.

[8] See, for example, *J. Organomet. Chem.* **1995**, *500*; published by Elsevier Science, Amsterdam/Lausanne/Oxford.

[9] These journals were started in **1975** (Elsevier) and **1982** (American Chemical Society), respectively.

[10] (a) C. A. Tolman, *Chem. Soc. Rev.* **1972**, *1*, 337; (b) M. Tsutsui, R. Ugo (Eds.), *Fundamental Research in Homogeneous Catalysis*, Plenum, New York, **1977**; (c) J. K. Kochi, *Organometallic Mechanisms and Catalysis*, Academic Press, New York, **1978**; (d) J. Halpern, *Pure Appl. Chem.* **1983**, *55*, 99.

[11] F. R. Hartley, *Supported Metal Complexes. A New Generation of Catalysts*, Reidel, Dordrecht, **1985**.

[12] (a) W. A. Herrmann, M. Elison, J. Fischer, Ch. Köcher, G. R. J. Artus, *Angew. Chem.* **1995**, *107*, 2602; *Angew. Chem., Int. Ed. Engl.* **1995**, *34*, 3005; (b) Hoechst AG (W. A. Herrmann, M. Elison, J. Fischer, Ch. Köcher), DE 4.447.066, 4.447.067, 4.447.068 (**1994**).

[13] U. Romano, A. Esposito, F. Maspero, C. Neri, M. G. Clerici, *Chim. Ind. (Milan)* **1990**, *72*, 610.

[14] Review: S. L. Scott, J.-M. Basset, G. P. Niccolai, C. C. Santini, J.-P. Candy, Ch. Lecuyer, F. Quignard, A. Choplin, *New J. Chem.* **1994**, *18*, 115.

[15] J.-M. Basset, personal communication to the editor (W. A. Herrmann); *Vth Königstein/ Kreuth Conference on Organometallic Chemistry*, Kreuth/Bavaria, Oct. 3–6, **1995**.

[16] D. Avnir, J. Blum, A. Rosenfeld, H. Schumann, H. Sertchook, S. Wernik, *Abstracts 9th Int. Symp. on Homogeneous Catalysis*, Jerusalem, Israel, **1994**, p. 144.

[17] G. Wulf, *Angew. Chem.* **1995**, *107*, 1958; *Angew. Chem., Int. Ed. Engl.* **1995**, *34*, 1812.

[18] G. A. Melson (Ed.), *Coordination Chemistry of Macrocyclic Compounds*, Plenum, New York, **1979**.

[19] G. Ertl, H. Knözinger, J. Weitkamp (Eds.), *Handbook of Heterogeneous Catalysis*, VCH, Weinheim, **1997**; especially Chapters 2.1.4, 2.3.5, 2.3.6, 4.9, 4.11, 5.3.3, 11.2.1, B/4.5, B/4.15.

[20] D. E. De Vos, F. Thibault-Starzyk, P. P. Knops-Gerrits, R. F. Parton, P. A. Jacobs, *Macromol. Symp.* **1994**, *80*, 157.

[21] I. T. Jarváth, J. Rábai, *Science* **1994**, *266*, 72.

[22] Cf. Ref. [19], Chapter 9.3.

[23] H. Vahrenkamp, *Adv. Organomet. Chem.* **1983**, *22*, 169.

[24] Recent monograph: G. Schmid (Eds.), *Clusters and Colloids*, VCH, Weinheim, **1994**.

[25] (a) H. Bönnemann, W. Brijoux, R. Brinkmann, E. Dinjus, T. Jouen, B. Korall, *Angew. Chem.* **1991**, *103*, 1344; *Angew. Chem., Int. Ed. Engl.* **1991**, *30*, 1312; (b) H. Bönnemann, W. Brijoux, R. Brinkmann, R. Fretzen, T. Joussen, R. Köppler, B. Korall, P. Neiteler, J. Richter, *J. Mol. Catal.* **1994**, *86*, 129; (c) L. N. Lewis, J. Stein, K. A. Smith in: *Progress in Organosilicon Chemistry* (Ed.: B. Marciniec, J. Chojonovski), Gordon and Breach, Langhorne, **1995**, p. 263; (d) T. J. Wehrli, A. Baiker, D. M. Monti, H. U. Blaser, *J. Mol. Catal.* **1990**, *61*, 207; (e) *Workshop on Bimetallic Effects in Chemistry*, European Science Foundation, Parma/Italy, April 26–29, **1995**.

[26] (a) Ref. [19], Chapter 4; (b) J. H. Sinfelt, *Bimetallic Catalysts – Discoveries, Concepts and Applications*, John Wiley, New York, **1983**.

[27] J. R. Anderson, M. Boudart, *Catalysis – Science and Technology*, Springer, Berlin, **1981**.

[28] (a) S. L. Stinson, *Chem. Eng. News*, Sept. 19, **1994**, 33; (b) S. L. Stinson, *ibid.* Oct. 9, **1995**, 44.

[29] C. Botteghi, S. Paganelli, A. Schionato, M. Marchetti, *Chirality* **1991**, *3*, 355.

[30] (a) L. S. Hegedus, *Transition Metals in the Synthesis of Complex Organic Molecules*, University Science Books, Mill Valley **1994** (German translation published by VCH, Weinheim, **1995**); (b) D. Seebach, *Angew. Chem.* **1990**, *102*, 1363; *Angew. Chem., Int. Ed. Engl.* **1990**, *29*, 1320.

[31] H. Brunner, W. Zettlmeier, *Handbook of Enantioselective Catalysis*, VCH, Weinheim, **1993**.

[32] (a) Y. Orito, S. Imai, S. Niwa, *J. Chem. Soc. Jpn.* **1979**, 1118; (b) H.-U. Blaser, H. P. Jalett in: *Heterogeneous Catalysis and Fine Chemicals*, Vol. III (Ed.: M. Guisnet), Elsevier, Amsterdam, **1993**, p. 139; (c) B. Minder, T. Mallat, A. Baiker, G. Wang, T. Heinz, A. Pfaltz, *J. Catal.* **1995**, *154*, 371; (d) K. E. Simons, G. Wang, T. Heinz, T. Giger, T. Mallat, A. Pfaltz, A. Baiker, *Tetrahedron: Asymmetry* **1995**, *6*, 505.

[33] (a) R. Noyori, *Science* **1990**, *248*, 1194; *Chem. Soc. Rev.* **1989**, *18*, 187; (b) W. A. Herrmann, R. Eckl, unpublished results, **1996**; (c) N. Sakai, S. Mano, K. Nozaki, H. Takaya, *J. Am. Chem. Soc.* **1993**, *115*, 7033; (d) N. Sakai, K. Nozaki, H. Takaya, *J. Chem. Soc., Chem. Commun.* **1994**, 395; (e) T. Higashizima, N. Sakai, S. Mano, K. Nozaki, H. Takaya, *Tetrahedron Lett.* **1994**, *35*, 2023; (f) T. Horiuchi, T. Ohta, K. Nozaki, H. Takaya, *J. Chem. Soc., Chem. Commun.* **1996**, 155; (g) T. Nanno, N. Sakai, K. Nozaki, H. Takaya, *Tetrahedron: Asymmetry* **1995**, *6*, 2583.

[34] A. Togni, L. M. Venanzi, *Angew. Chem.* **1994**, *106*, 517; *Angew. Chem., Int. Ed. Engl.* **1994**, *34*, 497.

[35] (a) A. Pfaltz, *Mod. Synth. Methods* **1989**, *5*, 199; (b) see Section 3.1.7 of this book; (c) T. Aratani, *Pure Appl. Chem.* **1985**, *57*, 1839; (d) M. P. Doyle, in: *Catalytic Asymmetric Synthesis* (Ed.: I. Ojima), VCH Publishers, Weinheim, New York, **1993**, p. 63.

[36] (a) E. N. Jacobsen, L. Deng, Y. Furukawa, E. Martinez, *Pure Appl. Chem.* **1994**, *50*, 4323; (b) E. N. Jacobsen in: *Catalytic Asymmetric Synthesis* (Ed.: I. Ojima), VCH, New York **1993**, pp. 159–202; (c) B. D. Brandes, E. N. Jacobsen, *Tetrahedron Lett.* **1995**, *36*, 5123; (d) B. D. Brandes, E. N. Jacobsen, *J. Org. Chem.* **1994**, *59*, 4378.

[37] (a) M. Kitamura, S. Suga, R. Noyori, *J. Am. Chem. Soc.* **1986**, *108*, 6071; (b) M. Kitamura, S. Okada, S. Suga, R. Noyori, *ibid.* **1989**, *111*, 4028; (c) K. Soai, T. Shibata, H. Morioka, K. Choji, *Nature* **1993**, *378*, 767.

[38] (a) K. T. Wan, M. E. Davis, *Nature (London)* **1994**, *370*, 449; I. T. Horváth, F. Joó (Eds.), *Aqueous Organometallic Chemistry and Catalysis*, Kluwer Academic, Dordrecht, **1995**; (b) M. Sawamura, K. Kitayama, Y. Ito, *Tetrahedron: Asymmetry* **1993**, *4*, 1529.

[39] (a) E. J. Roskamp, S. F. Pedersen, *J. Am. Chem. Soc.* **1987**, *109*, 3152; *ibid.* **1987**, *109*, 6551; (b) J. B. Hartung, jr., S. F. Pedersen, *ibid.* **1989**, *111*, 5468.

[40] (a) H. B. Kagan, J. L. Namy, *Tetrahedron* **1986**, *42*, 6573; (b) H. B. Kagan, *New J. Chem.* **1990**, *14*, 453; (c) J. A. Soderquist, *Aldrichim. Acta* **1991**, *24*, 24; (d) G. A. Molander, *Chem. Rev.* **1992**, *92*, 29.

[41] See, for example: H. B. Kagan, *Bull. Soc. Chim. Fr.* **1988**, 846.

[42] J. D. Atwood, *Mechanisms of Inorganic and Organometallic Reactions*, Brooks/Cole, Monterey/Canada, **1985**.

[43] (a) J. F. Waller, *J. Mol. Catal.* **1985**, *31*, 123; (b) G. W. Parshall, *Organometallics* **1987**, *6*, 687; (c) E. Drent, *Pure Appl. Chem.* **1990**, *62*, 661.

[44] (a) W. A. Herrmann, Ch. Broßmer, K. Öfele, C.-P. Reisinger, T. Priermeier, M. Beller, H. Fischer, *Angew. Chem.* **1995**, *107*, 1989; *Angew. Chem., Int. Ed. Engl.* **1995**, *35*, 1844; (b) M. Beller, H. Fischer, W. A. Herrmann, K. Öfele, Ch. Broßmer, *Angew. Chem.* **1995**, *107*, 1992; *Angew. Chem., Int. Ed. Engl.* **1995**, *35*, 1846; (c) J. J. Masters, J. T. Link, L. B. Snyder, W. B. Young, S. J. Danishefsky, *Angew. Chem.* **1995**, *107*, 1886; *Angew. Chem., Int. Ed. Engl.* **1995**, *35*, 1723; (d) Review: K. C. Nicolaou, R. K. Guy, *Angew. Chem.* **1995**, *107*, 2247; *Angew. Chem., Int. Ed. Engl.* **1995**, *35*, 2079.

[45] See, for example: (a) T. Ziegler, *Chem. Rev.* **1991**, *91*, 651; (b) J. K. Labanowski, J. W. Andzelm (Eds.), *Density Functional Methods in Chemistry*, Springer, New York, **1991**; (c) O. D. Häberlein, N. Rösch, *J. Phys. Chem.* **1993**, 4970; (d) N. Rösch, S. Köstlmeier, H. Bock, W. A. Herrmann, *Organometallics* **1996**, *15*, 1872; (e) G. Frenking in B. Cornils, W. A. Herrmann, R. Schlögl, C.-H. Wong (Eds.), *Catalysis from A to Z*, Wiley-VCH, Weinheim, **2000**.

[46] J. S. Giovannetti, Ch. M. Kelly, C. R. Landis, *J. Am. Chem. Soc.* **1993**, *115*, 4040.

[47] (a) M. W. Balakos, S. S. C. Chuang, *J. Catal.* **1995**, *151*, 253, 266; (b) R. M. Deshpande, R. V. Chaudhari, *Ind. Eng. Chem. Res.* **1988**, *27*, 1996; (c) S. S. Divekar, R. M. Deshpande, R. V. Chaudhari, *Catal. Lett.* **1993**, *21*, 191.

[48] R. V. Gholap, O. M. Kut, J. R. Bourne, *Ind. Eng. Chem. Res.* **1992**, *31*, 1597, 2446.

[49] BP Chemicals Ltd. (M. J. Baker, J. R. Dilworth, J. G. Glenn, N. Wheatley), EP 0.632.006 (1994).

[50] D. Astruc, *Electron Transfer and Radical Processes in Transition Metal Chemistry*, VCH, Weinheim, **1995**.

[51] G. W. Parshall, S. D. Ittel, *Homogeneous Catalysis*, 2nd ed., John Wiley, New York, **1992**.

[52] (a) K. Eller, H. Schwarz, *Chem. Rev.* **1991**, *91*, 1121; (c) B. S. Freiser, *Acc. Chem. Res.* **1994**, *27*, 353; (c) D. Schröder, H. Schwarz, *Angew. Chem.* **1995**, *107*, 2126; *Angew. Chem., Int. Ed. Engl.* **1995**, *34*, 1973.

[53] B. Brielbeck, E. Spika, M. Frede, E. Steckhan, *BIOforum* **1994**, *17*, 22.

[54] (a) C. G. Hill, *Activation and Functionalization of Alkanes*, Wiley, New York, **1989**; (b) R. G. Bergman, *ACS Adv. Chem. Ser.* **1992**, *230*, 211; (c) R. H. Crabtree, *Chem. Rev.* **1995**, *95*, 987.

[55] Review: A. Behr, *Angew. Chem.* **1988**, *100*, 681; *Angew. Chem., Int. Ed. Engl.* **1988**, *27*, 661.

[56] Monographs: (a) A. Behr, *Carbon Dioxide Activation by Metal Complexes*, VCH Weinheim, **1988**; (b) W. Keim (Ed.), *Catalysis in C_1-Chemistry*, D. Reidel, Dordrecht, **1983**.

[57] (a) R. G. Bergman, B. A. Arndtzen, *Science* **1995**, *270*, 1970; (b) P. L. Watson, G. W. Parshall, *Acc. Chem. Res.* **1985**, *18*, 51; (c) M. Gozin, A. Weisman, Y. Ben-David, D. Milstein, *Nature* **1993**, *364*, 699; see also: W. D. Jones, *ibid.* **1993**, *364*, 676; (d) M. Gozin, M. Aizenberg, S.-Y. Liou, A. Weisman, J. Ben-David, D. Milstein, *Nature* **1994**, *370*, 42; (e) S.-Y. Liou, M. Gozin, D. Milstein, *J. Am. Chem. Soc.* **1995**, *117*, 9774; (f) M. Hudlicky, *Chemistry of Organic Fluorine Compounds*, 2nd edition, p. 175ff, Prentice-Hall, New York **1992**; (g) M. Aizenberg, D. Milstein, *Science* **1994**, *265*, 359; (h) M. Aizenberg, D. Milstein, *J. Am. Chem. Soc.* **1995**, *117*, 8674;

(i) J.-M. Basset, V. Dufaud, unpublished results **1995/95**; (k) J. Corker, F. Lefevre, Ch. Lécuyer, V. Dufaud, F. Quignard, A. Choplin, J. Evans, J.-M. Basset, *Science* **1996**, *271*, 966; (l) J.-M. Basset, V. Dufaud, FR. 9.508.552 (July 13, **1995**).

[58] Ruhrchemie AG (B. Cornils, W. Konkol, H. W. Bach, G. Dämbkes, W. Gick, E. Wiebus, H. Bahrmann), DE 3.415.968 (1984); EP 0.160.249 (1985).

[59] Argonne National Laboratory (J. W. Rathke, R. J. Klinger), US 5.198.589 (1994).

[60] R. G. Jessop, T. Ikariya, R. Noyori, *Nature (London)* **1994**, *368*, 231; *ibid.* **1995**, *269*, 1065; *Chem. Rev.* **1995**, *95*, 259. See also: W. Leitner, *Angew. Chem.* **1995**, *107*, 2391; *Angew. Chem., Int. Ed. Engl.* **1995**, *34*, 2187.

[61] (a) R. A. Sheldon, *Top. Curr. Chem.* **1993**, *164*, 21; (b) R. A. Sheldon, *CHEMTECH* **1991**, 566; (c) R. A. Sheldon, J. Dakka, *Catal. Today* **1994**, *19*, 215; (d) Hoechst AG (W. A. Herrmann, W. Adam, R. W. Fischer, J. Lin, Ch. R. Saha-Möller, J. D. G. Correia), DE 4.419.799 (1994); (e) W. A. Herrmann, W. Adam, J. Lin, Ch. R. Saha-Möller, R. W. Fischer and J. D. G. Correia, *Angew. Chem.* **1994**, *106*, 2545; *Angew. Chem. Int. Ed. Engl.* **1994**, *33*, 2475; (f) W. Adam, W. A. Herrmann, Ch. R. Saha-Möller, M. Shimizu, *J. Mol. Catal.* **1995**, *97*, 15.

[62] Monographs: (a) G. Strukul (Ed.), *Catalytic Oxidations with Hydrogen Peroxide as Oxidant*, Kluwer Academic, Dordrecht, **1992**; (b) L. I. Simandi, *Catalytic Activation of Dioxygen by Metal Complexes*, Kluwer Academic, Dordrecht, **1992**; (c) R. A. Sheldon, J. K. Kochi, *Metal-Catalyzed Oxidations of Organic Compounds*, Academic Press, London, **1981**; (d) D. H. R. Barton, A. E. Martell, D. T. Sawyer (Eds.), *The Activation of Dioxygen and Homonuclear Catalytic Oxidation*, Plenum, New York, **1993**.

[63] Review: W. A. Herrmann, *J. Organomet. Chem.* **1995**, *500*, 149.

[64] See, for example: C. Botteghi, G. Del Ponte, M. Narchetti, S. Paganelli, *J. Mol. Catal.* **1994**, *93*, 1.

[65] (a) J.-L. Couturier, Ch. Paillet, M. Leconte, J.-M. Basset, K. Weiss, *Angew. Chem.* **1992**, *104*, 622; *Angew. Chem., Int. Ed. Engl.* **1992**, *31*, 628; (b) J.-L. Couturier, K. Tanaka, M. Leconte, J.-M. Basset, *Angew. Chem.* **1993**, *105*, 99; *Angew. Chem.,Int. Ed. Engl.* **1993**, *32*, 112.

[66] (a) Shell (E. Drent), EP 121.965 (1984); *Chem. Abstr.* **1985**, *102*, 46423; (b) E. Drent, J. A. M. van Broekhoven, M. J. Doyle, *J. Organomet. Chem.* **1991**, *417*, 235; (c) E. Drent, P. H. M. Budzelaar, *Chem. Rev.* **1996**, *96*, 663; (d) K. Nozaki, N. Sato, H. Takaya, *J. Am. Chem. Soc.* **1995**, *117*, 9911; (e) E. Drent, P. Arnoldy, P. H. M. Budzelaar, *J. Organomet. Chem.* **1994**, *475*, 57.

[67] (a) BASF AG (G. Witzel et al.), DE 843.876 (1951); (b) B. Cornils, H. Bahrmann, *Chem.-Ztg.* **1980**, *104*, 39; (c) J. F. Knifton, *Catal. Today* **1993**, *18*, 355.

[68] H. Sinn, W. Kaminsky, *Adv. Organomet. Chem.* **1980**, *18*, 99, and references cited therein.

[69] Reviews: (a) H.-H. Brintzinger, D. Fischer, R. Mülhaupt, B. Rieger, R. Waymouth, *Angew. Chem.* **1995**, *107*, 1255; *Angew. Chem., Int. Ed. Engl.* **1995**, *34*, 143; (b) H. Cherdron, M.-J. Brekner, F. Osan, *Angew. Makromol. Chem.* **1994**, *223*, 121; (c) G. Fink, R. Mülhaupt, H.-H. Brintzinger (Eds.), *Ziegler Catalysts*, Springer, Berlin, **1995**; (d) A. M. Thayer, *Chem. Eng. News*, **1995**, Sept. 11, p. 15.

[70] (a) F. Küber, *New Scientist*, Aug. **1993**, 28; (b) F. Küber, M. Aulbach, *Chem. uns. Zeit* **1994**, *28*, 197.

[71] (a) G. W. Coates, R. M. Waymouth, *Science* **1995**, *267*, 217; (b) W. A. Herrmann, W. Baratta, E. Herdtweck, *Angew. Chem.* **1996**, *108*, 2098; *Angew. Chem., Int. Ed. Engl.* **1996**, *35*, 1951.

[72] O. Nuyken et al., unpublished results, **1995**.

[73] (a) W. A. Herrmann, S. Bogdanovic, R. Poli, T. Priermeier, *J. Am. Chem. Soc.* **1994**, *116*, 4989; (b) S. Bogdanovic, Ph. D. Thesis, Technische Universität München, **1994**.

[74] R. R. Schrock, *Pure Appl. Chem.* **1994**, *66*, 1447.

[75] C. P. Snow, *The Two Cultures: And a Second Look*, Cambridge University Press, London, **1959**.

[76] (a) M. de Sousa Healy, A. J. Rest, *Adv. Inorg. Radiochem.* **1978**, *21*, 1; (b) J. Steinke, D. C. Sherrington, I. R. Dunkin, *Adv. Polym. Sci.* **1995**, *123*, 81.

[77] E. Montflier, G. Fremy, Y. Castanet, A. Montreux, *Angew. Chem.* **1995**, *107*, 2450; *Angew. Chem., Int. Ed. Engl.* **1995**, *34*, 2269.

[78] (a) R. A. Lerner, S. J. Benkovic, P. G. Schultz, *Science* **1991**, *252*, 659; (b) J. D. Stewart, L. J. Liotta, S. J. Benkovic, *Acc. Chem. Res.* **1993**, *26*, 396; (c) B. L. Iverson, *CHEMTECH* **1995**, *25* (June), 17; (d) E. Keinen, S. C. Sinha, D. Shabat, H. Itzhaky, J.-L. Reymond, in Ref. [16], p. 383.

[79] (a) M. T. Reetz, S.-H. Kyung, C. Bolm, T. Zierke, *Chem. Ind. (London)* **1986**, 824; (b) U. Koert, *Nachr. Chem. Tech. Lab.* (Weinheim) **1995**, *43*, 1068.

[80] G. Dyker, *Angew. Chem.* **1995**, *107*, 2407; *Angew. Chem., Int. Ed. Engl.* **1995**, *34*, 2223.

[81] E. Monflier, G. Fremy, Y. Castanet, A. Mortreux, *Angew. Chem.* **1995**, *107*, 2450; *Angew. Chem., Int. Ed. Engl.* **1995**, *34*.

[82] P. G. Schultz, *Angew. Chem.* **1989**, *101*, 1336; *Angew. Chem., Int. Ed. Engl.* **1989**, *28*, 1283.

[83] M. T. Reetz, S. A. Quaiser, *Angew. Chem.* **1995**, *107*, 2461; *Angew. Chem., Int. Ed. Engl.* **1995**, *34*, 2240.

[84] (a) H. König, R. Rogi-Kohlenprath, H. Weber in *Chiral Reactions in Heterogeneous Catalysis* (Eds.: G. Jannes, V. Dubois), Plenum, **1995**, p. 135; (b) C. Exl, I. Francesconi, H. Hönig, R. Rogi-Kohlenprath, *Preprints 8th Int. Symp. on the Relations between Homogeneous and Heterogeneous Catalysis*, Balatonfüred, **1995**.

[85] (a) J. M. Lehn, *Angew. Chem.* **1988**, *100*, 91; *Angew. Chem., Int. Ed. Engl.* **1988**, *27*, 89; (b) D. J. Cram, *Angew. Chem.* **1988**, *100*, 1041; *Angew. Chem., Int. Ed. Engl.* **1988**, *27*, 1009.

[86] (a) C. J. Li, *Tetrahedron Lett.* **1995**, *36*, 517; (b) R. Sjöholm, R. Rairama, M. Ahonen, *J. Chem. Soc., Chem. Commun.* **1994**, 1217.

[87] P. Maitlis, H. Long, Z.-Q. Wang, M. L. Turner, Ref. [16], p. 3.

[88] H. Heinemann, *CHEMTECH* **1971**, *1* (5), 286.

[89] A. Müller, H. Reuter, S. Dillinger, *Angew. Chem.* **1995**, *107*, 2505; *Angew. Chem., Int. Ed. Engl.* **1995**, *35*, 2328.

[90] (a) L. K. Johnson, Ch. M. Killian, M. Brookhart, *J. Am. Chem. Soc.* **1995**, *117*, 6414; (b) L. K. Johnson, S. Mecking, M. Brookhart, *ibid.* **1996**, *118*, 267.

[91] J. Haggin, *Chem. Eng. News* **1996**, *74* (6), p. 6.

[92] W. A. Herrmann (Ed.), *Topics in Current Chemistry*, Vol. 179, Springer, Berlin, **1996**.

[93] B. Cornils, W. A. Herrmann (Eds.), *Aqueous-Phase Organometallic Catalysis*, Wiley-VCH, Weinheim, **1998**.

[94] Cf. Ref. [19], Section 2.5.

[95] Review: F. Agbosson, J.-F. Carpentier, A. Mortreux, *Chem. Rev.* **1995**, *95*, 2485.

[96] M. J. Cannarsa, *Chem. Ind.* **1996**, 374.

[97] T. Shibata, K. Choji, H. Morioka, T. Hayase, K. Soai, *J. Chem. Soc., Chem. Commun.* **1996**, 751.

[98] R. R. Bader, P. Baumeister, H.-U. Blaser, *Chimia* **1996**, *50*, 99; Ciba-Geigy AG (P. Baumeister, G. Seifert, H. Steiner), EP 584.043 (1992).

[99] (a) U. Bremberg, M. Larhed, C. Moberg, A. Hallberg, *J. Org. Chem.* **1999**, *64*, 1082; (b) U. Bremberg, S. Lutsenko, N.-F. K. Kaiser, M. Larhed, A. Hallberg, C. Moberg, *Synthesis* **2000**, *1004*.

[100] N.-F. K. Kaiser, U. Bremberg, M. Larhed, Ch. Moberg, A. Hallberg, *Angew. Chem.* **2000,** *112*, 3742; *Angew. Chem., Int. Ed.* **2000**, *39*, 3596.

[101] H. Ihee, J. Cao, A. H. Zewail, *Angew. Chem., Int. Ed.* **2001**, *40*, 1334.

[102] B. Jandeleit, D. J. Schaefer, T. S. Powers, H. W. Turner, W. H. Weinberg, *Angew. Chem.* **1999,** *111*; *Angew. Chem., Int. Ed.* **1999**, *38*, 2494.

[103] F. Hollmann, A. Schmid, E. Steckhan, *Angew. Chem.* **2001**, *113*, 190; *Angew. Chem., Int. Ed.* **2001**, *40*, 169.

[104] W. A. Herrmann, J. P. Zoller, R. W. Fischer, *J. Organomet. Chem.* **1999**, *579,* 404.

[105] W. A. Herrmann, Th. Weskamp, J. P. Zoller, R. W. Fischer, *J. Mol. Catal. A: Chemical* **2000**, *153,* 49.

[106] (a) P. G. Jessop, T. Ikariya, R. Noyori, *Chem. Rev.* **1999**, *99,* 475; (b) J. L. Kendall, D. A. Canelas, J. L. Young, J. M. Defimore, *Chem. Rev.* **1999**, *99,* 543.

[107] A. Fürstner, D. Koch, K. Langemann, W. Leitner, C. Six, *Angew. Chem.* **1997**, *109*, 2562; *Angew. Chem., Int. Ed. Engl.* **1997**, *36*, 2466.

[108] A. Fürstner, *Angew. Chem.* **2000**, *112*, 3140; *Angew. Chem., Int. Ed.* **2000**, *39*, 3012.

[109] P. G. Jessop, Y. Hsiao, T. Ikariya, R. Noyori, *J. Am. Chem. Soc.* **1996**, *118*, 344.

[110] W. Kläui, J. Bongards, G. J. Reiß, *Angew. Chem.* **2000**, *112*, 4077; *Angew. Chem., Int. Ed.* **2000**, *39*, 3894.

[111] W. Leitner, *Adv. Organomet. Chem.* **2000**, *14,* 809.

[112] R. A. Brown, P. Pollet, E. McKoon, Ch. A. Eckert, Ch. L. Liotta, P. G. Jessop, *J. Am. Chem. Soc.* **2001**, *123,* 1254.

[113] Short review: P. Wasserscheid, W. Keim, *Angew. Chem.* **2000**, *112,* 3926; *Angew. Chem., Int. Ed.* **2000**, *39*, 3772.

[114] See, for example: Solvent Innovation GmbH: http://www.solvent-innovation.com.

[115] (a) W. A. Herrmann, V. P. W. Böhm, *J. Organomet. Chem.* **1999**, *572*, 141; (b) L. Xu, W. Chen, I. Xiao, *Organometallics* **2000**, *19*, 1123; (c) A. I. Carmichael, M. I. Earle, I. D. Holberg, P. B. McCormac, K. R. Seddon, *Org. Lett.* **1999**, *1*, 997; (d) V. P. W. Böhm, W. A. Herrmann, *Chem. Eur. J.* **2000**, *6*, 1017.

[116] P. Wasserscheid, M. Eichmann, *Proc. 3rd Int. Symp. Catal. In Multiphase Reactors,* Naples, **2000**, pp. 249–261.

[117] (a) Ch. Döbler, G. Mehltretter, M. Beller, *Angew. Chem.* **1999**, *111*, 3211; *Angew. Chem. Int. Edit. Engl.* **1999**, *38*, 3026; (b) Ch. Döbler, G. Mehltretter, U. Sundermeier, M. Beller, *J. Amer. Chem. Soc.* **2000**, *122*, 10289.

[118] S. L. Aeilts, D. R. Cefalo, P. J. Bonitatebus Jr., J. H. Houser, A. H. Hoveyda, R. R. Schrock, *Angew. Chem., Int. Ed.* **2001**, *40*, 1452.

[119] J. F. Hartwig et al., *J. Am. Chem. Soc.* **2000**, *122*, 9546.

[120] A. Togni et al., *J. Am. Chem. Soc.* 1997, *119,* 10857; Lonza AG, EP 0.909.762.

[121] Reviews: (a) W. A. Herrmann, Ch. Köcher, *Angew. Chem.* **1997**, *109*, 2256; *Angew. Chem., Int. Ed. Engl.* **1997**, *36*, 2162; (b) W. A. Herrmann, *Angew. Chem.* **2002**, in press; *Angew. Chem., Int. Ed.* **2002**, in press; (c) T. Weskamp, V. P. W. Böhm, W. A. Herrmann, *J. Organomet. Chem.* **2000**, *600*, 12; (d) W. A. Herrmann, T. Weskamp, V. P. W. Böhm, *Adv. Organomet. Chem.* **2002**, in press.

[122] (a) T. Weskamp, F. J. Kohl, D. Gleich, W. A. Herrmann, *Angew. Chem.* **1999**, *111*, 2573; *Angew. Chem., Int. Ed.* **1999**, *38*, 2416; (b) M. S. Sanford, M. Ulman, R. H. Grubbs, *J. Amer. Chem. Soc.* **2001**, *123*, 749.

[123] V. P. W. Böhm, T. Weskamp, Ch. W. K. Gstöttmayr, W. A. Herrmann, *Angew. Chem.* **2000**, *112*, 1672; *Angew. Chem., Int. Ed.* **2000**, *39*, 1602.

[124] Review: R. Anwander, in *Lanthanides: Chemistry and Use in Organic Synthesis* (Ed.: S. Kobayashi), Springer, Berlin, **1999**, *Vol. 2,* pp. 1–62.

[125] R. Anwander, O. Runte, J. Eppinger, G. Gerstberger, E. Herdtweck, M. Spiegler, *J. Chem. Soc. Dalton Trans.* **1998**, *847.*

[126] (a) T. Skrydstrup, *Angew. Chem.* **1997,***109*, 355; *Angew. Chem., Int. Ed. Engl.* **1997**, *36*, 345; (b) H. B. Kagan, J.-L. Namy, in Ref. [124], pp. 155–198.

[127] (a) M. A. Giardello, V. P. Conticelli, L. Brard, M. R. Gagne, T. J. Marks, *J. Am. Chem. Soc.* **1994**, *116*, 10241; (b) Y. Li, P.-F. Fu, T. J. Marks, *Organometallics* **1999**, *13*, 439.

[128] (a) Review: J. Yasuda, in Ref. [124], pp. 255–283; (b) H. Yasuda, H. Yamamoto, K. Yokota, S. Miyake, A. Nakamura, *J. Am. Chem. Soc.* **1992**, *114*, 4908; (c) H. Yasuda, E. Ihara, *Advan. Polym. Sci.* **1997**, *133*, 53.

[129] J. M. J. Fréchet, M. Kawa, *Chem. Mater.* **1998**, *10*, 286.

[130] Reviews: (a) T. J. Marks, *Acc. Chem. Res.* **1992**, *25*, 57; (b) W. M. H. Sachtler, Z. Zhang, *Adv. Catal.* **1993**, *39*, 129.

[131] C. Piguet, G. Bernardinelli, G. Hopfgartner, *Chem. Rev.* **1997**, *97*, 2005.

[132] H. W. Bohnen, B. Cornils, *Adv. Catal.* **2002**, in press.

[133] *Angew. Chem., Int. Ed. Engl.* **1997**, *36*, 1431 and **2001**, *40*, 4422; *Nature* **1999**, *401*, 254 and **2000**, *404*, 982; *Science* **1998**, *279*, 1021.

[134] Keyword "history of catalysis" in B. Cornils, W. A. Herrmann, R. Schlögl, C.-H. Wong (Eds.), *Catalysis from A to Z*, Wiley-VCH, Weinheim, **2000**.

[135] W. A. Herrmann, M. Mühlhofer, T. Strassner, unpublished results **2001**; *Angew. Chem.* **2002**, in press (review article on *N*-heterocyclic carbenes in catalysis).

[136] V. P. W. Böhm, C. W. K. Gstöttmayr, T. Weskamp, W. A. Herrmann, *Angew. Chem., Int. Ed.* **2001**, *40*, 3387.

[137] H. Arakawa et al., *Chem. Rev.* **2001**, *101*, 953.

Index

al tagging.